Engineering Economy

E. PAUL DeGARMO

Registered Professional Engineer
Professor of Industrial Engineering and Mechanical Engineering Emeritus
University of California, Berkeley

WILLIAM G. SULLIVAN

Registered Professional Engineer
Professor of Industrial Engineering
Arizona State University, Tempe

JOHN R. CANADA

Registered Professional Engineer
Professor of Industrial Engineering
North Carolina State University at Raleigh

ENGINEERING ECONOMY

Seventh Edition

Macmillan Publishing Company

New York

Collier Macmillan Publishers

London

Earlier editions entitled *Introduction to Engineering
Economy,* by E. P. DeGarmo and B. M. Woods,
copyright 1942 and 1953 by Macmillan Publishing Co.,
Inc. Earlier editions entitled *Engineering Economy* ©
1960 and copyright © 1967, 1973 and © 1979 by
Macmillan Publishing Co., Inc.

Macmillan Publishing Company
866 Third Avenue, New York, New York 10022

Collier Macmillan Canada, Inc.

Library of Congress Cataloging in Publication Data

DeGarmo, E. Paul (date)
   Engineering economy.
   Bibliography: p.
   Includes index.
   1. Engineering economy.   I. Sullivan, William G.,
1942-   .   II. Canada, John R.   III. Title.
TA177.4.D43   1984   658.1'55   83-704
ISBN 0-02-328600-8

Printing: 3 4 5 6 7 8   Year: 4 5 6 7 8 9 0 1

ISBN   0-02-328600-8

PREFACE

Engineering economy deals with the systematic evaluation of the equivalent worth of benefits resulting from a proposed engineering or business venture in relation to the costs associated with the undertaking. The basic purpose of this book is to provide a sound understanding of concepts and principles of engineering economy and to develop proficiency with methods for making rational decisions regarding problems likely to be encountered in professional practice. Accordingly, the book is intended primarily as a text and basic reference for a first formal course in engineering economy. It should also be useful in business-related courses concerned with the economic analysis of alternative plans of action.

The Seventh Edition features several notable changes compared to the previous edition. In general, most chapters have been expanded to include new topical areas and important changes that affect engineering economy studies. Furthermore, additional solved exercises are used to augment each chapter, and the majority of problems at the end of each chapter are new. Specifically, Chapter 4 has been increased to include geometric gradients and time-varying interest rates. A discussion of the mid-year cash flow convention is also given in Chapter 4. Chapter 5 provides expanded coverage of equivalent worth and rate of return methods for assessing the economic worth of a proposed alternative. A computer program written in BASIC is also introduced in Chapter 5 (and described in Appendix C) to assist with the solution of problems encountered throughout the text. Chapter 6 extends the methods of Chapter 5 to the analysis of multiple alternatives.

The subject of risk and uncertainty has been enlarged in Chapter 7 to include a variety of commonly-used techniques from industry and government. Chapter 8 deals with cost estimating and inflation, and it represents a significant expansion upon earlier treatments of these subjects. Two other key chapters, "Depreciation" (Chapter 9) and "The Effects of Income Taxes" (Chapter 10), have been extensively modified to include the provisions of the Economic Recovery Tax Act of 1981. The subject of replacement analysis (Chapter 11) includes depreciation and income tax provisions of Chapters 9 and 10 in addition to the treatment of inflation. In Chapter 13 the revenue requirements methodology is described and illustrated in more detail, and Chapter 14 contains expanded coverage of capital budgeting topics. Throughout the book, case studies are inserted at strategic points to reinforce important concepts.

For the many suggestions from colleagues, students (especially Saran Harinsuta), and practicing engineers who have used the previous editions, we express our deep appreciation. We hope they will find this edition to be equally helpful.

E. PAUL DEGARMO
WILLIAM G. SULLIVAN
JOHN R. CANADA

CONTENTS

PART 1

BACKGROUND AND TOOLS FOR ENGINEERING ECONOMIC ANALYSIS

CHAPTER 1

Introduction

This book is concerned with the evaluation of alternative uses of capital in engineering and business projects. An outstanding phenomenon of present-day industrialized civilizations is the extent to which engineers and business managers, through the use of capital, are able to multiply the effectiveness of efforts to harness resources to satisfy the needs and wants of people. Consequently, capital, in the form of money for people, machines, and materials, is an economic necessity in virtually all engineering and business projects.

In the United States an average of over $25,000 of capital is required for each worker in industry. In the more highly mechanized or automated industries, the investment per worker often exceeds $100,000. Naturally, those who supply the capital, and who ultimately control the spending of it, are concerned that it be used to best advantage. Likewise, those who design projects and make the managerial decisions as to whether they should be undertaken and how they should be operated are equally concerned that the available capital be used effectively.

In our capitalistic economy the success of engineering and business projects is more commonly measured in terms of financial efficiency than in any other way. It is unlikely that a project will achieve maximum financial success unless it is also properly planned and operated with respect to its technical, social, and financial requirements. Because engineers are most likely to understand the technical requirements of a project, they are very frequently called upon to make a study combining the technical and financial details, as well as social and aesthetic values, of a project and thus to provide the analysis upon which they or others can base sound managerial decisions.

Engineers play a unique and important role in the conception of new ideas and projects that will require the expenditure of capital to reach the hardware or operational stage. In fact, it is the necessity for considering properly such a combination of factors that distinguishes in large part the work of engineers from that of theoretical scientists.

ENGINEERING ECONOMY AND DECISIONS AMONG ALTERNATIVES

Engineering economy is a body of knowledge devoted to the systematic evaluation of the net worth of benefits resulting from proposed engineering and business ventures in relation to the expenditures associated with those undertakings. Accordingly, economic analyses that primarily involve engineering and technical projects commonly are called *engineering economy* studies.

In recent years the techniques that were originally developed in the field of engineering economy have been expanded and adapted for use in a much broader spectrum of business situations. Consequently, broader titles frequently are attached to these studies, as, for example, economic analyses for decision making. There is no clear line of demarcation between the various types of studies, and the matter of name is of little consequence.

Virtually all engineering problems can be solved in more than one way. Most projects can be carried out in more than one way. Almost all business decisions involve doing one thing or another, even though one alternative is merely to do nothing or to maintain the status quo. Clearly, decisions involve alternative courses of action that must be well defined and feasible before the economic merits of competing opportunities for investment can be properly evaluated. Moreover, *engineering economy studies deal with the differences in economic results among alternatives*. This concept of economic differences among alternatives is basic and most important in making engineering economy studies; if there is no alternative, there is no need to make an economy study.

The scope and importance of modern engineering projects make it essential that all the significant factors involved in the economy of an undertaking not only be considered but also handled in an accurate and correct manner so that the results will be satisfactory from all the viewpoints touched by the project.

In short, the alternative chosen should never be a matter of guess. Hunches and intuition are not reliable enough. Because much of the required information in economy studies is based on estimates of future conditions, one might be tempted to conclude that a decision can be based largely on guesswork. Estimates made after careful study and based on all the available information should be considerably better than haphazard guesses. The accuracy of an estimate,

4

and indeed the entire economy study, is usually limited only by the amount of time and effort that one is willing to devote to making the analysis.

ENGINEERING AND MANAGEMENT

Business continues to become more technical. Consequently, engineers play an increasingly important role in management. More and more decision making in government and industry is done by engineers. Some of these decisions are made primarily on the basis of the economic factors involved. More often there are many other factors that must be weighed, and sometimes these may prevail over purely economic considerations. When managers are not engineers, they increasingly call upon engineers to make technical-economic analyses and to provide the data and recommendations upon which managerial decisions can be based. In such situations an engineer is essentially in the position of a consultant to management and must combine technical and economic knowledge to provide sound conclusions and recommendations. With recent developments in mathematical, statistical, and computer techniques which permit the quantitative handling of increasingly complex economic problems, the engineer has an opportunity to play an even more important role in the decision-making process. Not only does the engineer have the mathematical and scientific background for understanding and using such techniques, but he or she has the engineering background that enables one to recognize the practical limitations of these techniques and the effect of the lack of information that usually exists in real situations. He or she is thus in a position to make the necessary compromises and adjustments that will enable a realistic, although possibly not perfect, solution to be achieved.

MEASURES OF FINANCIAL PRODUCTIVITY

In most technical courses studied by the engineer or engineering student, one is concerned with obtaining the most productive utilization of materials, labor, equipment, and energy. The degree of effectiveness of the utilization is often measured by the familiar equation

$$\text{efficiency} = \frac{\text{output}}{\text{input}} \tag{1-1}$$

When dealing with phenomena in the physical world, the engineer knows that efficiency can never exceed 100%. However, when dollars are considered, a different situation exists. This may be simply expressed as

$$\text{financial productivity} = \frac{\text{total dollars income}}{\text{total dollars spent}} \tag{1-2}$$

It may readily be seen that, unless the financial productivity can and does exceed 100%, a project is not desirable from a purely financial viewpoint. *It should be recognized that a vital component of a firm's overall productivity is the financial productivity resulting from the allocation of its capital.*

Engineers will immediately recognize that Equation 1-1 can be used in at least two ways. The first, and less common, use involves the *total* output and input over the life of a project. The second, and more frequent, use involves *instantaneous* values of output and input. The use of electrical meters to determine the input and output of an electric motor is an example of this. The latter type of efficiency is more useful, inasmuch as we do not wish to wait until a piece of equipment has worn out at the end of its life before being able to determine what its efficiency has been. In the same manner, a shorter period of time is used in obtaining a more usable measure of financial productivity.

A commonly used measure of financial productivity for a given year is

$$\text{annual rate of return} = \frac{\text{annual net profit}}{\text{initial capital investment}} \tag{1-3}$$

which is expressed in percent. Obviously, there will be a difference in the annual rate of return depending upon whether the annual net profit is calculated before or after income taxes have been considered. Thus it is wise to specify whether a computed rate of return is before or after taxes. Other, more useful, definitions of rate of return for *extended* periods of time are provided in a subsequent chapter.

Although the annual rate of return undoubtedly is the most universally used measure of financial productivity, we should be aware that alternative measures are commonly used in certain situations. In certain instances *least cost,* either investment cost or annual cost, may be used as a measure of financial productivity. However, in most cases the annual cost must include profit (return) on the invested capital in order to be meaningful. Other measures that can be used in some cases are *profit per sales dollar* and *number of years before total profits exceed total investment.* However, these measures must be used very carefully to avoid making incorrect recommendations. It should be noted that more theoretically correct methods of measuring financial productivity are emphasized in subsequent chapters. Chapter 4 discusses time-value-of-money concepts, which form the foundation for the methods that are discussed in Chapter 5. The matter of evaluating multiple alternatives in view of the time value of money is then described in Chapter 6.

RISK AND UNCERTAINTY

In practically all engineering economy studies, estimated values of the required data will be affected by such things as a changing economic environment, biases in the development and interpretation of data, and lack of experience with similar projects due to the "one-of-a-kind" nature of many engineering projects. Because of variations that are bound to occur in study conditions, risk and uncertainty surround the results of an engineering economy analysis and should be explicitly taken into account if at all possible. In this book "risk" and "uncertainty" are used interchangeably.

Procedures are presented in Chapter 7 to aid the analyst in determining how uncertainty in various study parameters affects the measure of financial productivity being utilized to judge the merits of a proposed project. When uncertainty in a parameter causes a reversal in preference for an alternative, particular care can be given to estimating that parameter and evaluating the relative uncertainties which exist among competing alternatives.

The major sources of uncertainty in an analysis are typically expense and revenue estimates, life of the venture, and the effects of inflation. In this regard, Chapter 8 addresses cost estimating and inflation so that these important topics can be highlighted in the hope that some of the uncertainty in an engineering economy study can be reduced.

NONMONETARY ATTRIBUTES AND MULTIPLE OBJECTIVES

Few decisions, either personal or business, are made solely on the basis of financial considerations. Furthermore, the financial productivity of a project may be affected to a considerable extent by *nonmonetary attributes* and *multiple objectives*. If a person goes to a store to buy a new suit of clothes and finds that one in plain black can be obtained for $10 less than one of comparable quality in attractive colors, he probably will not make the decision solely on the basis of price. He may also consider the fabric and cut of the clothing and have objectives in mind other than keeping warm. The financial well-being of the manufacturer of the garments will, without question, be affected by the desires of customers for color, style, fabric, and so on. Similarly, satisfactory decisions and recommendations regarding the feasibility of engineering projects must take into account all attributes, monetary and nonmonetary, that will affect the undertaking, as well as the multiple objectives that often must be considered. These subjects are explored in Chapter 7.

Some of the most common nonmonetary (also called intangible and irreducible) attributes that must be considered are general business conditions, social human values, consumer likes and dislikes, and governmental regulations. All businesses operate within an economic system that functions in accordance with certain general rules. Whether a product or service is socially desirable, acceptable, or needed can easily determine the success or failure of a venture. Similarly, whether a product or service meets the existing likes and dislikes of the public, which may change with time, may be all-important in the success or failure of an enterprise.

Although the primary focus of this book is on the correct use of techniques for considering economic or monetary desirability, it should be emphasized that decisions between alternatives involve objectives other than those which can be reasonably reduced to monetary terms. For example, a limited list of objectives other than profit maximization or cost minimization that may be important to a firm would include:

1. Improvement of the safety record to a certain level.
2. Minimization of risk of monetary loss.
3. Maximization of service quality.
4. Minimization of cyclic fluctuations in production.
5. Maintenance of a desired public image.
6. Reduction of pollutants to a certain level.
7. Maximization of employee satisfaction.

Economic analyses formally provide for those objectives or factors that can be reduced to monetary terms. The results of these analyses should then be weighed together with other objectives or factors before a final decision is made. There exist analytical methods to accomplish this if the appropriate estimates can be made. In general, one should pursue formal analysis as far as is technically and economically feasible, and then depend on the judgment and experience of the decision maker and his staff to consider nonmonetary factors and other objectives not formally included in the analysis.

EQUITY AND DEBT CAPITAL

Capital that is used for financing engineering and business ventures may be classified in two fundamental categories. *Equity capital* is that owned by those individuals (or organizations) who have invested their capital in a business venture in the hope of receiving a profit. *Debt capital,* often called *borrowed capital,* is obtained from lenders for use in a venture. In return, the lenders receive interest from the borrowers, but normally they do not receive any other benefits that may accrue from the use of the capital in the venture it finances. On the other hand, neither do lenders participate as fully in the risks of the venture. Thus the return to the owners of equity capital is profit; the return to the lenders of borrowed capital is interest.

THE DECISION-MAKING PROCESS

Rational decision making is a complex process. Following is a list of six interrelated phases which make up that process, together with brief comments on each:

1. *Recognition of the problem:* A problem is any condition or circumstance that may impede one from achieving his goals or objectives. It may merely be an opportunity for improvement.
2. *Definition of the goal or objective:* This may be specific, such as to produce X products; or general, such as to reduce costs or to increase profits.
3. *Identification of feasible alternatives:* This phase often is not given sufficient attention. Unless the best alternative is considered, the result of comparisons will always be suboptimal.
4. *Selection of a criterion for judging which alternative is best:* For studies

in which only monetary factors need to be considered, a single economy study criterion such as net present worth is sufficient. Otherwise, multiple criteria, to correspond with objectives, should be used.

5. *Prediction of the outcomes for each alternative:* This estimating phase is difficult and crucial to the soundness of the analysis.

6. *Choice of the best alternative to achieve the objective:* This wraps it up, normally.

Because economy studies are made for the purpose of providing information on which decisions can be made concerning the investment or use of capital, they should satisfy the following requirements:

1. Economy studies should be based on consideration of all available factors, both monetary and nonmonetary. Multiple objectives should be incorporated into the analysis when they are apparent.

2. Economy studies should be made relative to a stated viewpoint: for example, from the viewpoint of the stockholder or of the organization's customers.

3. Because economy studies deal with future events that require some estimates to be made, these estimates should be made intelligently, in the light of experience and sound judgment.

4. Insofar as possible, decisions should be based on differences that exist among alternatives.

5. The risks and uncertainties inherent to each alternative should be recognized and their effects considered.

6. Some valid measure of financial productivity, such as the internal rate of return or the net present worth, should be determined.

7. In the interest of cooperation and better understanding, studies should make use of the same factors that will be used to judge the worthiness of the investment after it has been made.

8. A recommended course of action, together with the reasons for the recommendation, should be clearly stated.

Studies made in this manner will assure that those who use them for decision making will have the necessary facts in a readily understandable form. It is essential that economy study results be clearly and accurately communicated to decision makers.

THE RELATIONSHIP OF ECONOMY STUDIES AND ACCOUNTING

Engineering economy studies are made for the purpose of determining whether capital should be invested in a project or whether it should be utilized in a different manner than it presently is being used. Economy studies always deal, at least for one of the alternatives being considered, with something that currently is not being done. Economy studies thus provide information upon which investment and managerial decisions about future operations can be based. Thus the engineering economic analyst might be termed an *alternatives fortune-teller*.

After a decision to invest capital in a project has been made and the capital has been invested, those who supply and manage the capital want to know the financial results. Therefore, procedures are established so that financial events relating to the investment can be recorded and summarized and financial productivity determined. At the same time, through the use of proper financial information, controls can be established and utilized to aid in guiding the venture toward the desired financial goals. General accounting and cost accounting are the procedures that provide these necessary services in a business organization. Accounting studies thus are concerned with *past* and *current* financial events. Thus the accountant might be termed a *financial historian*.

The accountant is somewhat like a data recorder in a scientific experiment. Such a recorder reads the pertinent gauges and meters and records all the essential data during the course of an experiment. From these it is possible to determine the results of the experiment and to prepare a report. Similarly, the accountant records all significant financial events connected with an investment, and from these data he or she can determine what the results have been and can prepare financial reports. Just as an engineer can, by taking cognizance of what is happening during the course of an experiment and making suitable corrections, gain more information and better results from the experiment, managers must also rely on accounting reports to make corrective decisions in order to improve the current and future financial performance of the business.

Accounting is generally a source of much of the past financial data that are needed in making estimates of future financial conditions. Accounting is also a prime source of data for *postmortem,* or after-the-fact, analyses that might be made regarding how well an investment project has turned out compared to the results that were predicted in the economy study.

A proper understanding of the origins and meaning of accounting data is needed in order to properly use or not use that data in making projections into the future and in comparing actual versus predicted results.

ACCOUNTING FUNDAMENTALS

Accounting is often referred to as the language of business. Engineers should make serious efforts to learn about a firm's accounting practice so that they can better communicate with top management. This section contains an extremely brief and simplified exposition of the elements of financial accounting in recording and summarizing transactions affecting the finances of the enterprise. These fundamentals apply to any entity (such as an individual, corporation, governmental unit, etc.), called here a ''firm.''

All accounting is based on the *fundamental accounting equation,* which is

$$\text{assets} = \text{liabilities} + \text{ownership} \tag{1-4}$$

where ''assets'' are those things of monetary value that the firm *possesses,* ''liabilities'' are those things of monetary value that the firm owes, and ''ownership'' is the worth of what the firm owes to its stockholders (also referred to as "equities," "net worth," etc.). For example, typical accounts in each term of Equation 1-4 are as follows:

Asset Accounts	=	Liability Accounts	+	Ownership Accounts
Cash		Long-term debt		
Receivables		Payables		Capital stock
Inventories		Short-term debt		
Equipment				Retained earnings
Buildings				(income retained in the firm)
Land				

The fundamental accounting equation defines the format of the *balance sheet*, which is one of the two most common accounting statements, and which shows the financial position of the firm at any given point in time.

Another important, and rather obvious, accounting relationship is

$$\text{revenues} - \text{expenses} = \text{profit (or loss)} \qquad (1\text{-}5)$$

This relationship defines the format of the *income statement* (also commonly known as a "profit-and-loss statement"), which summarizes the revenue and expense results of operations *over a period of time*. Equation 1-4 can be expanded to take account of profit defined in Equation 1-5:

$$\text{assets} = \text{liabilities} + (\text{beginning ownership} + \text{revenue} - \text{expenses}) \qquad (1\text{-}6)$$

Profit is the increase in money value, resulting from a firm's operations, which is available for distribution to stockholders. It therefore represents the return on owner's invested capital.

A useful analogy is that a balance sheet is like a "snapshot" of the firm at an instant in time, while an income statement is a summarized "moving picture" of the firm over an interval of time. It is also useful to note that a revenue serves to increase the ownership amount for a firm, while an expense serves to decrease the ownership amount for a firm.

To illustrate the workings of accounts in reflecting the decisions and actions of a firm, suppose that an individual decides to undertake an investment opportunity and that the following sequence of events occurs over a period of 1 year:

1. Organize XYZ firm and invest $3,000 cash as capital.
2. Purchase equipment for a total cost of $2,000 by paying cash.
3. Borrow $1,500 through note to bank.
4. Manufacture year's supply of inventory through the following:
   (a) Pay $1,200 cash for labor.
   (b) Incur $400 account payable for material.
   (c) Recognize the partial loss in value (depreciation) of the equipment amounting to $500.
5. Sell on credit all goods produced for year, 1,000 units at $3 each. Recognize that the accounting cost of these goods is $2,100, resulting in an increase in equity (through profits) of $900.

6. Collect $2,200 of account receivable.
7. Pay $300 account payable and $1,000 of bank note.

A simplified version of the accounting entries recording the same information in a format that reflects the effects on the fundamental accounting equation (with a " + " denoting an increase and a " − " denoting a decrease) is shown in Table 1-1. A summary of results is shown in Figure 1-1.

It should be noted that the profit for a period serves to increase the value of the ownership in the firm by that amount. Also, it is worth noting that the net cash flow from operation of $700 (= $2,200 − $1,200 − $300) is not the same as profit. This was recognized in transaction 4(c), in which capital consumption (depreciation) for equipment of $500 was declared. Depreciation serves to convert part of an asset into an expense which is then reflected in a firm's profits, as seen in Equation 1-5. Thus the profit was $900, or $200 more than the net cash flow. For purposes of financial accounting, revenue is recognized when it is earned and expenses are recognized when they are incurred.

One very important indicator of after-the-fact financial performance that can be obtained from Figure 1-1 is "annual rate of return," defined in Equation 1-3. If the invested capital is taken to be the owners' (equity) investment, the annual rate of return at the end of this particular year can be found to be $900/$3,900 = 23%.

XYZ FIRM
BALANCE SHEET
AS OF DEC. 31, 19xx

Assets		Liabilities and Ownership	
Cash	$2,200	Bank note	$ 500
Accounts receivable	800	Accounts payable	100
Equipment	1,500	Equity	3,900
TOTAL	$4,500	TOTAL	$4,500

XYZ FIRM
INCOME STATEMENT
FOR YEAR ENDING DEC. 31, 19xx

Operating revenues (Sales)		$3,000
Operating costs (Inventory depleted)		
Labor	$1,200	
Material	400	
Depreciation	500	
		$2,100
Net income (Profits)		$ 900

FIGURE 1-1 Balance sheet and income statement as a result of transactions shown in Table 1-1.

TABLE 1-1 Accounting Effects of Transactions—XYZ Firm

	Account			Transaction					Balances at End of Year
		1	2	3	4	5	6	7	
Assets {	Cash	+$3,000	-$2,000	+$1,500	-$1,200		+2,200	-$1,300	+$2,200
	Account receivable					+$3,000	-2,200		+ 800
	Inventory				+ 2,100	- 2,100			0
	Equipment		+ 2,000		- 500				+ 1,500
									$4,500
equals									
Liabilities {	Account payable				+ 400			- 300	+ 100
	Bank note			+ 1,500				- 1,000	+ 500
plus									
Ownership {	Equity	+ 3,000				+ 900			+ 3,900
									$4,500

13

Financial statements are usually most meaningful if figures are shown for 2 or more years (or other reporting period such as quarters or months), or for two or more individuals or firms. In so doing the figures can be used to reflect trends or comparative financial indications which are useful in enabling investors and management to determine the effectiveness of investments *after* they have been made.

COST ACCOUNTING

Cost accounting, or management accounting, is a phase of accounting that is of particular importance to the economy-study analyst because it is concerned principally with decision making and control in a firm. Consequently, cost accounting is the source of much of the cost data that are needed in making economy studies. Modern cost accounting may satisfy any or all of the following objectives:

1. Determination of the actual cost of products or services.
2. Provision of a rational basis for pricing goods or services.
3. Provision of a means for controlling expenditures.
4. Provision of information on which operating decisions may be based and by means of which operating decisions may be evaluated.

Although the basic objectives of cost accounting are simple, the exact determination of costs usually is not. As a result, some of the procedures used are arbitrary devices that make it possible to obtain reasonably accurate answers for most cases but which may contain a considerable percentage of error in other cases.

THE ELEMENTS OF COST

One of the first problems in cost accounting is that of determining the elements of cost that arise in the production of an article or the rendering of a service. A study of how these costs occur gives an indication of the accounting procedure that must be established to give satisfactory cost information. Also, an understanding of the procedure that is used to account for these costs makes it possible to use them more intelligently and correctly.

From an engineering and managerial viewpoint in manufacturing enterprises, it is common to consider the general elements of cost to be *direct materials, direct labor,* and *overhead.* Such terms as *burden* and *indirect costs* are often used synonymously with overhead, and overhead costs are often divided into several subcategories.

Ordinarily, the materials that can be conveniently and economically charged directly to the cost of the product are called direct materials. Several guiding principles are used when we decide whether a material is classified as a direct material. In general, direct materials should be readily measurable, be of the same quantity in identical products, and be used in economically significant amounts. Those materials that do not meet these criteria are classified as *indirect materials* and are a part of the charges for overhead. For example, the exact amount of glue used in making a chair would be difficult to determine. Still more difficult would be the measurement of the exact amount of coal that was used to produce the steam that generated the electricity that was used to heat the glue. Some reasonable line must be drawn beyond which no attempt is made to measure directly the material that is used for each unit of production.

Labor costs also are ordinarily divided into *direct* and *indirect* categories. Direct labor costs are those which can be conveniently and easily charged to the product or service on which worked. Other labor costs, such as for supervisors and material handlers, are charged as indirect labor and are thus included as part of overhead costs. It is often imperative to know what is included in direct labor and direct material cost data before attempting to use them in economy studies.

In addition to indirect materials and indirect labor there are numerous other cost items that must be incurred in the production of products or the rendering of services. Property taxes must be paid; accounting and personnel departments must be maintained; buildings and equipment must be purchased and maintained; supervision must be provided. It is essential that these necessary *overhead* costs be allocated to each unit produced in proper proportion to the benefits received. Proper allocation of these overhead costs is not easy, and some factual, yet reasonably simple, method of allocation must be used.

As might be expected where solutions attempt to meet conflicting requirements, such as exist in overhead-cost allocation, the resulting procedures are empirical approximations that are accurate in some cases and less accurate in others.

There are many methods of allocating overhead costs among the products or services produced. The most commonly used methods involve allocation in proportion to direct labor cost, or direct labor hours, or direct materials cost, or sum of direct labor and direct materials cost, or machine hours. In each of these methods it is necessary to know what the total of the overhead costs has been if postmortem costs are being determined, or to estimate what the total overhead costs will be if predicted or standard costs are being determined. In either case, *the total overhead costs will be associated with a certain level of production.* This is an important condition that should always be remembered when we are dealing with unit-cost data. They can only be correct for the conditions for which they were determined.

To illustrate one method of allocation of overhead costs, consider the method which assumes that overhead is incurred in direct proportion to the cost of direct labor used. With this method the overhead rate (overhead per dollar of direct labor) and the overhead cost per unit would be

$$\text{overhead rate} = \frac{\text{total overhead in dollars for period}}{\text{direct labor in dollars for period}}$$

$$\text{overhead cost per unit} = \text{overhead rate} \times \text{direct labor cost per unit} \quad (1\text{-}7)$$

Suppose that for a coming period (say quarter) the total overhead cost is expected to be $100,000 and the total direct labor cost is expected to be $50,000. From this, the overhead rate = $100,000/$50,000 = $2 per dollar of direct labor cost. Suppose further that for a given unit of production (or job) the direct labor cost is expected to be $60. From Equation 1-7, the overhead cost for the unit of production would be 60 × $2 = $120.

This method obviously is simple and easy to apply. In many cases it gives quite satisfactory results. However, in many other instances, it gives only very approximate results because some items of overhead, such as depreciation and taxes, have very little relationship to labor costs. Quite different total costs may be obtained for the same product when different procedures are used for the allocation of overhead costs. The magnitude of the difference will depend on the extent to which each method produces or fails to produce results that are in accord with the facts. The choice of an overhead allocation method should be based on what seems most reasonable under the circumstances.

THE USE OF ACCOUNTING COSTS
IN ECONOMY STUDIES

When we recognize that accounting costs are linked to a definite set of conditions, and that they are the result of certain arbitrary decisions concerning overhead-cost allocation, it is apparent that they should not be used without modification in cases where the conditions are different from those for which they were determined. Economy studies invariably deal with situations that now are *not* being done. Thus ordinary accounting costs normally cannot be used, without modification, in economy studies. However, if we understand how the accounting costs were determined, we should be able to break them down into their component elements and then, often, we find that these cost elements will supply much of the cost information that is needed for an economy study. Thus an understanding of the basic objectives and procedures of cost accounting will enable the economy-study analyst to make best use of available cost information and to avoid needless work and serious mistakes.

One should not assume that the figures contained in accounting reports are absolutely correct and indicative, even though they have been prepared with the utmost care by highly professional accountants. This is because accounting procedures often must include certain assumptions that are based on subjective judgment. For example, the years of life on which depreciation expense for a particular asset is based has to be determined or assumed and the estimate may turn out to have caused unrealistic depreciation expenses and book values in accounting reports. Also, there are many accepted practices in accounting that may provide unrealistic information for management control purposes. For example, the net book value of an asset is generally declared in the balance sheet at the original first cost price minus any accumulated depreciation, even though it may be recognized that the true value of the asset at a particular time is far above or below this reported book value.

THE FLOW OF CAPITAL WITHIN A FIRM

Figure 1-2 is a schematic diagram of the flow of capital within a typical firm. It shows how funds are generated with which the firm continues as a going concern from year to year. New debt and equity capital is initially obtained from external sources so that investments in buildings, land, and equipment can be made and supported with sufficient amounts of working capital to meet payrolls, build inventories, and so forth. As gross revenues (R) are produced through the sale of goods and/or services, they are reduced by operating ex-

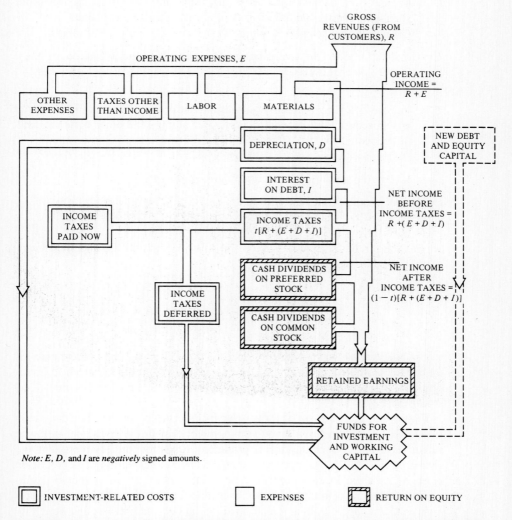

Note: E, D, and I are *negatively* signed amounts.

INVESTMENT-RELATED COSTS   EXPENSES   RETURN ON EQUITY

FIGURE 1-2 Flow of capital within a firm during a year. (Courtesy of the Commonwealth Edison Company, Chicago, Ill.)

penses (E) in arriving at operating income. Then it can be seen that operating income has subtracted from it depreciation and interest on debt capital to obtain net income before income taxes (NIBT), which is often referred to as "profit before income taxes."

Two classes of profit are of concern in economy studies. The first is profit before income taxes; the other is profit after income taxes. Operating expenses and investment-related costs may conveniently be divided into several classes as follows:

E = out-of-pocket expenses; these are current and tend to be proportional to the extent of the business activity and thus can be controlled to a degree

D = depreciation; recognition of cost due to loss in value of assets over time, such as property, buildings, or equipment; as will be explained in Chapter 9, the annual magnitude of this cost can only be estimated

I = interest paid for the use of borrowed capital: a financial cost due to the use of some borrowed money; if only equity capital is used, this cost does not exist

T = income taxes; costs to an organization that obviously are not ordinary costs because they depend on profits remaining after other costs are paid or accounted for (the effect of income taxes in engineering economy studies is the topic of Chapter 10)

t = income tax rate used for computing income taxes

Using this notation, we can write Equation 1-5 in a more specific form:

$$\text{NIBT} = R + (E + D + I) \tag{1-8}$$

where NIBT represents profits before income taxes and E, D, and I are negatively signed amounts. Similarly,

$$\text{NIAT} = \text{NIBT} - T \tag{1-9}$$

$$= R + (E + D + I) - t\,[R + (E + D + I)]$$

$$= (1 - t)[R + (E + D + I)] \tag{1-10}$$

where NIAT represents profits after income taxes. Because the interest paid for the use of borrowed capital is included as a cost, the profits indicated in Equations 1-8 to 1-10 are those belonging to the owners of equity capital (i.e., stockholders).

Figure 1-2 indicates that profits after income taxes are distributed as cash dividends on preferred stock and common stock, with the remainder going to retained earnings. Once the firm is profitable, funds for further investment and increased working capital are accumulated from retained earnings, depreciation reserves, deferred income taxes, and new equity–debt capital as shown. Additional investments can then be made to enable the firm to prosper over time.

Investment capital is thus transformed into goods and services that hopefully result in an excess of revenues over expenses. This excess must be sufficient to cover investment-related costs such as depreciation, interest on debt, income taxes, and profits to pay for the use of equity capital.

PROBLEMS

1-1 Discuss the concept of financial productivity and explain why it is different from thermal efficiency, for example.

1-2 Explain why differences between alternatives are of major interest in engineering economy studies.

1-3 Why do engineers get involved with economy studies? Why would it not be better for an accountant to perform economy studies? Explain the differences in viewpoint here.

1-4 Describe three situations in which monetary comparisons among alternatives could be less important than nonmonetary differences among them.

1-5 What is the difference between equity capital and debt capital? What constitutes "net worth" in the fundamental accounting equation? Give examples.

1-6 In your own words, explain the difference between the basic purposes of engineering economy and accounting.

1-7 Explain why profit is essential in business ventures.

1-8 Will the increased use of automation increase the importance of engineering economy studies? Why or why not?

1-9 What are some of the basic causes of uncertainty in economy studies?

1-10 How would you attempt to deal with multiple objectives in an engineering economy problem?

1-11 What is rational decision making, and why is flipping a coin not necessarily a rational way to make a decision?

1-12 List the six phases of the decision-making process and explain why cycling back through the process may be required in some situations.

1-13 What would occur if, through legislation, all profit were eliminated from a certain industry? Would this necessarily assure a decrease in the selling price of the product of this industry?

1-14 Explain why ordinary accounting data may not be useful in an economy study.

1-15 Explain the differences in the purposes of the two primary financial statements (balance sheets and income statements).

1-16 You have started your own business producing precision-wound induction coils. At the end of the first year the following sequence of events has occurred.
(1) Organized firm and invested $2,200 as capital.
(2) Purchased equipment worth $1,500 financed by a note with the bank.
(3) Manufactured year's inventory by:
    (a) $1,200 labor expense.
    (b) $500 materials expense on account.
    (c) $450 depreciation
(4) All of the 5,000 units produced were sold on account for $0.60 apiece.
(5) Collected $2,000 of accounts receivable.
(6) Paid $500 on accounts payable and $750 on the bank note.
Prepare (a) a balance sheet at the end of the year and (b) an income statement for the year.

1-17 Refer to Figure 1-1.
(a) Compute the ratio of cash and accounts receivable to the bank note, and comment on what this ratio might tell you.
(b) Also determine the ratio of total operating costs to operating revenues. Is a high ratio desirable?
(c) Compute the ratio of profit to operating revenues. Why is this probably a less valid measure of financial productivity than the annual rate of return in this situation?

1-18 Does a company that makes a single, multipart product need a good cost accounting system? Why?

1-19 What are the three common elements of cost in cost accounting? Why must some cost accountants also be able to do cost estimating work?

1-20 One product of a multiproduct company was produced by a group of four skilled and experienced workers, who were paid $8 per hour. They worked 8 hours per day, 250 days per year, in an area of 400 square feet in a single-story building of light construction. They required very little supervision, and the labor turnover was almost zero. The cost of the material in the product was $8 per unit. The equipment used by these workers cost $3,200 and had an expected life of 4 years. It consumed 1 kilowatt of electricity per hour. The company followed a policy of charging overhead to all its products on the basis of 125% of direct labor cost. The company was considering dropping this item from its product line because the selling price was only slightly greater than the computed product cost. Comment on this situation.

1-21 Under normal conditions, the Apex Company makes products A, B, and C in such quantities that the direct labor costs per day are $400, $500, and $600 for the three products, respectively, and the total overhead is just absorbed (covered) when it is charged at the rate of 150% of direct labor. Because of a change in market conditions, the sales of product C have been reduced by 50% and those of product A have doubled. The total overhead has not changed. To what percentage of direct labor cost must the overhead charge be reduced to avoid overcharging for overhead? (Assume that direct labor and direct material costs are directly proportional to output.)

1-22 List four basic sources of funds that a company can use for capital investment and working capital.

1-23 In a certain year, MNX Enterprises has a gross revenue of $7,400,000. Its operating expenditures amount to a total of $4,250,000 for the year. Depreciation and interest are $1,200,000 and $420,000, respectively. If income taxes paid during the year amount to $460,000, answer the following:
(a) What is the operating income for the year?
(b) What is the profit before income taxes?
(c) What is the profit after income taxes?

CHAPTER 2

The Economic Environment and Cost Concepts

There are certain economic principles that frequently must be taken into account in economy studies. In many instances, economic concepts aid materially in managerial decision making.

In general terms, economics deals with the interactions between people and wealth. Because people, as individuals, are not all alike as to their reactions, the subject of economics necessarily must deal with these interactions in generalized terms. Also, in most instances we cannot quantify the results of these interactions in absolute, monetary terms. Thus they must be dealt with as "non-monetary" or "intangible" considerations. Nevertheless, their effects on the financial aspects of a venture can be just as real and certain as can the law of gravity on the physical aspects of a venture. The purpose of this chapter is to examine, briefly, some of these basic economic concepts and to indicate how they may be factors for consideration in making economy studies and managerial decisions.

CONSUMER AND PRODUCER GOODS AND SERVICES

The goods and services that are produced and utilized may be divided conveniently into two classes. *Consumer goods and services* are those products or services that are directly used by people to satisfy their wants. Food, clothing, homes, cars, television sets, haircuts, opera, and medical services are examples. The producers and vendors of consumer goods and services must be aware of, and are subject to, the whims and caprices of the people to whom their products are sold. At the same time, the demand for such goods and services is directly related to people and in many cases, as will be discussed later, may be determined with considerable certainty.

Producer goods and services are used to produce consumer goods and services or other producer goods. Machine tools, factory buildings, buses, and farm machinery are examples. Although, in the long run, producer goods serve to satisfy human wants, they are the means to that end. Thus the amount of producer goods needed is determined indirectly by the amount of consumer goods or services that are demanded by people. However, because the relationship is much less direct than for consumer goods and services, the demand for, and production of, producer goods may greatly precede or lag behind the demand for the consumer goods that they will produce. For example, a company may build and equip a factory to manufacture a product for which there presently is no demand, in the hope that it will be able to create such a demand through advertising and sales effort. On the other hand, a company may wait until the demand for an existing product is well established before deciding to build facilities to produce the product. Thus the decision to buy or build producer goods is not a direct result of the demand for the consumer goods or services they can produce, but is related directly to management's estimate of the profit potential of the facilities. This is quite a different situation from that for many consumer goods—food, for example—where the consumption cannot be deferred indefinitely.

It can be seen that if an economy study involves investment in production facilities, the problem of estimating the demand for the product or service will be quite different, depending on whether producer goods or services or consumer goods or services are involved. Furthermore, the effects of variations in economic conditions could be quite different for the two cases.

MEASURES OF ECONOMIC WORTH

Goods and services are produced, and desired, because directly or indirectly they have *utility*—the power to satisfy human wants and needs. Thus they may be used or consumed directly, or they may be used to produce other goods or services that may, in turn, be used directly. Utility most commonly is measured in terms of *value*, expressed in some medium of exchange as the *price* that must be paid to obtain the particular item.

Much of our business activity revolves around increasing the utility (value) of materials and products by changing their form or location. Thus iron ore, worth only a few dollars per ton, may be increased in value to several dollars a pound by processing it and combining it with suitable alloying elements and converting it into razor blades. Similarly, snow, worth almost nothing when high in distant mountains, can be made quite valuable when it is delivered in melted form several hundred miles away to dry southern California.

NECESSITIES, LUXURIES, AND PRICE–DEMAND

Goods and services may be divided into two types, *necessities* and *luxuries*. Obviously, these terms are relative, because for most goods and services what one person may consider to be a necessity may be considered by another to be a luxury. Economic status is an important factor in one's views regarding luxuries and necessities. Other factors also may be determining. For example, a man living in one community may find that an automobile is an absolute necessity for him to be able to get to and from his place of employment. If the same man lived and worked in a different city, he might have adequate public transportation available, and an automobile would be strictly a luxury. Also, the classification of goods and services into luxuries and necessities is less easy in the case of producer goods than for consumer goods. However, it is clear that for all goods and services there is a relationship between the price that must be paid and the quantity that will be demanded, or purchased.

This general relationship is depicted in Figure 2-1; as the selling price is increased, there will be less demand for the product; and as the selling price is decreased, the demand will increase. The relationship between price and demand can be expressed as a linear function:

$$P = a - bD \qquad \text{for } 0 \le D \le \frac{a}{b} \qquad (2\text{-}1)$$

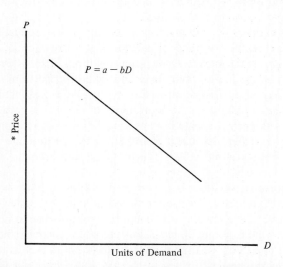

FIGURE 2-1 General price–demand relationship. Note that "price" is considered to be the independent variable but is shown as the vertical axis. This convention is commonly used by economists; because it may be somewhat confusing to engineers, who usually show the independent variable as the abscissa, an asterisk has been added before the independent variable.

where a is the intercept on the price axis and $-b$ is the slope. Thus b is the amount by which demand increases for each unit decrease in P. Both a and b are constants. It follows, of course, that

$$D = \frac{a - P}{b} \tag{2-2}$$

Although Figure 2-1 illustrates the general relationship between price and demand, it would be expected that this relationship would be different for necessities and luxuries. Consumers can readily forgo the consumption of luxuries if the price is greatly increased, but they find it more difficult to reduce their consumption of true necessities. Also, they will use the money saved by not buying luxuries to pay the increased cost of the necessities. Figure 2-2 shows how the demand curves for luxuries and necessities might differ.

The extent to which price changes influence demand varies according to the *elasticity* of the demand. The demand for products is said to be *elastic* when a decrease in the selling price results in a considerable increase in sales. On the other hand, if a change in selling price produces little or no effect on the demand, the demand is said to be *inelastic*. It is clear, from Figure 2-2, that luxury items have greater elasticity of demand than do necessities.

COMPETITION

Because economic laws are general statements regarding the interaction of people and wealth, they will be affected by the economic environment in which the people and the wealth exist. Most general economic principles are stated for situations in which *perfect competition* exists.

Perfect competition occurs in a situation in which any given product is supplied by a very large number of vendors and there is no restriction against additional vendors entering the market. Under such conditions, there is assurance of complete freedom on the part of both buyer and seller. Actually, of course, perfect competition may never exist because of a multitude of factors that impose some degree of limitation upon the actions of buyers or sellers, or both. However, with conditions of perfect competition assumed, it is easier to formulate general economic laws. When deviations from perfect competition are known to exist, their probable economic effects can be taken into account, at least approximately.

The existing competitive situation is an important factor in most economy studies. It will have a very real effect upon decisions that are made. Unless information is available to the contrary, it should be assumed that competition does or will exist, and the resulting effects should be taken into account.

Monopoly is at the opposite pole from perfect competition. A perfect monopoly exists when a unique product or service is available from only a single vendor and that vendor can prevent the entry of all others into the market. Under such conditions the buyer is at the complete mercy of the vendor as to the availability and price of the product. Actually, there seldon is a perfect monopoly. This is due to the fact that few products are so unique that substitutes cannot

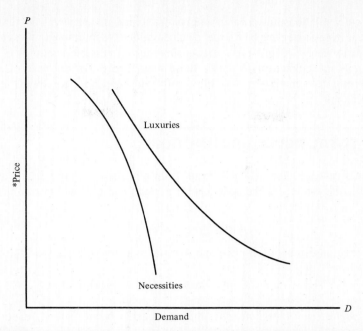

FIGURE 2-2   Generalized price–demand relationship for luxuries and necessities. Note that "price" is considered to be the independent variable but is shown as the vertical axis.

be used satisfactorily, or to the fact that governmental regulations prohibit monopolies if they are unduly restrictive.

A monopoly may be of great benefit to a producer in that he may be able to control the supply and the price to provide the maximum profit. Although this might result in high prices for the product, higher *total* profits may be obtained from lower prices and wider distribution. Under some conditions, a monopoly may avoid costly duplication of facilities and thus make possible lower prices for products and services. This situation is recognized by governing bodies in granting public utilities exclusive rights to render service in a given territory. Such a practice, for example, avoids having two electric power companies duplicate power-distribution lines and equipment in the same city. When vendors are granted such a monopolistic position, the governmental body also regulates the rates that the utility can charge for its services to assure that the customers are not overcharged. Thus governmental regulation takes the place of competition in determining prices.

An *oligopoly* exists when there are so few suppliers of a product or service that action by one will almost inevitably result in similar action by the others. Thus if one of the only three gasoline stations in an isolated town raises the price of its product by 1 cent per gallon, the other two would probably do the same because they could do so and still retain their previous competitive positions.

It is apparent that many conditions between those of perfect competition, perfectly monopoly, and oligopoly can and do exist. Thus it is to be expected that *actual* relationships among supply, price, and demand will be somewhat different from those derived for the ideal conditions. Also, it would be extremely difficult to state or depict such relationships for all possible conditions. Therefore, it is most practical to derive these relationships for the case of perfect competition; the changes that would result from nonperfect conditions are usually apparent.

THE TOTAL-REVENUE FUNCTION

The total revenue, TR, that will result from a business venture during a given period is the product of the selling price per unit and the number of units sold. Thus

$$TR = price \times demand \tag{2-3}$$

If the relationship between price and demand as given in Equation 2-1 is used,

$$TR = (a - bD)D = aD - bD^2 \qquad for\ 0 \le D \le \frac{a}{b} \tag{2-4}$$

If we neglect any cost functions, the relationship between total revenue and demand *for the condition expressed in Equation 2-4* may be represented by the curve shown in Figure 2-3. Under these conditions, the maximum total revenue also would produce maximum total profit. From calculus the demand, D^*, that will produce maximum total revenue can be obtained by solving

$$\frac{d\,TR}{dD} = a - 2bD = 0 \tag{2-5}$$

Thus†

$$D^* = \frac{a}{2b} \tag{2-6}$$

†D^* means the optimal value of D.

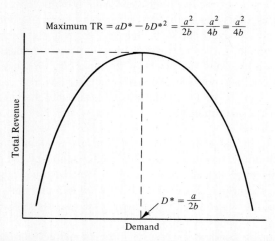

Maximum $TR = aD^* - bD^{*2} = \dfrac{a^2}{2b} - \dfrac{a^2}{4b} = \dfrac{a^2}{4b}$

Total Revenue

$D^* = \dfrac{a}{2b}$

Demand

FIGURE 2-3 Total-revenue function as a function of demand.

For example, if the equation for price is given by $50,000 - 200D$, the demand, D^*, that maximizes total revenue is equal to $50,000/400 = 125$ units. It must be emphasized that, because of cost–volume relationships, most businesses would not obtain maximum profits by maximizing revenue. Thus the cost–volume relationship must be considered and related to revenue.

At this point, attention is called to the derivative of the total revenue with respect to volume (demand), $d\text{TR}/dD$, which is called the *incremental* or *marginal revenue*; this will be discussed later in more detail.

COST–VOLUME RELATIONSHIPS

In virtually all businesses, there are certain costs that remain constant over a wide range of activity as long as the business does not permanently discontinue operations. Such costs as property taxes, interest on borrowed capital, and many of the overhead costs are of this type. Such costs commonly are referred to as *fixed costs*. There are other costs, however, that vary more or less directly with the volume of output, and these are called *variable costs*. These include the costs of such items as materials, labor, and power.† Using this concept of fixed and variable costs, their relationship to the total-cost function is portrayed in Figure 2-4.‡ Thus at any demand D,

$$C_{\text{total}} = C_F + C_v \tag{2-7}$$

where C_F and C_v denote fixed and variable costs, respectively. For the linear relationship, assumed here,

$$C_v = (v)(D) \tag{2-8}$$

where v is the variable cost per unit.

†Later in this chapter, a more detailed description of fixed and variable costs is provided.

‡In actual cases the variable costs frequently are not exactly proportional to volume, and thus the total-cost line is not exactly linear. However, in most cases the actual total-cost line does not deviate much from a straight line, and a straight line relationship is used here for simplification.

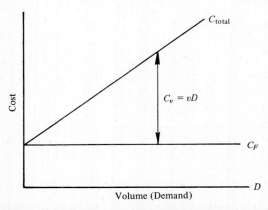

FIGURE 2-4 Typical fixed, variable, and total costs as functions of volume.

When the total revenue–demand relationship, as depicted in Figure 2-3, and the total cost–demand relationship, as depicted in Figure 2-4, are combined, Figure 2-5 results for the case where $(a - v) > 0$, and the volumes for which profit and loss result are clearly evident. We obviously are interested in the condition for which maximum profit will be obtained. First,

$$\text{profit} = \text{total revenue} - \text{total cost.}$$

$$= (aD - bD^2) - (C_F + vD)$$

$$= -C_F + (a - v)D - bD^2 \qquad \text{for } 0 \le D \le \frac{a}{b}$$

We now take the first derivative with respect to D and set it equal to zero:

$$\frac{d(\text{profit})}{dD} = a - v - 2bD = 0$$

The value of D that maximizes profit is

$$D^* = \frac{a - v}{2b} \qquad\qquad (2\text{-}9)$$

If $(a - v) \le 0$, the profit will be maximized when $D = 0$.†

To illustrate Equation 2-9, suppose again that $P = 50,000 - 200D$, where D is demand per month and P is the price in dollars. The fixed costs are \$500 per month and variable costs amount to \$5,000 per unit. The number of units that maximizes monthly profits is

$$D^* = \frac{50,000 - 5,000}{400} = 112.5$$

†This relationship simply says that if the variable cost per unit is equal to or greater than the price that will produce *no* demand, nothing should be produced; thus losses are minimized.

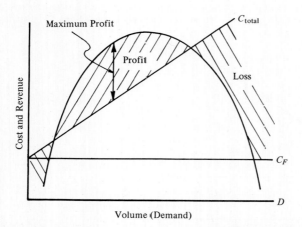

FIGURE 2-5 Combined cost and revenue functions as functions of volume, and their effect on profit.

BACKGROUND AND TOOLS

and the corresponding maximum profit is

$$-500 + (45,000)112.5 - 200(112.5)^2 = \$2,530,750$$

Of course, D^* would probably be rounded to 112 or 113 in an actual production problem.

The condition for maximum profit as expressed in Equation 2-9 becomes more meaningful and useful when it is related to the incremental revenue, Equation 2-5. If we substitute Equation 2-9, the condition for maximum profit, into Equation 2-5, we obtain

$$\frac{d\text{TR}}{dD} = a - \frac{2b(a-v)}{2b} = v \tag{2-10}$$

Because v is the variable cost per unit, Equation 2-10 means that in order to obtain maximum profit, we should increase the output as long as the incremental revenue exceeds the incremental production cost, stopping when they are equal.

THE LAW OF SUPPLY AND DEMAND

As was discussed previously and illustrated in Figure 2-1, under competitive conditions there is a relationship between the price customers must pay for a product and the amount that they will buy. There is a similar relationship between the price at which a product can be sold and the amount that will be made available. If the price they can get for their products is high, more producers will be willing to work harder, or perhaps risk more capital, in order to reap the greater reward. If the price they can obtain for their products declines, they will not produce as much because of the smaller reward that they can obtain for their labor and risk. Some will stop producing and turn their efforts to other endeavors. This relationship between price and the volume of product produced can be portrayed by the curve shown in Figure 2-6.

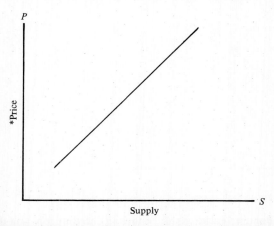

FIGURE 2-6   General price–supply relationship. Note that "price" is considered to be the independent variable but is shown as the vertical axis.

If Figures 2-1 and 2-6 are combined, Figure 2-7 results. Figure 2-7 illustrates the basic economic *law of supply and demand,* which states that under conditions of perfect competition, the price at which a given product will be supplied and purchased is the price that will result in the supply and the demand being equal.

Because many economy studies deal with investments that will increase the amount of a given product that will be available on the market, we are interested in what will happen to the selling price of the commodity under the proposed conditions. If a producer is willing to supply additional volume of a product to the market at existing prices, this means that a new price–supply condition has been created. As shown in Figure 2-8, at the original price P_1, an additional amount of a product is made available and a new price–supply curve exists. Because there has been no change in the price–demand relationship meanwhile, the intersection of the new supply curve results in a new and lower price, P_2, corresponding to the new demand, D_2. Obviously, a reverse situation results from a decrease in the amount of a product offered to the market at a given price.

THE LAW OF DIMINISHING RETURNS

The law of diminishing returns is another basic economic principle that frequently is a factor in economic decision making and helps to provide an understanding of conditions that often are encountered in economy studies. All en-

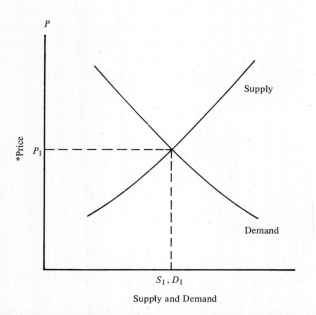

FIGURE 2-7 Price–supply–demand relationship, showing equal supply and demand at a given price. Note that "price" is considered to be the independent variable but is shown as the vertical axis.

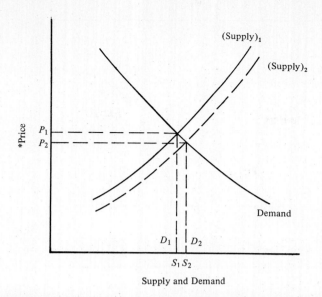

Price
P_1
P_2

(Supply)$_1$

(Supply)$_2$

Demand

D_1 D_2

$S_1 S_2$

Supply and Demand

FIGURE 2-8 Price–supply–demand relationship, showing how the addition of supply, at a given price, will cause a new and lower price to be established. Note that "price" is considered to be the independent variable but is shown as the vertical axis.

gineers and engineering students have encountered the results of this principle, but often they do not realize that what they have encountered is related to this law. Figure 2-9, for example, is such a case. Basically, the law refers to the amount of *extra output* that results from successive *additions* of equal units of *variable* input to some *fixed* (or limited) amount of some *other input factor*. The law may be stated as follows: When one or more of the input factors of production is limited, either by absolute quantity or by increasing cost, adding more of the variable input factors will result in an output's being reached beyond which such additions will result in a less-than-proportionate increase in output.

For an engineering problem, Table 2-1 shows the data from which Figure 2-9 was obtained. It will be noted that for the early increases in input, up to an input of 4.0 kilowatts (kW), the actual increase in output is greater than proportional;† beyond this point, the output is less than proportional. In this case the fixed input factor is the electric motor. As more and more of the variable factor—electrical energy—is added, each kilowatt has a smaller portion of the motor on which it can act, thus reaching an ultimate point beyond which the efficiency decreases.

As mentioned previously, the results of the principle of diminishing returns are found with great frequency. Some examples are all types of prime movers, office and apartment buildings, and production facilities where equipment cannot be expanded in capacity. In each case there is an input factor that is limited by

---

†In this range each successive increase of 0.5 kilowatt in input results in a greater increment of output than did the previous 0.5 kilowatt.

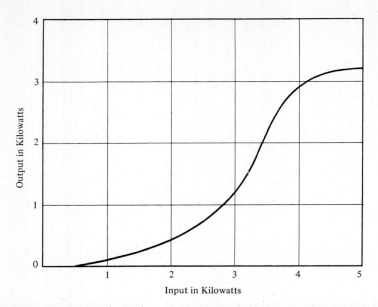

FIGURE 2-9   Input—output curve for a $3\frac{1}{2}$-horsepower constant-speed motor.

increasing cost or absolute quantity. It should also be recognized that capital and markets may also constitute limited input factors because of the increasing cost of obtaining them. Thus diminishing returns is an important consideration in many investment decisions.

UNIT COST FUNCTIONS

Most business and engineering projects are designed to operate at certain capacities. The level of output at which they are operated usually has a marked effect upon their economy and profitability as well as upon their physical efficiency. The principle of diminishing returns plays an important part in such situations because some of the input factors of production are limited.

To study the effect of diminishing returns upon unit cost where two factors of production are involved and one is fixed, examine Table 2-2. The first two columns of figures are the same as in Table 2-1 and concern the same test of a constant-speed electric motor. The fixed factor in this case is the motor. The third column shows the number of units of the variable factor that would be required so that 1 unit (1.0 kilowatt) of output will be obtained. Thus, in line B, since 1.0 kilowatt of input produced an output of only 0.1 kilowatt, to obtain 1.0 kilowatt of output 10 kilowatts of input would be required.

In the fourth column similar data are tabulated for the fixed factor, the motor. Thus, in line C, since one motor produced only 0.25 kilowatt of output, four motors would be required to produce 1.0 kilowatt of output if all were operated under the assumed conditions of load and efficiency.

TABLE 2-1 Load Test Data for $3\frac{1}{2}$-Horsepower Constant-Speed Motor Showing Proportional and Actual Output per Unit of Input

(1)	(2) Input (kW)		(3) Actual Output (kW)	(4) Proportional Output (kW)	(5) Output per Unit of Input, (3) ÷ (2)
A	0.5		0.00	—	0.000
B	1.0 ⎫	*increase*	0.10 × *150%* = ──	—	0.100
C	1.5 ⎭	*of 50%*	0.25 × *133%* = ──	0.15	0.167
D	2.0 ⎫	*increase*	0.41	0.33	0.205
E	2.5 ⎭	*of 33%*	0.70	0.51	0.280
F	3.0		1.20	0.84	0.400
G	3.5		2.20	1.40	0.628
H	4.0		2.92	2.52	0.730
I	4.5 ⎫	*increase*	3.15 × *111%* = ──	3.28	0.700
J	5.0 ⎭	*of 11%*	3.20	3.50	0.640

In order to determine the monetary cost of obtaining 1 unit of output, we must know the monetary value of the required units of input. In this case the variable input factor, electric energy, was assumed to cost 2 cents per kilowatthour. The cost of the fixed factor—the motor—was more difficult to determine. This cost is made up of the depreciation of the motor, rent on the building where it is housed, taxes, insurance, and other overhead costs. In this case it was assumed that the total of these costs was 8 cents per hour for one motor.

In this table it will be noted that the lowest total cost occurs at I, where the output is 3.15 kilowatts. In Table 2-1 the maximum mechanical efficiency occurred at H, where the output was only 2.92 kilowatts. The maximum output, however, is at J and is 3.20 kilowatts. This is a situation that is frequently difficult for engineers to accept. If maximum financial effectiveness (lowest unit

TABLE 2-2 Monetary Cost of Units of Output of a $3\frac{1}{2}$-Horsepower Constant-Speed Motor

	Input (kW)	Output (kW)	Input per Unit of Output		Monetary Unit Cost		Total Unit Cost (cents)
			Variable Factor	Fixed[a] Factor	Fixed Factor (at 8 cents)	Variable Factor (at 2 cents)	
A	0.5	0.00	—	—	—	—	—
B	1.0	0.10	10.00	10.00	80.00	20.00	100.00
C	1.5	0.25	6.00	4.00	32.00	12.00	44.00
D	2.0	0.41	4.88	2.44	19.52	9.76	29.28
E	2.5	0.70	3.57	1.43	11.44	7.14	18.58
F	3.0	1.20	2.50	0.83	6.64	5.00	11.64
G	3.5	2.20	1.59	0.45	3.60	3.18	6.78
H	4.0	2.92	1.37	0.34	2.72	2.74	5.46
I	4.5	3.15	1.43	0.32	2.56	2.86	5.42
J	5.0	3.20	1.56	0.31	2.48	3.12	5.60

[a]Number of motors to give 1.0-kW output.

cost) is desired, the motor should not be operated at either the point of maximum mechanical efficiency or maximum output. If a machine is to operate continuously at a fixed load, it would be desirable to have the three points of maximum output coincide. On the other hand, if the machine were to operate at three-quarters load 90% of the time, it would be desirable to have the points of maximum mechanical and financial effectiveness near 75% of capacity.

Situations of the type just discussed produce average unit-cost functions of the general type shown in Figure 2-10. Up to a point, as greater use is made of a production facility the cost per unit decreases until a minimum is reached. Expansion of production beyond this point results in greater cost per unit of output, because of the diminishing-returns principle. This condition sometimes is referred to as *the economy of mass production*. So we can continue to obtain lower unit cost with increasing volume, technological innovation must be achieved so as to break the restraints of limited input factors. We can summarize the situation by saying that *the average unit cost* will continue to decrease as long as it exceeds the incremental cost. If the incremental cost exceeds the existing average unit cost, it is apparent that the average unit cost will be increased if volume is increased. It then can be generalized that minimum average cost will be achieved when it is equal to the incremental cost. Because of their importance to engineering economy studies, incremental costs are discussed in more detail later in this chapter.

BREAKEVEN CHARTS

The total-revenue function expressed in Equation 2-4 and illustrated in Figure 2-3 was a generalized relationship involving a straight line relationship between price and demand, as illustrated in Figure 2-1. For a specific business the revenue function is the product of price and volume sold and thus is essentially a straight line over a considerable range of volume change. For this condition the relationship between volume and fixed costs, variable costs, and revenue is shown in Figure 2-11. These commonly are called *breakeven charts*. Breakeven charts

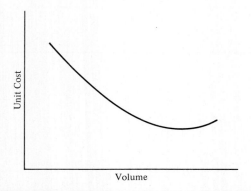

FIGURE 2-10   Unit-cost function—variation of average unit cost with volume.

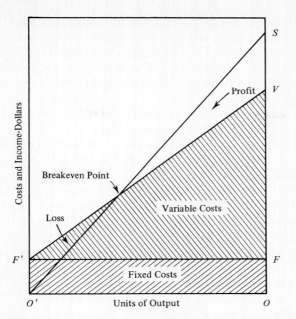

FIGURE 2-11   Typical breakeven chart for a business enterprise.

are very useful in portraying and understanding the effects of variations in fixed and variable costs on the profitability of a business. Thus they may be used to portray the effects of proposed changes in operational policy.

In breakeven charts the fixed costs, variable costs, and revenue are plotted against output, either in units, dollar volume, or percent of capacity. Thus in Figure 2-11 the line $F'F$ represents the fixed costs of production. The line $F'V$ shows the variation in total variable cost with production; because its starting point is at F', it actually represents the sum of all production costs. The gross revenue from sales is represented by the line $O'S$. Because $F'V$ represents the total costs of production and $O'S$ the total revenue from sales, the intersection of these two lines is often called the *breakeven point*. When units of output during a year are denoted Q, total revenues equal total costs at the breakeven point, Q':

$$Q'(\text{selling price/unit} - \text{variable cost/unit}) = \text{fixed cost/year} \quad (2\text{-}11)$$

Thus the breakeven point becomes

$$Q' = \frac{\text{fixed cost/year}}{\text{selling price/unit} - \text{variable cost/unit}} \quad (2\text{-}12)$$

With the rate of production at the breakeven point, the business will make no profit and have no loss. If the production rate is greater than that at the breakeven point, a profit will result. When the rate is less than that at the breakeven point, a loss will be sustained. By representing the revenue and costs of a business in this manner, we can easily determine the possibility of profit for any rate of production. Because these breakeven charts show the relationship between revenue and costs for all possible volumes of activity, they are, in effect, continuous income (profit and loss) statements.

One of the most satisfactory uses of breakeven charts is to show the relative effects of changes in fixed and variable costs of a business. This is illustrated in Figure 2-12. Figure 2-12a shows the breakeven chart for a certain business that has a sales capacity of $300,000 per year. For the values of revenue and cost shown, the breakeven point occurs at 50% of sales capacity. For any volume of sales over $150,000, the business will make a profit.

The chart in Figure 2-12b is drawn to determine what the effect will be if the variable costs are decreased 10%, with all other factors remaining constant. The profits are increased by $20,000 when the business is operated at 100% of sales capacity. However, a more significant effect is the change in the breakeven point. The 10% decrease in variable costs lowers the breakeven point to approximately 41% of sales capacity. This means that not only will the company make greater profits if the plant is operated at capacity, but, more important, some profit will be earned if operations are greater than 41% rather than 50% of capacity. In times of recession it is usually more important that a business be able to earn some profit, or not incur a loss, when operating at partial capacity than it is for it to be able to earn a very large profit if operated at a capacity that is entirely beyond the realm of possibility.

The third breakeven chart, Figure 2-12c was drawn to determine to what extent the breakeven point will be shifted if the same saving of $20,000 is effected out of the fixed costs. The solid lines in Figure 2-12c give the solution to this question. It is shown that this change in fixed costs will lower the breakeven point from 50% to approximately 29% of sales capacity. This makes the importance of controlling fixed costs very apparent. The saving of $20,000 in fixed costs lowers the breakeven point nearly 12% more than would be the case if the same saving were made in variable costs at maximum sales capacity.

Unfortunately, it is usually easier to effect savings in variable costs than in fixed costs. We might wish to know what saving in fixed costs would give the same result as a greater economy in variable costs, and the dashed lines in Figure 2-12c provide this information. A vertical line XE is drawn upward from

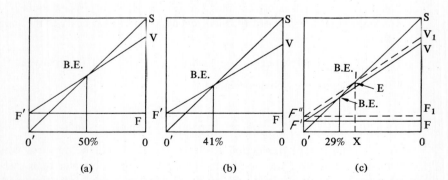

FIGURE 2-12 Effect of changes in fixed and variable costs on the location of the breakeven point. Maximum sales capacity (100%) corresponds to $300,000 per year. (a) Before change, (b) with a 10% decrease in variable costs, and (c) with a $20,000 decrease in fixed costs.

a point corresponding to 41% of sales capacity until it intersects the revenue line $O'S$ at point E. Through point E a line $F''V_1$ is drawn parallel to $F'V$. $F''F_1$ is the fixed-cost line that will give the required breakeven point of 41% of sales capacity. The fixed costs are determined by the ordinate $O'F''$, corresponding to \$42,000. This means that in this case a saving of only \$8,000 in fixed costs will lower the breakeven point as much as a saving of \$20,000 in variable costs. Thus the effect of fixed costs upon the breakeven point is evident.

It should be noted that in each case illustrated in Figure 2-12b and c the profit at 100% of capacity is the same. However, the percentage of capacity at which *some* profit can be earned differs. It should also be remembered that the changes in percentages apply only to the conditions of the example given, although other cases will show a similar change.

COST CONCEPTS FOR ECONOMIC ANALYSIS

The word *cost* has many meanings in many different settings. The cost concept that should be used in an economic evaluation depends on the decision to be made. Financial records resulting from the firm's accounting function aim at describing what has happened in the past, whereas useful decision-making concepts of cost attempt to project what is expected to happen in the future as a result of implementing alternative courses of action. Indeed, different types of cost are present in various kinds of management and engineering problems. The following sections contain descriptions of cost concepts that are commonly encountered throughout this book.

INVESTMENT COST

The investment cost, or *first cost,* is the capital required to get a project started. It is typically a single cash flow, or possibly a series of cash flows, that must be made on the front end of an activity's life span. For instance, the investment cost of acquiring a new automobile is the sum of the down payment, taxes, and other dealer's charges involved in obtaining the title for the automobile. In the case of a construction project that takes several years to complete, the investment cost is usually spread over time in the form of periodic progress payments. It is often true that the magnitude of a proposed project's first cost exceeds the amount of capital that is on hand or that can be borrowed, thus making the alternative infeasible.

FIXED, VARIABLE, AND INCREMENTAL COSTS

Fixed costs are those costs associated with a new or existing project that are essentially unaffected by changes in activity level over the normal range of operation of the project. Typical fixed costs include insurance and taxes on buildings, administrative salaries, license fees, and interest costs of borrowed capital.

Of course, any cost is subject to change, but fixed costs tend to remain constant during normal operating conditions of a firm. When large changes in usage of a company's resources are encountered or when plant expansion or shutdown are involved, one would expect fixed costs to be affected.

Variable costs are groups of costs that vary proportionately to changes in the activity level of a new or existing project. Different types of costs will increase by a certain amount as the output of a firm increases by some number of units. Examples of variable costs include direct labor and materials, direct utilities, sales commissions, shipping charges, and social security taxes. If one is making an economic analysis of a proposed change, it should be remembered that in most cases only the variable costs need to be considered, since prospective differences between alternatives need to be taken into account.

An *incremental cost*, or an *incremental revenue*, refers to the additional cost, or revenue, that will result from increasing the output of a system by one or more unit(s). Incremental costs are also referred to as marginal costs for purposes of this book. Reference is frequently made to incremental costs in connection with an increase in cost resulting from a unit increase in some factor of interest. For instance, the incremental cost per mile for driving an automobile may be $0.27, but this depends on numerous considerations, such as total mileage driven during the year, age of the automobile, and so forth. Also it is common to read of the "incremental cost of producing a barrel of oil" and the "incremental cost to the State for educating a student." As these examples would indicate, incremental costs (and revenues) are quite difficult to determine in practice.

Because differences in costs and revenues between alternatives are relevant in decision making, variable and incremental quantities are involved in engineering economy studies. Variable costs are typically encountered when alternatives deal with large-scale changes in activity levels, while incremental costs are relevant to studies of "with or without" situations such as that illustrated in the following simple example.

Four college students wish to go home for Christmas vacation, a distance of 400 miles each way. One has an automobile and agrees to take the other three if they will pay the cost of operating the car. When they return from the trip the owner presents each of them with a bill for $102.42, stating that he has kept careful records of the cost of operating his car and has found that, based on his average yearly mileage of 15,000 miles, his cost per mile is $0.384. The three students declare that they feel the charge is too high and ask to see his cost figures. He shows them the following list:

Item	Cost per Mile
Gasoline	$0.120
Oil and lubrication	0.021
Tires	0.027
Depreciation	0.150
Insurance and taxes	0.024
Repairs	0.030
Garage	0.012
TOTAL	$0.384

The three riders, on the other hand, claim that only the costs for gasoline, oil and lubrication, tires, and repairs are a function of mileage driven and thus could be caused by the trip. Since these four costs total only $0.198 per mile, and thus $158.40 for the 800-mile trip, the share for each student would be $158.40/3 = $52.80. Obviously, the opposing views are substantially different. The question is: Which, if either, is correct?

In this instance, assume that the owner of the automobile agreed to accept $52.80 per person from the three riders, based on the costs that were purely incremental for the Christmas trip. That is, the $52.80 per person is the "with" a trip cost relative to the "without" alternative. What would happen if the three students, because of the low costs, returned and proposed another 800-mile trip the following weekend? And what if there were several more such trips on subsequent weekends? Quite clearly what started out to be a temporary change in operating conditions—from 15,000 miles per year to 15,800 miles—soon would become a standard operating condition of 18,000 or 20,000 miles per year. On this basis it would not be valid to compute increment cost per mile as $0.198. A more valid incremental cost would be obtained by computing the total annual cost if the car were driven, say, 18,000 miles and, by subtracting the total cost for 15,000 miles of operation, determining the cost of 3,000 additional miles of operation. From this value an increment cost per mile for abnormal mileage could be obtained. In this instance the total cost for 15,000 miles of driving per year was

$$15,000 \times \$0.384 = \$5,760$$

If the cost of 18,000 miles per year of service, which is due to increased depreciation, repairs, reduced gasoline mileage, and so on, turned out to be $6,570, it is evident that the cost of 3,000 miles of additional driving was $810. The corresponding incremental cost per mile would be $0.270. Therefore, if the owner of the car suspected that several weekend trips might be made, he would be on much safer economic ground to quote an incremental cost of $0.270 per mile for even the first trip.

A MAKE-VERSUS-BUY EXAMPLE

It is probable that more mistakes have been made in economic decisions because of the improper use of unit costs based on accounting records than due to any other single cause. Here the relationship of fixed costs, incremental costs, and unit costs is of the utmost importance. Because unit costs are based on some level of activity, a number of arbitrary cost allocations are used in their determination, such as in the allocation of overhead. If operations are to be carried out at a different level, the unit costs are bound to be inaccurate. Since a large number of cost studies involve changes from existing conditions, it is at once apparent that the use of unit costs may lead to considerable error.

An excellent example of how a wrong decision may result from improper use of unit costs in an economy study is found in the following case. A large manufacturing plant consisted of 11 departments. In one of these automobile batteries were produced. The equipment for producing the hard-rubber covers for the cells of these batteries occupied about 100 square feet in one corner of this department. This operation required only one workman 4 hours per shift and was under the supervision of a person who also supervised a number of other operations. The daily production of cell covers was 576. The total cost of operation according to accounting records was as follows:

Labor	$12.00
Material	8.64
Overhead	8.20
TOTAL	$28.84

Unit cost = $28.84/576 = $0.05 per cover

The cell cover operation was the cause of considerable noise, and the percentage of defects was high. It required considerable attention by the foreman and was a general source of grief.

A new foreman was appointed for this department, and he soon became aware of the troubles connected with the making of the cell covers. He happened to have a friend who was the manager of another company that specialized in the production of hard rubber and plastic products. In discussing their problems together, the new foreman mentioned his troubles with the cell cover operation and the fact that it cost $0.05 to produce each cover. His friend thought the price seemed rather high and later made an offer to produce the cell covers from the existing molds for $0.035 each. The department manager computed that the saving involved through having the cell covers manufactured outside the plant would be as follows:

Present cost for 576 cell covers	$28.84
Cost if made outside (576 × $0.035)	20.16
DAILY SAVING	$ 8.68

This appeared to him to be a very worthwhile saving, inasmuch as it could be obtained with no investment of additional capital. As a result the change was made, and the cell covers were made by the outside company.

After the change had been in effect a little over a month, the cost department made a check on the savings that actually resulted from the change. The investigation showed that the actual saving in overhead was only $0.30 per day, which represented a saving in insurance. The molding press had to be kept available, since it served as a standby for a second press that was required for another operation. It was further found that a portion of the material that had been used for the cell covers was a waste product of another operation in the factory and that the actual saving in material was only $6.10 instead of $8.64. This waste product had been included in the unit-cost determination at $2.54 per day, because it would have cost this much to produce this material if it had not been available as scrap. As a result the company was now paying for the 576 cell covers the following:

Remaining overhead	$ 7.90
Waste material	2.54
To outside company	20.16
TOTAL	$30.60

Instead of a saving of $8.68 per day there was actually a loss of $30.60 − $28.84, or $1.76. Arrangements were made immediately to have the molds returned to the plant, and cell covers were again produced within the factory.

A little thought makes it apparent why the use of unit costs failed to give correct results. The unit cost of $0.05 per cover was based on the assumption that 576 covers *were to be produced* each day. If the covers were not produced, the unit cost had no meaning, since it did not apply to this set of conditions. We might even do a bit of computation and say that the unit cost of *not producing* 576 cell covers each day would be

$$\frac{\$7.90 + \$2.54}{576} = \$0.018 \text{ per cell cover}$$

It becomes apparent from the foregoing example that serious errors may result from the use of unit costs. For economy studies of this type we must determine the incremental cost of producing the products. The actual (incremental) costs that were incurred when the cell covers were produced, but that would not exist if they were not manufactured, would be as follows:

Incremental labor	$12.00
Incremental material	6.10
Incremental overhead (insurance)	0.30
TOTAL INCREMENTAL COST	$18.40

Dividing $18.40 by 576 gives a unit incremental cost of $0.032 for producing the cell covers. In other words, the actual cost of producing each cell cover, *compared to not making any cell covers,* was slightly less than $0.032. This result makes it obvious that an actual loss would result from paying $0.035 each for cell covers produced by an outside company. The department foreman had made the common mistake of using unit costs for a purpose for which they were never intended and of failing to determine the facts that would reveal the incremental costs for the alternatives of producing or not producing the cell covers.

SUNK COSTS

Another type of cost that often must be considered in economy studies of going concerns is a *sunk cost*. Sunk costs are different from other costs considered in engineering economy studies in that they are costs of the past, rather than of the future. Virtually all economy studies deal with future costs, and this fact should immediately indicate that sunk costs have no place in them. Yet many misconceptions have existed, and still do exist, concerning sunk costs, and there is a necessity for recognizing them so that they may be handled properly. Sunk costs may be defined in several ways, such as:

1. Any past expenditure (or commitment to expend).
2. The unrecovered balance (or book value) of an investment.
3. Capital already invested that cannot be retrieved.

The principle of sunk costs may be illustrated by the following simple example. Suppose that Joe College finds a motorcycle on Saturday that he likes and pays $40 "down payment," which will be applied toward the $1,300 purchase price but which must be forfeited if he decides not to take the cycle. Over the weekend, Joe finds another motorcycle, which he considers equally desirable, for a purchase price of $1,230. For purposes of deciding which cycle to purchase, the $40 is a sunk cost and thus would not enter into the decision. The decision then boils down to paying $1,300 − $40 = $1,260 for the first motorcycle versus $1,230 for the second motorcycle.

As another example of a sunk cost, assume that you bought a stock several years ago for $10 a share. The stock performed badly, and 1 year ago it was selling for $1. You have kept the stock and now it is selling for $3. Your question is: Should you sell or keep the stock? You might reason that selling would result in a ($10 − $3)/$10 = 70% loss, or in a ($3 − $1)/$1 = 200% gain over the low selling price. Actually, both the $10 and the $1 are in the past and are thus sunk costs. The valid viewpoint for answering the question is to consider the likely future performance of $3 in the stock compared to the same $3 in some other investment. Accordingly, the stockholder should guard against the tendency in such situations to "throw good money after bad money."

A classical example of a sunk cost occurs in the replacement of assets. Suppose that a piece of equipment, which originally cost $50,000, presently has an accounting book value of $20,000 and can be salvaged now for $5,000. For purposes of an economic analysis, the $50,000 is a sunk cost. However, the viewpoint can be taken that the sunk cost should be considered to be the difference between the accounting book value and the present realizable salvage value. According to this viewpoint, the sunk cost is $20,000 minus $5,000, or $15,000. Neither the $50,000 nor the $15,000 should be considered in an economic analysis, except for the manner in which the $15,000 affects income taxes, as will be discussed in Chapter 10.

In summary, sunk costs result from past decisions and therefore are irrelevant in the consideration of alternatives that affect the future. Another way to view sunk costs is that they are *common to all alternatives* that influence future cost patterns and hence do not affect relevant differences among them. Even though it is emotionally difficult to do so at times, sunk costs should be ignored except possibly to the extent that their existence assists one in better anticipating what will happen in the future.

OPPORTUNITY COSTS

An *opportunity cost* is a cost which, although hidden or implied, is incurred because of the use of limited resources in such a manner that the opportunity to use those resources to monetary advantage in an alternative use is forgone.

As an example suppose that a project involves the use of firm-owned warehouse space that is presently vacant. The cost for that space which should be charged to the project in question should be the income or savings that possible alternative uses of the space may bring to the firm. In other words, the cost for the space for purposes of an economy study should be the *opportunity cost* of the space. This may be more than or less than the average cost of that space which might be obtained from accounting records.

As another example, consider a student who could earn $8,000 for working during a year and who chooses instead to go to school and spend $3,000 to do so. The total cost of going to school for that year is $11,000: $3,000 cash outlay and $8,000 for income forgone. (*Note*: This neglects the influence of income taxes and assumes that the student has no earning capability while in school.)

Opportunity Cost in Determination of Interest Rates (Mininum Attractive Returns) for Economic Analyses

An extremely important use of the opportunity cost principle is in the determination of the interest rate (minimum attractive return) to use in analyses of proposed capital investment projects. *The proper interest rate is not just the amount that would be paid for the use of borrowed money, but is rather the opportunity cost, that is, the return forgone or expense incurred because the money is invested in that project rather than in possible alternative projects.* Suppose that a firm always has available certain investment opportunities, such as expansion or bond purchases, which will earn a minimum of, say, $X\%$ per year. This being the case, the firm would be unwise to invest in other alternative projects earning less than $X\%$. The $X\%$ may be thought of as the opportunity cost of not investing in the readily available alternatives.

In economy studies it is necessary to recognize the time value of money irrespective of how the money is obtained, whether it be through debt financing, through owners' capital, or through reinvestment of earnings generated by the firm. Interest on project investments is a cost in the sense of an opportunity forgone—an economic sacrifice of a possible income that might have been obtained by use of that money elsewhere.

Opportunity Cost in Replacement Analysis

As another illustration of the opportunity cost principle, suppose that a firm is considering replacing an existing piece of equipment which originally cost $50,000, presently has an accounting book value of $20,000, and can be salvaged now for $5,000. For purposes of an economic analysis of whether or not to replace the existing piece of equipment, the investment in that equipment should be considered as $5,000; for by keeping the equipment, the firm is giving up the *opportunity* to obtain $5,000 from its disposal. This principle is elaborated on in Chapter 11.

CONSIDERATION OF INFLATION IN ECONOMIC ANALYSES

One increasingly important factor in economic analyses is *inflation,* which is the increase in the number of units of currency (dollars, for example) necessary to buy a given unit of goods or service. The economic analysis methods and examples explained in the next several chapters may seem to ignore inflation, but this is not necessarily true. All the methods can be used so as to correctly consider inflation provided that the types of cash flow estimates and interest rates (minimum attractive rates of return) are consistent with respect to either including or not including inflation. Inflation and two methods of explicitly dealing with it in economy studies are described in depth in Chapter 8.

CASH COSTS VERSUS BOOK COSTS

Costs that involve payments of cash or increases in liability are called *cash costs* to distinguish them from noncash (book) costs. Other common terms for cash costs are "out-of-pocket costs" or costs that are "cash flows." Book costs are costs that do not involve cash payments, but rather represent the amortization of past expenditures for items of lengthy durability. The most common example of book costs is a depreciation charge for the use of assets such as plant and equipment. In economic analyses, only those costs that are cash flows or potential cash flows need to be considered. Depreciation, for example, is not a cash flow and is important only in the way it affects income taxes, which are cash flows. Determination of depreciation and income taxes are considered in Chapters 9 and 10, respectively.

PROBLEMS

2-1   Describe the difference between consumer goods and producer goods. Why is it more difficult to estimate the demand for producer goods?

2-2   Discuss the concept of economic utility. How does the utility of necessities differ from the utility of luxuries for low-income groups in this country?

2-3   Explain why the elasticity of demand for a luxury product (e.g., tennis rackets) may be relevant in an economy study of an investment in proposed equipment for manufacturing this product.

2-4   How would the elasticity of demand for necessities differ from the elasticity of demand for luxuries? Draw a simple graph to illustrate your answer.

2-5   Explain why perfect competition is in reality an ideal that is difficult to attain in the United States. List several business situations in which perfect competition is approached.

2.6   Differentiate between monopoly, oligopoly, and perfect competition. For which situation are most general economic principles stated? Is a monopoly ever desirable for the economic welfare of the public?

2-7   In the context of industrial production, define and illustrate a fixed cost and a variable cost. List as many cost components of each as you can.

2-8   Explain how the law of supply and demand would operate if a bumper crop of corn were produced this year (assume that the federal government does not buy any surpluses).

2-9   Company ABC has established that the relationship between sales price for an item and the quantity sold per month is roughly

$$D = 780 - 10P$$

where D is the demand (quantity sold) per month and P is the price in dollars. The company's fixed costs are $800 per month and the variable costs are $30 per unit produced. What number of units, D, should be produced and sold to maximize the net profit?

2-10 Your company estimates that the relationship between unit price and demand per year for a given product is approximated as $P = \$100.00 - \$0.10D$. The company can produce the product by expanding fixed costs of $17,500 per year and variable costs of $40.00 per unit. What number of units D should the company produce and sell to maximize net profit?

(a) Work out the complete solution by differential calculus starting with the formula for profit.

(b) Solve for approximate answer graphically.

2-11 A firm estimates that its total-cost and total-revenue functions are as follows:

$$C_{total} = C_F + vD, \qquad revenue = aD - bD^2$$

for $0 \le D \le a/b$, otherwise zero. When $C_F = \$500.00$, $v = \$2.00$, $a = \$4.00$, and $b = \$0.001$, find the production volume that maximizes profit, and the maximum profit.

2-12 A *breakeven point* in Figure 2-11 was defined as

$$Q' = \frac{fixed\ costs/year}{selling\ price/unit - variable\ cost/unit}$$

Suppose that the ABC Corporation has a sales capacity of $1,000,000 per month. Its fixed costs are $350,000 per month, and the variable costs over a considerable range of volume, are $0.50 per dollar of sales.

(a) What is the breakeven point volume?

(b) What would be the effect of decreasing the variable costs by 25% if the fixed costs thereby increased by 10%?

(c) What would be the effect if the fixed costs were decreased by 10% and the variable costs increased by the same percentage?

2-13 Suppose that the selling price per unit in Problem 2-10 has been set at $75.00. The fixed cost per year is $17,500 and the variable cost per unit is $40.00. What is the breakeven point of operations for this company?

2-14 A company estimates that as it increases its sales by decreasing the selling price of its product, the revenue changes by the relationship

$$revenue = aD - bD^2$$

where D represents the units of demand per month, with $0 \le D \le a/b$. The company has fixed costs of $1,000 per month, and the variable costs are $4 per unit. If $a = \$6$ and $b = \$0.001$, determine the sales volume for maximum profit and the maximum profit per month.

2-15 A company produces and sells a product and thus far has been able to control the volume of the product just about as the company desires by varying the selling price. The company is seeking to maximize its net profit. It has concluded that the relationship between price and demand, per month, is approximately $D = 500 - 5P$, where P is the price in dollars. The company's fixed costs are $1,000 per month and the variable costs are $20 per unit. What number of units, D, should be produced and sold to maximize the net profit? (Obtain the answer both mathematically and graphically.)

2-16 Describe the relationship between the law of diminishing returns and unit cost functions such as that of Figure 2-10.

2-17 A certain item can be readily purchased from a local vendor for $0.50 per unit. The shop foreman in your company has proposed manufacturing the item in an idle part of the production area. He has computed that labor, materials, and overhead per unit would be $0.15, $0.20, and $0.15, respectively. However, he contends that overhead should not be included in the manufactured cost so that it is less expensive to make the item compared to purchasing it. Do you agree with the foreman's analysis? What other factors might have a bearing on this decision?

2-18   What is an opportunity cost and how will it be used in various engineering economy studies? What are a few of the opportunity costs of taking a job in Buffalo, New York, instead of a job in Manhattan, Kansas, that would have paid $2,000 per year less?

2-19   What is an incremental, or differential, cost? Give an example.

2-20   Why should sunk costs be ignored in economy studies when income taxes are not considered? Make up an example of a sunk cost that you encounter every day.

CHAPTER 3

Selections in Present Economy

This chapter is devoted to economy studies for cases in which interest and money–time relationships do not need to be considered. Chapter 4 and subsequent chapters concentrate on economy studies in the more general (and somewhat more difficult) cases in which alternatives have differing amounts of money at different points in time, so that the time value of money must be taken into account.

Studies that do not require consideration of time–money relationships are sometimes called *present-economy studies* and generally involve selection between alternative designs, materials, or methods. The effects of interest do not enter into the solution of these problems because of one or more of the following conditions:

1. There is no investment of capital; only out-of-pocket costs are involved. As an example, should a businessman make a trip by air, requiring 3 hours of his time, or by train, requiring 12 hours? Here there is only a balancing of the difference in fares against the value of his time.
2. After any first cost is paid, the long-term costs will be the same for all alternatives, or will be proportional to the first cost. Thus the alternative that

has the lowest first cost also will be the most economical in the long run. An example of this type may be found in the construction of a highway overpass. Whether conventional or prestressed-concrete construction is used, the maintenance costs would be the same and the ownership costs would be proportional to the first costs.

3. The alternatives will have essentially identical results regardless of the capital investment involved. For example, in the making of a certain screw-machine product, it is found that either free-cutting aluminum or screw-machine steel can be used and can be machined in the same amount of time. The decision will be determined by the costs of the two materials.

In none of these cases is time a significant factor. The first case involves selection between alternative methods; the second involves selection between alternative designs; the third concerns selection between materials. Because most parts, products, and structures can be designed in several ways to achieve a given function, and because so many different materials and processes are available for use, it is inevitable that many present-economy studies must be made.

Such studies tend to have certain characteristics. First, they usually are relatively simple in form. Second, the required decision usually is quite easily made, once the cost data have been organized. Third, the major effort in such studies is in the collection or computation of the various cost elements. In this regard, when there is an absence of desired, factual cost information—and this too often is the case—a considerable amount of experience may be necessary so that accurate estimates can be made. Otherwise, important, but seemingly insignificant, items may be overlooked.

Another problem that often faces one in present-economy studies is the fact that a large and sometimes almost unlimited number of alternatives may be available for study. This is particularly true in connection with designs and methods. In some cases, if it were desired the analyst could originate new designs almost without limit. However, at some point he or she must recognize that the basic objective is to arrive at an economic decision. This means that a cutoff point must be set regarding the number of alternatives to be considered. Clearly, we do not wish to miss considering those alternatives that offer the prospect of real economic advantage. On the other hand, we do not wish to devote much time to making detailed analyses of alternatives that quickly can be determined as not offering much prospect for economic gain.

Handling such situations requires the exercise of good judgment. One should quickly gain a general idea of the probable economic advantages that appear to be possible in those alternatives that seem to have definite differences. This usually will require rather brief, preliminary economy studies to be made of a number of alternatives. Such studies will not be highly accurate, but they will give the analyst an idea of the general magnitude of probable economic differences. Thus they provide a rather good guide as to which alternatives should be dropped from consideration and which should be studied in detail. This commonsense approach is merely the application of two very basic principles relating to engineering practice: (1) there is no virtue in spending $10 to save $1; and (2) sooner or later a decision must be made so that the work can proceed.

THE NECESSITY FOR EQUIVALENT RESULTS

One important factor that must be kept in mind in present-economy studies is that the results of various alternatives being evaluated must be substantially *equivalent*.† For example, if we were considering the alternatives of digging a ditch by pick and shovel or by a ditchdigging machine and found that the narrowest ditch that could be dug by the machine was 2 feet wider than was wanted, the results could hardly be said to be equivalent. Thus we must be certain that each alternative being considered will produce results that are satisfactory. This concept of equivalence does not mean that the results must always be identical. If it did not matter that the ditch were somewhat wider than a certain minimum, the extra width produced by the machine would be neither an advantage nor a disadvantage. Frequently, materials or methods must be compared that are not in all ways identical but that may be used so as to produce equivalent results. For example, a mild-steel bolt with a cross-sectional area of 1.0 square inch has an ultimate strength in tension of approximately 60,000 pounds. A nickel–steel alloy bolt of the same size would have a strength of 90,000 pounds. If strength were a factor in the design, it would not be correct to compare the cost of mild-steel and nickel–steel bolts of the same size. Yet a fair selection could be made between a mild-steel bolt with a cross-sectional area of 1.0 square inch and a nickel–steel bolt having a cross section of only 0.667 square inch, inasmuch as both have the same strength, provided that tensile strength is the sole design criterion.

Although there must be at least *minimal* equivalence among the alternatives being considered, this does not mean that one alternative cannot exceed another in a requisite property. For example, if it were found, in the example just cited, that the same size bolts of mild-steel and nickel–steel could be purchased for the same price, in most instances the extra strength of the nickel–steel bolt would be no detriment, and this fact would be no bar to its adoption. But at the same time we should not assign added economic advantage to a superior quality that cannot in fact be used or put to actual economic gain.

SELECTION AMONG MATERIALS

One of the most frequent types of present-economy studies is that in which selection must be made among materials. As more new and specialized materials become available, such problems become more numerous. These studies are frequently complicated by the fact that materials usually differ from each other in more than one property. For example, consider the choice between using ordinary low-carbon steel and a low alloy/high yield strength steel in an appli-

†Here "equivalent" should be taken to mean that the economic results have roughly the same effect or meaning to the firm.

cation where *yield strength is the design criterion upon which the selection must be based*. The properties of the two materials are as follows:

	Low-Carbon Steel	Low Alloy/High Yield Strength Steel
Yield strength (psi)	32,000	52,000
Ultimate strength (psi)	64,000	75,000
Cost per pound	$0.24	$0.30

These materials may be compared on the basis of the cost to obtain 1 pound of yield strength in the following manner:

Low-carbon steel:

$$\frac{\$0.24}{32,000} = \$0.00000750$$

Low alloy/high yield strength steel:

$$\frac{\$0.30}{52,000} = \$0.00000577$$

Thus for this application the low alloy/high yield strength steel would be less expensive.

In making selections between two or more materials, we must make certain that the proper criterion is used. For example, if we were selecting between steel and aluminum for a given product, both rigidity and strength might be of importance. It would therefore be necessary to consider both the strength and the modulus of elasticity of each material in order to arrive at a proper decision. Because materials usually differ with respect to more than one property, we must be certain that true equivalence is achieved.

TOTAL COST IN MATERIAL SELECTION

In a large proportion of cases, economic selection between materials cannot be based solely on the costs of the materials. Very frequently, a change in materials will affect the processing costs, and often shipping costs may be altered. A good example of this is the part illustrated in Figure 3-1. The part was produced in considerable quantities on a high-speed turret lathe, using 1112 screw-machine steel costing $0.30 per pound. A study was made to determine whether it might be cheaper to use brass screw stock, costing $1.40 per pound. Since the weight of steel required per piece was 0.0353 pound and of brass was 0.0384 pound, the material cost per piece was $0.0106 for steel and $0.0538 for brass. However, when the manufacturing and standards departments were consulted, it was found that, although 57.1 parts per hour were being produced using steel, if brass were used, the output would be 102.9 parts per hour. Inasmuch as the machine operator was paid $7.50 per hour and the overhead cost for the turret

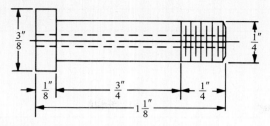

FIGURE 3-1 Small screw-machine product.

lathe was $10.00 per hour, the total-cost comparison for the two materials was as follows:

	1112 Steel			Brass		
Material	$0.30 × 0.0353	=	$0.0106	$1.40 × 0.0384	=	$0.0538
Labor	$7.50/57.1	=	0.1313	$7.50/102.9	=	0.0729
Overhead[a]	$10.00/57.1	=	0.1751	$10.00/102.9	=	0.0972
	TOTAL COST PER PIECE		$0.3170			$0.2239

Saving per piece by use of brass = $0.3170 − $0.2239 = $0.0931

[a]A given overhead rate applied to different alternatives without modification may be invalid for economic analyses even though useful in after-the-fact accounting allocations for whatever alternative is used. Unless explicitly stated otherwise, any given overhead rate is assumed to be applicable to all alternatives considered in the examples of this chapter and the remainder of the book.

Because a large number was made each year, the saving of $93.10 per thousand was a substantial amount. It also is clear that costs other than the cost of material were of basic importance in the economy study.

Later history of this same product illustrates the fact that shipping costs also must often be considered in the selection between materials. After the part had been made from brass stock for several years, it was found that it was desirable to supply domestic and foreign assembly plants of the company by using air freight as a shipping medium. This led to a study of the possible use of a heat-treated aluminum alloy. This material cost $0.85 per pound and the cost of heat-treating each part, at an outside plant, was $0.018. Production studies indicated that the aluminum alloy could be machined at the same speeds as the brass stock.

The specific gravities of the brass and aluminum alloy are 8.7 and 2.75, respectively, and the raw and finished weights of the parts were as follows:

	Brass (lb)	Aluminum Alloy (lb)
Raw material	0.0384	(0.0384)(2.75/8.7) = 0.01213
Finished part	0.0150	(0.0150)(2.75/8.7) = 0.00474

Consequently, the comparative costs, including shipping at $3.00 per pound of finished part, were as follows:

	Brass	Aluminum Alloy
Material	$0.0538	$0.0103
Labor	0.0729	0.0729
Heat treatment	—	0.0180
Overhead[a]	0.0972	0.0972
Shipping	0.0450	0.0142
TOTAL COST PER PIECE	0.2689	0.2126

[a]See the footnote to the table on page 51.

A decision was made to use the aluminum alloy for those parts that were to be shipped by air to the subsidiary plants and for the parts that were consumed at the local plant. There was no advantage to using brass when shipping costs could be omitted.

Care should be taken in making economic selections between materials to assure that any differences in yields or resulting scrap are taken into account. Very commonly, alternative materials do not come in the same stock sizes, such as sheet sizes and bar lengths. This may considerably affect the yield obtained from a given weight of material; similarly, the resulting scrap may differ for different materials. This factor can have serious economic implications when one of the materials is considerably more costly than another. Determination of these effects is an illustration of where experience may be most helpful.

ECONOMY OF LOCATION

Many problems of location, particularly those of alternative locations for manufacturing plants, warehouses, retail stores, and so on, involve time because of differences in construction costs, taxes, insurance rates, and operating costs. Such examples will be considered in later chapters. However, some location problems are not affected by time. The following is a typical example.

In connection with surfacing a new highway, the contractor has a choice of two sites on which to set up his asphalt-treating equipment. He will pay a subcontractor $0.20 per cubic yard per mile for hauling the mixed material from the mixing plant to the job site. Factors relating to the two sites are as follows:

	Site A	Site B
Average hauling distance	3 miles	2 miles
Monthly rental of site	$100	$500
Cost to set up and remove equipment	$1,500	$2,500

If site B is selected, there will be an added charge of $64 per day for a flagman.

The job involves 50,000 cubic yards of mixed material. It is estimated that 4 months (17 weeks of 5 working days per week) will be required for the job.

The two sites may be compared in the following manner:

	Site A		Site B	
Rental	4 × $100 =	$ 400	4 × $500 =	$ 2,000
Setup cost	=	1,500	=	2,500
Flagman			5 × 17 × $64 =	5,440
Hauling	50,000 × $0.20 × 3 =	30,000	50,000 × $0.20 × 2 =	20,000
TOTAL		$31,900		$29,940

Thus site B is clearly less costly. There are some location problems that can be handled as present-economy studies but that are of such complexity that they can be analyzed more readily by the application of linear programming procedures. These are discussed in Chapter 17.

PROBLEMS INVOLVING MAKEREADY AND PUT-AWAY TIMES

There are a substantial number of operations that involve a combination of makeready and put-away times that precede and follow the production portion of the cycle. Such problems frequently do not have a least-cost or maximum-profit alternative within practicable limits. However, proper analysis can provide information on which reasonable decisions can be based.

This type of problem can be illustrated by the situation where typed material is duplicated in large quantities with stencil masters. First, the duplication machine must be inked. After the run is completed, the stencil is removed from the duplicator, cleaned, and stored. For a particular case, this makeready and put-away time is 11 minutes. Once the duplicator has been started, it can turn out the work at the rate of 0.6 second per copy, or 100 copies per minute. Consequently, the total time required to produce any number of copies is as shown in Figure 3-2.

It is apparent that the total time required is the sum of a fixed amount (11 minutes) plus a variable amount equal to $0.01N$, where N is the number of copies. It is also evident that the operator cost per copy produced will be affected by the number of copies that are run each time the duplicator is set up. If the operator of the machine receives $3.60 per hour, the operator cost per copy, as a function of the number of copies in a single run, would vary as follows:

Copies per Run	Time Required (min)	Operator Cost per Copy
10	11.1	$0.06660
50	11.5	0.01380
100	12.0	0.00720
200	13.0	0.00390
300	14.0	0.00280
500	16.0	0.00192
1,000	21.0	0.00126

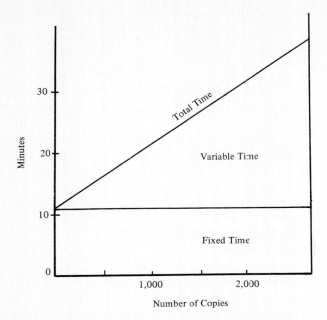

FIGURE 3-2 Time required to set up and mimeograph copies of materials.

These results are shown in graphical form in Figure 3-3.

It becomes apparent from an examination of Figure 3-3 that neither this curve nor the data given in the table provide an answer as to how many copies should be run in a single setup of the duplicator. However, they do supply information upon which a decision may be based. It is apparent that we should endeavor not to run less than about 200 copies at each setup in order to avoid the steep portion of the cost curve. However, before we can arrive at a reasonable decision, a quick determination should be made of the approximate amount of money that may be saved by running various quantities. For example, if 1,000 copies were made in five runs of 200 each, the total cost would be $3.90. If the entire 1,000 copies were run at one time, the cost would be $1.26—a possible saving of $2.64. Should this be considered to be a worthwhile amount (and storage and inventory costs are neglected), we might decide to run the 1,000 copies in one lot. On the other hand, if it were stated that changes in method should not be made unless a saving of at least, say, $4.00 would result, a change from 200 to 1,000 copies per run would not be justified. Thus in problems of this type, we must decide what magnitude of saving (or cost or profit) is significant. When this is coupled to cost versus run size data, such as portrayed in Figure 3-3, a rational decision can be reached.

It must be emphasized that the foregoing analysis assumes that the amount of money tied up in unused product between runs is negligible and that the productive value of the machine during the time required for setup may be neglected. In many cases these factors cannot be neglected. Under such conditions, time must be taken into account, and such problems will be considered in Chapter 15.

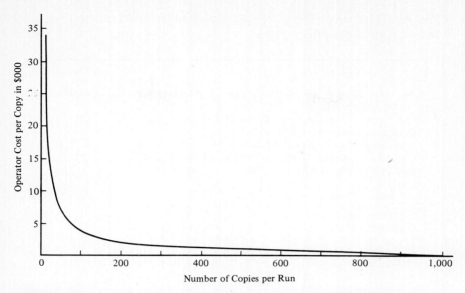

FIGURE 3-3   Cost for mimeographing runs of various numbers of copies.

ALTERNATIVE MACHINE SPEEDS

Machines frequently can be operated at different speeds, resulting in different rates of product output. However, this usually results in different frequencies of machine downtime to permit servicing or maintaining the machine, such as resharpening or adjusting tooling. Such situations lead to economy studies to determine the optimum or preferred operating speed.

A simple example of this type involved the planing of lumber. Lumber put through the planer increased in value by $0.02 per board foot. When the planer was operated at a cutting speed of 5,000 feet per minute, the blades had to be sharpened after 2 hours of operation, and lumber could be planed at the rate of 1,000 board feet per hour. When the machine was operated at 6,000 feet per minute, the blades had to be sharpened after $1\frac{1}{2}$ hours of operation, and the rate of planing was 1,200 board feet per hour. Each time the blades were changed, the machine had to be shut down for 15 minutes. The blades, unsharpened, cost $5 per set and could be sharpened 10 times before having to be discarded. Sharpening cost $1 per set. The crew that operated the planer changed and reset the blades. At what speed should the planer be operated?

Because the labor cost for the crew would be the same for either speed of operation, and because there was no discernible difference in wear upon the planer, these factors did not have to be included in the study.

In problems of this type, the operating time plus the delay time due to the necessity for tool changes constitute a cycle of time that determines the output from the machine. The time required for the complete cycle determines the number of cycles that can be completed in a period of available time—for example, 1 day—and a certain portion of each complete cycle is productive.

The actual productive time will be the product of the productive time per cycle and the number of cycles per day.

	Value per Day
At 5,000 feet per minute	
Cycle time = 2 hours + 0.25 hour = 2.25 hours	
Cycles per day = 8 ÷ 2.25 = 3.555	
Value added by planing = 1,000 × 3.555 × 2 × $0.02 =	$142.20
Cost of resharpening blades = 3.555 × $1 = $3.56	
Cost of blades = 3.555 × $5/10 = 1.78	
TOTAL COST	5.34
Net increase in value per day	$136.86
At 6,000 feet per minute	
Cycle time = 1.5 hours + 0.25 hour = 1.75 hours	
Cycles per day = 8 ÷ 1.75 = 4.57	
Value added by planing = 4.57 × 1.5 × 1,200 × $0.02 =	$164.50
Cost of resharpening blades = 4.57 × $1 = $4.57	
Cost of blades = 4.57 × $5/10 = 2.29	
TOTAL COST	6.86
Net increase in value per day	$157.64

Thus it was more economical to operate at the higher speed, in spite of the more frequent sharpening of blades that was required.

It should be noted that this analysis assumes that the added production can be used. If, for example, the maximum production needed is equal to or less than that obtained by the slower machine (1,000 × 3.555 cycles × 2 hours = 7,110 board feet per day), then the value added would be the same for each speed, and the decision then should be based on which speed minimizes total cost.

This type of study is of great importance in connection with metal-cutting machine tool operations. Changes of cutting speeds can have a great effect on tool life. In addition, because the cost of machine tools and wage rates has increased, it is important that productivity be maintained at as high a level as possible. Under these conditions, it has frequently been found that increased cutting speeds give greater overall economy, even though the cutting-tool life is considerably less than was accepted practice in former years. This is particularly true if rapid means can be devised for changing tools when required.

PROCESSING MATERIALS HAVING LIMITED AND UNLIMITED SUPPLY

Some present-economy studies involve alternative methods of processing raw materials that sometimes are limited in quantity. When this restriction is present, certain precautions must be observed that need not be considered when the supply is unlimited.

As an example of this type, consider two methods that are available for processing a certain ore to recover the contained metal. The ore, as it comes from a certain mine, contains 22% of the metal. Two methods of treatment for the ore are available. Each uses the same equipment, but they involve the use of different chemicals and grinding to different degrees of fineness. To process a ton of ore by method A costs $14.35 and recovers 84% of the contained metal. Processing by method B costs $16.70 per ton of ore, and 91% of the metal is recovered. If the recovered metal can be sold for $0.06 per pound, which method of processing should be used?

There is only a certain amount of ore in the mine; thus the supply of raw material obviously is limited. However, the fact that a different total amount of metal will be recovered and sold when the two different processes are used must be taken into account in the economy study. Such problems can be handled in either of two ways.

Probably the most foolproof method is to base the calculations on a unit of raw material—in this case a ton of ore. This procedure automatically puts each alternative on the proper basis.

Method A:

$$profit \ per \ ton \ of \ ore = 2{,}000 \times 0.22 \times 0.84 \times \$0.06 - \$14.35$$
$$= \$7.83$$

Method B:

$$profit \ per \ ton \ of \ ore = 2{,}000 \times 0.22 \times 0.91 \times \$0.06 - \$16.70$$
$$= \$7.32$$

Thus, although method B would result in more metal being recovered and sold, the profit obtained would not be as great as if method A were used.

A second method of working this type of problem is either to determine the cost of producing a unit of output—1 pound of metal in this case—by each method of processing, or to determine the profit obtained from the sale of 1 pound. However, we must then take into account the fact that, because of the different recoveries from the same amount of ore, the two methods will not produce the same total number of pounds of the product.

When the supply of the raw material is not limited, the procedure for determining which method of processing will give the lowest cost for producing a unit of output is a satisfactory, and usually somewhat simpler, procedure. Since there is no limitation on the amount that can be produced, we need be concerned only with obtaining each unit of product at the least cost. For the example above, the cost per pound of metal produced would be

$$Method \ A: \quad \$14.35/2{,}000 \times 0.22 \times 0.84 = \$0.0388$$

$$Method \ B: \quad \$16.70/2{,}000 \times 0.22 \times 0.91 = \$0.0417$$

Thus method A is the least costly, which in this case happens to correspond with the choice based on profit per ton of ore.

THE PROFICIENCY OF LABOR

Workers of different proficiencies frequently produce different amounts of output in a given time. Through the use of various types of incentives, either monetary or nonmonetary, workers of all types may be induced to develop and exhibit different degrees of proficiency, resulting in various levels of output. Present-economy studies are useful in establishing and maintaining the proper incentives.

As an example of this type, consider the case of a product that was made by hand in a small factory. The workers were paid $0.40 per acceptable piece produced. It was found that if a worker produced 80 pieces per day, 5% would be rejected. If 90 pieces were produced per day, 10% would be rejected, and at a rate of 100 pieces per day, 20% would be rejected. The cost for materials was $0.50 per piece, and the materials in any rejected products had to be thrown away. There was a fixed overhead expense of $10.00 per day per worker, regardless of considerable change in output. Three questions arose concerning the situation: (1) at which of the three outputs did the workers make the highest wages; (2) at which output did the factory achieve the lowest unit cost; and (3) was some adjustment in the wage-payment situation desirable?

The wages obtained by the workers for the three output rates were as follows:

	Pieces Produced per Day		
	80	90	100
Rejects	4	9	20
Acceptable pieces per day (N)	76	81	80
Earnings ($N \times \$0.40$)	$30.40	$32.40	$32.00

It was thus evident that it was to the workers' advantage to produce at the rate of 90 pieces per day.

From the viewpoint of unit cost, the situation was as follows:

	Rate per Day		
	80	90	100
Material cost	$40.00	$45.00	$50.00
Labor cost	30.40	32.40	32.00
Overhead	10.00	10.00	10.00
TOTAL	$80.40	$87.40	$92.00
Cost per acceptable piece	$1.058	$1.079	$1.150

The calculations showed that the lowest unit cost would be obtained at an output rate of 80 pieces per day. It was thus evident that the degree of proficiency that was likely to result from the incentive wage system would not be best from the viewpoint of the company. A change in the wage structure was therefore desirable.

ECONOMIC SELECTION OF A BEAM

A typical design selection may be found in the choice of wooden beams used to support a floor that will sustain a uniformly distributed load of 2,200 pounds over a span length of 12 feet. Using structural-grade Douglas fir and limiting the deflection to $\frac{1}{300}$ of the span, the analyst found that there are three common sizes that will meet the required conditions—4 by 8 inches, 3 by 10 inches, and 2 by 12 inches.

Obviously, beams of these three sizes do not contain the same number of board feet of material. Also, when the cost is determined, it is learned that these three sizes do not cost quite the same per board foot. The 4- by 8-inch and 3- by 10-inch sizes list at $175 per thousand board feet, and the 2- by 12-inch size can be obtained for only $150 per thousand board feet. A final cost comparison of the beams is as follows:

Nominal Size (in.)	Cost per 1000 Board Feet	Board Feet per Beam	Cost per Beam
4 by 8	$175	$(4 \times 8 \times 144)/144 = 32$	$5.60
3 by 10	175	$(3 \times 10 \times 144)/144 = 30$	5.25
2 by 12	150	$(2 \times 12 \times 144)/144 = 24$	3.60

It is thus clear that for this particular situation a design that utilizes 2- by 12-inch beams would be advantageous.

ECONOMIC SPAN LENGTH FOR BRIDGES

The design of a multiple-span bridge is an example that can be used to illustrate (1) the effect of design on costs, (2) the use of approximate cost curves, and (3) some of the limitations of such curves and data.

In many cases of design selection, some of the cost factors increase as certain design parameters are changed, and others decrease. The problem is one of selecting the design that affords the lowest total cost. Figure 3-4 shows the cost curves relating to such a condition. These cost curves show the cost per foot of structure for low-level combined bridges on sand foundations 200 feet deep. As might be expected, the cost of the substructure decreases somewhat as the span length increases. On the other hand, the cost of the steelwork increases quite rapidly as the span length increases. The greatest economy would be achieved by using span lengths of about 325 feet.

Now consider the application of these cost curves to the design of a bridge having a total length of 1,600 feet and the added requirement that at or adjacent to the center of the bridge there must be one span of at least 400 feet. Two designs are being considered: (1) four equal spans of 400-foot length, and (2)

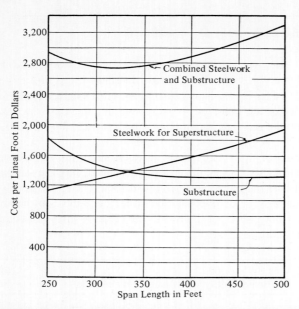

FIGURE 3-4 Costs per lineal foot of structure for low-level combined bridges on sand foundations 200 ft deep. (From J. A. L. Waddell, _Economics of Bridgework_ New York: John Wiley & Sons, Inc.)

one center span of 400 feet and two 300-foot spans on each side of the center span.

Directly applying the cost data contained in Figure 3-4, we obtain the following figures:

Alternative	Length of Spans (ft)	Number of Spans	Cost per Foot	Cost
1	400	4	$2,800	$4,608,000
2	400	1	2,880	1,152,000
	300	4	2,760	3,312,000
TOTAL COST				$4,464,000

It thus appears that alternative 2 will be more economical. However, before coming to a final decision, we should consider two factors related to the use of the cost curves of Figure 3-4. First, these cost curves assume that _all_ spans are of equal length. Thus, when these curves are used in estimating the cost for alternative 2, some error would be introduced. Second, it should be asked whether the construction methods and materials used in connection with determining the cost relationships portrayed in Figure 3-4 were the same as would be involved in building the bridge being considered. The effect of changed materials and methods could be considerable. Therefore, when using historical-cost data, either in tabular or curve form, considerable care and judgment must be exercised. For preliminary economy studies such data may be quite adequate; for final studies upon which important decisions are to be based, such data may not be sufficiently accurate.

THE RELATION OF DESIGN TOLERANCES AND QUALITY TO PRODUCTION COST

Design alternatives very commonly must be considered from the viewpoint of economy of manufacture. Some designs contain features that inherently are more costly to produce than others. Unnecessarily tight dimensional tolerances are prime culprits in this respect. Not only does increased accuracy cost money, but the relaxation of accuracy requirements may make it possible to use different and less expensive processes. Figure 3-5 shows a set of curves relating tolerance and increased cost for several casting methods.

An example of how relaxation of tolerances can affect costs is found in the

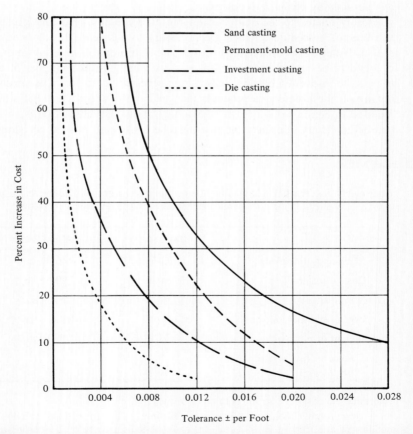

FIGURE 3-5   Relative costs for maintaining specified tolerances with various casting processes.

case of a sand-cast part that required considerable machining so that the specified tolerances could be obtained. The cost for the finished part was as follows:

Cost of casting	$ 9.50
Tooling cost, per part	0.50
Machining	8.25
TOTAL	$18.25

When the tolerance requirements were reconsidered, it was found that by relaxing the tolerances the part could be used as cast if it were made by die casting, except for tapping some holes. An economy study of the die cast part showed the cost to be as follows:

Cost of casting	$2.25
Tooling cost, per part	1.75
Machining	3.10
TOTAL	$7.10

Thus, in making design studies and economy studies of alternative designs, we should not overlook the possibility of making more economical production methods possible by changes in design.

Closely related to the matter of tolerances is the cost of quality. Inspection and control procedures, which are necessary to ensure quality as well as the cost of better quality itself, add to the cost of a product. In many instances it is found to be more economical to permit a small, assured number of defects to pass the inspection barriers and pay the necessary costs for replacements and repairs on these defective products than to pay the added cost that would be required to eliminate all defects. Obviously, customer satisfaction is an important nonmonetary consideration that must be considered in such decisions.

PROBLEMS

3-1   After machining, the finished volume of a certain metal part is 0.17 cubic inch. Data for two types of metal being considered for manufacturing the part are as follows:

	Brass	Aluminum
Machining time/piece	0.64 min	0.42 min
Cost of material	$0.96/lb	$0.52/lb
Scrap value	$0.24/lb	Negligible
Cost of machine operator	$12.00/hr	$12.00/hr
Weight of material	0.31 lb/in.^{3}	0.10 lb/in.^{3}
Volume of raw materials for the part	0.30 in.^{3}	0.45 in.^{3}

Determine the cost per part for both types of material.

3-2   For a certain application, a stainless steel, having a yield strength of 117,000 psi, a modulus of elasticity of 29,000,000 psi, and a specific gravity of 7.6 can be used. It costs $2.80 per pound. As an alternative, an aluminum alloy, having a yield strength of 67,000 psi, a modulus of elasticity of 10,000,000 psi, and a specific gravity of 2.8 could be used. This alloy costs $1.90 per pound. If the selection can be based solely on the yield strength, which material would be more economical?

3-3 Two machines can be operated at different speeds. The faster machine has to be shut down more frequently to change cutting tools. The production rate, duration of production period between tool changes, and time required to change tools are as follows:

Machine	Production Rate (pieces per hour)	Production Time (hr)	Tool Change Time (hr)
A	4,000	20	1.5
B	5,000	15	2.0

The machine operator receives $8.50 per hour and the tool changer receives $10.00 per hour. Find the labor cost per piece produced for each machine. (*Hint:* The machine operator's wages are paid while the machine is producing *and* while the tools are being changed. The tool changer's wages are paid only for time involved in changing tools.)

3-4 Fiber X has 40% greater heat insulating value than fiber Y. A particular design calls for a 6-inch thickness of fiber X or its equivalent. If fiber X costs $7.50 per cubic foot and fiber Y costs $5.00 per cubic foot, which is more economical? Remember to make the comparison on an equivalent basis.

3-5 An operator of a fleet of diesel trucks has been using oil-filter cartridge A on each truck. This type of filter cartridge costs $5.00 per unit, and it has been replaced each 5,000 miles, with 1 quart of oil being added each 1,000 miles thereafter until the next oil change. A salesperson for filter cartridge B, which costs only $2.00, claims that if this unit is used and replaced each 2,000 miles, with the oil also being changed, no oil will have to be added between changes, and oil costing only $0.90 per quart can be used in place of the presently used oil, which costs $1.25 per quart. If the motors require 6 quarts of oil when the oil is changed, which type of filter would be more economical?

3-6 A company finds that, on the average, two of its engineers spend 60 hours each month in flying time, using commercial airlines, in making service calls to customers' plants, always traveling together. It finds that the cost for airline tickets, airport buses, car rental, and so on, costs approximately $2,000 per person per month. An air charter service offers to supply a small business jet and pilot on 24-hour notice at a cost of $1,200 per month plus $125 per hour of flying time and $25 per hour for waiting time on the ground at the destination. It states that experience for similar situations has shown that total travel time will be reduced by 50% by using the charter service. The company estimates that the cost of car rental at destinations probably would amount to about $250 per month if the charter service is used, and the average "waiting time" will be about 40 hours per month. It also estimates that each engineer's time is worth $40 per hour to the company. Should the charter service be used?

3-7 For a construction job, a contractor can haul rock by means of an ordinary 10-cubic-yard dump truck, or he can use a combination rig where a 10-cubic-yard truck also pulls an 8-cubic yard trailer. When this combination is used, the trailer slides into the empty dump truck body for the return trip. The total equivalent operating cost, including driver, amortization of capital, interest on the investment, fuel, and maintenance, is $12.50 per hour for the dump truck and $17.50 per hour for the tandem rig. For the required 10-mile haul, the loading, unloading, hauling, and return times are as follows:

	Time (min)	
	Conventional	Double Rig
Loading	3	5
Unloading	4	10
Hauling	24	28
Return	20	24

(a) Which type of equipment will be more economical?

(b) If hauling and return times are directly proportional to the haul distance, for what length of haul would the two types of equipment provide equal costs?

3-8   A duplicating service will make electrostatic copies of documents up to $8\frac{1}{2}$ by 11 inches in size for 7 cents per copy for the first 10 and 4 cents per copy for all additional. It also will reproduce such documents by offset printing at a cost of 30 cents for the master and 2 cents per copy, with a minimum quantity of 30 being required.

(a) Assuming that each process gives satisfactory copies, which will be more economical for 25 copies?

(b) At what quantity are the two processes equal in cost?

3-9   An automatic machine can be operated at three speeds with the following results:

Speed	Output (pieces per hour)	Time Between Tool Grinds (hr)
A	400	15
B	480	12
C	540	10

A set of unground tools costs $150 and can be ground 20 times. The cost of each grinding is $25. The time required to change and reset the tools is $1\frac{1}{2}$ hours, and such changes are made by a tool-setter who is paid $5.75 per hour. Overhead on the machine is charged at the rate of $3.75 per hour, including tool-change time. All output produced can be used. At which speed should the machine be operated?

3-10   A recent engineering graduate was given the job of determining the best production rate for a new type of casting in a foundry. After experimenting with many combinations of hourly production rates and total production cost per hour, he summarized his findings:

Total Cost/Hour	$1,000	$2,600	$3,200	$3,900	$4,700
Castings Produced/Hour	100	200	300	400	500

The engineer then talked to the firm's marketing specialist, who provided these estimates of selling price per casting as a function of production output:

Selling Price/Casting	$20.00	$17.00	$16.00	$15.00	$14.50
Castings Produced/Hour	100	200	300	400	500

(a) What production rate would you recommend to maximize total profits?

(b) How sensitive is the rate in part (a) to changes in total production cost per hour?

3-11   A manufacturer of wooden shipping containers has lost the business of a major customer who, for economic reasons, has converted to the use of foamed plastic containers for his product, which is shipped by air freight. This customer ships three sizes of product, weighing 25, 50, and 100 pounds, at a rate of $1 per pound. The company pays $30, $42, and $50 for the three sizes of plastic containers, which weigh 7, 12, and 25 pounds, respectively. The manufacturer of the wooden containers, which weigh four times as much as the plastic containers, is attempting to regain the lost business by pricing her product so as to be competitive. The prices for the three sizes have been $8, $12, and $22, respectively.

(a) At what prices would she have to sell her products in order to compete?

(b) She estimates that decreasing the weight of the wooden containers by one-third could be achieved without causing the selling price to be increased more than 25% over the present price. Would this procedure produce favorable results?

3-12 In the design of an automobile engine part, an engineer has a choice of either a steel casting or an aluminum alloy casting. Either material provides the same service. However, the steel casting weighs 8 ounces as compared with 5 ounces for the aluminum casting. Every pound of extra weight in the automobile has been assigned a penalty of $6 to account for increased fuel consumption during the expected life of the car. The steel casting costs $3.20 per pound while the aluminum alloy can be cast for $7.40 per pound. Machining costs per casting are $5.00 for steel and $4.20 for aluminum. Which material should the engineer select and what is the difference in unit costs?

3-13 Seawater contains 2.1 pounds of magnesium per ton. By using the processing method A, 85% of the metal can be recovered at a cost of $3.25 per ton of water pumped and processed. If process B is used, 70% of the available metal is recovered, at a cost of only $2.60 per ton of water pumped and processed. The two processes are substantially equal as to investment costs and time requirements.

(a) If the extracted metal can be sold for $2.40 per pound, which processing method should be used?

(b) At what selling price for the metal would the two processes be equally economical?

3-14 One method for developing a mine will result in the recovery of 62% of the available ore deposit and will cost $23 per ton of material removed. A second method of development will recover only 50% of the ore deposit, but it will cost only $15 per ton of material removed. Subsequent processing of the removed ore recovers 300 pounds of metal from each ton of processed ore and costs $40 per ton of ore processed. The ore contains 300 pounds of metal per ton, which can be sold for $0.80 per pound. Which method for developing the mine should be used?

3-15 The assembly of a certain component for television sets has been done by two skilled workers, each of whom is paid $5.00 per hour and can produce 10 units per hour (as a group). Of the 10 units, 1 will be defective and will have to be discarded at a material loss of $2.50. It has been proposed that this operation could be done by three persons of lesser skill, who would be paid only $4.00 per hour, and who, as a group, could complete 13 units per hour. It is estimated that the number of defects per hour would not be changed if this alternative procedure is used. Which procedure would you recommend?

3-16 In cooperation with a rehabilitation program of a veterans' hospital, a company has offered to employ a certain number of partially disabled veterans on the basis that they will be paid an hourly, or piece, wage which will permit the company to obtain the same unit cost as it achieves with nondisabled workers. On a certain job, it has been paying regular workers $5.20 per hour and production has been 40 units per hour, of which 4% are rejected as defective. The direct-material cost is $2.00 per unit and is not recoverable from defective units. A test has shown that a certain type of trained but partially disabled worker can produce 32 units per hour, with only 1% defective. The overhead cost of the space and equipment used is $2.50 per hour per worker.

(a) What hourly pay rate should be established for the disabled workers?

(b) Comment on the resulting situation.

3-17 The bottler of a popular soft drink can purchase glass bottles for $0.04 each or all-aluminum cans for $0.05 each. If the aluminum cans are used, she believes she can achieve a saving of $0.005 per can in shipping and handling costs, owing to the reduced weight. In addition, by setting up a reclamation drive, she believes she can recover about 50% of the cans by offering to pay $0.10 per pound for returned cans at four return centers which she will establish in the area her company serves. The used cans weigh 20 to the pound, and she can sell the used cans to the can manufacturer for $0.15 per pound. She estimates that the overhead cost of operating the return centers will not exceed 20% of the cost of repurchasing the used cans. In addition, she believes that she will obtain favorable response from ecology-minded customers. Which type of beverage container would you recommend that she use?

CHAPTER 4

Interest and Money—
Time Relationships

The majority of economy studies involve commitment of capital (expressed in money) for a period of time such that the effect of time on the money must be considered. In this regard it is widely recognized that a dollar today is worth more than a dollar one year from now because of the *interest* it could earn. Consequently, money has a "time value." It is the purpose of this chapter to provide an understanding of the return to capital (i.e., interest) and to show how calculations relating to its proper consideration in engineering economy studies are made.

THE RETURN TO CAPITAL

In privately financed sectors of a capitalistic economy, suppliers of capital expect to receive "rent" from the use of their money. This rent is called interest and is referred to by economists as the return to capital. It thus represents payment for the use of capital. Even when the capital is supplied by governmental agencies, there is a requirement for most types of projects that a return for the use of the capital be paid. This is because the government either borrows the capital directly from the public or obtains it from the public through taxation, which implies that the individual taxpayer could have obtained a return from the capital had the government permitted the person to keep it.

There are several reasons why interest is essential and must be considered in engineering economy studies and economic decision making. First, interest pays the supplier for forgoing the use of his or her money (or property) during the time the user has it. Second, the interest is payment for the risk the supplier takes in permitting another person, or an organization, to use his or her capital. Third, the fact that the supplier can earn interest by making capital available acts as an incentive to accumulate capital and to make it available. *Whenever capital is required in engineering and business projects, it is essential that proper consideration be given to the cost of capital in economy studies.* Consequently, maximizing the productive use of capital must correctly take into account this expected return over the period of time that the capital has been committed.

WHEN MUST INTEREST AND PROFIT BE CONSIDERED?

If capital for financing an enterprise must be borrowed, money that is paid for its use is called interest. On the other hand, if a person or corporation owns sufficient capital to finance a proposed project, the return that accrues is called *profit*. When owner's capital is the sole source of funds used to sponsor a project, there is no borrowed money in the true meaning of the term and there is no interest expense. However, in this case if the owner of the capital decides to invest it in the proposed venture, he or she must forgo using it for some other profitable purpose—even if the other purpose is merely leaving it in a bank where it would draw interest.

In typical situations an investor must decide whether the expected profit, usually measured in terms of an annual rate of return on investment, is sufficient to justify investment of capital in the proposed venture. Although no interest cost is involved, if the capital is invested in the project the investor would expect, as a minimum, to receive profit at least equal to the amount the person has sacrificed by not using it in some other equally attractive and available opportunity. This profit, which is lost or forgone, is called the *opportunity cost* of using capital in the proposed venture. Throughout the book we shall refer to this cost as an annual ''minimum attractive rate of return'' required by investors. Thus whether equity or borrowed capital is involved, there is a cost for the capital used in the sense that the venture must provide a return on capital sufficient to make it profitable to suppliers of the capital.

To determine whether the expected capital return (profitability) is sufficient, it usually is necessary to compare the expected rate of profit with the rate that could be obtained from using the same capital in some other manner. Thus a return (profit) to owners of capital is a factor that must be considered in a large proportion of economy studies, although it is not a cash cost in the normal sense when equity capital is used.

ORIGINS OF INTEREST

Like taxes, interest has existed from the time of earliest recorded human history. Records reveal its existence in Babylon in 2000 B.C. In the earliest instances interest was paid in money for the use of grain or other commodities that were borrowed; it was also paid in the form of grain or other goods. Many of the existing interest practices stem from the early customs in the borrowing and repayment of grain and other crops.

History also reveals that the idea of interest became so well established that a firm of international bankers existed in 575 B.C., with home offices in Babylon. Its income was derived from the high interest rates it charged for the use of its money for financing international trade.

Throughout early recorded history, typical annual rates of interest on loans of money were in the neighborhood of 6 to 25%, although legally sanctioned rates as high as 40% were permitted in some instances. The charging of exhorbitant interest rates on a loan was termed "usury," and prohibition of usury is found in the Law of Moses.

During the Middle Ages, interest taking on loans of money was generally outlawed on scriptural grounds. In 1536 the Protestant theory of what comprised usury was established by John Calvin and it refuted the notion that interest was unlawful. Consequently, interest taking again became viewed as an essential and legal part of doing business, and this eventually led to the publication of interest tables which became available to the public.

SIMPLE INTEREST

When the total interest earned or charged is directly proportional to the amount of the loan (principal), the interest rate, and the number of interest periods for which the principal is committed, the interest and interest rate are said to be *simple*.

When simple interest is applicable, the total interest, I, earned or paid may be computed by the formula

$$I = (P)(N)(i) \qquad (4\text{-}1)$$

where P = principal amount lent or borrowed
N = number of interest periods (e.g., years)
i = interest rate per interest period

Thus, if $100 is loaned for 3 years at a simple interest rate of 10% per annum, the interest earned will be

$$I = \$100 \times 0.10 \times 3 = \$30$$

The total amount owed at the end of 3 years would be $100 + $30 = $130.

Simple interest is not used frequently in commercial practice in modern times, but it is of importance to contrast simple interest with compound interest, which is explained below.

COMPOUND INTEREST

Whenever the interest charge for any interest *period* (a *year*, for example) is based on the remaining principal amount plus any accumulated interest charges up to the beginning of that period, the interest is said to be *compound*. The effect of compounding of interest can be shown by the following table for $100 loaned for three periods at an interest rate of 10% compounded per period.

Period	(1) Amount Owed at Beginning of Period	(2) = (1) × 10% Interest Charge for Period	(3) = (1) + (2) Amount Owed at End of Period
1	$100.00	$10.00	$110.00
2	110.00	11.00	121.00
3	121.00	12.10	133.10

Thus $133.10 would be due for repayment at the end of the third period. If the length of a period is 1 year, the $133.10 at the end of three periods (years) can be compared directly with the $130.00 given earlier for the same problem with simple interest. The difference is due to the effect of *compounding,* which essentially is the calculation of interest on previously earned interest. This difference would be much greater for larger amounts of money, higher interest rates, or greater numbers of years. Compound interest is much more common in practice than is simple interest and is used throughout the remainder of this book.

EQUIVALENCE

The concept of equivalence was introduced in Chapter 3. There it was stated that alternatives should be compared insofar as possible when they produce similar results, serve the same purpose, or accomplish the same function. This is not always possible to achieve in some types of economy studies as we shall see later, but now our attention is directed at answering the question: How can alternatives for providing the same service or accomplishing the same function be compared when interest is involved over extended periods of time? Thus we should consider the comparison of alternative options, or proposals, by reducing them to an *equivalent basis* that is dependent on (1) the interest rate, (2) the amounts of money involved, (3) the timing of the monetary receipts and/or disbursements, and (4) the manner in which the interest, or profit, on invested capital is repaid and the initial capital recovered.

To understand better the mechanics of interest and to expand on the notion of economic equivalence, we consider a situation in which we borrow $8,000 and agree to repay it in 4 years at an annual interest rate of 10%. There are many ways in which the principal of this loan (i.e., $8,000) and the interest on it can be repaid. For simplicity, we have selected four plans to demonstrate the idea of economic equivalence. In each the interest rate is 10% and the original amount borrowed is $8,000; thus the primary differences among plans rest with items (3) and (4) above. The four plans are shown in Table 4-1, and it will soon be apparent that all are equivalent at an interest rate of 10% per year.

It can be seen in plan 1 that $2,000 of the loan principal is repaid at the end of years 1 through 4. As a result, the interest we repay at the end of a particular year is affected by how much we still owe on the loan at the beginning of that year. Our end-of-year payment is just the sum of $2,000 and interest paid on the beginning-of-year amount owed.

Plan 2 indicates that none of the loan principal is repaid until the end of the

TABLE 4-1 Four Plans for Repayment of $8,000 in 4 Years with Interest at 10%

(1) Year	(2) Amount Owed at Beginning of Year	(3) = 10% × (2) Interest Owed for Year	(4) = (2) + (3) Total Money Owed at End of Year	(5) Principal Payment	(6) = (3) + (5) Total End-of-Year Payment
\multicolumn{6}{c}{PLAN 1: AT END OF EACH YEAR PAY $2,000 PRINCIPAL PLUS INTEREST DUE}
1	$8,000	$ 800	$8,800	$2,000	$ 2,800
2	6,000	600	6,600	2,000	2,600
3	4,000	400	4,400	2,000	2,400
4	2,000	200	2,200	2,000	2,200
		$2,000 (total interest)		$8,000	$10,000 (total amount repaid)
\multicolumn{6}{c}{PLAN 2: PAY INTEREST DUE AT END OF EACH YEAR AND PRINCIPAL AT END OF 4 YEARS}
1	$8,000	$ 800	$8,800	$ 0	$ 800
2	8,000	800	8,800	0	800
3	8,000	800	8,800	0	800
4	8,000	800	8,800	8,000	8,800
		$3,200 (total interest)		$8,000	$11,200 (total amount repaid)
\multicolumn{6}{c}{PLAN 3: PAY IN 4 EQUAL END-OF-YEAR PAYMENTS}
1	$8,000	$ 800	$8,800	$1,724	$ 2,524
2	6,276	628	6,904	1,896	2,524
3	4,380	438	4,818	2,086	2,524
4	2,294	230	2,524	2,294	2,524
		$2,096 (total interest)		$8,000	$10,096 (total amount repaid)
\multicolumn{6}{c}{PLAN 4: PAY PRINCIPAL AND INTEREST IN ONE PAYMENT AT END OF 4 YEARS}
1	$ 8,000	$ 800	8,800	$ 0	$ 0
2	8,800	880	9,680	0	0
3	9,680	968	10,648	0	0
4	10,648	1,065	11,713	8,000	11,713
		$3,713 (total interest)		$8,000	$11,713 (total amount repaid)

fourth year. Our interest cost each year is $800, and it is repaid at the end of years 1 through 4. Since interest does not accumulate in either plan 1 or plan 2, compounding of interest is not present. Notice that $3,200 in interest is paid with plan 2, whereas only $2,000 is paid in plan 1. The reason, of course, is that we had use of the $8,000 principal for 4 years in plan 2 but, on average, had use of much less than $8,000 in plan 1.

Plan 3 requires that we repay equal end-of-year amounts of $2,524 each. Later in this chapter we will show how the $2,524 per year is computed. For purposes here, the student should observe that the four end-of-year payments in plan 3 completely repay the $8,000 loan principal with interest at 10%.

Finally, plan 4 shows that no interest and no principal are repaid for the first 3 years of the loan period. Then at the end of the fourth year, the original loan principal plus accumulated interest for the 4 years is repaid in a single lump-sum amount of $11,712.80 (rounded in Table 4-1 to $11,713). Both plans 3 and 4 involve compound interest, even though it may not be readily apparent in plan 3. The total amount of interest repaid in plan 4 is the highest of all plans considered. Not only was the principal repayment in plan 4 deferred until the end of year 4, but we also deferred all interest until that time. If annual interest rates are rising above 10%, can you see that plan 4 causes bankers to turn gray-haired rather quickly?

This brings us back to the notion of economic equivalence. If interest rates remain constant at 10% for the plans shown in Table 4-1, all four plans are equivalent. That is, we would be indifferent about whether the principal is repaid early in the loan's life (e.g., plans 1 and 3) or repaid at the end of year 4 (e.g., plans 2 and 4). Economic equivalence is established, in general, when we are indifferent between a future payment, or series of future payments, and a present sum of money.

To see *why* the four plans in Table 4-1 are equivalent at 10%, we could plot the amount owed at the beginning of each year (column 2) versus the year. The area under the resultant curve represents the "dollar-years" that the money is owed. For example, the dollar-years for plan 1 equals 20,000, which is obtained from this graph:

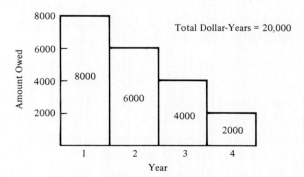

When total dollar-years are calculated for each plan and divided into total interest paid over the 4 years (the sum of column 3), one discovers that the ratio is constant:

Plan	Area Under Curve (dollar-years)	Total Interest Paid	Ratio of Total Interest to Dollar-Years
1	$20,000	$2,000	0.10
2	32,000	3,200	0.10
3	20,960	2,096	0.10
4	37,130	3,713	0.10

Because the ratio is constant at 0.10 for all plans, we can deduce that all repayment methods considered in Table 4-1 are equivalent even though each involves a different total end-of-year payment in column 6. Dissimilar dollar-years of borrowing, by itself, does not necessarily allow one to conclude that different loan repayment plans *are* equivalent or *are not* equivalent. In summary, equivalence is established when total interest paid, divided by dollar-years of borrowing, is a constant ratio among financing plans.

One last important point to emphasize is that the loan repayment plans of Table 4-1 are equivalent only at an interest rate of 10%. If these plans are evaluated with methods presented later in Chapter 4 at interest rates other than 10%, one plan can be identified that is superior to the other three. For instance, when $8,000 has been lent at 10% interest and subsequently the cost of borrowed money increases to 15%, the lender would prefer plan 1 in order to recover his or her funds quickly so that they might be reinvested at higher interest rates.

NOTATION AND CASH FLOW DIAGRAMS

The following notation is utilized for compound interest calculations:

i = effective interest rate per interest period

N = number of compounding periods

P = present sum of money; the *equivalent* worth of one or more cash flows at a relative point in time called the present

F = future sum of money; the *equivalent* worth of one or more cash flows at a relative point in time called the future

A = end-of-period cash flows (or *equivalent* end-of-period values) in a uniform series continuing for a specified number of periods

These and other commonly used symbols and terminology that appear throughout this book are defined in Appendix A.

The use of cash flow (time) diagrams is strongly recommended for situations in which the analyst needs to clarify or visualize what is involved when flows of money occur at various times. Cash flows are dollar transactions that "trade hands" or represent opportunities during whatever study period is being con-

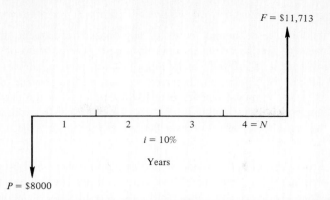

FIGURE 4-1   Cash Flow Diagram for Plan 4 of Table 4-1 (Lender's Viewpoint).

sidered for an alternative. Indeed, the usefulness of the cash flow diagram for economic analysis problems is analogous to the use of the free-body diagram for mechanics problems.

Figure 4-1 shows a cash flow diagram for plan 4 of Table 4-1, and Figure 4-2 depicts the cash flows of plan 3. These two figures also illustrate the definition of the above symbols and their placement on a cash flow diagram. Notice that all cash flows have been placed at the end of the year to correspond with the convention used in Table 4-1.

The cash flow diagram employs several conventions:

1. The horizontal line is a *time scale* with progression of time moving from left to right. The period (or year) labels are applied to intervals of time rather than points on the time scale. Note, for example, that the end of period 2 is coincident with the beginning of period 3. Only if specific dates are employed should the points in time rather than intervals be labeled.
2. The arrows signify cash flows. If a distinction needs to be made, downward arrows represent disbursements (negative cash flows or cash outflows) and upward arrows represent receipts (positive cash flows or cash inflows).
3. The cash flow diagram is dependent on point of view. For example, the situations shown in Figures 4-1 and 4-2 were based on cash flow as seen by the lender. If the direction of all arrows had been reversed, the problem would have been diagrammed from the borrower's viewpoint.

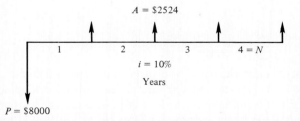

FIGURE 4-2   Cash Flow Diagram for Plan 3 of Table 4-1 (Lender's Viewpoint).

EXAMPLE 4-1

Before evaluating the economic merits of a proposed investment, the XYZ Corporation insists that its engineers develop a cash flow diagram of the proposal. An investment of $10,000 can be made that will produce uniform annual revenue of $5,310 for 5 years and then have a positive salvage value of $2,000 at the end of year 5. Annual disbursements will be $3,000 at the end of each year for operating and maintaining the project. Draw a cash flow diagram and determine the cumulative cash flow over the 5-year life of the project.

Solution

The initial investment of $10,000 and annual disbursements of $3,000 are cash outflows, while annual revenues and the salvage value are cash inflows (see Ex. 4-1 figure).

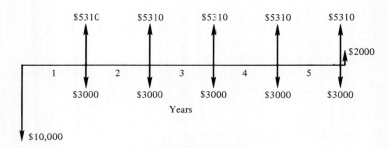

EXAMPLE 4-1

Notice that the beginning of a given year is the end of the preceding year. For example, the beginning of year 1 is the end of year 0. Cumulative cash flow is shown in this tabulation.

End of Year	Net Cash Flow	Cumulative Cash Flow
0	− $10,000	− $10,000
1	+ 2,310	− 7,690
2	+ 2,310	− 5,380
3	+ 2,310	− 3,070
4	+ 2,310	− 760
5	+ 4,310	+ 3,550

When monetary returns each year to owners of the $10,000 capital are considered, the economic attractiveness of the project can be determined. This general subject involves different methods of compounding interest (or forgone profits) and also different conventions and assumptions concerning the timing of cash flows. The remainder of Chapter 4 addresses these and other related matters.

INTEREST FORMULAS FOR DISCRETE COMPOUNDING AND DISCRETE CASH FLOWS

Table 4-2 provides a summary of the six most common discrete compound interest factors and utilizes notation of the preceding section. These factors are derived and explained by example problems in the following sections. The formulas are for *discrete compounding,* which means that the interest is compounded at the end of each finite-length period, such as a month or a year. Furthermore, the formulas also assume discrete (i.e., lump-sum) cash flows spaced at equal time intervals on a cash flow diagram. Discrete compound interest factors are given in Appendix E, where the assumption is made that i remains constant during the N compounding periods. Interest formulas for continuous compounding will be discussed at the end of this chapter.

INTEREST FORMULAS RELATING PRESENT AND FUTURE WORTHS OF SINGLE CASH FLOWS

Figure 4-3 shows a cash flow diagram involving a present single sum, P, and a future single sum, F, separated by N periods with interest at $i\%$ per period. Two formulas relating these amounts are presented next.

TABLE 4-2 **Discrete Compounding Interest Factors and Symbols[a]**

To Find:	Given:	Factor by Which to Multiply "Given"	Factor Name	Factor Functional Symbol[b]
For single cash flows:				
F	P	$(1 + i)^N$	Single payment compound amount	$(F/P, i\%, N)$
P	F	$\dfrac{1}{(1 + i)^N}$	Single payment present worth	$(P/F, i\%, N)$
For uniform series (annuities):				
F	A	$\dfrac{(1 + i)^N - 1}{i}$	Uniform series compound amount	$(F/A, i\%, N)$
P	A	$\dfrac{(1 + i)^N - 1}{i(1 + i)^N}$	Uniform series present worth	$(P/A, i\%, N)$
A	F	$\dfrac{i}{(1 + i)^N - 1}$	Sinking fund	$(A/F, i\%, N)$
A	P	$\dfrac{i(1 + i)^N}{(1 + i)^N - 1}$	Capital recovery	$(A/P, i\%, N)$

[a]i, effective interest rate per interest period; N, number of interest periods; A, uniform series amount (occurs at the end of each interest period); F, future worth; P, present worth.
[b]The functional symbol system is used throughout this book.

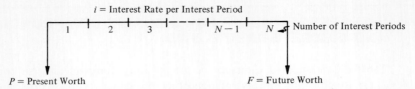

i = Interest Rate per Interest Period

Number of Interest Periods

P = Present Worth F = Future Worth

FIGURE 4-3   General Cash Flow Diagram Relating Present Worth and Future Worth of Single Payments.

Finding *F* When Given *P*

If an amount of P dollars exists at a point in time and $i\%$ is the interest (profit or growth) rate per period, the amount will grow to a future amount $F = P + Pi = P(1 + i)$ by the end of one period; by the end of two periods, the amount will grow to $P(1 + i)(1 + i) = P(1 + i)^2$; by the end of three periods, the amount will grow to $P(1 + i)^2(1 + i) = P(1 + i)^3$; and by the end of N periods the amount will grow to

$$F = P(1 + i)^N \qquad (4\text{-}2)$$

The quantity $(1 + i)^N$ is commonly called the *single payment compound amount factor*. Numerical values for this factor are given in the second column of the tables in Appendix E, for a wide range of values of i and N. In this book we shall use the functional symbol $(F/P, i\%, N)$ for $(1 + i)^N$. Hence Equation 4-2 can be expressed as

$$F = P(F/P, i\%, N) \qquad (4\text{-}3)$$

where the factor in parentheses is read "find F given P at $i\%$ interest per period for N interest periods." Note that the sequence of F and P in F/P is the same as in the initial part of Equation 4-3 where the unknown quantity, F, is placed on the left-hand side of the equation. This sequencing of letters is true of all functional symbols used in this book, which makes them easy to remember.

An example of this compounding situation was given earlier as plan 4 of Table 4-1. With Equation 4-2, the total amount of principal and interest to be repaid at the end of year 4 is easily determined:

$$F = \$8,000(1 + 0.10)^4 = \$8,000(1.4641) = \$11,712.80$$

Another example of this type of problem, together with a cash flow diagram and solution, is given in the first part of the body of Table 4-3. Note in Table 4-3 that for each of the six common discrete compound interest circumstances covered, two problem statements are given: (a) in borrowing–lending terminology, and (b) in equivalence terminology, but that they both represent the same cash flow situation. Indeed, there are generally many ways in which a given cash flow situation can be expressed.

In general, a good way to interpret a relationship such as Equation 4-3 is that the calculated amount, F, at the point in time at which it occurs, *is equivalent to* (has the same effect or value as) the known value, P, at the point in time at which it occurs, for the given interest or profit rate, i.

Finding *P* When Given *F*

From Equation 4-2, $F = P(1 + i)^N$. Solving this for P gives the relationship

$$P = F\left(\frac{1}{1 + i}\right)^N = F(1 + i)^{-N} \qquad (4\text{-}4)$$

The quantity $(1 + i)^{-N}$ is called the *single payment present worth factor*. Numerical values for this factor are given in the third column of the tables in Appendix E for a wide range of values of i and N. We shall use the functional symbol $(P/F, i\%, N)$ for this factor. Hence

$$P = F(P/F, i\%, N) \qquad (4\text{-}5)$$

Examples of this type of problem, together with a cash flow diagram and solution, are given in the second part of the body of Table 4-3.

INTEREST FORMULAS RELATING A UNIFORM SERIES (ANNUITY) TO ITS PRESENT AND FUTURE WORTHS

Figure 4-4 shows a general cash flow diagram involving a series of uniform receipts, each of amount A, occurring at the end of each period for N periods with interest at $i\%$ per period. Such a uniform series is often called an *annuity*. It should be noted that the formulas and tables below are derived such that A occurs at the end of each period, and thus:

1. P (present worth) occurs one interest period before the first A (uniform payment).
2. F (future worth) occurs at the same time as the last A, and N periods after P.

The timing relationship for P, A, and F can be observed in Figure 4-4. Four formulas relating A to F and P are developed in Fig. 4-4.

Finding *F* When Given *A*

If A dollars exist at the end of each period for N periods and $i\%$ is the interest (profit or growth) rate per period, the future worth, F, at the end of the Nth

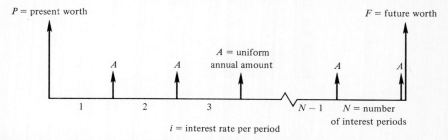

FIGURE 4-4 General Cash Flow Diagram Relating Uniform Series (Ordinary Annuity) to its Present Worth and Future Worth.

TABLE 4-3 Discrete Compounding Example Problems and Solutions

Example Problems (all using interest or profit rate of $i = 10\%$ compounded annually)

To Find:	Given:	(a) In Borrowing–Lending Terminology:	(b) In Equivalence Terminology:	Diagram	Solution
For single cash flows:					
F	P	A firm borrows $1,000 for 8 years. How much must it repay in a lump sum at the end of the eighth year?	What is the equivalent worth at the end of 8 years of $1,000 at the beginning of those 8 years?	$P = \$1{,}000$ $i = 10\%$ $1 \qquad 8$ $F = ?$	$F = P(F/P, 10\%, 8)$ $= \$1{,}000(2.1436)$ $= \$2{,}143.60$
P	F	A firm desires to have $1,000 6 years from now. What amount should be deposited now to provide for it?	What is the equivalent present worth of $1,000 6 years from now?	$F = \$1{,}000$ $i = 10\%$ $1 \qquad 6$ $P = ?$	$P = F(P/F, 10\%, 6)$ $= \$1{,}000(0.5645)$ $= \$564.50$
For uniform series:					
F	A	If three annual deposits of $2,000 each are placed in an account, how much money has accumulated immediately after the last deposit?	What amount at the end of the third year is equivalent to three $2,000 end-of-year payments?	$F = ?$ $i = 10\%$ $1 \quad 2 \quad 3$ $A = \$2{,}000$	$F = A(F/A, 10\%, 3)$ $= \$2{,}000(3.3100)$ $= \$6{,}620$

78

P	A	How much should be deposited in a fund now to provide for nine end-of-year withdrawals of $100 each?	 $A = \$100$ 1 2 3 8 9 $i = 10\%$ $P = ?$ $P = A(P/A, 10\%, 9)$ $= \$100(5.7590)$ $= \$575.90$
A	F	What uniform annual amount should be deposited each year in order to accumulate $10,000 at the time of the fifth annual deposit?	 $F = \$10,000$ $i = 10\%$ 1 2 3 4 5 $A = ?$ $A = F(A/F, 10\%, 5)$ $= \$10,000(0.1638)$ $= \$1,638$
A	P	What is the size of 10 equal annual payments to repay a loan of $1,000? The first payment is due 1 year after receiving loan.	 $P = \$1,000$ $i = 10\%$ 1 2 3 9 10 $A = ?$ $A = P(A/P, 10\%, 10)$ $= \$1,000(0.1627)$ $= \$162.70$

period is obtained by summing the future worths of each of the payments of amount A. Thus, if A_1 is the payment at the end of the first period, A_2 is the payment at the end of the second period, . . . , A_{N-1} is the payment at the end of the $(N - 1)$th period, and A_N is the payment at the end of the Nth period, then

$$F = A_1(1 + i)^{N-1} + A_2(1 + i)^{N-2} + A_3(1 + i)^{N-3} + \cdots$$
$$+ A_{N-1}(1 + i)^1 + A_N(1 + i)^0$$
$$= A[(1 + i)^{N-1} + (1 + i)^{N-2} + (1 + i)^{N-3} + \cdots$$
$$+ (1 + i)^1 + (1 + i)^0]$$

The term in brackets is a geometric series having a common ratio of $(1 + i)^{-1}$, which reduces to

$$F = A\left[\frac{(1 + i)^{N-1} - (1 + i)^{-1}}{1 - (1 + i)^{-1}}\right]$$

By multiplying numerator and denominator by $(1 + i)$, this reduces to

$$F = A\left[\frac{(1 + i)^N - 1}{i}\right] \qquad (4\text{-}6)$$

The quantity $\{[(1 + i)^N - 1]/i\}$ is called the *uniform series compound amount factor*. Numerical values for this factor are given in the fourth column of the tables in Appendix E for a wide range of values of i and N. We shall use the functional symbol $(F/A, i\%, N)$ for this factor. Hence Equation 4-6 can be expressed as

$$F = A(F/A, i\%, N) \qquad (4\text{-}7)$$

Examples of this type of problem, together with a cash flow diagram and solution, are given in the third row of the body of Table 4-3.

Finding *P* When Given *A*

From Equation 4-2, $F = P(1 + i)^N$. Substituting for F in Equation 4-6, one determines that

$$P(1 + i)^N = A\left[\frac{(1 + i)^N - 1}{i}\right]$$

Dividing both sides by $(1 + i)^N$,

$$P = A\left[\frac{(1 + i)^N - 1}{i(1 + i)^N}\right] \qquad (4\text{-}8)$$

Thus Equation 4-8 is the relation for finding the equivalent present worth (as of the beginning of the first period) of a uniform series of end-of-period cash flows of amount A. The quantity in brackets is called the *uniform series present worth factor*. Numerical values for this factor are given in the fifth column of the tables in Appendix E for a wide range of values of i and N. We shall use the functional symbol $(P/A, i\%, N)$ for this factor. Hence

$$P = A(P/A, i\%, N) \qquad (4\text{-}9)$$

Examples of this type of problem, together with a cash flow diagram and solution, are given in the fourth row of the body of Table 4-3.

Finding *A* When Given *F*

Taking Equation 4-6 and solving for A, one finds that

$$A = F\left[\frac{i}{(1 + i)^N - 1}\right] \qquad (4\text{-}10)$$

Thus Equation 4-10 is the relation for finding the amount, A, of a uniform series of cash flows occurring at the end of each of N interest periods which would be equivalent to (have the same value as) its future worth, F, occurring at the end of the last period. The quantity in brackets is called the *sinking fund factor*. Numerical values for this factor are given in the sixth column of the tables in Appendix E for a wide range of values of i and N. We shall use the functional symbol $(A/F, i\%, N)$ for this factor. Hence

$$A = F(A/F, i\%, N) \qquad (4\text{-}11)$$

Examples of this type of problem, together with a cash flow diagram and solution, are given in the fifth row of the body of Table 4-3.

Finding *A* When Given *P*

Taking Equation 4-8 and solving for A, one finds that

$$A = P\left[\frac{i(1 + i)^N}{(1 + i)^N - 1}\right]. \qquad (4\text{-}12)$$

Thus Equation 4-12 is the relation for finding the amount, A, of a uniform series of cash flows occurring at the end of each of N interest periods which would be equivalent to (have the same value as) the present worth, P, occurring at the beginning of the first period. The quantity in brackets is called the *capital recovery factor*.† Numerical values for this factor are given in the seventh column of the tables in Appendix E for a wide range of values of i and N. We shall use the functional symbol $(A/P, i\%, N)$ for this factor. Hence

$$A = P(A/P, i\%, N) \qquad (4\text{-}13)$$

An example that utilizes the equivalence between a present lump-sum amount and a series of equal uniform annual amounts starting at the end of year 1 and

---

†The capital recovery factor is more conveniently expressed as $i/[1 - (1 + i)^{-N}]$ when computing this factor with a hand-held calculator.

continuing through year 4 is provided in Table 4-1 as plan 3. By using Equation 4-13, the equivalent value of A that repays the $8,000 loan plus 10% interest per year over 4 years is

$$A = \$8,000(A/P, 10\%, 4) = \$8,000(0.3155) = \$2,524$$

The entries in columns 3 and 5 of plan 3 in Table 4-1 can now be better understood. Interest owed at the end of year 1 equals $8,000(0.10), and therefore the principal repaid out of the total end-of-year payment of $2,524 is the difference of $1,724. At the beginning of year 2, the amount of principal owed is $8,000 - $1,724 = $6,276. Interest owed at the end of year 2 is $6,276(0.10) \cong $628, and the principal repaid at that time is $2,524 - $628 = $1,896. After performing these calculations for years 3 and 4, the remaining entries in plan 3 are obtained.

A graphical summary of plan 3 is given in Figure 4-5. Here it can be seen that 10% interest is being paid on the beginning-of-year amount owed and that year-end payments of $2,524 which consist of interest and principal bring the amount owed to $0 at the end of the fourth year. (The exact value of A is $2,523.77 and produces an exact value of $0 at the end of 4 years.) It is important to note that all the uniform series interest factors in Table 4-2 involve the same concept as the one illustrated in Figure 4-5.

Another example of a problem where we desire to compute an equivalent value for A, from a given value of P and a known interest rate and number of compounding periods, is given in the last row of the body of Table 4-3.

DEFERRED ANNUITIES (UNIFORM SERIES)

All annuities (uniform series) discussed to this point involved the first cash flow being made at the end of the first period, and they are called *ordinary annuities*. If the cash flow does not begin until some later date, the annuity is known as a

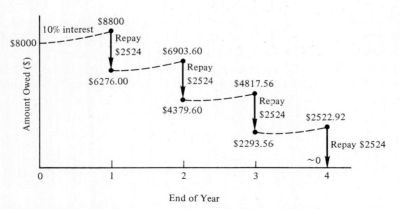

FIGURE 4-5   Relationship of Cash Flows for Plan 3 of Table 4-1 to Repayment of the $8,000 Loan Principal.

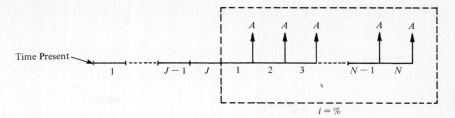

FIGURE 4-6 General Cash Flow Representation of a Deferred Annuity (Uniform Series).

deferred annuity. If the annuity is deferred J periods, the situation is as portrayed in Figure 4-6. It should be noted in Figure 4-6 that the entire framed ordinary annuity has been moved forward from "time present" or "time 0" by J periods. It must be remembered that in an annuity deferred for J periods the first payment is made at the end of the $(J + 1)$ period, assuming that all periods involved are equal in length.

The present worth at the end of period J of an annuity with cash flows of amount A is, from Equation 4-9, $A(P/A, i\%, N)$. The present worth of the single amount $A(P/A, i\%, N)$ as of time 0 will then be $A(P/A, i\%, N)(P/F, i\%, J)$.

EXAMPLE 4-2

To illustrate the discussion above, suppose that a father, on the day his son is born, wishes to determine what lump amount would have to be paid into an account bearing interest at 12% compounded annually to provide payments of $2,000 on each of the son's 18th, 19th, 20th, and 21st birthdays.

Solution

The problem is represented in Figure 4-7. One should first recognize that an ordinary annuity of four payments of $2,000 each is involved, and that the present worth of this annuity occurs at the 17th birthday when a $(P/A, i\%, N)$ is utilized. It often is helpful to use a subscript with P or F to denote the point in time. Hence

$$P_{17} = A(P/A, 12\%, 4) = \$2,000(3.0373) = \$6,074.60$$

Note the dashed arrow in Figure 4-7, denoting P_{17}. Now that P_{17} is known, the next step is to calculate P_0. With respect to P_0, P_{17} is a future worth, and hence it could also be denoted F_{17}. Money at a given point in time, such as end of

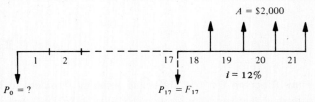

FIGURE 4-7 Cash Flow Diagram of the Deferred Annuity Problem in Example 4-2.

period 17, is the same regardless of whether it is called a present worth or a future worth. Hence

$$P_0 = F_{17}(P/F, 12\%, 17) = \$6,074.60(0.1456) = \$834.46$$

which is the amount that the father would have to deposit. ∎

EXAMPLE 4-3

As an addition to the problem in Example 4-2, suppose that it is desired to determine the equivalent worth of the four \$2,000 payments as of the son's 24th birthday. Physically, this could mean that the four payments never were withdrawn or that possibly the son took them and immediately redeposited them in an account also earning interest at 12% compounded annually. Using our subscript system, we desire to calculate F_{24} as shown in Figure 4-8.

Solution

One way to work this is to calculate

$$F_{21} = A(F/A, 12\%, 4) = \$2,000(4.7793) = \$9,558.60$$

To determine F_{24}, F_{21} becomes P_{21}, and

$$F_{24} = P_{21}(F/P, 12\%, 3) = \$9,558.60(1.4049) = \$13,428.88$$

Another quicker way to work the problem is to recognize that the $P_{17} = \$6,074.60$ and $P_0 = \$884.46$ are each equivalent to the four \$2,000 payments. Hence one can find F_{24} directly given P_{17} or P_0. Using P_0, we obtain

$$F_{24} = P_0(F/P, 12\%, 24) = \$884.46(15.1786) = \$13,424.86$$

which checks closely with the previous answer. The two numbers differ by \$4.02, which can be attributed to round-off error in the interest factors. ∎

UNIFORM SERIES WITH BEGINNING-OF-PERIOD CASH FLOWS

It should be noted that up to this point all the interest formulas and corresponding tabled values for uniform series have assumed *end-of-period* cash flows. These

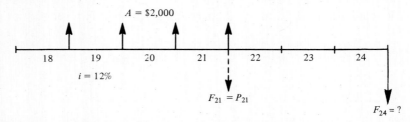

FIGURE 4-8   Cash Flow Diagram for the Deferred Annuity Problem in Example 4-3.

same tables can be used for cases in which beginning-of-period cash flows exist merely by remembering that:

1. P (present worth) occurs one interest period before the first A (uniform payment amount).
2. F (future worth) occurs at same time as the last A, and N periods after P.

EXAMPLE 4-4

Figure 4-9 is a cash flow diagram depicting a uniform series of five beginning-of-period cash flows of $100 each. Thus the first cash flow is at the beginning of the first period (time 0), and the fifth is at the beginning of the fifth period, which coincides with the end of the fourth period (time 4). If the interest or profit rate is 10%, it is desired to find (a) the worth of the uniform series at the beginning of the first period, P_0, and (b) the worth at the end of the fifth period, F_5.

Solution

(a) There are several ways to work these types of problems. To find P_0, one way is first to calculate

$$P_{-1} = A(P/A, 10\%, 5) = \$100(3.7908) = \$379.08$$

Note that this P is at time -1 because that is one period before the first A cash flow. Also note that the interest factor is for five periods because there were five payments. Next, the problem is to find P_0 given P_{-1}. In this case P_0 becomes a future worth and could be denoted F_0. Hence

$$P_0 = F_0 = P_{-1}(F/P, 10\%, 1) = \$379.08(1.100) = \$416.99$$

Alternatively, one could directly utilize the Appendix E interest factors to determine P_0 in this manner:

$$P_0 = \$100 + \$100(P/A, 10\%, 4)$$

$$= 100 + 100(3.1699)$$

$$= \$416.99$$

(b) To find F_5 without use of the fact that $P_0 = \$416.99$, the first logical step is to calculate

$$F_4 = A(F/A, 10\%, 5) = \$100(6.1051) = \$610.51$$

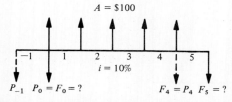

FIGURE 4-9   Cash Flow Diagram of Uniform Series with Beginning-of-Period Payments in Example 4-4.

Note that the F is at time 4 because that is at the same time as the last A cash flow. Also note that the interest factor is again for five periods, corresponding to the number of cash flows. Next, the problem is to find F_5 given F_4. In this case F_4 becomes a present worth and could be denoted P_4. Hence

$$F_5 = P_4(F/P, 10\%, 1) = \$610.51(1.10) = \$671.56 \qquad \blacksquare$$

Another, easier way to determine F_5 would be to start with \$379.08 as of time -1, or \$416.99 as of time 0, and to calculate the future worth at time 5. Indeed, once the equivalent worth of one or more cash flows is found as of a certain point in time, the equivalent worth of those same cash flows can be found as of any other point in time as long as the interest or profit rate, i, is known.

UNIFORM SERIES WITH MIDDLE-OF-PERIOD CASH FLOWS

Rather than using end-of-period or beginning-of-period cash flow conventions, some companies and individuals prefer to assume that discrete cash flows occur at the middle of each time period. Fortunately, the $(P/A, i, N)$, $(F/A, i, N)$, and $(P/F, i, N)$ interest factors of Appendix E can be utilized in the case of middle-of-period cash flows simply by *multiplying* the Appendix E interest factor by a "half-period correction" factor. When i is the effective interest rate per period, the half-period correction factor (HPC) to apply to interest factors in Appendix E is

$$HPC = \sqrt{1 + i}$$

When the $(A/F, i, N)$, $(A/P, i, N)$, and $(F/P, i, N)$ factors are involved, the Appendix E values are *divided* by the half-period correction factor to obtain midperiod interest factors.

EXAMPLE 4-5
Figure 4-10a is a cash flow diagram of an annuity consisting of four middle-of-year amounts of \$200. If $i/\text{year} = 10\%$, determine (a) the equivalent present

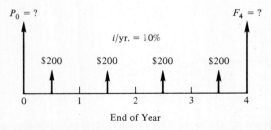

FIGURE 4-10a   Cash Flow Diagram of a Uniform Series of Midyear Amounts in Example 4-5.

worth of the uniform series at the beginning of the first year, and (b) the equivalent future worth at the end of the fourth year.

Solution
(a) The half-period correction factor equals $\sqrt{1.10} = 1.04881$. Therefore, the equivalent value of P_0, utilizing the $(P/A, 10\%, 4)$ factor from Appendix E, can be determined in two steps:

$$P_{-\frac{1}{2}} = \$200(P/A, 10\%, 4) = \$633.98$$

$$P_0 = P_{-\frac{1}{2}}(\text{HPC}) = \$664.92$$

(b) The equivalent value of F_4 is determined with a $(F/A, 10\%, 4)$ factor from Appendix E and a half-period correction factor of $\sqrt{1.10}$:

$$F_4 = \$200(F/A, 10\%, 4)(1.04881)$$

$$= \$200(4.641)(1.04881)$$

$$= \$973.50 \qquad \blacksquare$$

EXAMPLE 4-6
If \$1,500 is to be received at the end of 5 years when i per year is 15%, what middle-of-year uniform amount would be equivalent to this future amount? The cash flow diagram is shown in Figure 4-10b.

Solution
The half-period correction is $\sqrt{1.15} = 1.0724$, so the following relationships, utilizing the appropriate Appendix E factor, can be used to solve for the middle-of-year equivalent A:

$$F_{4\frac{1}{2}}(\text{HPC}) = F_5 \qquad \text{or} \qquad F_5\left(\frac{1}{\text{HPC}}\right) = F_{4\frac{1}{2}}$$

$$A = F_{4\frac{1}{2}}(A/F, 15\%, 5)$$

$$= \frac{\$1,500}{1.0724}(0.1483) = \$207.43 \qquad \blacksquare$$

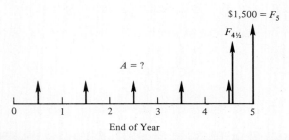

FIGURE 4-10b   Cash Flow Diagram for Mid-Year Amounts of Example 4-6.

EQUIVALENT PRESENT WORTH, FUTURE WORTH, AND ANNUAL WORTH

The reader should now be comfortable with equivalence problems that involve discrete compounding of interest and discrete cash flows. All compounding of interest takes place once per time period (e.g., a year) and to this point cash flows also occur once per time period. The following example demonstrates several equivalence calculations based on the commonly used end-of-year cash flow convention.

EXAMPLE 4-7

Figure 4-11 depicts an example problem with a series of year-end cash flows extending over 8 years. The amounts are $100 for the first year, $200 for the second year, $500 for the third year, and $400 for each year from the fourth through the eighth. These could represent something like the expected maintenance expenditures for a certain piece of equipment or payments into a fund. Note that the payments are shown at the end of each year, which is a standard assumption for this book and for economic analyses in general unless one has information to the contrary. It is desired to find the equivalent (a) present worth, (b) future worth, and (c) annual worth of these cash flows if the annual interest rate is 20%.

Solution

(a) To find the equivalent present worth, P_0, one needs to sum the worth of all payments as of the beginning of the first year (time 0). The required movements of money through time are shown graphically in Figure 4-11a.

$$
\begin{aligned}
P_0 = F_1(P/F, 20\%, 1) &\quad = \$100(0.8333) &\quad = \$ \quad 83.33 \\
+ F_2(P/F, 20\%, 2) &\quad + \$200(0.6944) &\quad + \quad 138.88 \\
+ F_3(P/F, 20\%, 3) &\quad + \$500(0.5787) &\quad + \quad 289.35 \\
+ A(P/A, 20\%, 5) &\quad + \$400(2.9906) &\quad + \quad \underline{692.26} \\
\times (P/F, 20\%, 3) &\quad \times (0.5787) &\quad \\
&\quad &\quad \$1{,}203.82
\end{aligned}
$$

(b) To find the equivalent future worth, F_8, one can sum the worth of all payments as of the end of the eighth year (time 8). Figure 4-11b indicates these movements of money through time. However, since the equivalent present worth is already known to be $1,203.82, one can calculate directly

$$F_8 = P_0(F/P, 20\%, 8) = \$1{,}203.82(4.2998) = \$5{,}176.19$$

(c) The equivalent annual worth of the irregular series can be calculated directly from either P_0 of F_8, as follows:

$$A = P_0(A/P, 20\%, 8) = \$1.203.82(0.2606) = \$313.73$$

or

$$A = F_8(A/F, 20\%, 8) = \$5{,}176.19(0.0606) = \$313.73$$

The computation of A from P_0 and F_8 is shown in Figure 4-11c. Thus one finds that the irregular series of payments shown in Figure 4-11 is equivalent to

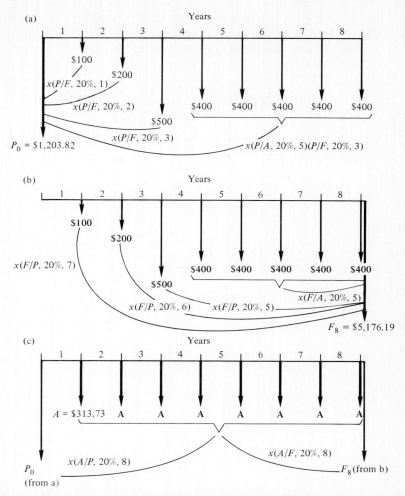

FIGURE 4-11   Example 4-7 for Calculating the Equivalent Present Worth, Future Worth, and Annual Worth.

$1,203.82 at time 0, $5,176.19 at time 8, or a uniform series of $313.73 at the end of each of 8 years. ∎

INTEREST FORMULAS RELATING AN ARITHMETIC GRADIENT SERIES TO ITS PRESENT AND ANNUAL WORTHS

Some economic analysis problems involve receipts or disbursements that are projected to increase or decrease by a uniform *amount* each period, thus constituting an arithmetic series. For example, maintenance and repair expenses on specific equipment may increase by a relatively constant amount each period.

Figure 4-12 is a cash flow diagram of a series of end-of-period payments increasing by a constant amount, G, each period. The G is known as the *gradient amount*. Note that the timing of payments on which the derived formulas and tabled values are based is as follows:

End of Year	Payments
1	0
2	G
3	$2G$
.	.
.	.
.	.
$N - 1$	$(N - 2)G$
N	$(N - 1)G$

Finding *P* When Given *G*

The present worth, P, of the series shown in Figure 4-12 is

$$P = G\left[\frac{1}{(1 + i)^2}\right] + 2G\left[\frac{1}{(1 + i)^3}\right]$$

$$+ \cdots + (N - 2)G\left[\frac{1}{(1 + i)^{N-1}}\right]$$

$$+ (N - 1)G\left[\frac{1}{(1 + i)^N}\right]$$

$$P = G \times \frac{1}{i}\left[\frac{(1 + i)^N - 1}{i(1 + i)^N} - \frac{N}{(1 + i)^N}\right] \tag{4-14}$$

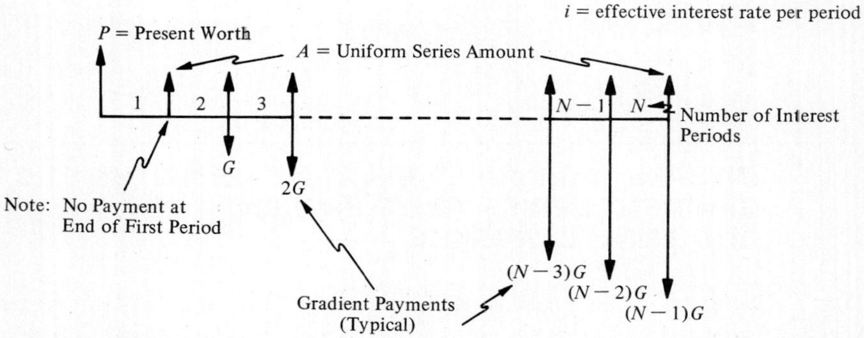

FIGURE 4-12 General Cash Flow Diagram for an Arithmetic Gradient Series of *G* Dollars Per Period.

The term

$$\frac{1}{i}\left[\frac{(1 + i)^N - 1}{i(1 + i)^N} - \frac{N}{(1 + i)^N}\right]$$

is called the *gradient to present worth conversion factor*. It also can be expressed as $(1/i)[(P/A, i\%, N) - N(P/F, i\%, N)]$. Numerical values for this factor are given in Table 16 of Appendix E for a wide range of i and N values. We shall use the functional symbol $(P/G, i\%, N)$ for this factor. Hence

$$P = G(P/G, i\%, N) \qquad (4\text{-}15)$$

Finding *A* When Given *G*

To obtain a uniform series of amount A that is equivalent to the arithmetic gradient series shown in Figure 4-12, one need only multiply the present worth in Equation 4-15 by $(A/P, i\%, N)$. Hence

$$A = P(A/P, i\%, N) = G(P/G, i\%, N)(A/P, i\%, N)$$

$$= G \times \frac{1}{i}\left[\frac{(1 + i)^N - 1}{i(1 + i)^N} - \frac{N}{(1 + i)^N}\right]\left[\frac{i(1 + i)^N}{(1 + i)^N - 1}\right]$$

$$= G \times \left[\frac{1}{i} - \frac{N}{(1 + i)^N - 1}\right] \qquad (4\text{-}16)$$

The term in brackets is called the *gradient to uniform series conversion factor*. Numerical values for this factor are given in Table 17 of Appendix E for a wide range of i and N values. We shall use the functional symbol $(A/G, i\%, N)$ for this factor. Hence

$$A = G(A/G, i\%, N) \qquad (4\text{-}17)$$

Again, note that the use of these gradient conversion factors calls for there being no payment at the end of the first period.

EXAMPLE 4-8

As an example of the straightforward use of the gradient conversion factors, suppose that certain end-of-year expenses are expected to be \$1,000 for the second year, \$2,000 for the third year, and \$3,000 for the fourth year, and that if interest is 15% per year, it is desired to find the equivalent (a) present worth at the beginning of the first year, and (b) uniform annual worth at the end of each of the 4 years.

Solution
It can be observed that this schedule of expenses fits the model of the arithmetic gradient formulas with $G = \$1,000$ and $N = 4$. Note there was no payment at the end of the first period.

(a) The present worth can be calculated as

$$P_0 = G(P/G, 15\%, 4) = \$1,000(3.79) = \$3,790$$

(b) The uniform annual worth can be calculated from Equation 4-17 as

$$A = G(A/G, \ 15\%, \ 4) = \$1,000(1.3263) = \$1,326.30$$

Of course, once the present worth was known, the uniform annual worth could have been calculated as

$$A = P_0(A/P, \ 15\%, \ 4) = \$3,790(0.3503) = \$1,326.30 \qquad \blacksquare$$

EXAMPLE 4-9

As a further example of the use of arithmetic gradient formulas, suppose that one has payments as follows:

End of Year	Payment
1	$5,000
2	6,000
3	7,000
4	8,000

and that one wishes to calculate their equivalent present worth at $i = 15\%$ using arithmetic gradient interest formulas.

Solution

The schedule of payments is depicted in the top diagram of Figure 4-13. The bottom two diagrams of Figure 4-13 show how the original schedule can be broken into two subschedules of payments, a uniform series of $5,000 payments plus an arithmetic gradient of $1,000 that fits the general gradient model for

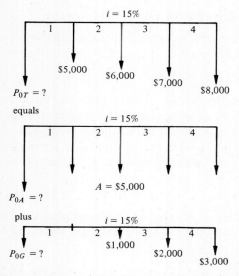

FIGURE 4-13   Example 4-9 Involving an Increasing Arithmetic Gradient.

which factors are tabled. The summed present worths of these two subschedules equal the present worth of the original problem. Thus, using the symbols shown in Figure 4-13, we have

$$P_{0T} = P_{0A} + P_{0G}$$

$$= A(P/A,\ 15\%,\ 4) + G(P/G,\ 15\%,\ 4)$$

$$= \$5,000(2.8550) + \$1,000(3.79) = \$14,275 + 3,790 = \$18,065$$

The uniform annual worth of the original payments could be calculated with Equation 4-17 as follows:

$$A_T = A + A_G$$

$$= \$5,000 + \$1,000(A/G,\ 15\%,\ 4) = \$6,326.30$$

A_T is equivalent to P_{0T} because $\$6,326.30(P/A,\ 15\%,\ 4) = \$18,061$, which is the same value obtained above (subject to round-off error). ■

EXAMPLE 4-10

For another example of the use of arithmetic gradient formulas, suppose that one has payments which are timed in exact reverse of the payments depicted in Example 4-9. The top diagram of Figure 4-14 shows these payments as follows:

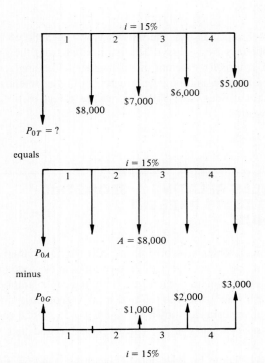

FIGURE 4-14   Example 4-10 Involving a Decreasing Arithmetic Gradient.

End of Year	Payment
1	$8,000
2	7,000
3	6,000
4	5,000

Solution

The bottom two diagrams of Figure 4-14 show how these payments can be broken into two subschedules of payments. It must be remembered that the arithmetic gradient formulas and tables provided are for increasing gradients only. Hence one must subtract an *increasing* gradient of payments that *did not* occur. Thus

$$P_{OT} = P_{OA} - P_{OG}$$

$$= A(P/A, \ 15\%, \ 4) - G(P/G, \ 15\%, \ 4)$$

$$= \$8,000(2.8550) - \$1,000(3.79) = \$22,840 - 3,790 = \$19,050$$

Again, the uniform annual worth of the original decreasing series of payments could be calculated by the same rationale, or by finding A, given that the equivalent present worth is $19,050. ∎

Note from Examples 4-9 and 4-10 that the present worth of $18,065 for an increasing arithmetic gradient series of payments is different from the present worth of $19,050 for an arithmetic gradient of payments of like amounts but reversed timing. This difference would be even greater for higher interest rates and gradient payments, but it does exemplify the marked effect of timing of payments on equivalent worths.

INTEREST FORMULAS RELATING A GEOMETRIC GRADIENT SERIES TO ITS PRESENT AND ANNUAL WORTH

Some economic equivalence problems involve projected cash flow patterns that are increasing at a constant *rate, f,* each period. A fixed amount of a commodity that inflates in price at a constant rate each year is a typical example. The resultant end-of-period cash flow pattern is referred to as a geometric gradient series and has the general appearance shown in Figure 4-15. Notice that *the initial cash flow in this series occurs at the end of period 1* and that $A_k = (A_{k-1})(1 + f), \ 1 \le k \le N$.

Each term in Figure 4-15 could be discounted, or compounded, at interest rate i per period to obtain a value of P or F, respectively. However, this becomes quite tedious for large N and it is convenient to have a single equation instead.

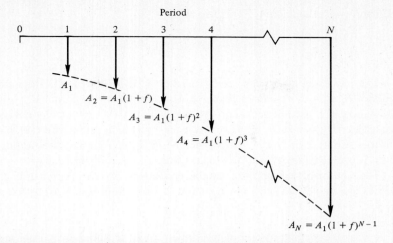

FIGURE 4-15 Cash Flow Diagram for a Geometric Gradient Series.

To develop a compact expression for P at interest rate i per period for the cash flows of Figure 4-15, consider the following summation:

$$P = \sum_{k=1}^{N} A_k(1 + i)^{-k} = \sum_{k=1}^{N} A_1(1 + f)^{k-1}(1 + i)^{-k}$$

or

$$P = \frac{A_1}{1 + f} \sum_{k=1}^{N} \left(\frac{1 + f}{1 + i}\right)^k \qquad (4\text{-}18)$$

When $i \neq f$, we let

$$i_r = \frac{1 + i}{1 + f} - 1$$

which can also be written as $i_r = (i - f)/(1 + f)$. In the situation where $i \neq f$, Equation 4-18 can thus be rewritten as

$$P = \frac{A_1}{1 + f} \sum_{k=1}^{N} \left(\frac{1 + i}{1 + f}\right)^{-k}$$

$$= \frac{A_1}{1 + f} \sum_{k=1}^{N} (1 + i_r)^{-k}$$

$$= \frac{A_1}{1 + f} \, (P/A, \, i_r, \, N)† \qquad (4\text{-}19)$$

Equation 4-19 makes use of the fact that

$$(P/A, \, i_r, \, N) = \sum_{k=1}^{N} (1 + i_r)^{-k} = \sum_{k=1}^{N} (P/F, \, i_r, \, k)$$

†When f exceeds i, i_r is negative and the above summation is valid only when N is finite.

When $i = f$, Equation 4-19 reduces to

$$P = \frac{A_1}{1 + f} (P/A, 0\%, N) = \frac{NA_1}{1 + f} \qquad (4\text{-}20)$$

The interested reader can verify Equation 4-20 by applying L'Hôpital's Rule to the $(P/A, i_r, N)$ factor in Equation 4-19 and taking the limit as $i_r \rightarrow 0$.

Values of i_r used in connection with Equation 4-19 are typically not included in tables in Appendix E. Because i_r is usually a noninteger interest rate, resorting to the definition of a $(P/A, i_r, N)$ factor (see Table 4-2) and substituting terms into it is a satisfactory way to obtain values of these interest factors.

The end-of-period uniform annual equivalent, A, of a geometric gradient series can be determined from Equation 4-19 (or Equation 4-20) as follows:

$$A = P(A/P, i, N) \qquad (4\text{-}21)$$

The year 0 equivalent value of this annuity, which increases at a constant rate of $f\%$ per period, is A_0 and equals

$$A_0 = P(A/P, i_r, N) \qquad (4\text{-}22)$$

The difference between A and A_0 can be seen in Figure 4-16. Finally, the future equivalent of this geometric gradient series is

$$F = P(F/P, i, N) \qquad (4\text{-}23)$$

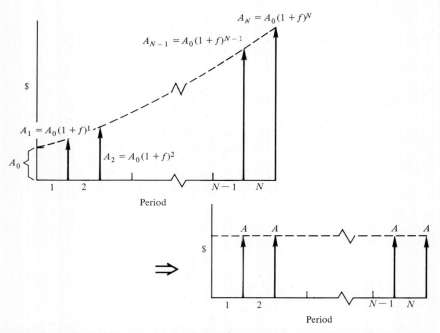

FIGURE 4-16 Graphical Interpretation of A and A_0 Equivalents of a Geometric Gradient Series when $f > 0$.

EXAMPLE 4-11

Consider the end-of-year geometric gradient series below and determine its P, A, A_0, and F equivalent values. The rate of increase is 20% per year after the first year, and the interest rate is 25%.

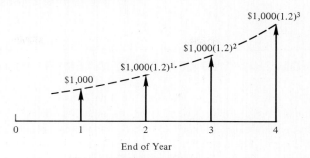

End of Year

EXAMPLE 4-11

Solution

$$P = \frac{\$1,000}{1.2}(P/A, \frac{25\% - 20\%}{1.20}, 4) = \$833.33 \left(P/A, 4.167\%, 4 \right)$$

$$= \$833.33(3.6157) = \$3,013.08$$

$$A = \$3,013.08(A/P, 25\%, 4) = \$1,275.86$$

$$A_0 = \$3,013.08(A/P, 4.167\%, 4)$$

$$= \$3,013.08\left[\frac{0.04167(1.04167)^4}{(1.04167)^4 - 1} \right] = \$833.34$$

$$F = \$3,013.08(F/P, 25\%, 4) = \$7,356.15$$ ∎

EXAMPLE 4-12

A heat pump is being considered as a replacement for an existing electric resistance furnace. Based on winter heating requirements, it has been estimated that $600 per year will be saved through reduced electricity bills. These savings have been computed at *present* electricity rates. If electricity is expected to increase at an average rate of 14% per year into the foreseeable future, how much can we justify spending now for the heat pump if the interest rate is 12% per year? Assume that the life of the heat pump is 15 years and that its salvage value at this time is negligible.

Solution

Present savings are $600 per year, so savings by the end of year 1 will be $600(1.14) = $684. With end-of-year cash flow convention, the maximum amount that can be justified now for the purchase of this heat pump is

$$P \le \frac{\$684}{1.14} \left(P/A, \frac{12\% - 14\%}{1.14}, 15 \right)$$

$$P \le \$600(P/A, -1.75\%, 15)$$

or
$$P \leq \$600\left[\frac{(0.9825)^{15} - 1}{-0.0175(0.9825)^{15}}\right]$$
$$\leq \$600(17.326)$$
$$\leq \$10,395$$ ∎

NOMINAL AND EFFECTIVE INTEREST RATES

Very often, the interest period, or time between successive compounding, is something less than 1 year. It has become customary to quote interest rates on an annual basis, followed by the compounding period if different from 1 year in length. For example, if the interest rate is 5% per interest period and the interest period is 6 months, it is customary to speak of this rate as ''10% compounded semiannually.'' The basic annual rate of interest is known as the *nominal rate*, 10% in this case. A nominal interest rate is represented by r. The actual annual rate on the principal is not 10% but something greater, because of the compounding that occurs twice during a year. For instance, consider $100 to be invested at a nominal rate of 10% compounded semiannually. The interest earned during the year would be as follows:

First 6 months:
$$I = \$100 \times 0.05 = \$5$$

Total principal and interest at beginning of the second period:
$$P + Pi = \$100 - \$5 = \$105$$

Interest earned during second 6 months
$$\$105 \times 0.05 = \$5.25$$

Total interest earned during year:
$$\$5.00 + \$5.25 = \$10.25$$

Actual annual interest rate:
$$\frac{\$10.25}{\$100} \times 100 = 10.25\%$$

This can be shown graphically as follows:

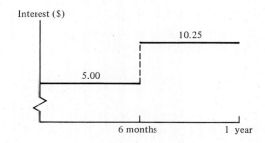

This actual or exact rate of interest earned on the principal during 1 year is known as the *effective rate*. It should be noted that effective interest rates always are expressed on an annual basis, unless specifically stated otherwise. In this text the interest rate per period is designated by i and the nominal interest rate per year by r. In most engineering economy studies in which compounding is annual, $i = r$, of course. In general terms,

$$\text{effective rate, } i = (1 + r/M)^M - 1$$
$$= (F/P, r/M, M) - 1 \qquad (4\text{-}24)$$

where M is the number of compounding periods per year.

The effective rate of interest is useful for describing the compounding effect of interest earned on interest within 1 year. Table 4-4 shows effective rates for various nominal rates and compounding periods. As a matter of interest, the federal "truth in lending" law requires a statement regarding the annual percentage rate (APR) being charged in contracts involving borrowed money. The APR is a nominal interest rate.

The reader will now realize that in Examples 4-5 and 4-6 an assumption was made regarding an effective interest rate per 6 months which was compounded to an effective interest rate per year: $i/\text{year} = (1 + i/6 \text{ months})^2 - 1$, or $1 + i/6 \text{ months} = \sqrt{1 + i/\text{year}}$.

INTEREST PROBLEMS WITH COMPOUNDING MORE OFTEN THAN ONCE PER YEAR

Single Amounts

If a nominal interest rate is quoted and the number of compounding periods per year and number of years are known, any problem involving calculating future worths or present worths can be calculated by straightforward use of Equations 4-3 and 4-5, respectively.

TABLE 4-4 Effective Interest Rates for Various Nominal Rates and Compounding Periods

Compounding Period	Number of Periods per Year, M	Effective Rate (%) for Nominal Rate of:		
		6%	12%	24%
Annually	1	6.00	12.00	24.00
Semiannually	2	6.09	12.36	25.44
Quarterly	4	6.14	12.55	26.25
Bimonthly	6	6.15	12.62	26.53
Monthly	12	6.17	12.68	26.82
Continuously	∞	6.18	12.75	27.12

EXAMPLE 4-13

For example, $100.00 is invested for 10 years at 6% compounded quarterly. How much is it worth at the end of the 10th year?

Solution

There are four compounding periods per year, or a total of $4 \times 10 = 40$ periods. The interest rate per interest period is $6\%/4 = 1.5\%$. When the values are used in Equation 4-3, one finds that

$$F = P(F/P, 1.5\%, 40) = \$100.00(1.814) = \$181.40$$

Alternatively, the effective interest rate from Equation 4-24 is 6.14%. Therefore, $F = \$100.00(F/P, 6.14\%, 10) = \$181.40.$ ∎

Uniform Series and Gradient Series

When there is more than one compounded interest period per year, the formulas and tables for uniform series and gradient series can be used as long as there is a cash flow at the end of each interest period, as shown in Figures 4-4 and 4-12 for uniform series and gradient series, respectively.

EXAMPLE 4-14

Suppose that one has a beginning indebtedness of $10,000 which is to be repaid by equal end-of-month installments for 5 years with interest at 12% compounded monthly. What is the amount of each payment?

Solution

The number of installment payments is $5 \times 12 = 60$, and the interest rate per month is $12\%/12 = 1\%$. When these values are used in Equation 4-13, one finds that

$$A = P(A/P, 1\%, 60) = \$10,000(0.0222) = \$222$$ ∎

EXAMPLE 4-15

Suppose that certain operating expenditures were expected to be 0 at the end of the first 6 months, $1,000 at the end of the second 6 months, and to increase by $1,000 at the end of each 6-month period thereafter for a total of 4 years. It is desired to find the equivalent uniform payment at the end of each of the eight 6-month periods if interest is 20% compounded semiannually.

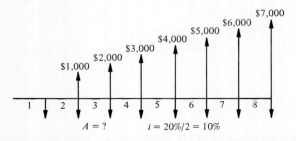

FIGURE 4-17 Example Arithmetic Gradient with Compounding More Often than Once Per Year in Example 4-15.

Solution

A cash flow diagram is shown in Figure 4-17, and the solution is

$$A = G(A/G,\ 10\%,\ 8) = \$1{,}000(3.0045) = \$3{,}004.50 \qquad \blacksquare$$

INTEREST PROBLEMS WITH UNIFORM CASH FLOWS LESS OFTEN THAN COMPOUNDING PERIODS

EXAMPLE 4-16

Suppose that there exists a series of 10 end-of-year payments of $1,000 each and it is desired to compute the equivalent worth of those payments as of the end of the 10th year if interest is 12% compounded quarterly. The problem is depicted in Figure 4-18.

Solution

Interest is $12\%/4 = 3\%$ per quarter, but the uniform series cash flows are not at the end of each quarter. Hence one must make special adaptations to fit the interest formulas to the tables provided. To solve this type of problem, an equivalent cash flow must be computed for the time interval that corresponds to the stated compounding frequency, or an effective interest rate must be determined for the interval of time separating cash flows.

One useful adaptation procedure is to take the number of compounding periods over which a cash flow occurs and convert the cash flow into its equivalent uniform end-of-period series. The upper cash flow diagram in Figure 4-19 shows this approach applied to the first year (four interest periods) in the example of Figure 4-18. The uniform end-of-quarter payment which is equivalent to the $1,000 at the end of the year with interest at 3% per quarter can be calculated as

$$A = F(A/F,\ 3\%,\ 4) = \$1{,}000(0.2390) = \$239$$

Thus $239 at the end of each quarter is equivalent to $1,000 at the end of each year. This is true not only for the first year but also for each of the 10 years under consideration. Hence the original series of 10 end-of-year payments of $1,000 each can be converted to a problem involving 40 end-of-quarter payments of $239 each, as shown in the lower cash flow diagram of Figure

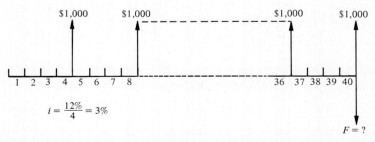

FIGURE 4-18 **Uniform Series with Payments Less Often than Compounding Periods in Example 4-16.**

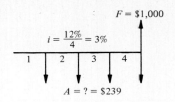

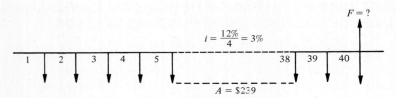

FIGURE 4-19   First Adaptation to Solve Example 4-16.

4-19. The worth at the end of the tenth year (40th quarter) may then be computed as

$$F = A(F/A, \ 3\%, \ 40) = \$239(75.4012) = \$18,021 \quad \blacksquare$$

In general, if i is the interest rate per interest period and there is a uniform payment, X, at the end of each kth interest period, then the equivalent payment, A, at the end of each interest period is

$$A = X(A/F, \ i\%, \ k) \tag{4-25}$$

By similar reasoning, if i is the interest rate per interest period and there is a uniform payment, X, at the *beginning* of each kth interest period, then the equivalent payment, A, at the end of each interest period is

$$A = X(A/P, \ i\%, \ k) \tag{4-26}$$

Another adaptation procedure for handling uniform series with cash flows less often than compounding periods is to find the exact interest rate for each time period separating cash flows and then to straightforwardly apply the interest formulas and tables for that exact interest rate. For Example 4-16, interest is 3% per quarter and payments occur each year. Hence the interest rate to be found is the exact rate each year, or the *effective rate*. The effective rate of 3% per quarter (12% nominal) can be found from Equation 4-24 to be

$$\left(1 + \frac{0.12}{4}\right)^4 - 1 = (F/P, \ 3\%, \ 4) - 1 = 0.1255$$

Hence the original problem in Figure 4-18 can now be expressed as shown in Figure 4-20. The future worth of this series can then be found as

$$F = A(F/A, \ 12.55\%, \ 10) = \$1,000(F/A, \ 12.55\%, \ 10) = \$18,022$$

Because interest factors are not commonly tabled for $i = 12.55\%$, one must compute the $(F/A, \ 12.55\%, \ 10)$ factor by substituting $i = 0.1255$ and $N = 10$ into its algebraic equivalent, $[(1 + i)^N - 1]/i$.

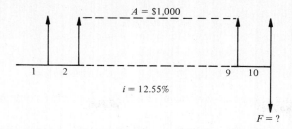

FIGURE 4-20 Second Adaptation to Solve Example 4-16.

Substitution into the algebraic equivalent will give the exact answer (same as the $18,021 by the first adaptation method in Figure 4-19 except for any round-off error). Linear interpolation may also be utilized as a good approximation in most problems.

Yet another adaptation for problems involving uniform series cash flows that occur less often than compounding periods is to treat each cash flow as a single sum. This is usually unsatisfactory because of the number of computations involved, but it is well to recognize the possibility. Thus the solution to Example 4-16 can be computed by recognizing that the first payment is compounded 36 interest periods at 3% per period, the second payment is compounded 32 interest periods, and so on. Hence the problem can be solved as

$$F = \$1,000[(F/P, 3\%, 36) \quad + (F/P, 3\%, 32), + \cdots$$
$$+ (F/P, 3\%, 4) + (F/P, 3\%, 0)]$$
$$= \$1,000(2.8983 + 2.5751 + \cdots + 1.1255 + 1.00)$$

The exact answer of $18,021 would be obtained from this calculation if there were no interpolation or round-off error.

INTEREST PROBLEMS WITH UNIFORM CASH FLOWS OCCURRING MORE OFTEN THAN COMPOUNDING PERIODS

EXAMPLE 4-17

Suppose that an individual insists on making weekly payments of $50 into a savings account upon which nominal interest of 10% is compounded quarterly. What is the equivalent lump-sum future amount in the account at the end of 2 years? There are several assumptions that might be made in working this problem, and two of them are shown below.

Solution

Assumption 1: There is no interest computed on any amount except that which is in the account by the end of each quarter. In this case the amount deposited each quarter totals $50(13) = $650. Because the nominal interest per quarter is 2.5%, the compound amount at the end of 2 years is

$$F = \$650(F/A, 2.5\%, 8) = \$5,678.48$$

Assumption 2: There is an appropriate nominal interest rate per week, r_{52}, that compounds to 2.5% each quarter. This rate is

$$(1 + r_{52})^{13} - 1 = 0.025 \qquad \text{or} \qquad r_{52} = 0.0019$$

Therefore, the future amount at the end of 2 years with weekly compounding at r_{52} per week is

$$F = \$50(F/A, 0.19\%, 104)$$

$$= \$5,743.33$$

Of the two assumptions, probably the first is the more realistic and is recommended for solving this type of problem. ∎

INSTALLMENT FINANCING

When a series of deferred equal periodic cash flows is substituted for a single (lump-sum) cash amount, as when merchandise such as an automobile is purchased, a modification of the ordinary annuity frequently is used. A finance charge is made upon the total amount owed at the beginning of the loan instead of only upon the unpaid balance. Such a charge is, of course, not in accord with the true nature and definition of interest. To see what the true interest rate being charged really is in such cases, consider the following example.

EXAMPLE 4-18

The Fly-by-Night finance company advertises a "bargain 6% plan" for financing the purchase of automobiles. To the amount remaining to be paid through installment payments, 6 percent is added for each year in which money is owed. This total is divided by the number of months over which the payments are to be made, and the result is the amount of the monthly payments. For example, a woman purchases a $10,000 automobile under this plan and makes an initial cash payment of $2,500. She wishes to pay the balance in 24 monthly payments. What will be the amount of each payment, and what rate of interest does she actually pay?

Solution

Purchase price	= \$10,000
− Initial payment	= 2,500
= Balance due, (P)	= 7,500
+ 6% finance charge = 0.06×2 years $\times \$7,500$ =	900
= Total to be paid	= 8,400
∴ Monthly payments = $\$8,400/24$, (A)	= \$ 350

Because there are to be 24 payments of $350 each, made at the end of each month, these constitute an annuity at some unknown rate of interest that should be computed only upon the unpaid balance. Therefore,

$$P = A(P/A, i\%, N)$$

$$\$7,500 = \$350(P/A, i\%, N)$$

$$(P/A, i\%, N) = \frac{\$7,500}{\$350} = 21.43$$

By examination of the interest tables for P/A factors for $N = 24$ which comes closest to 21.43, one finds that $(P/A, \frac{3}{4}\%, 24) = 21.8891$ and $(P/A, 1\%, 24) = 21.2434$.

Linear interpolation gives

$$i\% = 1\% - \frac{21.43 - 21.2434}{21.8891 - 21.2434}(1\% - \tfrac{3}{4}\%) = 0.93\%$$

Since payments are monthly, 0.93% is the interest rate per month. The nominal rate (APR) paid on the borrowed money is $0.93\%(12) = 11.16\%$ compounded monthly. This corresponds to an effective annual interest rate of $(1 + 0.0093)^{12} - 1 \cong 12\%$. What appeared at first to be a real bargain turns out to involve effective annual interest at recent typical rates for automobile loans. ■

EXAMPLE 4-19

A small company needs to borrow $160,000. The local (and only) banker makes this statement: "We can loan you $160,000 at a very favorable rate of 12% per year for a 5-year loan. However, to secure this loan you must agree to establish a checking account with us in which the *minimum* average balance is $32,000. In addition, your interest payments are due at the end of each year and the principal will be repaid in a lump-sum amount at the end of year 5." What is the true effective interest rate being charged?

Solution

The cash flow diagram from the banker's viewpoint has this appearance:

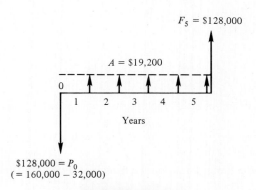

EXAMPLE 4-19

The interest rate that establishes equivalence among cash flows can easily be computed:

$$P_0 = F_5(P/F, i, 5) + A(P/A, i, 5)$$

$$\$128,000 = \$128,000(P/F, i, 5) + \$19,200(P/A, i, 5)$$

If we try $i = 15\%$, we discover that $\$128,000 = \$128,000$. Therefore, the true effective interest rate is 15%. Because $P_0 = F_5$ and A is constant, the true interest rate can be computed more easily: $i = A/P = \$19,200/\$128,000 = 0.15$. ■

DISCOUNT

Two types of transactions are sometimes encountered in which *discount* is involved. In the first the holder of a negotiable paper, such as a bond or a sales contract that is not due and payable until some future date, desires to exchange the paper for immediate cash. To do this, he will accept a sum of cash smaller in amount than the face value of the paper. The difference between the present worth (the amount received for the paper in cash) and the worth of the paper at some time in the future (the face value of the paper or principal) is known as the discount for the period involved.

The second type of transaction occurs in many bank loans. As an example, a woman may wish to borrow \$100 from a bank for 1 year at an interest rate of 12%. The bank computes the interest, 12%, and *deducts* this amount from the \$100, giving the borrower only \$88. The borrower signs a note promising to repay \$100 at the end of a year. The \$12 that was deducted represents interest paid in advance. It also represents the difference between the present worth of the note—the \$88 received by the borrower—and the worth of the note at the end of 1 year—\$100. It is therefore the discount for the period involved. The *rate of discount* is defined as the discount of one unit of principal for one unit of time. If the rate of discount is designated by d, it is equal to the difference between 1 and its present worth:

$$d = 1 - \frac{1}{1 + i} \tag{4-27}$$

and

$$i = \frac{d}{1 - d} \tag{4-28}$$

For the example cited, the discount was \$12. The rate of discount was

$$d = \frac{\$12}{\$100} = 0.12 = 12\%$$

The interest rate, based on the principal actually received by the borrower, was

$$i = \frac{0.12}{0.88} = 0.1364 = 13.64\%$$

EXAMPLE 4-20

A bond has a future value of $1,000 (its face value) in 8 years and pays interest of $100 at the end of each of the 8 years. It can now be purchased at a discounted price of $900. What is the true rate of interest being earned on the bond if it is purchased for $900?

Solution

The true interest rate being earned on this bond can be calculated as follows:

present worth of cash outflow = present worth of cash inflow

or

$$\$900 = \$100(P/A, i, 8) + \$1,000(P/F, i, 8)$$

By trial and error, the value of i is found to be quite close to 12%. ■

INTEREST RATES THAT VARY WITH TIME

When the interest rate on a loan can vary with, for example, the Federal Reserve Board's discount rate, it is necessary to take this into account when determining the future worth of the loan. It is becoming commonplace to see interest-rate "escalation riders" on some types of loans. Example 4-21 demonstrates how this situation is treated.

EXAMPLE 4-21

A person has made an arrangement to borrow $1,000 now and another $1,000 two years hence. The entire obligation is to be repaid at the end of 4 years. If the projected interest rates in years 1, 2, 3, and 4 are 10%, 12%, 12%, and 14%, respectively, how much will be repaid as a lump-sum amount at the end of 4 years?

Solution

This problem can be solved by compounding the amount owed at the beginning of each year by the interest rate that applies to each individual year and repeating this process over the 4 years to obtain the total future worth:

$$F_1 = \$1,000(F/P, 10\%, 1) = \$1,100$$

$$F_2 = \$1,100(F/P, 12\%, 1) = \$1,232$$

$$F_3 = (\$1,232 + \$1,000)(F/P, 12\%, 1) = \$2,500$$

$$F_4 = \$2,500(F/P, 14\%, 1) = \$2,850$$

To obtain the present worth of a series of future amounts, the procedure above would be utilized with a sequence of single-payment present worth factors. In general, the present worth of a cash flow occurring at the end of period n can be computed with this formula, where i_k is the interest rate for the kth period:

$$P = \frac{F_n}{\prod_{k=1}^{n} (1 + i_k)}$$

For instance, if $F_4 = \$1,000$ and $i_1 = 10\%$, $i_2 = 12\%$, $i_3 = 13\%$, and $i_4 = 10\%$,

$$P = \$1,000(P/F, 10\%, 1)(P/F, 12\%, 1)(P/F, 13\%, 1)(P/F, 10\%, 1)$$

$$= \$1,000/[(1.10)(1.12)(1.13)(1.10)] = \$653 \qquad \blacksquare$$

INTEREST FACTOR RELATIONSHIPS

The following relationships among the six basic discrete compounding interest factors should be recognized:

$$(P/F, i\%, N) = \frac{1}{(F/P, i\%, N)} \qquad (4\text{-}29)$$

$$(A/P, i\%, N) = \frac{1}{(P/A, i\%, N)} \qquad (4\text{-}30)$$

$$(A/F, i\%, N) = \frac{1}{(F/A, i\%, N)} \qquad (4\text{-}31)$$

$$(F/A, i\%, N) = (P/A, i\%, N)(F/P, i\%, N) \qquad (4\text{-}32)$$

$$(P/A, i\%, N) = \sum_{k=1}^{N} (P/F, i\%, k) \qquad (4\text{-}33)$$

$$(F/A, i\%, N) = \sum_{k=0}^{N-1} (F/P, i\%, k) \qquad (4\text{-}34)$$

$$(A/F, i\%, N) = (A/P, i\%, N) - i \qquad (4\text{-}35)$$

These same relationships exist among the corresponding continuous compounding interest factors discussed in the following sections.

INTEREST FORMULAS FOR CONTINUOUS COMPOUNDING AND DISCRETE CASH FLOWS

In most business transactions and economy studies, interest is compounded at the end of discrete periods of time and, as has been discussed previously, cash flows are assumed to occur in discrete amounts at the beginning, middle or end of such periods. *This practice will be used throughout the remaining chapters of this book.* However, it is evident that in most enterprises cash is flowing in and out in an almost continuous stream. Because cash, whenever available, can usually be used profitably, this situation, in effect, produces very frequent compounding of the interest earned. So that this condition can be accounted for, the concepts of continuous compounding and continuous cash flow are sometimes used in economy studies. Actually, the effects of these procedures compared to discrete compounding are rather small in most cases.

In the concept of continuous compounding it is assumed that cash payments occur once per year. but that compounding is continuous throughout the year. Thus, with a nominal rate of interest per year of r, if the interest is compounded M times per year, at the end of 1 year one unit of principal will amount to $[1 + (r/M)]^M$. If $M/r = p$, the foregoing expression becomes

$$\left[1 + \frac{1}{p}\right]^{rp} = \left[\left(1 + \frac{1}{p}\right)^p\right]^r$$

The limit of $(1 + 1/p)^p$ as p approaches infinity is e, the Naperian base of natural logarithms which equals 2.71828. . . . Thus the previous expression can be written as e^r. Consequently, the *continuous compounding compound amount factor* (*single cash flow*) at $r\%$ nominal interest for N years is e^{rN}. Using our functional notation, we express this as

$$(F/P, \underline{r}\%, N) = e^{rN} \tag{4-36}$$

Note that the symbol is directly comparable to that used for discrete compounding and discrete cash flows except that $\underline{r}\%$ is used to denote the nominal rate *and* the use of continuous compounding.

Since e^{rN} for continuous compounding corresponds to $(1 + i)^N$ for discrete compounding, e^r corresponds to $(1 + i)$. Equating the latter expression gives

$$i = e^r - 1 \tag{4-37}$$

By use of this relationship, the corresponding values of (P/F), (F/A), and (P/A) for continuous compounding may be obtained from Equations 4-4, 4-6, and 4-8, respectively, by substitution of $e^r - 1$ for i in these equations. Thus, for continuous compounding and discrete cash flows,

$$(P/F, \underline{r}\%, N) = \frac{1}{e^{rN}} = e^{-rN} \tag{4-38}$$

$$(F/A, \underline{r}\%, N) = \frac{e^{rN} - 1}{e^r - 1} \tag{4-39}$$

$$(P/A, \underline{r}\%, N) = \frac{1 - e^{-rN}}{e^r - 1} = \frac{e^{rN} - 1}{e^{rN}(e^r - 1)} \tag{4-40}$$

Values for $(A/P, \underline{r}\%, N)$ and $(A/F \underline{r}\%, N)$ may be derived through their inverse relationships to $(P/A, \underline{r}\%, N)$ and $(F/A, \underline{r}\%, N)$, respectively. All the continuous compounding, discrete cash flow interest factors and their uses are summarized in Table 4-5.

Because continuous compounding is used rather infrequently and is not used in subsequent problems in this text (except for limited exercises at the end of this chapter), detailed values for $(A/F, \underline{r}\%, N)$ and $(A/P, \underline{r}\%, N)$ are not given in the Appendix. However, the tables in Appendix F give values of $(F/P, \underline{r}\%, N)$, $(P/F, \underline{r}\%, N)$, $(F/A, \underline{r}\%, N)$, and $(P/A, \underline{r}\%, N)$ for a limited number of interest rates.

It is important to note that tables of interest and annuity factors for continuous compounding are tabulated in terms of nominal annual rates of interest.

TABLE 4-5 Continuous Compounding and Discrete Cash Flows Interest Factors and Symbols[a]

To Find:	Given:	Factor by Which to Multiply "Given"	Factor Name	Factor Functional Symbol
For single cash flows:				
F	P	e^{rN}	Continuous compounding compound amount (single cash flow)	$(F/P, \underline{r}\%, N)$
P	F	e^{-rN}	Continuous compounding present worth (single cash flow)	$(P/F, \underline{r}\%, N)$
For uniform series (annuities):				
F	A	$\dfrac{e^{rN} - 1}{e^{r} - 1}$	Continuous compounding compounding amount (uniform series)	$(F/A, \underline{r}\%, N)$
P	A	$\dfrac{e^{rN} - 1}{e^{rN}(e^{r} - 1)}$	Continuous compounding present worth (uniform series)	$(P/A, \underline{r}\%, N)$
A	F	$\dfrac{e^{r} - 1}{e^{rN} - 1}$	Continuous compounding sinking fund	$(A/F, \underline{r}\%, N)$
A	P	$\dfrac{e^{rN}(e^{r} - 1)}{e^{rN} - 1}$	Continuous compounding capital recovery	$(A/P, \underline{r}\%, N)$

[a]$\underline{r}$, nominal annual interest rate, compounded continuously; N, number of periods (years); A, uniform series amount (occurs at end of each year); F, future worth; P, present worth.

EXAMPLE 4-22

Suppose that one has a present amount of \$1,000 and it is desired to determine what uniform end-of-year payments could be obtained from it for 10 years if interest is 20% compounded continuously.

Solution

$$A = P(A/P, \underline{r}\%, N)$$

Since the (A/P) factor is not tabled for continuous compounding we substitute its inverse (P/A), which is tabled. Thus

$$A = P \times \frac{1}{(P/A, \underline{20}\%, 10)} = \$1,000 \times \frac{1}{3.9054} = \$256$$

It is interesting to note that the answer to the same problem except with discrete annual compounding is

$$A = P(A/P, i\%, N) = P(A/P, 20\%, 10)$$

$$= \$1,000(0.2385) = \$239 \qquad \blacksquare$$

110 BACKGROUND AND TOOLS

EXAMPLE 4-23

A person needs $12,000 immediately as a down payment on a new home. Suppose that he can borrow this money from his insurance company. He will be required to repay the loan in equal payments, made every 6 months over the next 8 years. The nominal interest rate being charged is 7% compounded continuously. What is the amount of each payment?

Solution

The nominal interest rate per 6 months is 3.5%. Thus A each 6 months is $12,000($A/P$, $r = \underline{3.5}$%, 16). By substituting terms in Equation 4-40 and then using its inverse, we determine the value of A per 6 months to be $997, that is, $A = \$12,000/ (P/A$, $r = \underline{3.5}$%, 16). ∎

INTEREST FORMULAS FOR CONTINUOUS COMPOUNDING AND CONTINUOUS CASH FLOWS

Continuous flow of funds means a series of cash flows occurring at infinitely short intervals of time; this corresponds to an annuity having an infinite number of short periods. This formulation could apply to companies having receipts and disbursements that occur frequently during each working day. In such cases the interest normally is compounded continuously. If the nominal interest rate per year is r and there are p payments per year, which amount to a total of one unit per year, then, by use of Equation 4-8, for 1 year the present worth at the beginning of the year is

$$P = \frac{1}{p}\left\{\frac{[1 + (r/p)]^p - 1}{r/p[1 + (r/p)]^p}\right\} = \frac{[1 + (r/p)]^p - 1}{r[1 + (r/p)]^p}$$

The limit of $[1 + (r/p)]^p$ as p approaches infinity is e^r. By letting the present worth of one unit per year, flowing continuously and with continuous compounding of interest, be called the *continuous compounding present worth factor* (*continuous, uniform cash flow over one period*),

$$(P/\overline{A},\ r\%,\ 1) = \frac{e^r - 1}{re^r} \qquad (4\text{-}41)$$

where $\overline{A}$ is the amount flowing continuously over 1 year (here $1). For $\overline{A}$ flowing each year over N years, as depicted in Figure 4-21,

$$(P/\overline{A},\ r\%,\ N) = \frac{e^{rN} - 1}{re^{rN}} \qquad (4\text{-}42)$$

which is the *continuous compounding present worth factor* (*continuous, uniform cash flows*).

Equation 4-41 can also be written

$$(P/\overline{A},\ r\%,\ 1) = e^{-r}\left[\frac{e^r - 1}{r}\right] = (P/F,\ r\%,\ 1)\left[\frac{e^r - 1}{r}\right]$$

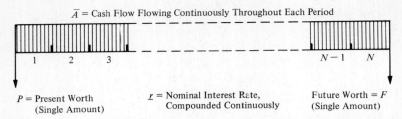

$\overline{A}$ = Cash Flow Flowing Continuously Throughout Each Period

| 1 | 2 | 3 | | | N − 1 | N |

P = Present Worth r = Nominal Interest Rate,   Future Worth = F
(Single Amount)   Compounded Continuously   (Single Amount)

FIGURE 4-21 **General Cash Flow Diagram for Continuous Compounding, Continuous Cash Flows.**

Because the present worth of 1 per year, flowing continuously with continuous compounding of interest, is $(P/F, r\%, 1)(e^r − 1)/r$, it follows that $(e^r − 1)/r$ must also be the compound amount of 1 per year, flowing continuously with continuous compounding of interest. Consequently, the *continuous compounding compound amount factor* (*continuous, uniform cash flow over 1 year*) is

$$(F/\overline{A}, r\%, 1) = \frac{e^r − 1}{r} \tag{4-43}$$

For N years,

$$(F/\overline{A}, r\%, N) = \frac{e^{rN} − 1}{r} \tag{4-44}$$

Equation 4-44 can also be developed by integration in this manner:

$$F = \overline{A} \int_0^N e^{rt}\, dt = \overline{A} \left(\frac{1}{r}\right) \int_0^N r e^{rt}\, dt$$

or

$$F = \frac{\overline{A}}{r}(e^{rt}) \Big|_0^N = \overline{A}\left[\frac{e^{rN} − 1}{r}\right]$$

This is the *continuous compounding compound amount factor* (*continuous uniform cash flows for N years*).

Values of $(P/\overline{A}, r\%, N)$ and $(F/\overline{A}, r\%, N)$ are given in the tables in Appendix F for various interest rates. Values for $(\overline{A}/P, r\%, N)$ and $(\overline{A}/F, r\%, N)$ can readily be obtained through their inverse relationship to $(P/\overline{A}, r\%, N)$ and $(F/\overline{A}, r\%, N)$, respectively. A summary of these factors and their use is given in Table 4-6.

EXAMPLE 4-24

What will be the worth at the end of 5 years of a uniform, continuous cash flow, at the rate of $500 per year for 5 years, with interest compounded continuously at the nominal annual rate of 8%?

Solution

$$F = \overline{A}(F/\overline{A}, 8\%, 5) = \$500 \times 6.1478 = \$3074$$

TABLE 4-6 Continuous Compounding Continuous Uniform Cash Flows Interest Factors and Symbols[a]

To Find:	Given:	Factor by Which to Multiply "Given"	Factor Name	Factor Functional Symbol
F	$\overline{A}$	$\dfrac{e^{rN} - 1}{r}$	Continuous compounding compounding amount (continuous, uniform cash flows)	$(F/\overline{A},\ \underline{r}\%,\ N)$
P	$\overline{A}$	$\dfrac{e^{rN} - 1}{re^{rN}}$	Continuous compounding present worth (continuous, uniform cash flows)	$(P/\overline{A},\ \underline{r}\%,\ N)$
$\overline{A}$	F	$\dfrac{r}{e^{rN} - 1}$	Continuous compounding sinking-fund (continuous, uniform cash flows)	$(\overline{A}/F,\ \underline{r}\%,\ N)$
$\overline{A}$	P	$\dfrac{re^{rN}}{e^{rN} - 1}$	Continuous compounding capital recovery (continuous, uniform cash flows)	$(\overline{A}/P,\ \underline{r}\%,\ N)$

[a]$\underline{r}$, nominal annual interest rate, compounded continuously; N, number of periods (years); A, amount of money flowing continuously and uniformly during each period; F, future worth; P, present worth.

It is interesting to note that if this cash flow had been in year-end amounts of $500 with discrete compounding at 8%, the amount would have been

$$F = A(F/A,\ 8\%,\ 5) = \$500 \times 5.8666 = \$2,933$$

If the year-end payments had occurred with 8% interest compounded continuously, the amount would have been

$$F = A(F/A,\ \underline{8}\%,\ 5) = \$500 \times 5.9052 = \$2,953 \qquad \blacksquare$$

EXAMPLE 4-25

What is the future worth of $10,000 per year that flows continuously for $8\frac{1}{2}$ years if the nominal interest rate is 10% per year? Continuous compounding is utilized.

Solution

There are 17 6-month periods in $8\frac{1}{2}$ years, and the r per 6 months is 5%. The $\overline{A}$ every 6 months is $5,000, so $F = \$5,000(F/\overline{A},\ \underline{5}\%,\ 17) = \$133,964.50$.

$\blacksquare$

PROBLEMS

4-1   What lump-sum amount of interest will be paid on a $1,000 loan that was made on August 1, 1983, and repaid on November 1, 1988, with ordinary simple interest at 10% per year?

4-2   How much interest would be *payable each year* on a loan of $2,000 at an annual interest rate of 12%? How much interest would have been paid over an 8-year period?

4-3 In Problem 4-2, if the interest had not been paid each year but had been allowed to compound, how much interest would be due to the lender as a lump sum at the end of the eighth year? How much extra interest is being paid here (as compared to Problem 4-2) and what is the reason for the difference?

4-4 Draw a cash flow diagram for $10,500 being loaned out at a simple interest rate of 15% per annum over a period of 6 years. How much interest would be repaid as a lump-sum amount at the end of the sixth year?

4-5 A future amount, F, is equivalent to $1,500 now when 8 years separates the amounts and the annual interest is 12%. What is the value of F?

4-6 A present obligation of $12,000 is to be repaid in equal uniform annual amounts, each of which includes repayment of the debt (principal) and interest on the debt, over a period of 6 years. If the interest rate per year is 10%, what is the amount of the annual repayment?

4-7 Suppose that the $12,000 in Problem 4-6 is to be repaid at a rate of $2,000 per year plus the interest that is owed and based on the beginning of year unpaid principal. Compute the total amount of interest repaid in this situation and compare it with that of Problem 4-6. Why are the two amounts different?

4-8 A person desires to accumulate $2,500 over a period of 7 years so that a cash payment can be made for a new roof on a summer cottage. To have this amount when it is needed, annual payments will be made to a savings account that earns 8% annual interest per year. How much must each annual payment be? Draw a cash flow diagram.

4-9 Mrs. Green has just purchased a new car for $12,000. She makes a down payment of 30% of the negotiated price and then makes payments of $303.68 each month thereafter for 36 months. Furthermore, she believes the car can be sold for $3,500 at the end of 3 years. Draw a cash flow diagram of this situation from Mrs. Green's viewpoint.

4-10 If $25,000 is deposited now into a savings account that earns 12% per year, what uniform annual amount could be withdrawn at the end of each year for 10 years so that nothing would be left in the account after the tenth withdrawal?

4-11 It is estimated that a certain piece of equipment can save $6,000 per year in labor and materials costs. The equipment has an expected life of 5 years and no salvage value. If the company must earn a 20% rate of return on such investments, how much could be justified now for the purchase of this piece of equipment? Draw a cash flow diagram.

4-12 A young woman, 22 years old, has just graduated from college. She accepts a good job and desires to establish an Individual Retirement Account (I.R.A.). At the end of each year thereafter she plans to deposit a tax-free amount of $2,000 into the I.R.A. The annual interest rate on the account is expected to average 12% into the foreseeable future. How old will she be when the I.R.A. has an accumulated (compounded) value of $1,000,000?

4-13 Determine the present equivalent and annual equivalent value of the following cash flow pattern when $i = 8\%$ per year.

End of Year	0	1	2	3	4	5	6	7
Amount ($)	−1,500	+500	+500	+500	+400	+300	+200	+100

4-14 Maintenance costs for a new bridge with an expected 50-year life are estimated to be $1,000 each year for the first 5 years, followed by a $10,000 expenditure in the fifteenth year and a $10,000 expenditure in year 30. If $i = 10\%$ per year, what is the equivalent uniform annual cost over the entire 50-year period?

4-15 Equal end-of-year payments of $263.80 each are being made on a $1,000 loan at 10% effective interest per year.
(a) How many payments are required to repay the entire loan?
(b) Immediately after the second payment, what lump-sum payment would completely pay off the loan?

4-16   John Q. wants his estate to be worth $65,000 at the end of 10 years. His net worth is now zero. He can accumulate the desired $65,000 by depositing $3,887 at the end of each year for the next 10 years. At what effective interest rate must his deposits be invested? Give answer to the nearest tenth of a percent.

4-17   Suppose that the parents of a young child decide to make annual payments into a savings account, with the first payment being made on the child's fifth birthday and the last payment being made on the fifteenth birthday. Then starting on the child's eighteenth birthday, the withdrawals shown below will be made. If the effective annual interest rate is 10% during this period of time, what is the amount of the annuity in years 5 through 15?

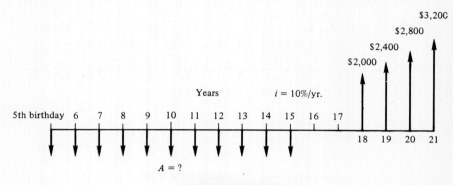

PROBLEM 4-17

4-18   Calculate the future worth at the end of 1986 at 8% compounded annually of this savings account:

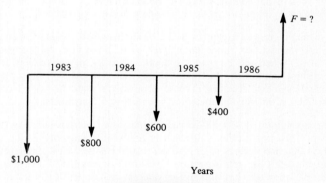

PROBLEM 4-18

4-19   Convert the following cash flow pattern to a uniform series of end-of-year costs over a 7-year period. Let $i = 12\%$.

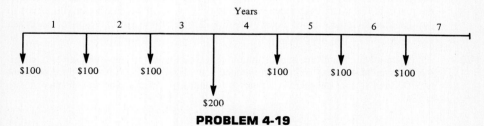

PROBLEM 4-19

4-20   A certain fluidized-bed combustion vessel has a first cost of $100,000, a life of 10 years, and negligible salvage value. Annual costs of materials, maintenance, and electric power for the vessel are expected to total $8,000. A major relining of the combustion vessel will occur during the fifth year at a cost of $20,000, during this year the vessel will *not* be in service. If the interest rate is 15% per year, what is the lump-sum equivalent cost of this project at the present time ($t = 0$)? Assume that a beginning-of-year cash flow convention is being utilized.

4-21   A certain government agency utilizes midperiod cash flow convention in its engineering economy studies. Project R-127 has this estimated cash flow pattern:

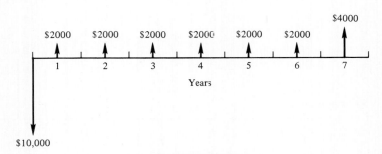

PROBLEM 4-21

If $i = 10\%$ per year, answer the following questions.
(a) What is the *present equivalent* value of this project at the beginning of year 1?
(b) What is the *annual equivalent* value of the project at the end of the 7 years shown?
(c) What is the *future equivalent* value at the end of year 7?

4-22   A woman arranges to repay a $1,000 bank loan in 10 equal payments at a 10% effective annual interest rate. Immediately *after* her third payment she borrows another $500, also at 10%. When she borrows the $500, she talks the banker into letting her repay the remaining debt of the first loan and the entire amount of the second loan in 12 equal annual payments. The first of these 12 payments would be made 1 year after she receives the $500. Compute the amount of each of the 12 payments.

4-23   An expenditure of $20,000 is made to modify a materials-handling system in a small job shop. This modification will result in first-year savings of $2,000, second-year savings of $4,000 and savings of $5,000 per year thereafter. How many years must the system last if a 25% per year return on investment is required? The system is tailor-made for this job shop and has no salvage value at any time. Use beginning-of-year cash flow convention to solve this problem.

4-24   You purchase special equipment that reduces defects by $10,000 per year on item A. This item is sold on contract for the next 5 years. After the contract expires, the special equipment will save approximately $2,000 per year for 5 more years. You assume that the machine has no salvage value at the end of 10 years. How much can you afford to pay for this equipment now, if you require a 25% return on your investment? All cash flows are end-of-year amounts.

4-25   Find the equivalent uniform annual amount that is equivalent to a gradient series in which the first annual payment is $500, the second annual payment is $600, the third payment is $700, and so on, and there is a total of 10 annual payments. The annual interest rate is 8%.

4-26   Find the equivalent value of Q in this cash flow diagram:

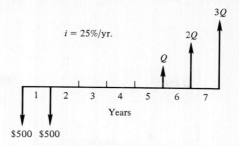

PROBLEM 4-26

4-27   The heat loss through the exterior walls of a certain poultry processing plant is estimated to cost the owner $3,000 next year. A salesman from Superfiber Insulation, Inc., has told you, the plant engineer, that he can reduce the heat loss by 80% with the installation of $15,000 worth of Superfiber now. If the cost of heat loss rises by $200 per year (gradient) after next year and the owner plans to keep the present building for 10 more years, what would you recommend if the cost of money is 12% per year?

4-28   Solve for the value of G below so that the left-hand cash flow diagram is equivalent to the one on the right. Let $i = 10\%$ per year.

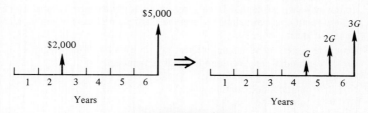

PROBLEM 4-28

4-29   What value of N comes closest to making the left cash flow diagram equivalent to the one on the right-hand side? Let $i = 15\%$ per year.

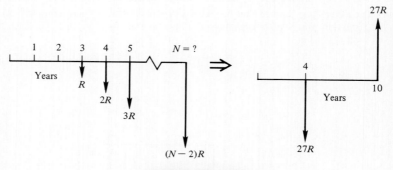

PROBLEM 4-29

4-30 A geometric gradient that increases at $f = 6\%$ per year for 15 years is shown below. The annual interest rate is 12%. What is the present equivalent value of this gradient?

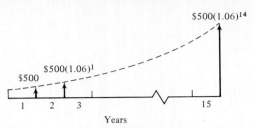

PROBLEM 4-30

4-31 An electronic device is available that will reduce this year's labor costs by $10,000. The equipment is expected to last for 8 years. If labor inflates at an average rate of 7% per year and the interest rate is 15% per year, answer these questions:
  (a) What is the maximum amount that we could justify spending for the device?
  (b) What is the uniform annual equivalent value (A) of labor costs over the 8-year period?
  (c) What annual "year 0" amount (A_0) that inflates at 7% per year is equivalent to the answer in part (a)?

4-32 You are the manager of a large crude oil refinery. As part of the refining process, a certain heat exchanger (operated at high temperatures and with abrasive material flowing through it) must be replaced every year. The replacement and downtime cost in the first year is $75,000. It is expected to increase due to inflation at a rate of 8% per year for 5 years at which time this particular heat exchanger will no longer be needed. If the company's cost of capital is 18% per year, how much could you afford to spend for a higher-quality heat exchanger so that these annual replacement and downtime costs could be eliminated?

4-33 Compute the effective annual interest rate in each of these situations:
  (a) 10% nominal interest, compounded semiannually.
  (b) 10% compounded quarterly.
  (c) 10% compounded continuously.
  (d) 10% compounded weekly.

4-34 **(a)** A certain savings and loan association advertises that they pay 8% interest, compounded quarterly. What is the *effective* interest rate per annum? If you deposit $5,000 now and plan to withdraw it in 3 years, how much would your account be worth at that time?
  (b) If in part (a) you decide to deposit $800 every year for 3 years, how much could be withdrawn at the end of the third year? Suppose that, instead, you deposit $400 every 6 months for 3 years, now what would the accumulated amount be?

4-35 You have just learned that the ABC Corporation has a "zero coupon" bond that costs $350 now and 8 years later pays a lump-sum amount of $1,000. The cash flow diagram looks like this:

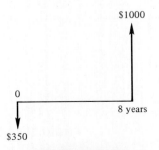

PROBLEM 4-35

If you bought one of these bonds, what interest rate would you earn on your investment? Calculate your answer to the nearest $\frac{1}{10}$ of 1%.

4-36 Exactly 140 years ago, my great-grandmother deposited $200 in a large New York bank as part of a savings plan. She forgot all about the deposit during her lifetime. Two years ago the bank notified my family that the account was worth $191,516. What was the effective annual interest rate in this situation?

4-37 A local bank offers a "Vacation Made Easy" plan as follows. Each participant in the plan deposits an amount of money, A, at the end of each week for 50 weeks with no interest being paid by the bank. Then the bank makes the 51st and 52nd payments and returns to each participant a grand total of 52A at the end of week 52 that is used to pay for the vacation. What is the true interest rate per year being earned *by participants* in this plan? Assume that an opportunity exists for weekly compounding if participants elect another investment plan elsewhere.

4-38 An individual approaches the Loan Shark Agency for $1,000 to be repaid in 24 monthly installments. The agency advertises an interest rate of $1\frac{1}{2}$% per month. They proceed to calculate the size of his monthly payment in the following manner:

Amount requested	$1,000
Credit investigation	25
Credit risk insurance	5
TOTAL	$1,030

Interest: $(1030)(24)(0.015) = \$371$
Total owed: $\$1,030 + 371 = \$1,401$
Payment: $\dfrac{\$1,401}{24} = \$58.50$

What effective annual interest rate is the individual paying?

4-39 In a certain foreign country, a man who wanted to borrow $10,000 for a 1-year period was informed he would only have to pay $2,000 in interest (i.e., a 20% interest rate). The lender stated that the total owed to him, $12,000, would be repaid at the rate of $1,000 per month for 12 months. What was the true effective interest being charged in this transaction? Why is it greater than the apparent rate of 20%?

4-40 Suppose that you borrow $500 from the Easy Credit Company with the agreement to repay it over a 3-year period. Their stated interest rate is 9% per year. They show you the following items in determining the monthly payment.

Principal	$500
Total interest: 0.09(3 years)($500)	135
Loan application fee	15

They ask you to pay the interest immediately, so you leave with $365 in your pocket. Your monthly payment is calculated as follows:

$$\frac{\$500 + \$15}{36} = \$14.30/\text{month}$$

What is the effective interest rate per year?

4-41 A company purchased a fleet of trucks for $156,000. Payment was to be made by an immediate cash payment of $10,000 and 12 month-end payments of $12,972 each. Another dealer offered to finance the $156,000 purchase at an interest rate of $1\frac{1}{2}$% per month on the unpaid balance. Which offer should the company have accepted?

4-42 How long does it take a given amount of money to triple itself if the money is invested at a nominal rate of 15%, compounded monthly?

4-43 Determine the present equivalent value of $400 paid every 3 months over a period of 7 years in each of these situations:
  (a) The interest rate is 12%, compounded annually.
  (b) The interest rate is 12%, compounded quarterly.
  (c) The interest rate is 12%, compounded continuously.

4-44 How many deposits of $200 must a person make at the end of each month if she desires to accumulate $11,500 for a new automobile? The savings account pays 9% interest compounded monthly.

4-45 The membership dues of a professional society are $47 per year. However, this society encourages its members to prepay their dues by the following incentive. If the member prepays his dues for N years in a lump-sum amount, the total amount due to the society is $47 + \$40(N - 1)$. Otherwise, the member would pay $47 per year at the beginning of the year for membership privileges. If an individual estimates that his personal before-tax rate of return ought to be 10%, how many years in advance should he pay his dues?

4-46 Determine the equivalent uniform semiannual cost of the following cash flow diagram. The interest rate is 10% compounded semiannually.

Six-Month Periods

$G = \$100/6$ Months

PROBLEM 4-46

4-47 If the nominal interest rate is 10% and compounding is semiannual, what is the present worth of the following receipts?

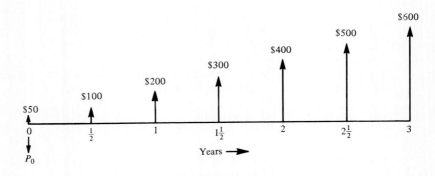

Years →

PROBLEM 4-47

4-48 Find the value of A that is equivalent to the gradient shown below if the nominal interest rate is 12% compounded monthly.

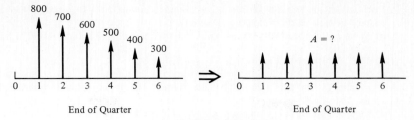

4-49 (a) Monthly amounts of $200 each are deposited into an account that earns 12% nominal interest, compounded quarterly. After 48 deposits of $200 each, what is the *future equivalent* worth of the account? State your assumptions.
(b) A "Christmas Plan" requires deposits of $10 per week for 52 weeks each year. The stated nominal interest rate is 20% compounded semiannually. What is the *present equivalent* value of the plan (beginning of week 1)?

4-50 A $20,000 ordinary life insurance policy for a 22-year-old female can be obtained for annual premiums of approximately $250. This type of policy (ordinary life) would pay a death benefit of $20,000 in exchange for annual premiums of $250 that are paid during the lifetime of the insured person. If the average life expectancy of a 22-year-old female is 77 years, what interest rate establishes equivalence between cash outflows and inflows for this type of insurance policy? Assume that all premiums are paid on a beginning-of-year basis and that the last premium is paid on the female's 76th birthday.

4-51 On January 1, 1980, a government bond is purchased for $9,400. The face value of the bond is $10,000 and 8% nominal interest is paid on the face value four times each year. Thus every 3 months the bondholder receives $200 as a cash payment. When the bond matures on January 1, 1990, $10,000 is paid to the bondholder. What is the effective rate of interest being earned in this situation?

4-52 What is the annual equivalent worth of $125,000 now, when 18% nominal interest per year is compounded monthly? Let $N = 15$ years.

4-53 Suppose that you have a money market certificate earning an average annual rate of interest, which varies over time as follows:

Year k	1	2	3	4	5
i_k	14%	12%	10%	10%	12%

If you invest $5,000 in this certificate at the beginning of year 1 and do not add or withdraw any money for 5 years, what is the value of the certificate at the end of the fifth year?

4-54 Determine the present equivalent value of this cash flow diagram when the annual interest rate (i_k) varies as indicated.

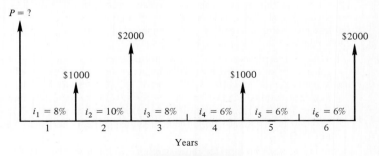

PROBLEM 4-54

4-55 If a nominal interest rate is 8% is compounded continuously, determine the unknown quantity in each of the following situations.
   (a) What uniform end-of-year amount for 10 years is equivalent to $8,000 at the end of year 10?
   (b) What is the present equivalent value of $1,000 per year for 12 years?
   (c) What is the future worth at the end of the sixth year of $243 payments every 6 months during the 6 years? The first payment occurs 6 months from the present and the last occurs at the end of the sixth year.
   (d) Find the equivalent lump-sum amount at the end of year 9 when $P_0 = \$1,000$ and a nominal interest rate of 8% is compounded continuously.

4-56 Find the value of the unknown quantity in the diagram when $\underline{r} = 10\%$ continuously compounded.

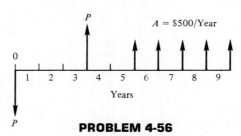

PROBLEM 4-56

4-57 A man deposited $2,000 in a savings account when his son was born. The nominal interest rate was 8% per year, compounded continuously. On the son's 18th birthday, the accumulated sum is withdrawn from the account. How much would this accumulated amount be?

4-58 Find the value of P in this cash flow diagram:

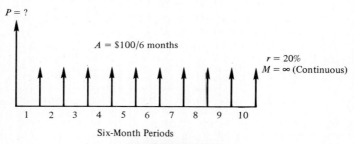

PROBLEM 4-58

4-59 A person needs $12,000 immediately as a down payment on a new home. Suppose that he can borrow this money from his company credit union. He will be required to repay the loan in equal payments, *made every 6 months* over the next 8 years. The annual interest rate being charged is 10% compounded continuously. What is the amount of each payment?

4-60 **(a)** What is the present worth of a uniform series of annual payments of $3,500 each for 5 years if the interest rate, compounded continuously, is 10%?
   (b) What is the future worth in 6.5 years of $12,000 deposited now in a savings account paying a nominal interest rate of 7% per annum? Assume continuous compounding.
   (c) An amount of $7,000 is invested in a certificate of deposit (C.D.) and will be worth $16,000 in 9 years. What is the continuously compounded nominal interest rate for this C.D.?

4-61 **(a)** Many persons prepare for retirement by making monthly contributions to a savings program. Suppose that $100 is set aside each month and invested in a savings account that pays 12% interest each year, compounded continuously. Determine the accumulated savings in this account at the end of 30 years.

(b) In part (a), suppose that an annuity will be withdrawn from savings that have been accumulated at the end of year 30. The annuity will extend from the end of year 31 to the end of year 40. What is the value of this annuity if the interest rate and compounding frequency in part (a) do not change? Graphically, here is the question:

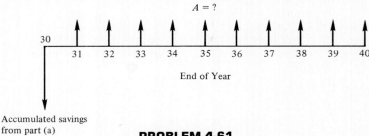

End of Year

Accumulated savings
from part (a)

PROBLEM 4-61

4-62   (a) What is the present worth on February 1, 1984, of $10,243 that is received on August 1, 1989? The nominal rate of interest is 10% compounded continuously.

(b) What is the future worth of a continuous funds flow amounting to $10,500 per year when $r = 15\%$, $M = \infty$, and $N = 12$ years?

(c) If the nominal interest rate is 10% per year, continuously compounded, what is the future value of $10,000 per year flowing continuously for 8.5 years?

(d) Let $\bar{A} = \$7859$ per year with $r = 15\%$, $M = \infty$. How many years will it take to have $1 million in this account?

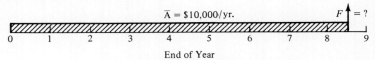

End of Year

PROBLEM 4-62(c)

4-63   (a) Determine the equivalent uniform series of annual *beginning-of-year* payments for the following cash flow pattern. The nominal interest rate is 8% compounded continuously.

(b) For how many years must an investment of $63,000 provide a continuous flow of funds at the rate of $16,000 per year so that a nominal interest rate of 10%, continuously compounded, will be earned?

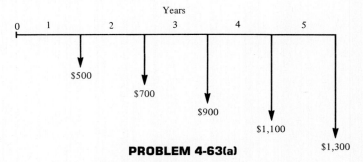

PROBLEM 4-63(a)

4-64   What is the present value of the following continuous funds flow situations?

(a) $1,000,000 per year for 4 years at 10% compounded continuously.

(b) $6,000 per year for 10 years at 8% compounded annually.

(c) $500 per quarter for 6.75 years at 12% compounded continuously.

APPLICATIONS OF ENGINEERING ECONOMY

Basic Methods for Making Economy Studies

All engineering economy studies of capital projects should consider the return that a given project will or should produce. The basic question this book addresses is whether a proposed capital investment and its associated expenditures can be recovered over time *in addition to* a return on the capital that is sufficiently attractive in view of risks involved and alternative uses of limited funds. Hence the interest and money–time relationships of Chapter 4 emerge as an essential ingredient in answering this question.

Because patterns of capital investment, revenue or savings cash flows, and cost cash flows can be quite different in various projects, there is no single method for making economy studies that is ideal for all cases. Consequently, several methods commonly are used in practice. All will produce equally satisfactory results and will lead to the same decision in cases where the inherent assumptions of each method are applicable.

The main objective of this chapter is to demonstrate the mechanics of six basic methods for making economy studies in view of the time value of money and to describe briefly the underlying assumptions and interrelationships of these methods.

All discussions and illustrations of economy study methods in this chapter and continuing through Chapter 8 are for *before-tax* studies. However, it is recognized that income taxes are an ever-present and somewhat unpleasant fact of life; most corporations pay out about half of their gross profits in the form of income taxes, both to State and federal governments. However, because each alternative use for available capital often is subject to the same income tax effects, in some economy studies one obtains the same decisions or relative rankings of alternatives without consideration of income taxes. Chapter 10 and parts of subsequent chapters show how to make *after-tax* studies by explicitly considering the effect of income taxes.

Economy studies can be made either from a project viewpoint or from the viewpoint of the owners of the organization. This difference occurs primarily when some borrowed capital is used to finance the project. If we make an economy study in which we are concerned only with the potential profitability of a project without consideration of the source of investment funds that would be required, we do not need to consider whether equity capital or borrowed capital is to be used. For most purposes such a viewpoint is quite satisfactory, and the appropriate interest rate to be used for discounting purposes is the organization's minimum attractive rate of return (M.A.R.R.). Usually, the M.A.R.R. is established in view of the opportunity cost of capital, which is the return forgone because money is invested in one particular project rather than in other feasible alternative projects. Additional problems associated with the acquisition and ownership of debt and/or equity capital are treated separately in Chapter 14.

BASIC METHODS

These six basic methods for making economy studies are discussed in this chapter.

Equivalent worth:

1. Present worth (P.W.).
2. Annual worth (A.W.).
3. Future worth (F.W.).

Rate of return:

1. Internal rate of return (I.R.R.).
2. External rate of return (E.R.R.).
3. Explicit reinvestment rate of return (E.R.R.R.).

The first three methods convert all cash flows into equivalent worths at some point or points in time using an annual interest rate (before taxes) equal to the minimum attractive rate of return. The last three methods are different ways to calculate an annual rate of profit or savings resulting from an investment so that this measure of merit can, in turn, be compared against the M.A.R.R. The "payback period" is a method that typically ignores the time value of money, and it is also briefly discussed. Another popular method is the benefit–cost technique. It will be discussed in connection with public sector economy studies in Chapter 12.

The present chapter concentrates on the correct use of the above six basic methods for analyzing single projects, and later chapters will show how to use each method for studies that involve multiple alternatives. *Unless specified otherwise, end-of-period cash flow convention and discrete compounding of interest are utilized throughout this and subsequent chapters.*

THE PRESENT WORTH METHOD

The present worth (P.W.) method is based on the concept of equivalent worth of all cash flows relative to some base or beginning point in time called the present. That is, all cash inflows and outflows are discounted to the base point at an interest rate that is generally the M.A.R.R.

The present worth of an alternative is a measure of how much money will have to be put aside now to provide for one or more future expenditures. It is assumed that such cash placed in reserve earns interest at a rate equal to a firm's M.A.R.R.

To find the P.W. of a series of cash receipts and/or disbursements, it is necessary to discount future amounts to the present by using an interest rate for the appropriate number of periods (years, for example) in the following manner.

$$\text{P.W.} = F_0(1 + i)^0 + F_1(1 + i)^{-1} + F_2(1 + i)^{-2} + \cdots$$

$$+ F_k(1 + i)^{-k} + \cdots + F_N(1 + i)^{-N} \quad (5\text{-}1)$$

where i = effective interest rate, or M.A.R.R., per compounding period
k = index for each compounding period ($0 \leq k \leq N$)
F_k = future cash flow at the end of period k
N = number of compounding periods

The relationship given in Equation 5-1 is based on the assumption of a constant interest rate throughout the life of a particular project. If the interest rate is assumed to change, the present worth must be computed in two or more steps as illustrated in Chapter 4.

The higher the interest rate and the further into the future a cash flow occurs, the lower is its present worth. This is shown graphically in Figure 5-1. The criterion for this method is that as long as the net present worth, N.P.W., (i.e. present equivalent of cash inflows minus cash outflows) is greater than or equal to zero, the project is economically justified; otherwise, it is not justified.

If receipts or savings are not known so that only cash outflows (disbursements) are relevant, the method is characterized by negative-valued present worth, and then it is often referred to as the present worth–cost (P.W.-C.) method.

EXAMPLE 5-1

An investment of $10,000 can be made in a project that will produce a uniform annual revenue of $5,310 for 5 years and then have a salvage value of $2,000. Annual disbursements will be $3,000 each year for operation and maintenance costs. The company is willing to accept any project that will earn 10% or more,

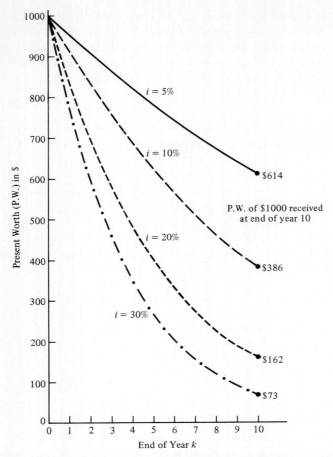

FIGURE 5-1 Present Worth of $1,000 Received in Year *k* at an Interest Rate of *i*%.

before income taxes, on all invested capital. Show whether this is a desirable investment by using the present worth method.

Solution

	Present Worth	
	Outflows	Inflows
Annual revenue: $5,310(P/A, 10%, 5)		+ $20,125
Salvage value: $2,000(P/F, 10%, 5)		+ 1,245
Investment	− $10,000	
Annual disbursements: $3,000(P/A, 10%, 5)	− 11,370	
TOTAL	− $21,370	+ $21,370
Net P.W.		$ 0

Since N.P.W. = $0, the project is shown to be barely justified. ∎

EXAMPLE 5-2

A piece of new equipment has been proposed by engineers to increase the productivity of a certain manual welding operation. The initial investment (first cost) is $25,000 and the equipment will have a salvage value of $5,000 at the end of its expected life of 5 years. Increased productivity attributable to the equipment will amount to $8,000 per year after extra operating costs have been subtracted from the value of the additional production. A cash flow diagram for this equipment is given in Figure 5-2. If the firm's minimum attractive rate of return (before income taxes) is 20%, is this proposal a sound one? Use the present worth method.

Solution

$$\text{Net P.W.} = \text{P.W. of cash receipts} - \text{P.W. of cash outlays}$$

or

$$\text{Net P.W.} = \$8,000(P/A, 20\%, 5) + \$5,000(P/F, 20\% \ 5) - \$25,000$$

$$= \$934.29$$

Because N.P.W. > 0, this equipment is economically justified. ∎

In Example 5-2 a table can be utilized to determine the cumulative present worth of cash flow through year k. A graph of cumulative present worth (Figure 5-3) serves to communicate clearly the behavior of time-related characteristics of the alternative under consideration.

End of Year k	Cash Flow	Present Worth of Cash Flow at 20%	Cumulative P.W. at 20% Through Year k	Cum. P.W. at 0% Through Year k
0	− $25,000	− $25,000	− $25,000	− $25,000
1	+ 8,000	+ 6,667	− 18,333	− 17,000
2	+ 8,000	+ 5,556	− 12,777	− 9,000
3	+ 8,000	+ 4,630	− 8,147	− 1,000
4	+ 8,000	+ 3,858	− 4,289	+ 7,000
5	+ 13,000	+ 5,223	+ 934	+ 20,000

The minimum attractive rate of return in Example 5-2 (and in other examples throughout this chapter) is to be interpreted as an effective interest rate (i) of 20% per year. Cash flows are discrete, end-of-year amounts. If *continuous compounding* had been specified for a nominal interest rate (r) of 20%, the net present worth would have been calculated by using Appendix F interest factors:

$$\text{N.P.W.} = -\$25,000 + \$8,000(P/A, \underset{\cdot}{r} = 20\%, 5)$$

$$+ \$5,000(P/F, \underset{\cdot}{r} = 20\%, 5)$$

$$= -\$25,000 + \$8,000(2.8551) + \$5,000(0.3679)$$

$$= -\$319.60$$

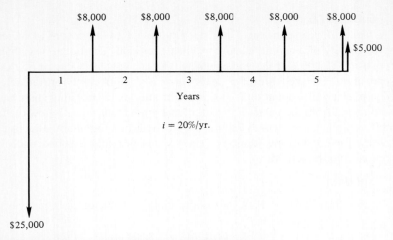

FIGURE 5-2 Cash Flow Diagram for Example 5-2.

Consequently, with continuous compounding the equipment would not be economically justifiable. The reason is that the higher effective interest rate ($e^{0.20} - 1 = 0.2214$) reduces the present worth of future positive cash flows but does not affect the present worth of the capital invested at the beginning of year 1.

When the life of an investment or the study period is infinite, a special case

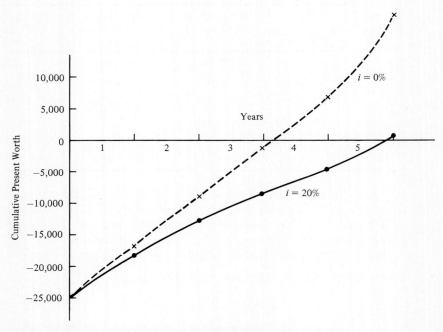

FIGURE 5-3 Graph of Cumulative Present Worth for Example 5-2.

of the present worth method is the *capitalized worth* method, which is discussed in Chapter 6.

THE ANNUAL WORTH METHOD

The term *annual worth* (A.W.) means a uniform annual series of dollar amounts for a certain period of time that is *equivalent* to a particular schedule of cash inflows (receipts or savings) and/or cash outflows (disbursements) under consideration. The net annual worth of a project is its annual equivalent receipts (R) minus annual equivalent expenses (E), less its annual equivalent capital recovery (C.R.) amount. An annual equivalent value of R, E, and C.R. is computed at the M.A.R.R. The study period is denoted by N, which is usually years. In equation form the net annual worth is

$$A.W. = R - E - C.R. \qquad (5\text{-}2)$$

The criterion for this method is that as long as the net annual worth (i.e., annual equivalent of inflows minus outflows) is ≥ 0, the project is economically justified; otherwise, it is not justified.

Calculation of Capital Recovery Cost

The capital recovery cost (C.R.) for a project is the equivalent uniform annual cost of the capital invested. It is an annual amount which covers the following two items:

1. Depreciation (loss in value of the asset).
2. Interest on invested capital (minimum attractive rate of return).

As an example, consider a machine or other asset that will cost $10,000, last 5 years, and have a salvage value of $2,000. Further, the interest on invested capital, i, is 10%.

It can be shown that no matter which method of calculating depreciation is used, the equivalent annual cost of the capital recovery is the same. For example, if straight line depreciation† is used, the equivalent annual cost of interest is calculated to be $710, as shown in Table 5-1. The annual depreciation cost by the straight line method is ($10,000 − $2,000)/5 = $1,600. The $710 added to $1,600 results in a calculated capital recovery cost of $2,310.

There are several convenient formulas by which capital recovery cost may be calculated to obtain the same answer as above. Probably, the easiest formula to understand involves finding the annual equivalent of the investment and then subtracting the annual equivalent of the salvage value. Thus

$$C.R. = P(A/P, i\%, N) - S(A/F, i\%, N) \qquad (5\text{-}3)$$

---

†Explanation of various methods of depreciation is given in Chapter 9.

TABLE 5-1   Calculation of Equivalent Annual Cost of Interest and Capital Recovery Cost Assuming Straight Line Depreciation

Year	Value of Investment at Beginning of Year	Straight Line Depreciation	Interest on Beginning-of-Year Investment at 10%	Present Worth of Interest at 10%		
1	$10,000	$1,600	$1,000	$1,000(P/F, 10%, 1)	=	$ 909
2	8,400	1,600	840	840(P/F, 10%, 2)	=	694
3	6,800	1,600	680	680(P/F, 10%, 3)	=	511
4	5,200	1,600	520	520(P/F, 10%, 4)	=	355
5	3,600	1,600	360	360(P/F, 10%, 5)	=	244
						$2,693

Annual equivalent interest = $2,693(A/P, 10%, 5) = $710
Total C.R. cost = annual depreciation + annual equivalent interest
$$= (\$10,000 - \$2,000)/5 + \$710 = \$2,310$$

where P = investment at beginning of life
S = salvage value at end of life
N = life of project

When Equation 5-3 is applied to the example in Table 5-1, the C.R. amount is

$$C.R. = \$10,000(A/P, 10\%, 5) - \$2,000(A/F, 10\%, 5)$$

$$= \$10,000(0.2638) - 2,000(0.1638) = \$2,310$$

Another way to calculate the C.R. cost is to add the annual sinking fund depreciation charge (or deposit) to the interest on the original investment (sometimes called minimum required profit). Thus

$$C.R. = (P - S)(A/F, i\%, N) + P(i\%) \tag{5-4}$$

When Equation 5-4 is applied to the example in Table 5-1, the C.R. amount is

$$C.R. = (\$10,000 - \$2,000)(A/F, 10\%, 5) + \$10,000(10\%)$$

$$= \$8,000(0.1638) + \$10,000(0.10) = \$2,310$$

Yet another very popular way to calculate the C.R. cost is to add the equivalent annual cost of the depreciable portion of the investment and the interest on the nondepreciable portion (salvage value). Thus

$$C.R. = (P - S)(A/P, i\%, N) + S(i\%) \tag{5-5}$$

Applied to the same example as above,

$$C.R. = (\$10,000 - \$2,000)(A/P, 10\%, 5) + \$2,000(10\%)$$

$$= \$8,000(0.2638) + \$2,000(0.10) = \$2,310$$

EXAMPLE 5-3
Considering the same project as in Example 5-1, show whether it is justified using the A.W. method.

Solution

	Annual Worth	
	Outflows	Inflows
Annual revenue		$5,310
Annual disbursements	− $3,000	
C.R. cost^a = ($10,000 − $2,000)(A/P, 10%, 5)		
+ $2,000(10%)	− 2,310	
TOTAL	− $5,310	+ $5,310
Net A.W.		$ 0

^{a}Uses Equation 5-5.

Since net A.W. = $0, the project earns exactly 10% and is thus barely justified. Of course, nonmonetary considerations would probably sway the decision either way. ∎

EXAMPLE 5-4

By using the annual worth method, determine whether the equipment described in Example 5-2 should be recommended.

Solution

The A.W. method applied to Example 5-2 yields the following:

$$\text{A.W.}(20\%) = \overbrace{\$8000}^{R-E} - \overbrace{[\$25,000(A/P, 20\%, 5) - \$5000(A/F, 20\%, 5)]}^{\text{C.R. amount (Eq. 5-3)}}$$

$$= \$8000 - (\$8359.49 - \$671.90)$$

$$= \$312.41$$

Because its net A.W. is positive, the equipment more than pays for itself over a period of 5 years while earning a 20% return per year on the unrecovered investment (e.g., see Table 5-1). In fact, the annual equivalent "surplus" is $312.41, which means that the equipment provided more than a 20% before-tax return on beginning-of-year unrecovered investment. This piece of equipment should be recommended as an attractive investment opportunity. ∎

The annual worth method is often called the "annual cost" (A.C.) method when only costs are involved. The object is then to select the least negative A.C. when several alternatives are being compared and when the "do-nothing" alternative is not an option.

Many decision makers prefer the annual worth method because it is relatively easy to interpret when one is accustomed to working with annual income statements and cash flow summaries.

THE FUTURE WORTH METHOD

Because a primary objective of all time value of money methods treated in this book is to maximize the future wealth of the owners of a firm, the future worth

(F.W.) criterion has become increasingly popular in recent years. With this method, the future worth of an alternative can be calculated in view of the M.A.R.R. and compared with the do-nothing option. If F.W. ≥ 0, the alternative would be recommended.

The future worth method for economy studies is exactly comparable to the present worth method except that all cash inflows and outflows are compounded forward to a reference point in time called the future.

EXAMPLE 5-5

Considering the same project as in Example 5-1, show whether it is justified using the F.W. method.

Solution

	Future Worth	
	Outflows	Inflows
Annual revenue: $5,310(F/A, 10\%, 5)$		$+\$32,420$
Salvage value		$+\quad 2,000$
Investment: $10,000(F/P, 10\%, 5)$	$-\$16,105$	
Annual disbursements: $3,000(F/A, 10\%, 5)$	$-\quad 18,315$	
TOTAL	$-\$34,420$	$+\$34,420$
Net F.W.		$\$\quad 0$

Again, the project is shown to be barely justified. ∎

EXAMPLE 5-6

Evaluate the future worth of the equipment described in Example 5-2. Show the relationship among the F.W., P.W., and A.W. for this example.

Solution

$$\text{F.W.} = -\$25,000(F/P, 20\%, 5) + \$8,000(F/A, 20\%, 5) + \$5,000$$

$$= \$2,324.80 > 0$$

Again, the equipment is a good investment. The F.W. is a larger number compared to the corresponding A.W. and P.W., but it can be shown to be equivalent.

$$\text{P.W.} = \$2,324.80(P/F, 20\%, 5) = \$934.29$$

$$\text{A.W.} = \$2,324.80(A/F, 20\%, 5) = \$312.41$$

These results were obtained in Examples 5-2 and 5-4, respectively. ∎

To this point the P.W., A.W., and F.W. methods have utilized a known and constant M.A.R.R. over the study period. Each method produces a measure of merit expressed in dollars and is equivalent to the other two. Another group of methods is now considered that produces an interest rate as its measure of merit. The three methods are internal rate of return (I.R.R.), external rate of return (E.R.R.), and explicit reinvestment rate of return (E.R.R.R.).

THE INTERNAL RATE OF RETURN METHOD

The internal rate of return (I.R.R.) method is the most general and widely used rate of return method for making economy studies. It commonly is called by several other names, such as investor's method, discounted cash flow method, receipts versus disbursements method, and profitability index.

This method solves for the interest rate that equates the present worth of an alternative's cash inflows (receipts or savings) to the present worth of cash outflows (expenditures, including investments). The resultant interest rate is termed the "internal rate of return" (I.R.R.). For a single alternative, the I.R.R. is not defined unless both receipts and disbursements are present in the cash flow pattern. Both the annual worth and future worth methods can also be used to determine the I.R.R. of an alternative.

Expressed in general, the I.R.R. is the $i'\%$† at which

$$\sum_{k=0}^{N} R_k(P/F, i'\%, k) = \sum_{k=0}^{N} E_k(P/F, i'\%, k) \qquad (5\text{-}6)$$

where R_k = net receipts or savings for the kth year

E_k = net expenditures including investments for the kth year

N = project life

Once i' has been calculated, it is then compared with the M.A.R.R. to assess whether the alternative in question is acceptable. If $i' \geq$ M.A.R.R. the alternative is acceptable; otherwise, it is not.

A popular variation of Equation 5-6 for computing the I.R.R. for an alternative is to determine the i' at which its *net* present worth is zero. In equation form, the I.R.R. is the value of i' at which

$$\sum_{k=0}^{N} R_k(P/F, i', k) - \sum_{k=0}^{N} E_k(P/F, i', k) = 0 \qquad (5\text{-}7)$$

For an alternative with a single investment at the present time followed by a series of positive cash flows over N, a graph of net present worth (N.P.W.) versus the interest rate typically has this general form:

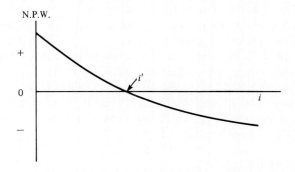

† i' is often used in place of i to mean the interest rate that is to be determined.

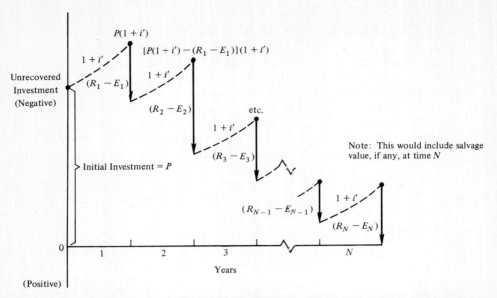

FIGURE 5-4 Unrecovered Investment Diagram That Illustrates the Internal Rate of Return.

The point at which N.P.W. = 0 on this graph defines i', which is the alternative's internal rate of return.

The value of i' can also be determined as the interest rate at which net F.W. = 0 or at which net A.W. = 0. For example, by setting net F.W. equal to zero, Equation 5-8 would result:

$$\sum_{k=0}^{N} R_k(F/P, i', N - k) - \sum_{k=0}^{N} E_k(F/P, i', N - k) = 0 \qquad (5\text{-}8)$$

Another way to interpret the I.R.R. is through an unrecovered investment diagram. Figure 5-4 shows how much of the original investment in an alternative is still to be recovered as a function of time. The downward arrows in Figure 5-4 represent annual returns, $(R_k - E_k)$ for $1 \le k \le N$, against the unrecovered investment and the dashed lines indicate the opportunity cost of interest, or profit, on the beginning of year investment balance. The I.R.R. is that value of i' in Figure 5-4 that causes the unrecovered investment balance to exactly equal 0 at the end of the study period (year N). It is important to notice that i' is calculated on the beginning of year *unrecovered* investment through the life of a project rather than on the total initial investment.

The method of solving Equations 5-6 to 5-8 normally involves trial-and-error calculations until the $i'\%$ is found or can be interpolated. Example 5-7 shows a typical solution using the common convention of "+" signs for cash inflows and "−" signs for cash outflows.

EXAMPLE 5-7
An investment of $10,000 can be made in a project that will produce a uniform annual revenue of $5,310 for 5 years and then have a salvage value of $2,000.

Annual disbursements will be $3,000 each year for operation and maintenance costs. The company is willing to accept any project that will earn 10% or more, before income taxes, on all invested capital. Determine whether it is justified using the I.R.R. method.

Solution

By writing an equation for net present worth and setting it equal to zero, the I.R.R. can be determined.

$$0 = -\$10,000 + (\$5,310 - \$3,000)(P/A, i'\%, 5) + \$2,000(P/F, i'\%, 5); \qquad i' = ?$$

If we did not already know the answer, we would probably try a relatively low i', such as 5%, and a relatively high i', such as 25%.

At $i' = 5\%$: $-\$10,000 + \$2,310(4.3295)$

$+ \$2,000(0.7835) = +\$1,568$

At $i' = 25\%$: $-\$10,000 + \$2,310(2.6893)$

$+ \$2,000(0.3277) = -\$3,132$

Since we have both a positive and a negative P.W. of net cash flow, the answer is bracketed. Linear interpolation can be used to find an approximation of the unknown $i'\%$ as shown in Figure 5-5. The answer, $i'\%$, can be obtained graphically to be approximately 12.0%, which is the $i\%$ at which the net P.W. = $0.

Linear interpolation for the answer, $i'\%$, can be accomplished by using the similar triangles dashed in Figure 5-5.

$$\frac{25\% - 5\%}{\$1,568 - (-\$3,132)} = \frac{i'\% - 5\%}{\$1,568 - \$0}$$

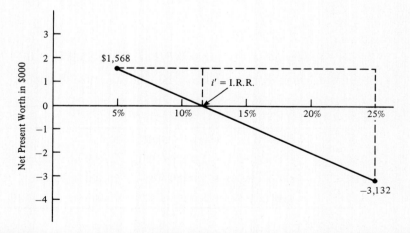

FIGURE 5-5 Use of Linear Interpolation to Find the Approximation of I.R.R. for Example 5-7.

or

$$i'\% = 5\% + \frac{\$1,568}{\$1,568 - (-\$3,132)}(25\% - 5\%)$$

$$= 5\% + 6.7\% = 11.7\%$$

The approximate solution above was merely to illustrate the trial-and-error process, together with linear interpolation. The error in this answer is due to nonlinearity of the net P.W. function and would be less if the range between interest rates used in the interpolation were smaller.

From the results of Examples 5-1, 5-3, and 5-5, we already know that the project is barely acceptable and that $i' = $ M.A.R.R. $= 10\%$. We can confirm this by trying $i = 10\%$ in the net P.W. equation, as follows:

At $i = 10\%$: $-\$10,000 + (\$5,310 - \$3,000)(P/A, 10\%, 5)$

$$+ \$2,000(P/F, 10\%, 5) = 0 \qquad \blacksquare$$

EXAMPLE 5-8

A piece of new equipment has been proposed by engineers to increase the productivity of a certain manual welding operation. The initial investment (first cost) is $25,000 and the equipment will have a salvage value of $5,000 at the end of its expected life of 5 years. Increased productivity attributable to the equipment will amount to $8,000/year after extra operating costs have been subtracted from the value of the additional production. A cash flow diagram for this equipment is given in Figure 5-2. Evaluate the internal rate of return of the proposed equipment. Is the investment a good one?

Solution
By utilizing Equation 5-7, the following expression is obtained:

$$\sum_{k=1}^{5} 8,000(P/F, i', k) + 5,000(P/F, i', 5) - 25,000 = 0$$

or

$$8,000(P/A, i', 5) + 5,000(P/F, i', 5) - 25,000 = 0$$

To solve this equation by trial and error, the following table is helpful.

i'	N.P.W.(i')
0.00	$8,000(5) + 5,000(1) - 25,000 = 20,000$
0.10	$8,000(3.7908) + 5,000(0.6209) - 25,000 = 8,430.90$
0.20	$8,000(2.9906) + 5,000(0.4019) - 25,000 = 934.30$
0.25	$8,000(2.6893) + 5,000(0.3277) - 25,000 = -1,847.10$

In graphical form the net P.W. has this appearance (top of p. 141).

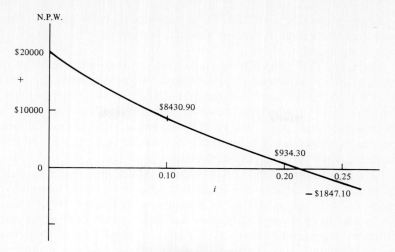

EXAMPLE 5-8(a)

By inspection, the value of i' where N.P.W. = 0 is about 22%. For most applications an i' value of 22% is accurate enough since our major concern is whether i' exceeds the M.A.R.R. A more exact value of i' can be determined by directly solving Equation 5-7:

$$0 = -25,000 + 8,000(P/A, i', 5) + 5,000(P/F, i', 5)$$

By repeated trial-and-error calculations, it can be determined that $i' = 21.577\%$. ∎

Below is the unrecovered investment diagram for Example 5-8.

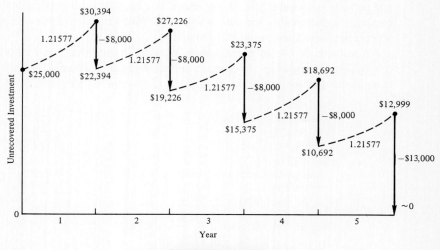

EXAMPLE 5-8(b)

Again, notice that $i' = 21.577\%$ is calculated on the beginning-of-year unrecovered investment and not on the total investment of $25,000 over the 5-year study period.

The A.W., P.W., and F.W. methods implicitly make the assumption that net receipts less disbursements (recovered funds) are reinvested by the firm at the M.A.R.R. However, the I.R.R. method is based on the assumption that funds are reinvested at i' rather than M.A.R.R. This assumption may not mirror reality in some problems, thus making I.R.R. an unacceptable method for analyzing engineering alternatives. This can usually be remedied by the E.R.R. and E.R.R.R. methods, which are both described later.

Other difficulties with the I.R.R. method include its computational intractability and the occurrence of multiple internal rates of return in some types of problems. A computational aid for determining the internal rate of return for a single alternative having both positive and negative cash flows or for cash flow differences between two alternatives is given in Appendix 5-A. Concerning the matter of multiple rates of return, a procedure for dealing with multiple rates is discussed and demonstrated in Appendix 5-B. Generally speaking, multiple rates are meaningless for decisionmaking purposes and another method of evaluation must be utilized.

Another possible drawback to the I.R.R. method is that it must be carefully applied and interpreted in the analysis of two or more alternatives when only one of them is to be selected (i.e., mutually exclusive alternatives). This is discussed further in Chapter 7. The key advantage of the method is its widespread acceptance by industry, where various types of rates of return and ratios are routinely used in making project selections.

Selecting Trial Rates of Return When Using the I.R.R. Method

Users of the I.R.R. method are often perplexed as to how to select an initial and perhaps subsequent trial rate of return to reduce the number of trials required to obtain an acceptably accurate answer. The following is an intuitive approach when calculations are performed manually.

An initial trial rate of return can be obtained by ignoring the time value of money and determining the average annual profit as a percent of the average investment. For the project in Example 5-1, this can be done as follows:

Cash inflow:	
Annual receipts: $5,310 × 5	$26,550
Salvage value	2,000
TOTAL	$28,550
Cash outflow:	
Annual disbursements: $3,000 × 5	−$15,000
Investment	− 10,000
NET CASH INFLOW (PROFIT)	$ 3,550

$$\text{Average profit per year} = \frac{\$3,550}{5} = \$710$$

$$\text{Average investment} = \frac{\$10,000 + \$2,000}{2} = \$6,000$$

$$\frac{\text{Average profit per year}}{\text{Average investment}} = \frac{\$710}{\$6,000} = 11.8\%$$

Thus one might start with a first trial rate of return of 12%. Substituting this into Equation 5-7 for finding the I.R.R. gives a negative N.P.W. It should be kept in mind that a lower interest rate will result in higher present worth factors for both single sums and uniform series; also, if the positive terms need to become larger so that the net present worth will be ≥ 0, the next trial rate should be lower.

It is recommended that the second trial rate be enough different from the first trial rate so that the answer will be bracketed between a positive and a negative net P.W. One can then interpolate to find the I.R.R. as demonstrated in Example 5-7.

A Computer Program for Selected Methods

To assist with the solution of engineering economy problems by methods discussed in this and subsequent chapters, a computer program is provided in Appendix C. The program is written in BASIC and is suitable for most personal computers. The results of running this computer program for Examples 5-2, 5-4, 5-6, and 5-8 are shown below.

Other capabilities of the program are described in Appendix C.

```
DO YOU WANT TO SEE INSTRUCTIONS FOR DATA INPUT
(TYPE YES OR NO)   ?NO
HOW MANY DIFFERENT CASH FLOW VALUES
 ?3
ENTER CASH FLOW, FIRST PERIOD, LAST PERIOD
 ?-25000,0,0
 ?8000,1,4
 ?13000,5,5

YEAR      CASH FLOW
   0       -25,000.00

   1         8,000.00
   2         8,000.00
   3         8,000.00
   4         8,000.00
   5        13,000.00
WANT TO MAKE ANY CORRECTIONS OR ADDITIONS (Y,N)
 ?N
WHAT WOULD YOU LIKE TO DO (TYPE
PW,FW,AW,IRR,SPP,DPP,TABLE,? OR END)
 ?PW
```

5-2   ENTER INTEREST RATE AS DECIMAL (FOR EXAMPLE
      ENTER 10% AS .1)
       ?.2

```
          DO YOU WANT CONTINUOUS (CON) OR DISCRETE (DIS)
          COMPOUNDING ?DIS
          =========> PW =   934.285
          ------------------------------------------------
          WANT ANOTHER RUN ? (Y,N) ?Y
          TYPE NEW,OLD OR CHANGE (TYPE '?' FOR
          EXPLANATION) ?OLD
          WHAT WOULD YOU LIKE TO DO (TYPE
          PW,FW,AW,IRR,SPP,DPP,TABLE,? OR END)
           ?AW
5-4       ENTER INTEREST RATE AS DECIMAL , FOR EXAMPLE
          ENTER 10% AS .1)
           ?.2
          DO YOU WANT CONTINUOUS (CON) OR DISCRETE (DIS)
          COMPOUNDING ?DIS
          =========> AW =   312.406
          ------------------------------------------------
          WANT ANOTHER RUN ? (Y,N) ?Y
          TYPE NEW,OLD OR CHANGE (TYPE '?' FOR
          EXPLANATION) ? OLD
          WHAT WOULD YOU LIKE TO DO (TYPE
          PW,FW,AW,IRR,SPP,DPP,TABLE,? OR END)
           ?FW
5-6       ENTER INTEREST RATE AS DECIMAL , FOR EXAMPLE
          ENTER 10% AS .1)
           ?.2
          DO YOU WANT CONTINUOUS (CON) OR DISCRETE (DIS)
          COMPOUNDING ?DIS
          =========> FW =   2324.8
          ------------------------------------------------
          WANT ANOTHER RUN ? (Y,N) ?Y
          TYPE NEW,OLD OR CHANGE (TYPE '?' FOR
          EXPLANATION) ?OLD
          WHAT WOULD YOU LIKE TO DO (TYPE
          PW,FW,AW,IRR,SPP,DPP,TABLE,? OR END)
           ?IRR
5-8       DO YOU WANT DISCRETE (DIS) OR CONTINUOUS (CON)
          INTEREST RATE ?DIS
          =========> IRR IS BETWEEN 21.577 AND 21.578 %
          ------------------------------------------------
          WANT ANOTHER RUN ? (Y,N) ?NO
```

THE EXTERNAL RATE OF RETURN METHOD

The reinvestment assumption of the I.R.R. method noted above may not be valid in an engineering economy study. For instance, if a firm's M.A.R.R. is

20% per year and the I.R.R. for a project is 42.4%, it may not be possible for the firm to reinvest net cash proceeds from the project at much more than 20%. This situation, coupled with the computational demands and possible multiple interest rates associated with the I.R.R. method, has given rise to other rate of return methods that can remedy some of these weaknesses.

The first method to be considered is the external rate of return (E.R.R.) method. It directly takes into account the external interest rate (e) at which net cash flows generated (or required) by a project over its life can be reinvested (or borrowed) outside the firm. If this external reinvestment rate happens to equal the project's I.R.R., then the E.R.R. method produces results identical to those of the I.R.R. method.

In general, all cash outflows are discounted to period 0 (the present) at $e\%$ per compounding period while all cash inflows are compounded to period N at $e\%$. The external rate of return is then the interest rate that establishes equivalence between the two quantities. In equation form, the E.R.R. is the $i'\%$ at which

$$\sum_{k=0}^{N} E_k(P/F, e\%, k)(F/P, i'\%, N) = \sum_{k=0}^{N} R_k(F/P, e\%, N - k) \quad (5\text{-}9)$$

where R_k = excess of receipts over disbursements in period k

E_k = excess of expenditures over receipts in period k

N = project life or number of periods for the study

e = external reinvestment rate per period

(not to be confused with the Naperian base)

Graphically, we have the following:

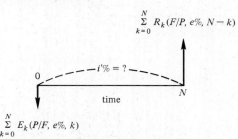

A project is acceptable when $i'\%$ of the E.R.R. method is greater than or equal to the firm's M.A.R.R.

The external rate of return method has two basic advantages over the I.R.R. method:

1. It usually can be solved for directly rather than by trial and error.
2. It is not subject to the possibility of multiple rates of return. (*Note:* The multiple rate of return problem with the I.R.R. method is discussed in Appendix 5-B.)

EXAMPLE 5-9

Referring to Example 5-8, suppose that e = M.A.R.R. = 20% per year. What is the alternative's external rate of return, and is the alternative acceptable?

Solution

By utilizing Equation 5-9, we have this relationship to solve for i'.

$$\$25,000(F/P, i'\%, 5) = \$8,000(F/A, 20\%, 5) + \$5,000$$

$$(F/P, i'\%, 5) = \frac{\$64,532.80}{25,000} = 2.5813$$

$$i' = 20.88\%$$

Because $i' >$ M.A.R.R., the alternative is barely justified. ∎

EXAMPLE 5-10

When $e = 15\%$ and M.A.R.R. $= 20\%$, determine whether the project whose cash flow diagram appears below is acceptable.

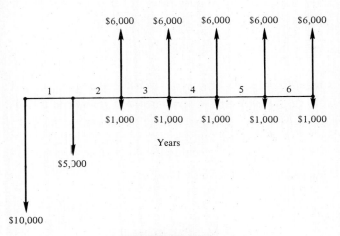

EXAMPLE 5-10

Solution

$$E_0 = \$10,000 \ (k = 0)$$

$$E_1 = \$5,000 \ (k = 1)$$

$$R_k = \$5,000 \text{ for } k = 2, 3, \ldots, 6$$

$$[10,000 + 5,000(P/F, 15\%, 1)](F/P, i'\%, 6)$$
$$= 5,000(F/A, 15\%, 5); \quad i' = 15.3\%$$

The i' is less than M.A.R.R. $= 20\%$, therefore, this project would be unacceptable according to the E.R.R. method. ∎

EXAMPLE 5-11

Considering the same project as in Example 5-7, determine whether it is justified using the E.R.R. method, assuming that funds can be reinvested at the M.A.R.R. $= 10\%$.

Solution

$$\$10,000(F/P, i'\%, 5) = (\$5,310 - \$3,000)(F/A, 10\%, 5) + \$2,000$$

$$= \$2,310(6.105) + \$2,000$$

$$(F/P, i'\%, 5) = 1.61$$

From interest table searching, one can find that the single-payment compound amount factor for $N = 5$ is 1.61 for $i = 10\%$. Thus 10% is the E.R.R., and the project is shown to be marginally justified.

Another way of finding the E.R.R. $= i'\%$ is to solve directly in the algebraic equivalent of the equation above. Thus

$$(F/P, i'\%, 5) = 1.61$$

$$(1 + i')^5 = 1.61$$

$$5 \ln (1 + i') = \ln 1.61$$

$$\ln (1 + i') = 0.0952$$

$$\ln^{-1} (1 + i') = 1.10$$

$$i = 0.10 = 10\%$$ ■

THE EXPLICIT REINVESTMENT RATE OF RETURN METHOD

The explicit reinvestment rate of return method (E.R.R.R.) overcomes some of the weaknesses of the I.R.R. method and is applicable to problems where there is a single lump-sum investment followed by uniform cash savings or net receipts at the end of each period throughout the life of a project.

This method is appealing because a rate of return is determined by dividing net annual "profit" (receipts less disbursements less annual equivalent replacement cost of the investment) by the amount of the initial investment. The annual replacement cost is the sinking fund equivalent of the original investment, less its salvage value, computed at $e\%$ per year. Thus the E.R.R.R. method assumes that a sinking fund in the amount of $(P - S)$ has been established for purposes of replacing the asset. Moreover, the value of e is the effective annual rate at which these recovered funds can be reinvested. Often e is equal to a firm's M.A.R.R.

In general,

$$\text{E.R.R.R.} = \frac{(R - E) - (P - S)(A/F, e\%, N)}{P} \qquad (5\text{-}10)$$

where terms are as defined previously and S equals the salvage value at the end of year N.

EXAMPLE 5-12
Based on the data of Example 5-8 and $e = 20\%$, is the alternative acceptable by the E.R.R.R. method?

Solution

$$\text{E.R.R.R.} = \frac{\$8{,}000 - \$20{,}000(A/F,\ 20\%,\ 5)}{\$25{,}000}$$

$$= \frac{\$8{,}000 - \$2687.59}{\$25{,}000}$$

$$= 0.2125 \quad \text{or} \quad 21.25\%$$

Because 21.25% > M.A.R.R., the alternative is justified.　■

SUMMARY COMPARISON OF ECONOMY STUDY METHODS

The reader may well have wondered why six different methods for economy studies have been presented in this chapter when any one of these methods will give a valid answer for the reinvestment assumption inherent in that method. The best answer is that preferences differ between analysts and decision makers. It must always be remembered that results of an engineering economy study should be communicated to decision makers in terms they can readily understand. Additionally, one method may be easier to use than another method because of the particular cash flow patterns involved.

Usually equivalence methods such as A.W., P.W., and F.W. are easier to use computationally than rate of return methods. However, many decision makers prefer to think and analyze study results in terms of rates of return. Business enterprises generally adopt one, or at most two, analysis techniques for broad categories of alternatives to be analyzed.

The three equivalence methods are directly related so that one amount can be obtained from the other by the following relations:

$$\text{A.W.} = \text{P.W.}(A/P,\ i\%,\ N) = \text{F.W.}(A/F,\ i\%,\ N) \qquad (5\text{-}11)$$

$$\text{P.W.} = \text{A.W.}(P/A,\ i\%,\ N) = \text{F.W.}(P/F,\ i\%,\ N) \qquad (5\text{-}12)$$

$$\text{F.W.} = \text{A.W.}(F/A,\ i\%,\ N) = \text{P.W.}(F/P,\ i\%,\ N) \qquad (5\text{-}13)$$

These methods all involve the built-in assumption that funds can be reinvested at $i\%$ per compounding period, which is normally the M.A.R.R. This same assumption is often utilized in the E.R.R. method. However, it is possible to consider the possibility that funds are reinvested at an external rate, $e \neq$ M.A.R.R. The E.R.R.R. method assumes that the reinvestment at M.A.R.R. (or e) applies only to recovered depreciation funds. The I.R.R. method, on the other hand, has the built-in assumption that all funds are reinvested at the particular I.R.R. rate computed for the project generating those funds.

In general, the numerical answer for a given project will be close for each of the rate of return methods. If the reinvestment rate is the same as the rate of return calculated, then I.R.R. = E.R.R. = E.R.R.R. This and other conditions relating answers by the three rate of return methods are summarized as follows:

If Reinvestment Rate Is:	Then E.R.R. and E.R.R.R. Are:
= I.R.R.	= I.R.R.
> I.R.R.	> I.R.R.
< I.R.R.	< I.R.R.

THE PAYBACK (PAYOUT) PERIOD METHOD

There is an economy study method that is often used by government and industry, but it can produce misleading results. It is definitely not recommended except as supplemental information in conjunction with one of the six methods previously discussed.

All methods presented thus far reflect the profitability of a proposed alternative. The payback method, which is often called the simple payout method, mainly indicates a project's liquidity rather than its profitability. Historically, the payback method has been used as a measure of a project's riskiness since liquidity deals with how fast an investment can be recovered. Quite simply, the payback method calculates the number of years required for positive cash flows to equal the initial investment. Hence the simple payback period is the smallest value of θ for which this relationship is satisfied under end-of-year cash flow convention:

$$\sum_{k=1}^{\theta} (R_k - E_k) - P \geq 0 \qquad (5\text{-}14)$$

The payback period, θ, ignores the time value of money *and* all cash flows that occur after θ. If this method were applied to the investment project in Example 5-7, the number of years for payout can be calculated to be $\$10,000/(\$5,310 - \$3,000) = 4.33$ years. When $k = N$, the salvage value is included in the determination of a payback period. As can be seen from Equation 5-14, the payback period does not indicate anything about the project desirability except the speed with which the investment will be recovered.

Sometimes the *discounted* payback period, θ', is calculated so that the time value of money is considered:

$$\sum_{k=1}^{\theta'} (R_k - E_k)(P/F, i\% \ k) - P \geq 0 \qquad (5\text{-}15)$$

where $i\%$ is the minimum attractive rate of return and θ' is the smallest value that satisfies Equation 5-15. This variation of the simple payback period is also called the "capital recovery period." However, neither payback period calculation necessarily includes cash flows occurring after θ (or θ'). Another serious defect is that they do not take into consideration the economic life of physical assets. Thus it is quite possible for one alternative that has a longer payout period than another to produce a higher rate of return (or net present worth) on the invested capital. The use of the payout period for making investment decisions should be avoided except as a measure of how quickly invested capital will be recovered, which is an indicator of project risk.

EXAMPLE 5-13

Considering the alternative presented in Example 5-8, calculate the simple payback period and the discounted payback period. The so-called "payback rate of return" equals $1/\theta$, so if M.A.R.R. $= 20\%$, the approximate minimum acceptable payback period is 5 years ($\frac{1}{5} = 0.20$). In view of a minimum acceptable payback period of 5 years, is this alternative economically attractive?

Solution
Payback period:

$$\sum_{k=1}^{\theta=3.13} (\$8,000)_k - \$25,000 \geq 0$$

Thus $\theta = 3.13$ years and the alternative is acceptable. (With end-of-year cash flow convention, this value of θ would be rounded up to 4 years.) Notice that the $5,000 salvage value was not considered.

Discounted payback period:

$$\sum_{k=1}^{\theta'=5} (\$8,000)_k(P/F, 20\%, k) + \$5,000(P/F, 20\%, 5) - \$25,000 > 0$$

or

$$\$8,000(P/A, 20\%, 5) + \$5,000(P/F, 20\%, 5) - \$25,000 > 0$$

at the project's expected life of $N = 5$ years. The alternative is barely acceptable with the discounted payback method since $\theta' = 5$ years. In this case, the salvage value was considered in determining the value of θ'. ∎

AN EXAMPLE OF A NEW VENTURE ANALYZED WITH VARIOUS METHODS

The following is an example of an economy study of a new venture involving a single, initial investment and uniform revenue and cost data. It thus represents a very simple type of basic, long-term investment decision problem. The reader is encouraged to use the computer program in Appendix C to evaluate this and other time-consuming problems with the methods described in Chapter 5.

EXAMPLE 5-14

Mr. Brown has an opportunity to purchase an apartment house that has just been completed. It is located in the suburban area of a medium-size city that contains a fair amount of industrial plants. It also is within walking distance of a large university. The apartment house is in the process of being rented and now is more than 80% occupied, with prospects of being fully rented within a few weeks. The purchase price would be $125,000, of which $50,000 represents the value of the land. The building consists of 10 four-room apartments, a small apartment for a caretaker, and garage space for 11 automobiles. From a study of similar apartment buildings, Mr. Brown estimates that each apartment can be rented for $325 per month, with at least 95% average occupancy at all times. Heat and water are included in the rental. The operating costs are estimated to be as follows:

Caretaker	$400 per month, plus his apartment
Fuel	$2,000 per year
Water	$150 per year
Maintenance and repair	Equal to 1 month's rental on each rental unit per year
Taxes	$4 per $100 of assessed value; assessed value will be approximately 30% of cost of building and land
Insurance	0.5% of first cost of building, per year
Agent's commission	$2\frac{1}{2}$% of gross rental revenue

At present Mr. Brown's capital is invested in bonds that yield approximately 12% before income taxes and he feels that a project of this risk should earn at least 15% before income taxes. He estimates that the economic life of the apartment house will be at least 40 years and that the $50,000 for land will be the only salvage value at the end of that time.

The revenue each year can be computed as $10 \times \$325 \times 12 \times 0.95 = \$37,050$. The out-of-pocket costs (disbursements) each year can be computed as follows:

Caretaker: $400 × 12	$ 4,800
Fuel	2,000
Water	150
Maintenance and repair: $325 × 10	3,250
Taxes: ($125,000/100) × 0.3 × $4	1,500
Insurance: $75,000 × 0.005	375
Agent's commission: $37,050 × 0.025	926
TOTAL	$13,001

Hence the net cash inflow (revenue minus disbursements) each year is $37,050 − $13,001 = $24,049.

By using the I.R.R. method, show whether Mr. Brown should purchase the apartment house.

Solution

The present worth can be written

$$-\$125,000 + \$24,049(P/A, i', 40)$$

$$+ \$50,000(P/F, i', 40) = 0$$

At $i' = 20\%$: $\quad -\$125,000 + \$24,049(4.9966)$

$$+ \$50,000(0.0007) = -4802$$

$$\text{At } i' = 19\%: \quad -\$125,000 + \$24,049(5.2582)$$
$$+ \$50,000(0.0010) = +1504$$

Since we are seeking the i' at which the present worth equation is zero, by linear interpolation we find that

$$i' = \text{I.R.R.} = 19.2\%$$

Since 19.2% is greater than 15%, the project is apparently worthy of investment. ∎

EXAMPLE 5-15

Analyze the same investment project as in Example 5-14 except use the E.R.R. method. Assume that he expects to reinvest annual net cash inflows in a money market account earning 15% before taxes.

Solution

By utilizing Equation 5-9, the E.R.R. for this project can be determined:

$$\$125,000(F/P, i', 40) = \$24,049(F/A, 15\%, 40) + \$50,000$$

$$(F/P, i', 40) = \frac{\$42,835,343}{\$125,000} = 342.68$$

$$(1 + i')^{40} = 342.68$$

$$\therefore i' = \text{E.R.R.} = 15.7\%$$

Since 15.7% > 15%, the project would be accepted with the E.R.R. method. If funds generated by the project can be reinvested at 15%, the E.R.R. method provides a more realistic assessment than the I.R.R. method in which the reinvestment rate is assumed to 19.3%. The difference in the I.R.R. and E.R.R. is rather pronounced when extended periods of time and relatively high interest rates are involved. ∎

EXAMPLE 5-16

Analyze the same investment project as in Example 5-14 but use the annual worth method. The M.A.R.R. is 15%.

Solution

With Equation 5-1, the annual worth can be calculated:

Annual revenue		$37,050
Annual costs (out-of-pocket)	$13,001	
Capital recovery cost at M.A.R.R. = 15%		
(From Equation 5-2, 5-3, or 5-4)	18,792	
TOTAL	$31,793	$37,050

Because the annual revenue of $37,050 exceeds the total annual costs of $31,793, the investment would be justified. The net A.W. is $37,050 − $31,793 = $5,257, which indicates a favorable project. ∎

EXAMPLE 5-17

Analyze the investment project in Example 5-14 with the present worth method at M.A.R.R. = 15%.

Solution

P.W. of cash inflow	
Revenue: $37,050(P/A, 15%, 40)	$246,078
Salvage of land: $50,000(P/F, 15%, 40)	187
TOTAL	$246,265
P.W. of cash outflow:	
Investment	$125,000
Out-of-pocket: $13,001(P/A, 15%, 40)	86,350
TOTAL	$211,350

Since $246,265 > $211,350, the net P.W. equals $34,915 and the project is again shown to be worthy of investment. ∎

DISCUSSION OF DECISION CRITERIA TO SUPPLEMENT AN ENGINEERING ECONOMY STUDY

In order to arrive at a decision on the project for which economy studies were made in Examples 5-14 through 5-17, Mr. Brown would have to consider a number of factors, somewhat as follows. The first item would undoubtedly be the revenue estimate. Some questions to be answered would be as follows:

1. Is the assumed rental rate reasonable, particularly in relationship to similar apartments?
2. Is the assumed occupancy rate a reasonable one for this type of housing?
3. How will possible changes in economic conditions affect the rentals, both as to price and occupancy rate?

It is evident that it might be quite easy to obtain reliable answers to the first two of these questions. Answers to the third would be less precise. However, it would be possible to predict the general effect of changing economic conditions. We would also have to take into account the other types and classes of rental housing available in the area. In addition, there would be the fact that if changing economic conditions forced a reduction in the monthly rental charge, some of the expenses also probably would be reduced. By such an analysis the validity of the estimated revenue data can be determined in any engineering economy study. In this particular case it should be possible to obtain revenue data that are reasonably accurate. In this study the amount of investment required is known exactly because the finished and already rented building is to be purchased at a stated price.

Maintenance and repair costs are difficult to estimate unless historical data are available from similar projects. Such data are usually available for apartment buildings. In the absence of such data, it is well to remember that estimates of these costs are frequently too low and that they usually increase as time passes. This may be of considerable importance when long study periods are involved.

It thus appears that in this case all the revenue and cost data can be determined with accuracy. The primary element of risk is due to the rather long, but not excessive, assumed life of the building. There is also the added factor that investment in the apartment would provide a hedge against inflation that does not exist in the bonds that Mr. Brown now owns. If he feels that the data have been determined carefully, he probably would make the investment.

A more complete discussion of how to include risk considerations and non-monetary ("intangible") decision criteria into engineering economy studies is presented in Chapter 7.

AN EXAMPLE OF A PROPOSED INVESTMENT TO REDUCE COSTS

EXAMPLE 5-18

A manufacturer of jewelry is contemplating the installation of a system that will recover a larger portion of the fine particles of gold and platinum that result from the various manufacturing operations. At the present time a little over $45,000 worth of these metals is being lost per year, and it is anticipated that, because of the growth of the company, this amount will increase by $5,000 each year for the next 10 years. The proposed system, involving a network of exhaust ducts and separators, will recover approximately two-thirds of the gold and platinum that otherwise would be lost. The complete installation would cost $140,000. The best estimates for the operating costs of the system, obtained from operations of similar systems, are $10,000 per year for operating expense, $1,800 per year for maintenance and repairs, and 2% of the first cost annually for property taxes and insurance. The company would require the investment to be recovered with interest in 10 years. The M.A.R.R. of the company, before taxes, is 15%. Should the recovery system be installed?

Solution

Such an investment is made to reduce some of the operating expenses, in this case the cost of the material used. Thus the saving (income) to be obtained by making an investment is almost entirely within the control of the investors. The company knows exactly what expenses have been. If the efficiency of the proposed equipment is known, the only factors that should affect the saving are the variation of production, operation, and maintenance expenses of the proposed equipment. In most cases of this type these items are known or may be predicted quite accurately. The company would have a good idea of how its volume might vary. Operation and maintenance expenses can usually be estimated accurately, especially if historical data are available on the proposed equipment.

Using the data given, we find that the cash flow situation and the internal rate of return calculation would be as shown in Table 5-2. In deciding whether or not the calculated internal rate of return of 16.8% is sufficient to justify investment, each factor that might contribute to the risk must be examined so that a measure of the total risk may be obtained. In this case it appears that the factors are quite well controlled or known. There is little reason to believe that much more risk would be involved than is present in all the normal operations of the company. Thus the company has its own experience to use as a basis of comparison. If the company is sound and its business quite stable, a return of 16.8% should be satisfactory, inasmuch as the M.A.R.R. for projects of this type is 15%. ■

It may be seen that when capital is invested in a going concern in order to bring about reduction in costs, the risk is usually easier to determine and is often much less than when entirely new enterprises are involved. As a result, the rate of return required for such investments is often lower.

A CASE STUDY OF A LARGE INDUSTRIAL INVESTMENT OPPORTUNITY

The following case study illustrates the before-tax cash flow analysis of a typical industrial plant expansion problem. It involves many types of cash flows and numerous considerations that can be readily expanded upon in later chapters. Particular attention is directed to variable, fixed and sunk costs that are present in addition to expenditures which are capitalized (e.g. investments) versus those charged against operations (e.g. raw materials costs).

EXAMPLE 5-19
I. Problem Statement

Product X-21, a food preservative, is manufactured in a facility with a nominal capacity of 180,000 pounds per year. The ultimate capacity of the facility can be increased, at some loss in cost efficiency, by process modifications and the use of extensive amounts of overtime, to 195,000 pounds per year. With the present marketing strategy, sales are expected to level off at about 190,000 pounds per year in 1985 and remain at that volume. Therefore, the existing capacity will be sufficient under the present marketing strategy.

However, an opportunity has become available to enter a new market for an X-21 product modified slightly to give it better warm weather stability. A German firm has a trade secret on the modifier process but is willing to enter into a nondisclosure agreement for a one-time fee of $75,000. A proposal has been submitted to management to expand the existing X-21 plant to permit the company to enter this new market. *Management has requested an economic analysis of this proposal and a recommendation.* Your assignment is to provide this recommendation.

TABLE 5-2 Tabular Determination of Internal Rate of Return[a] for Proposed Gold-Recovery System in Example 5-18

Year End, N	Investment	Recovery	Costs	Net Cash Flow	(P/F, 15%, N)	Net Present Worth at i = 15%	(P/F, 20%, N)	Net Present Worth at i = 20%
0	−$140,000	—	—	−$140,000	1.000	−$140,000	1.000	−$140,000
1	—	$33,333	−$14,600	+ 18,733	0.8696	+ 16,290	0.8333	+ 15,610
2	—	36,667	− 14,600	+ 22,067	0.7561	+ 16,685	0.6944	+ 15,323
3	—	40,000	− 14,600	+ 25,400	0.6575	+ 16,701	0.5787	+ 14,699
4	—	43,333	− 14,600	+ 28,733	0.5718	+ 16,430	0.4823	+ 13,858
5	—	46,667	− 14,600	+ 32,067	0.4972	+ 15,944	0.4019	+ 12,888
6	—	50,000	− 14,600	+ 35,400	0.4323	+ 15,303	0.3349	+ 11,855
7	—	53,333	− 14,600	+ 38,733	0.3759	+ 14,560	0.2791	+ 10,810
8	—	56,667	− 14,600	+ 42,067	0.3269	+ 13,752	0.2326	+ 9,785
9	—	60,000	− 14,600	+ 45,400	0.2843	+ 12,907	0.1938	+ 8,799
10	—	63,333	− 14,600	+ 48,733	0.2472	+ 12,047	0.1615	+ 7,870
TOTAL				+$197,333		+$ 10,619		−$ 18,503

[a]I.R.R. ≅ 15% + [$10,619/(10,619 + 18,503)] × (20% − 15%) ≅ 16.8%.

II. General Background Information

A. *Demand schedule:* The demand schedule for X-21 is as follows:

Year	Total Demand (lb)	Incremental Sales from This Project (lb)
1984	175,000	0
1985	190,000	0
1986	220,000	30,000
1987	250,000	60,000
1988–2000	260,000	70,000

Both the modified and regular grades of X-21 will be sold at the same price of $39.50 per pound. All demand over 190,000 pounds per year is attributable to the new market.

B. *Existing manufacturing facility:* The existing facility consists of two production lines, each with an ultimate capacity of 97,500 pounds per year each. A new line of the same capacity can be added in an existing building, which because of federal restrictions and its unique layout, cannot be used for any other purpose. That is, no cost need be allocated for the building space. The new line can be used for both regular X-21 and the modified X-21.

C. *Expansion plan:* The total installed cost of the new line is estimated at $430,000, assuming project operation by January 1, 1986. Of this total, engineering expenditures of $40,000 would be required in 1984 and all other capital costs would be incurred in 1985.

D. *Project life:* The company considers all projects of this type and degree of risk to have a 15-year life. Thus commercial operation of the new production line would begin in 1986 and terminate at the end of year 2000.

E. *Inflation:* The analysis will ignore the effects of inflation (later, in Chapter 8, this topic is considered).

F. *Federal and state income taxes:* The analysis will ignore all income taxes (later in Chapter 10 income taxes are considered).

G. *Interest rate and financing:* The minimum acceptable before-tax rate of return for projects with this degree of risk is 20% per year (effective). The financing decision for large projects of this type is discussed in Chapter 14.

H. *Other profitable company operations:* These will offset any short-term negative cash flows.

I. *Expenditures:* Those occurring throughout a year will be assumed for cash flow purposes to occur at the end of the year.

III. Cash Flow Data

A. *Capitalizable project expenditures:* A total of $430,000 will be required to install the new production line. In 1984, $40,000 is needed, and the remainder of $390,000 would be expected in 1985. (Capitalizable expenditures are those expenditures that must be depreciated using the methods discussed later in Chapter 9.)

B. Project expenditures chargeable to operations: Certain expenditures will result from the dismantlement and rearrangement of special equipment that will be required to fit the new line into the existing building. Estimated before-tax cost for this work is $48,600. (These expenditures would *not* be depreciated with methods of Chapter 9.)

C. Related projects and work orders: Various types of auxiliary equipment will have to be replaced at 5-year intervals. Therefore, equipment replacement expenditures of $120,000 are planned for 1990 and 1995.

D. Net change in working capital: All firms require capital for day-to-day operations, including funds to support prepaid expenses, inventories, and bank balances. Such *working capital* has associated with it an opportunity cost that in many cases is a substantial item in an economy study. Working capital caused by a project is normally treated as an asset whose first cost and salvage value are equal $(P = S)$. Thus the annual opportunity cost (i) of working capital in year k (WC_k) is iWC_k.

The change in working capital should be roughly proportional to the change in sales. Working capital in 1983 was approximately $200,000 when sales were 164,000 pounds. The following net changes in working capital are computed using this proportionality:

Year	Net Change from Previous Year
1984	$13,400
1985	18,300
1986	36,600
1987	36,600
1988	12,200
1989–2000	0

Only changes in 1986 and later are pertinent to the analysis.

E. Prepaid know-how: A single payment of $75,000 to the German firm is required before startup. This payment will be made at the end of 1985.

F. Pretax cash flows from commercial operation (incremental due to the new line):

1. *Fixed costs:* Fixed costs for supervisory salaries, general plant overhead, insurance, and property taxes will increase by $750,000 per year if this project is implemented. This increase will remain constant over the 15-year life of the projects.

2. *Variable costs:* These costs consist primarily of raw materials costs, direct labor, utilities, and the variable portion of maintenance costs. Variable costs are expected to be $9.18 per pound in 1986. As operating experience is gained, yields should increase, resulting in variable costs of $8.83 per pound in 1987 and $7.56 per pound in 1988 and thereafter.

3. *General and administrative (G and A) costs:* G and A costs are budgeted

for $900,000 in 1986 when the product is first introduced. After 1986, G and A costs are budgeted for $1,100,000 per year. This amounts to about 40% of sales revenue after sales have leveled out.

The following table summarizes the annual before-tax cash flow in years 1986–2000 attributable to this proposed project.

Year	(1) Fixed Cost (thousands)	(2) Variable Cost (thousands)	(3) G and A Costs (thousands)	Sales Revenues (thousands)	Sales Revenue Less Costs in Columns 1–3 (thousands)
1986	$750	$275.4	$ 900	$1,185	− $740.4
1987	750	529.8	1,100	2,370	− 9.8
1988–2000	750	529.2	1,100	2,765	+ 385.8

G. Introductory costs: Introductory costs will be relatively small. The only significant introductory cost is training personnel in modifications to the process and technical assistance during the startup. The total introductory cost in 1985 is estimated at $8,500.

IV. Analysis Results and Recommendations

A cash flow worksheet for this example is given in Table 5-3. From the before-tax cash flows in column 9, the project's internal rate of return can be calculated:

$$0 = -\$40,000 - \$522,100(P/F, i', 1) - \$777,000(P/F, i', 2) \cdots$$
$$+ \$471,200(P/F, i', 16)$$

By trial and error, $i' = 18.3\% <$ M.A.R.R. Because this determination is quite tedious, it is suggested that the computer program in Appendix C be employed to verify the project's I.R.R.

The before-tax M.A.R.R. is 20% and the net present worth at the end of 1984 is − $87,137. This may also be confirmed by utilizing the computer program in Appendix C. Thus the I.R.R. and P.W. criteria indicate that the company should not enter the new market for a modified X-21 product. Even if these criteria had signaled a favorable project (e.g., I.R.R. $\geq$ 20%), a detailed analysis of uncertainty and nonmonetary considerations should be performed before a decision is made. These topics are addressed in Chapter 7. ∎

PROBLEMS

Unless stated otherwise, discrete compounding of interest and end-of-period cash flows should be assumed in all problem exercises in the remainder of the book.

5-1 Uncle Wilbur's trout ranch is now for sale for $40,000. Annual property taxes, maintenance, supplies, and so on, are estimated to continue to be $3,000 per year. Revenues

TABLE 5-3 Cash Flow Worksheet for Example 5-19

End of Year	(1) Capitalizable Project Expenditures	(2) Expenditures Chargeable to Operations	(3) Related Projects and Work Orders	(4) Net Change in Working Capital	(5) Prepaid Know-How	(6) Total Investment Cash Flow: Cols. 1 + 2 + 3 + 4 + 5	(7) Before-Tax Cash Flows During Commercial Operation	(8) Introductory Costs	(9) Before-Tax Cash Flows for the Entire Project
1984	− $ 40,000					− $ 40,000			− $ 40,000
1985	− 390,000	− $48,600			− $75,000	− 513,600		− $8,500	− 522,100
1986				− $36,600		− 36,600	− $740,400		− 777,000
1987				− 36,600		− 36,600	− 9,800		− 46,400
1988				− 12,200		− 12,200	385,800		+ 373,600
1989							385,800		+ 385,800
1990			− $120,000			− 120,000	385,800		+ 265,800
1991							385,800		+ 385,800
1992							385,800		+ 385,800
1993							385,800		+ 385,800
1994							385,800		+ 385,800
1995			− 120,000			− 120,000	385,800		+ 265,800
1996							385,800		+ 385,800
1997							385,800		+ 385,800
1998							385,800		+ 385,800
1999							385,800		+ 385,800
2000				+ 85,400[a]		+ 85,400	385,800		+ 471,200

[a]This is return of working capital at the end of the project.

from the ranch are expected to be $10,000 next year and then decline by $500 per year thereafter through the tenth year. If you bought the ranch, you would plan to keep it only 5 years and at that time sell it for the value of the land, which is $15,000. If your desired annual rate of return is 12%, should you become a trout rancher? Use the present worth method.

5-2　A certain heat exchanger costs $30,000 installed and has an estimated life of 6 years. By the addition of certain auxiliary equipment when the heat exchanger is initially purchased, an annual saving of $1,000 in operating cost can be obtained and the estimated life of the heat exchanger can be doubled. Neglecting any salvage value for either plan and with effective annual interest at 8%, what *present* expenditure can be justified for the auxiliary equipment? Use a study period of 12 years.

5-3　(a) Machine XYZ is to be evaluated on the basis of the present worth method when the minimum attractive rate of return is 12%. Pertinent cost data are as follows:

	Machine XYZ
First cost	$13,000
Useful life	15 years
Salvage value	$ 3,000
Annual operating costs	$ 100
Overhaul—end of fifth year	$ 200
Overhaul—end of tenth year	$ 550

(b) Determine the capital recovery cost of machine XYZ by all three formulas presented in the text.

5-4　(a) A certain service can be performed satisfactorily by process R, which has a first cost of $8,000, an estimated service life of 10 years, no salvage value, and annual net receipts (revenues − disbursements) of $2,400. Assuming a minimum attractive rate of return of 18% before income taxes, find the annual worth of this process and specify whether you would recommend it.

(b) A compressor that costs $2,500 has a 5-year useful life and a salvage value of $1,000 after 5 years. At nominal interest of 10%, compounded quarterly, what is the *annual* capital recovery cost of the compressor?

5-5　(a) Determine the P.W. and F.W. of the following proposal when the M.A.R.R. is 15%.

	Proposal A
First cost	$10,000
Expected life	5 years
Salvage value[a]	−$ 1,000
Annual receipts	$ 8,000
Annual disbursements	$ 4,000

[a]A negative salvage value means there is a net cost to dispose of an asset.

(b) Determine the I.R.R. for proposal A. Is it acceptable?
(c) Determine the E.R.R.R. for proposal A when the external reinvestment rate is 15%.

5-6　Find the internal rate of return (I.R.R.) in each of these situations:

(a)

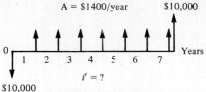

PROBLEM 5-6(a)

(b)

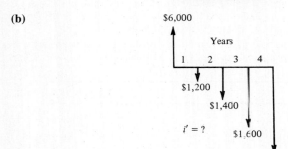

PROBLEM 5-6(b)

(c) You purchase a used car for $4,200. After you make a $1,000 down payment on the car, the salesperson looks in her *Interest Calculations Made Simple* handbook and announces: "The monthly payments will be $160 for the next 24 months and the first payment is due 1 month from now." (Draw a cash flow diagram.)

(d)

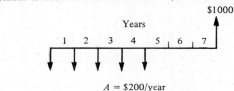

PROBLEM 5-6(d)

(e)

End of Year	Net Cash Flow
0	+$2,000
1	− 7,000
2	+ 2,000
3	+ 5,000

(f)

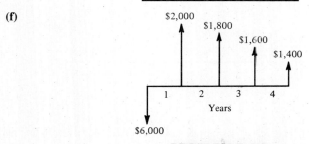

PROBLEM 5-6(f)

APPLICATIONS OF ENGINEERING ECONOMY

5-7   Rework parts (c), (e), and (f) of Problem 5-6 by using the E.R.R. method when e = M.A.R.R. = 8% per year.

5-8   Calculate the payback period and discounted payback period of parts (a) and (f) of Problem 5-6. Let M.A.R.R. be 8% per year.

5-9   Plot the net present worth of parts (a) and (f) of Problem 5-6 as a function of the interest rate. The M.A.R.R. is equal to 8%. Now plot the future worth of parts (a) and (f) versus i. What similarities do you see?

5-10   Draw an unrecovered investment diagram for part (a) of Problem 5-6 using the I.R.R. determined in that problem.

5-11   A small company purchased now for $23,000 will lose $1,200 each year the first 4 years. An additional $8,000 invested in the company during the fourth year will result in a profit of $5,500 each year from the 5th year through the 15th year. At the end of 15 years the company can be sold for $33,000.
(a) Determine the internal rate of return (I.R.R.).
(b) Determine the E.R.R. when e = 8%.
(c) Calculate the future worth if M.A.R.R. = 12%.

5-12   Consider these two machines:

	Machine A	Machine B
Initial Cost	$11,500	$8,000
Annual net cash flow	+$ 2,400	+$1,800
Life	6 years	6 years
Salvage value	0	0

(a) If M.A.R.R. = 8%, determine whether the machines are acceptable alternatives with the three rate of return methods. The reinvestment rate is equal to the M.A.R.R.
(b) Use the equivalent worth methods when M.A.R.R. = 8% to determine which machines are acceptable.

5-13   Evaluate machines A and B in Problem 5-12 by using the simple payback method. Is either one acceptable when the maximum permissible value of θ is 5 years?

5-14   Your boss has just presented you with a summary of projected costs and net annual receipts (before taxes) for a new product line. He asks you to calculate the before-tax internal rate of return for this investment opportunity. What would you present to your boss and how would you explain the results of your analysis? (It is widely known that the boss likes to see graphs of net present worth versus rate of return for this type of problem.)

End of Year	Net Cash Flow
0	− $450,000
1	− 42,500
2	+ 92,800
3	+ 386,000
4	+ 614,600
5	− 202,200

5-15   Rework Problem 5-14 with the E.R.R. method and an external reinvestment rate of 12%. Why are answers to the two problems different?

5-16 List the advantages and disadvantages of each of the six basic methods for performing engineering economy studies (A.W., P.W., F.W., I.R.R., E.R.R., E.R.R.R.)

5-17 A piece of equipment requires an investment of $10,000 and has a $2,000 salvage value at the end of its expected 9-year life. Savings attributable to the equipment will be $2,000 in the first year and will increase by $200 each year thereafter.
(a) Calculate the simple payout period for the equipment.
(b) Calculate the I.R.R. for the equipment.

5-18 List the major disadvantages of the payout method for making economic comparisons. In what situations might the payout method be appropriate?

5-19 The Anirup Food Processing Company is presently using an outdated method for filling 25-pound sacks of dry dog food. To compensate for weighing inaccuracies inherent to this packaging method, the process engineer at the plant has estimated that each sack is overfilled by $\frac{1}{8}$ pound on the average. A modern method of packaging is now available that would eliminate overfilling (and underfilling). The production quota for the plant is 300,000 sacks per year for the next 6 years, and a pound of dog food costs this plant $0.15 to produce. The present system has no salvage value and will last another 6 years, and the modern method has an estimated life of 4 years with a salvage value equal to 10% of its purchase price. It is also estimated that the present packaging operation will cost $2,100 per year more to maintain than the modern method. If the minimum attractive rate of return is 12% for this company, what amount could be justified for the purchase of the modern packaging method? (*Hint:* Find P at which net present worth of savings from the modern method equal zero.)

5-20 Evaluate the acceptability of the project below with all methods discussed in Chapter 5. Let M.A.R.R. = 15%, e = 12%, maximum acceptable θ = 5 years and maximum acceptable θ' = 6 years.

Project: R137-A   *Title:* Syn-Tree Fabrication

Description: Establish a production facility to manufacture synthetic palm trees for sale to resort areas in Alaska.

Cash Flow Estimates:

Year	Amount (thousands)
0	− $1,500
1	+ 200
2	+ 400
3	+ 450
4	+ 450
5	+ 600
6	+ 900
7	+ 1,100

5-21 In medieval Europe, usury was an unpardonable sin. However, loans could be made if they were to benefit the public. For these purposes a nominal interest rate of 12% was sanctioned by public officials. It is recorded that a Silas Smythe borrowed £90 from a local money merchant and agreed to repay the debt under the following conditions: £25 was repaid to the lender at the end of each of 4 years thereafter; and £20 was to be repaid at the end of the fifth year. This business venture was in the interest of the public. Assuming that interest was compounded continuously in medieval Europe, was the money merchant guilty of usury?

5-22 An insurance company is considering building a 25-unit apartment house in a growing town. Because of the long-term growth potential of the town, it is felt that the company could average 90% of full rent for the whole house each year. If the following items are reasonably accurate predictions, would you advise the company to make the investment if their invested capital is making 12%? Use the annual worth method.

Land investment	$50,000
Building investment	$225,000
Capital recovery period, N	20 years
Rent per unit per month	$350
Upkeep per unit per month	$35
Property taxes and insurance per year	10% of total investment

5-23 A company is considering constructing a plant to manufacture a proposed new product. The land costs $300,000, the building costs $600,000, the equipment costs $250,000, and $100,000 working capital is required. It is expected that the product will result in sales of $750,000 per year for 10 years, at which time the land can be sold for $400,000, the building for $350,000, the equipment for $50,000, and all of the working capital recovered. The annual out-of-pocket disbursements for labor, materials, and all other expenses are estimated to total $475,000. If the company requires a minimum return of 25% on projects of comparable risk, determine if it should invest in the new product line. Use the internal rate of return method.

5-24 In order to enter the market to produce a new toy for children, a manufacturer will have to make an immediate investment of $60,000 and additional investments of $5,000 at the end of 1 year and $3,000 more at the end of 2 years. Competing toys now are being produced by two large manufacturers. From a fairly extensive study of the market, it is believed that sufficient sales can be achieved to produce year-end, before-tax, net cash flows of:

Year	Cash Flow	Year	Cash Flow
1	− $10,000	6	+ $26,000
2	+ 5,000	7	+ 26,000
3	+ 5,000	8	+ 20,000
4	+ 20,000	9	+ 15,000
5	+ 21,000	10	+ 7,000

In addition, while it is believed that after 10 years the demand for the toy will no longer be sufficient to justify production, it is estimated that the physical assets would have a scrap value of about $8,000. If capital is worth not less than 12% before taxes, would you recommend undertaking the project? Make a recommendation with each of these methods: (a) future worth, and (b) discounted payback with the minimum acceptable payback of 5 years.

5-25 A prospective project requires an investment of $15,000, has an expected useful life of 10 years, and an expected salvage value of $4,000. The project should result in eliminating an existing out-of-pocket disbursement of $11,000 per year, but will require new out-of-pocket disbursements of $7,500 per year for maintenance, upkeep, and so on. If the firm expects to earn at least 15% nominal interest, compounded continuously, on projects of this nature, determine if the project is good using the (a) present worth method, and (b) internal rate of return method.

5-26 Insulation for steam pipes is estimated to reduce the fuel bill by 20%. The initial cost of the insulation is $1,500 installed. The fuel bill without the insulation is $1,800 per year. If the insulation is worthless after 8 years of use and a minimum return of 15% is desired, determine if the investment should be made. Use the internal rate of return method.

5-27 Joe Roe is considering establishing a company to produce impellers for water pumps. An investment of $100,000 will be required for the plant and equipment, and $15,000 will be required for working capital. It is expected that the property will last for 15 years, at which time only the working capital part of the investment can be recovered. It is estimated that sales will be $200,000 per year, and that operating expenses will be as follows:

Materials	$40,000 per year
Labor	$70,000 per year
Overhead	$10,000 + 10% of sales per year
Selling expense	$5,000 per year

Joe has a regular job paying $30,000 per year, but he will keep that job even if he establishes this company. If Joe expects to earn at least 15% on his capital, should this investment be made? Use the annual worth method.

5-28 The equipment in one department of a company is operating at only 75% of capacity because of the fact that the painting booths are overloaded. By building and equipping a new painting shed on adjoining land that is owned by the company, at a total cost of $14,000 the output could be stepped up to 100% (4,000 units per year). It is estimated that the useful life of the new paint shed and equipment would be 10 years. The old paint booths could be utilized for 10 more years. To operate at full capacity would require employing 2 more machinists and 2 painters at monthly salaries of $1,200 each. Indirect labor costs would be 10% of direct labor costs. Operation and maintenance costs per month on the new equipment are estimated to be $50 per 1,000 units processed. The cost for materials is $50 per unit. Annual cost for taxes and insurance on the new facilities would be 8% of the first cost. The sales department assures management that the full output of 4,000 units per year can be sold at the present selling price of $200 per unit. One half of the required capital would have to be borrowed at 12%. Equity capital of the company earns an average of 20% before income taxes. What would you recommend?

5-29 A machine that is not equipped with a brake "coasts" 30 seconds after the power is turned off upon completion of each piece, thus preventing removal of the work from the machine. The time per piece, exclusive of this stopping time, is 2 minutes. The machine is used to produce 40,000 pieces per year. The operator receives $6.50 per hour and the machine overhead rate is $4.00 per hour. How much could the company afford to pay for a brake that would reduce the stopping time to 3 seconds, if it would have a life of 5 years? Assume zero salvage value, capital worth 10%, and that repairs and maintenance on the brake would total not over $250 per year.

5-30 A meat-packing company is considering producing a new product, the entire domestic supply of which is now manufactured by three large companies. A new plant would be required that would cost $200,000 and be built on land that now is owned by the company but not used. The plant would have an expected life of 20-years. Annual costs for labor would be $60,000 and for material, $110,000. These would provide for an annual output of 600,000 pounds. This would constitute 25% of the present domestic consumption and would be 80% of the plant capacity. Advertising and other overhead expenses would amount to $50,000 per year. Taxes and insurance would total 3% of the value of the plant. The product is a very stable one and at present is selling for 55 cents per pound. Over a period of years the price has varied by plus or minus 15%.
(a) Would you recommend that the company go ahead on this project? Capital is available and earns not less than 15%.
(b) What is the minimum selling price for the product that would justify the investment?

5-31 The ABC Company recently started producing a new product, in addition to two others it has been producing for several years. One of the major parts of this new product now is being purchased from another company at a cost of $7.00 per unit. One of the officers of the ABC Company believes it would be advisable to purchase the required equipment, at a cost of $7,000, that would permit making this component. This equipment would have a capacity of 7,000 units per year and a useful life of at least 5 years. It could be installed in the existing plant provided a small storage shed were built at a cost of $2,000 to make the necessary floor space available. Material costs would be $1.10 per unit, and direct labor costs $2.40 per unit. Incremental overhead is 50% of direct labor cost. There would also have to be an added annual charge of 2% of the first cost of the storage shed to cover taxes and insurance. The company now is purchasing 4,000 of the parts per year.

(a) If capital is worth 15%, should the part be purchased or made in the plant?

(b) What volume would be required to justify purchasing the equipment? Use the annual worth method.

5-32   A company has the opportunity to take over a redevelopment project in an industrial area of a city. No immediate investment is required, but it must raze the existing buildings over a 4-year period and at the end of the fourth year invest $2,400,000 for new construction. It will collect all revenues and pay all costs for a period of 10 years, at which time the entire project, and the properties thereon, will revert to the city. The net cash flow is estimated to be as follows:

Year End	Net Cash Flow
1	+ $ 500,000
2	+ 300,000
3	+ 100,000
4	− 2,400,000
5	+ 150,000
6	+ 200,000
7	+ 250,000
8	+ 300,000
9	+ 350,000
10	+ 400,000

Tabulate net present worth versus the interest rate and determine whether multiple I.R.R.'s exist. If so, use the E.R.R. method to determine a rate of return when $e = 8\%$.

5-33   The prospective exploration for oil in the outer continental shelf by a small independent drilling company has produced a rather curious pattern of cash flows, as follows:

End of Year	Net Cash Flow
0	− $ 520,000
1–10	+ 200,000
10	− 1,500,000

The $1,500,000 expense at the end of year 10 will be incurred by the company in dismantling the drilling rig.

(a) Over the 10-year period, plot present worth versus the interest rate (i) in an attempt to discover whether multiple rates of return exist.

(b) Based on the projected net cash flows and results in part (a), what would you recommend regarding the pursuit of this project? Customarily, the company expects to earn at least 20% on invested capital before taxes.

5-34 A certain project has net receipts equaling $1,000 now, has costs of $5,000 at the end of the first year, and then earns $6,000 at the end of the second year.
(a) Show that multiple rates of return exist for this problem when using the I.R.R. method ($i' = 100\%, 200\%$).
(b) If an external reinvestment rate of 10% is available, what is the E.R.R. for this project?

CHAPTER 5 APPENDIX A: AIDS FOR CALCULATION OF THE INTERNAL RATE OF RETURN

The computations to determine the internal rate of return, I.R.R., can be rather laborious, particularly if the periodic cash flows do not follow some pattern to which tabled interest factors can be readily applied.

Figure 5-A-1 is a form that can be used to simplify the calculations if the analyst elects *not* to use the computer program in Appendix C. Net cash flows for each year of a project's life are entered in the column headed "CASH FLOW (0% Int. Rate)." These are the present worths at 0%. Each of these cash flows is then multiplied by the factor in the adjacent subcolumn labeled "Factor" [which are $(P/F, i\%, N)$ factors] and the result entered in the next subcolumn headed "Present Worth." These are the present worths at a 10% interest rate. The calculations are repeated for the 20%, 40%, and 60% interest rates as needed and the columns added to obtain the total present worth of disbursements (A), and the total present worth of receipts (B), for each trial interest rate. At each interest rate the total present worth of expenditures is divided by the total present worth of receipts and the result entered in the space designated "Ratio A/B."

The rate of return answer sought is the interest rate at which A equals B, or at which "ratio A/B" equals unity. If one of the interest rates used does not result in the unity ratio, we hope that the unity ratio will be bracketed so that the answer can be interpolated. To provide for ease of interpolation, a chart is provided in Figure 5-A-2.

To illustrate the use of these aids, a sample problem and the associated computations are shown in Table 5-A-1. For the sample problem, an initial investment of $60,000 is required, and an additional investment of $20,000 will be required at the end of the second year. The project would be terminated at the end of 10 years, at which time it is expected that a recovery of $20,000 will be obtained from the salvage value of the assets. The revenues and disbursements are estimated to be as shown for the various years.

The net cash flow amounts in the right-hand column of Table 5-A-1 can then be transferred to the "CASH FLOW (0% Int. Rate)" column of Figure 5-A-1. For computational simplicity, the "zero time" is taken to be the year of the last negative cash flow, which is the end of year 1987. Computations using Figure 5-A-1 result in ratios of present worth of disbursements to present worth of receipts (called "Ratio A/B") of 0.58 for 0% interest, 0.96 for 10% interest, and 1.51 for 20% interest. Hence the interest rate (I.R.R.), at which the ratio

TIMING		CASH FLOW (0% Int. Rate)		10% Int. Rate		20% Int. Rate		40% Int. Rate		60% Int. Rate	
Year	Period	Disbursement		Factor	Present Worth	Factor	Present Worth	Factor	Present Worth	Factor	Present Worth
	5th Yr			1.464		2.073		3.842		6.560	
1982	4th			1.333		1.728		2.744		4.100	
	3rd	$60,000		1.210	$72,000	1.440	$86,400	1.960		2.560	
83	2nd	5,000		1.100	5,500	1.200	6,000	1.400		1.600	
84	1st	13,000		1.000	13,000	1.000	13,000	1.000		1.000	
TOTALS (A)		78,000			90,500		105,400				

(BEFORE)

Cal. Year	Period	Receipt	Factor	Present Worth	Factor	Present Worth	Factor	Present Worth	Factor	Present Worth
85	1st Yr	$18,000	.909	$16,400	.833	$15,100	.714		.624	
86	2nd	22,000	.826	18,200	.694	15,300	.510		.300	
87	3rd	23,000	.751	17,400	.579	13,400	.364		.244	
88	4th	22,000	.683	15,100	.482	10,600	.260		.152	
89	5th	18,000	.621	11,200	.402	7,200	.186		.095	
90	6th	11,000	.565	6,200	.335	3,700	.133		.059	
91	7th	5,000	.513	2,560	.279	1,400	.095		.037	
92	8th	15,000	.466	7,000	.233	3,300	.067		.023	
	9th		.424		.194		.048		.015	
	10th		.385		.162		.035		.009	
	11th		.351		.135		.024		.006	
	12th		.318		.112		.018		.004	
	13th		.290		.094		.013		.002	
	14th		.263		.078		.009		.001	
	15th		.239		.065		.006			
	16th		.217		.054		.005			
	17th		.197		.045		.003			
	18th		.180		.038		.002			
	19th		.164		.031		.002			
	20th		.149		.026		.001			
	21st		.135		.022					
	22nd		.123		.018					
	23rd		.112		.015					
	24th		.102		.013					
	25th		.092		.011					
TOTALS (B)		$134,000		$94,060		$70,000				
RATIO A/B		0.58		0.96		1.51				

(AFTER ZERO TIME)

FIGURE 5-A-1 Tabular Determination of the Internal Rate of Return for a Proposed New Venture.

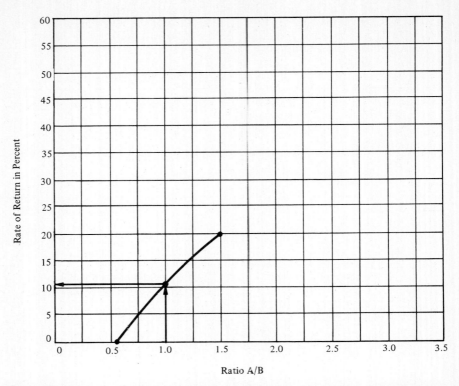

FIGURE 5-A-2 Rate of Return Interpolation Chart with Entries for the Example Project of Figure 5-A-1.

TABLE 5-A-1 Investments, Revenues, and Costs for a Proposed New Venture

End of Year	(A) Investment (−) or Recovery (+) of Capital	(B) Receipts (or Savings) (+)	(C) Disbursements (−)	(D) = (A) + (B) + (C) Net Cash Flow (+) or (−)
1985	− $60,000	—	—	− $60,000
1986	—	$20,000	$25,000	− 5,000
1987	− 20,000	35,000	25,000	− 13,000
1988	—	43,000	30,000	+ 18,000
1989	—	52,000	30,000	+ 22,000
1990	—	55,000	32,000	+ 23,000
1991	—	55,000	33,000	+ 22,000
1992	—	50,000	32,000	+ 18,000
1993	—	35,000	24,000	+ 11,000
1994	—	25,000	20,000	+ 5,000
1995	+ 20,000	10,000	15,000	+ 15,000

is 1.00, is bracketed between 10% and 20%. Linear interpolation gives the answer:

$$\text{I.R.R.} = i' = 10\% + \frac{1.00 - 0.96}{1.51 - 0.96}(20\% - 10\%) = 10.7\%$$

Figure 5-A-2 is a chart for graphical interpolation. To obtain the answer for the example in Table 5-A-1 and Figure 5-A-1, note that the three plotted points form a curvilinear relationship. Such graphical interpolation, if performed on large, accurately scaled paper, can result in greater accuracy than linear interpolation.

CHAPTER 5 APPENDIX B: THE MULTIPLE RATE OF RETURN PROBLEM WITH THE I.R.R. METHOD

Whenever the I.R.R. method is used and the cash flow reverses sign (from net outflow to net inflow or the opposite) more than once over the period of study, one should be alert to the rather remote possibility that either no interest rate or multiple interest rates may exist. Actually, the maximum number of possible rates of return for any given project is equal to the number of cash flow reversals over time. As an example, consider the following project for which the I.R.R. is desired.

EXAMPLE 5-B-1

Year	Net Cash Flow
0	+ $ 500
1	− 1,000
2	0
3	+ 250
4	+ 250
5	+ 250

Solution

Year	Net Cash Flow	P.W. at 35% Factor	P.W. at 35% Amount	P.W. at 63% Factor	P.W. at 63% Amount
0	+ $ 500	1.3500	+ $ 675	1.6300	+ $ 815
1	− 1,000	1.0000	− 1,000	1.0000	− 1,000
2	0				
3	+ 250	0.5487	+ 137	0.3764	+ 94
4	+ 250	0.4064	+ 102	0.2309	+ 58
5	+ 250	0.3011	+ 75	0.1417	+ 35
NET P.W.			Σ = 11		Σ = 2

Thus the present worth of the net cash flows equals zero for interest rates of about 35% and 63%. Whenever multiple answers such as this exist, it is likely that none is correct.

An effective way to overcome this difficulty and obtain a plausible answer is to manipulate cash flows as little as necessary so that there is only one reversal of the net cash flow over time. This can be done by using the M.A.R.R. or an external reinvestment rate to manipulate the funds, and then solving for the I.R.R. by using Equation 5-6 or 5-7. For the example, if the minimum attractive rate of return is 10%, the +$500 at year 0 can be compounded to year 1 to be $500(F/P, 10\%, 1) = +\$550$. This, added to the $-\$1,000$ at year 1, equals $-\$450$. The $-\$450$, together with the remaining cash flows, which are all positive, now fit the condition that there be only one reversal in the net cash flow over time. The return on invested capital at which the present worth of the net cash flows equals 0 can now be shown to be 19%, per the following table:

Year	Net Cash Flow	P.W. at 19% Factor	P.W. at 19% Amount
1	$-\$450$	1.0000	$-\$450$
3	$+\ 250$	0.7062	$+\ \ 177$
4	$+\ 250$	0.5934	$+\ \ 148$
5	$+\ 250$	0.4987	$+\ \ 125$
NET P.W.			$\Sigma = 0$

It should be noted that whenever a manipulation of net cash flows such as the above is done, the calculated rate of return will vary according to what cash flows are manipulated and at what interest rate. The less the manipulation and the closer the minimum rate of return to the calculated rate of return, the less the variation in the final calculated rate of return.

Probably the most straightforward way to overcome this multiple rate of return problem in general is to use the E.R.R. method. For Example 5-B-1, the E.R.R. equals 13% when the reinvestment rate (e) is 10%:

$$\$1000(F/P, i', 4) = 500(F/P, 10\%, 5) + 250(F/A, 10\%, 3)$$

$$(F/P, i', 4) = 1.632$$

$$i' = 0.13$$

This differs from the 19% return in the table above because the E.R.R. method makes the assumption that all positive cash flows are reinvested at 10%. The 19% return on invested capital assumes that funds can be reinvested at 19%.

EXAMPLE 5-B-2

Use the E.R.R. method to analyze the cash flow pattern shown on page 173. The internal rate of return is indeterminant (none exists) so that the I.R.R. is not a workable procedure. The external reinvestment rate is 12% and M.A.R.R. equals 15%.

Year	Net Cash Flow
0	+ $5000
1	− 7000
2	+ 2000
3	+ 2000

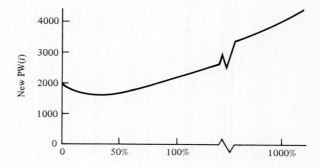

Solution

The E.R.R. method provides these results:

$$\$7,000(F/P, i', 2) = \$5,000(F/P, 12\%, 3) + \$2,000(F/P, 12\%, 1)$$
$$+ \$2,000$$
$$(F/P, i', 2) = 1.61$$
$$i' = 26.9\%$$

Thus the E.R.R. is greater than the M.A.R.R. Apparently, the project having this cash flow pattern would be acceptable. The P.W. at 15% is equal to $1,740.36.

CHAPTER 6

Selections Among Alternatives

Most engineering and business endeavors can be accomplished by more than one method or alternative. When the selection of one alternative excludes the choice of any other alternative being considered, the alternatives can be classified as *mutually exclusive*. Typically, the alternatives available require the investment of different amounts of capital, and the out-of-pocket costs (disbursements) usually will be different. Sometimes the alternatives may produce different revenues, and frequently the useful lives of the projects will be different. Because in such situations different levels of investment produce different economic outcomes, we must perform a study to determine how much capital should be invested.

Economy studies of mutually exclusive alternatives are the principal topic of this chapter. Studies of independent alternatives (opportunities) as well as sets of mutually exclusive, independent, and contingent projects are considered briefly in the last section of this chapter. All of the basic methods of Chapter 5 are applicable to these studies.

A BASIC PHILOSOPHY FOR STUDIES OF ALTERNATIVES

The decision regarding selection among several competing investment alternatives must take into account the fundamental purpose of a capital investment, which is to obtain at least the minimum acceptable return from *each* dollar invested. In actual practice, there are usually a limited number of investment alternatives to consider at one time so that the return associated with each increment, rather than each unit, of capital must be evaluated. The problem of deciding which mutually exclusive alternative should be selected is made easier if we adopt a simple policy: *The alternative that requires the minimum investment of capital and produces satisfactory functional results will always be chosen unless the incremental first cost associated with an alternative having a larger investment can be justified with respect to its incremental savings (or benefits).* Under this policy we consider the alternative that requires the least investment of capital to be the "baseline" alternative. Sometimes the baseline alternative is to "do nothing," or maintain the status quo. The next most capital intensive alternative would then be compared with it. By successive comparisons of the incremental investment required by the various alternatives, our policy would dictate that additional amounts of capital would *not* be justified unless the additional savings, or benefits, produce (for example) a sufficiently high rate of return on the added capital.

The investment of additional amounts of capital over the baseline alternative usually results in increased capacity, increased revenue, decreased operating expenses, or increased life. Before extra money is invested, it must be shown that each avoidable increment of capital can "pay its own way" relative to other available opportunities. In summary, if the extra return obtained by investing additional capital is better than could be obtained from investment of the same additional capital elsewhere, the investment probably should be made. If this is not the case we obviously would not invest more than the minimum amount required, including the possibility of nothing at all.

The philosophy above can be reduced to the following principles when comparing mutually exclusive alternatives by any rate of return (R.R.) method. Here the rate of return method can be either the I.R.R., E.R.R., or E.R.R.R. method.

1. Each increment of capital must justify itself by producing a sufficient R.R. on that increment.
2. Compare a higher investment alternative against a lower investment alternative only when the lower investment alternative is justified.
3. Choose the alternative that requires the largest investment *for which its increment of investment capital is justified* by savings in view of the minimum attractive rate of return.

This procedure assumes that the firm wants to invest capital as long as a sufficient rate of return is earned on each increment of capital. In general, a *sufficient rate of return* is one that is greater than or equal to the minimum attractive rate of return (M.A.R.R.).

When alternatives are compared by using equivalent worth methods (such as P.W.), the choices are completely consistent with those made when using rate of return methods. Thus the maximization of P.W., A.W., and F.W. leads to the identical choice among mutually exclusive alternatives as that resulting from the procedure described above for rate of return methods. This is shown to be true in examples that follow.

The aforementioned principles for comparing mutually exclusive alternatives

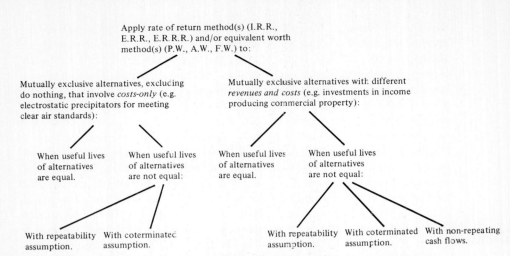

Apply rate of return method(s) (I.R.R., E.R.R., E.R.R.R.) and/or equivalent worth method(s) (P.W., A.W., F.W.) to:

Mutually exclusive alternatives, excluding do nothing, that involve *costs-only* (e.g. electrostatic precipitators for meeting clear air standards):

Mutually exclusive alternatives with different *revenues and costs* (e.g. investments in income producing commercial property):

When useful lives of alternatives are equal.

When useful lives of alternatives are not equal:

When useful lives of alternatives are equal.

When useful lives of alternatives are not equal:

With repeatability assumption.

With coterminated assumption.

With repeatability assumption.

With coterminated assumption.

With non-repeating cash flows.

FIGURE 6-1 Treatment of Mutually Exclusive Alternatives in Chapter 6.

with rate of return and equivalent worth methods are applied to a wide assortment of problems in this chapter. These methods provide consistent rank orderings of alternatives and conform to accepted business practices for evaluating capital investments. As mentioned above, alternatives may involve only costs or they may have different revenues and costs associated with them. Moreover, their useful lives may be identical or the alternatives could have dissimilar lives. When unequal lives are present, two distinctly different assumptions (repeatability and coterminated lives) are utilized to deal with this situation. An overview of the treatment of mutually exclusive alternatives in this chapter is shown in Figure 6-1.

ALTERNATIVES HAVING IDENTICAL (OR NO KNOWN) REVENUES AND LIVES

A good many economy studies are made of situations in which revenues from the various alternatives do not exist or may be assumed to be identical, and the useful lives are also the same. Such conditions lead to the simplest type of investment study among mutually exclusive alternatives, and it is illustrated by the left-hand side of Figure 6-1. Each of the six methods introduced in Chapter 5 is utilized in Example 6-1 for this frequently encountered type of problem.

EXAMPLE 6-1

A company is going to install a new plastic-molding press. Four different presses are available. The essential differences in first cost and operating expenditures are shown below.

	Press			
	A	B	C	D
Investment (installed)	$6,000	$7,600	$12,400	$13,000
Useful life	5 years	5 years	5 years	5 years
Annual disbursements				
Power	$ 680	$ 680	$1,200	$1,260
Labor	6,600	6,000	4,200	3,700
Maintenance	400	450	650	500
Property taxes and insurance	120	152	248	260
TOTAL ANNUAL DISBURSEMENTS	$7,800	$7,282	$6,298	$5,720

Each press will produce the same number of units. However, because of different degrees of mechanization, some require different amounts and grades of labor and have different operation and maintenance costs. None is expected to have a salvage value. Any capital invested in this company is expected to earn at least 10% before taxes. Which press should be chosen?

Solution of Example 6-1 by the Present Worth–Cost (P.W.-C.) Method

When alternatives for which revenues are not known are compared using the present worth method, that method is more commonly and descriptively called the present worth–cost (P.W.-C.) method. The alternative that has the minimum (least negative) P.W.-C. is judged to be the most desirable. Table 6-1 shows the analysis by the P.W.-C. method. The economic criterion is to choose that alternative with the minimum P.W.-C., which is press D. The order of desirability among the alternatives in *decreasing* order is press D, press B, press A, and press C. This rank ordering is identical for all methods considered when they are correctly applied. ■

TABLE 6-1 Comparison of Four Molding Presses by P.W.-C. Method

	Press			
	A	B	C	D
Present worth of:				
Investment	$ 6,000	$ 7,600	$12,400	$13,000
Disbursements:				
(annual disbursements)				
× (P/A, 10%, 5)	29,568	27,605	23,874	21,683
TOTAL P.W.-C.	$35,568	$35,205	$36,274	$34,683

Solution of Example 6-1 by the Annual Cost (A.C.) Method

Table 6-2 shows the analysis by the A.C. method. The economic criterion is to choose that alternative with the minimum A.C., which is press D. ■

TABLE 6-2 Comparison of Four Molding Presses by A.C. Method

	Press			
	A	B	C	D
Annual expenses:				
Disbursements	$7,800	$7,282	$6,298	$5,720
Capital recovery (depreciation + interest)				
= (investment) × (A/P, 10%, 5)	1,583	2,005	3,271	3,429
TOTAL A.C.	$9,383	$9,287	$9,569	$9,149

Solution of Example 6-1
by the Future Worth–Cost (F.W.-C.) Method

Table 6-3 shows the analysis by the F.W.-C. method. Once again, press D is shown to be the least costly, and the order of desirability of the four alternatives is the same as for the P.W. and A.W. methods. ∎

Solution of Example 6-1
by the Internal Rate of Return (I.R.R.) Method

Table 6-4 provides a tabulation of cash flows and a calculated I.R.R. for each *increment* of investment considered. *In this table and in subsequent studies of alternatives by a rate of return (R.R.) method, the symbol "Δ" is used to mean "incremental" or "change in," and the letters on the end of arrows indicate the projects for which the increment is considered. Thus the symbol A → B, for example, denotes "A is compared incrementally with B."*

Note that the alternatives are ordered according to increasing amounts of investment to facilitate step-by-step consideration of each increment of investment. Furthermore, an internal rate of return cannot be calculated for each individual press because all cash flows are negative.

The first increment subject to analysis is the $7,600 − $6,000 = $1,600 extra investment required for press B compared to press A. For this increment, annual disbursements are reduced or saved by $7,800 − $7,282 = $518. The I.R.R. on the incremental investment is the interest rate at which the present worth of the incremental net cash flows is zero. Thus

$$-\$1,600 + \$518(P/A, i', 5) = 0$$

$$(P/A, i', 5) = \$1,600/\$518 = 3.10$$

TABLE 6-3 Comparison of Four Molding Presses by F.W.-C. Method

	Press			
	A	B	C	D
Future worth of:				
Investment:				
(investment) × (F/P, 10%, 5)	$ 9,663	$12,240	$19,970	$20,937
Disbursements:				
(annual disbursements)				
× (F/A, 10%, 5)	47,620	44,457	38,450	34,921
TOTAL F.W.-C.	$57,283	$56,697	$58,420	$55,858

TABLE 6-4 Comparison of Four Molding Presses by I.R.R. Method

	Press			
	A	B	C	D
Investment	$6,000	$7,600	$12,400	$13,000
Annual disbursements	$7,800	$7,282	$6,298	$5,720
Useful life	5 years	5 years	5 years	5 years
Increment considered:		A → B	B → C	B → D
Δ Investment		$1,600	$4,800	$5,400
Δ Annual disbursements (savings)		$518	$984	$1,564
I.R.R. on Δ investment		18.6%	0.8%	13.8%
Is increment justified?		Yes	No	Yes

Interpolating between the tabled factors for the two closest interest rates, one can find that i' = I.R.R. = 18.6%. Since 18.6% > 10% (the M.A.R.R.) the increment A → B is justified.

The next increment subject to analysis is the $12,400 − $7,600 = $4,800 extra investment required for press C compared to press B. This increment results in annual disbursement reduction of $7,282 − $6,298 = $984. The I.R.R. on increment B → C can then be determined by finding the i' at which

$$-\$4,800 + \$984(P/A, i', 5) = 0$$

Thus i' = I.R.R. can be found to be approximately 0.8%. [*Note:* $(P/A, 0\%, 5) = 5.00$.] Since 0.8% < 10%, the increment B → C is *not* justified, and hence we can say that press C itself is not justified.

The next increment that should be analyzed is B → D, and not C → D. This is because press C has already been shown to be unsatisfactory, and hence it can no longer be a valid basis for comparison with other alternatives.† For B → D, the incremental investment is $13,000 − $7,600 = $5,400 and the incremental saving in annual disbursements is $7,282 − $5,720 = $1,564. The I.R.R. on increment B → D can then be determined by finding the i' at which

$$-\$5,400 + \$1,564(P/A, i', 5) = 0$$

Thus i' can be found to be 13.8%. Since 13.8% > 10%, the increment B → D is justified.

Based on the preceding analysis, press D would be the choice because it is the highest investment for which each increment of investment capital is justified. In choosing press D, one is actually accepting the first $6,000 investment as necessary without justification, the $1,600 increment A → B earning 18.6%, and the $5,400 increment B → D earning 13.8%.

Figure 6-2 depicts these results for all separable increments that comprise the total investment of $13,000 in press D. Note that this choice is reached without calculation of the I.R.R. on the total investment for press D or *any* of the individual presses. ■

†If the I.R.R. were computed for increment C → D, it would be in excess of 90%. This only means that press D is *very* attractive compared to press C, but since press C is not justified, this is not a valid basis for comparison.

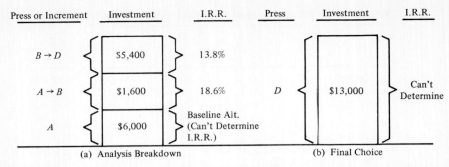

FIGURE 6-2 Representation of Increments and I.R.R. on Increments Considered in Justifying Press D in Example 6-1. (a) Analysis Breakdown, and (b) Final Choice.

The underlying rationale for the type of analysis above is that the firm wants to invest capital if and only if it is necessary or will earn at least the M.A.R.R., which in this case is 10%. Thus the firm supposedly has opportunities to invest elsewhere any capital not used for one of the presses in comparable risk projects where at least 10% per year can be earned.

Three errors commonly made in the type of analysis above are to choose the alternative (1) with the highest overall R.R. or (2) with the highest R.R. on an incremental investment or (3) with the largest investment that has an R.R. greater than or equal the M.A.R.R. None of these criteria is correct generally. For example, in the problem above, one might choose press B rather than press D because the I.R.R. for increment A → B is 18.6% and for increment B → D is 13.8%. However, the choice of press B means that one is forgoing the opportunity to invest an extra $5,400 to obtain D (i.e., investment in B → D) and thereby make 13.8% on that $5,400. If 10% is the return obtainable from alternative use of the money, clearly one should invest in D rather than B.

Solution of Example 6-1
by the External Rate of Return (E.R.R.) Method
The rationale and criteria for using the E.R.R. method to compare alternatives are the same as for the I.R.R. method. The only difference is in the calculation methodology. Table 6-5 provides a tabulation of the calculation and acceptability

TABLE 6-5 Comparison of Four Molding Presses by E.R.R. Method

	Press			
	A	B	C	D
Investment	$6,000	$7,600	$12,400	$13,000
F.W. of annual disbursements: (annual disbursements) × (F/A, 10%, 5)	$47,620	$44,457	$38,450	$34,921
Increment considered		A → B	B → C	B → D
Δ Investment		$1,600	$4,800	$5,400
Δ F.W. of annual disbursements (savings)		$3,163	$6,007	$9,536
E.R.R. on Δ investment		14.6%	4.6%	12.0%
Is increment justified?		Yes	No	Yes

of each increment of investment considered. The annual disbursements are compounded to the end of the 5-year life at e = M.A.R.R. = 10%.

To illustrate the procedure for finding the E.R.R. on a Δ investment, consider increment A $\rightarrow$ B. The formula is

$$1,600(F/P, i'\%, 5) = \$518(F/A, 10\%, 5)$$

$$(F/P, i'\%, 5) = [\$518(6.1051)]/1,600 = 1.977$$

Linear interpolation between 12% and 15% results in i' = 14.6%, which is the E.R.R. Using the M.A.R.R. = 10%, increment A $\rightarrow$ B is justified. Using similar calculations, increment B $\rightarrow$ C, earning 4.6%, is not justified and increment B $\rightarrow$ D, earning 12.0%, is justified. Hence press D is again, as anticipated, the choice. ∎

Solution of Example 6-1
by the Explicit Reinvestment Rate of Return (E.R.R.R.) Method

The rationale and criteria for using the E.R.R.R. method to compare alternatives are the same as for the I.R.R. method. The only difference is in the calculation methodology. Table 6-6 provides a tabulation of the calculation and acceptability of each increment of investment considered. Keeping in mind that the M.A.R.R. is 10%, increment A $\rightarrow$ B earning an E.R.R.R. of 16% is justified, increment B $\rightarrow$ C earning 4.1% is not justified, and increment B $\rightarrow$ D earning 12.5% is justified. Hence press D would be the choice, which of course is the same as when using the other methods.

To find the E.R.R.R. on the incremental investment for any two alternatives, one need only divide the incremental total annual expenses (including depreciation using the sinking fund method) by the incremental investment for those two alternatives. As an example, for increment A $\rightarrow$ B in Table 6-6, the incremental total annual expense is $8,783 - $8,527 = $256, and the incremental investment is $7,600 - $6,000 = $1,600. Hence the E.R.R.R. on increment A $\rightarrow$ B is $256/$1,600 = 16%. One can similarly calculate the E.R.R.R. on all other applicable incremental investments. ∎

TABLE 6-6   Comparison of Four Molding Presses by E.R.R.R. Method

	Press			
	A	B	C	D
Investment	$6,000	$7,600	$12,400	$13,000
Annual expenses				
Disbursements	7,800	7,282	6,298	5,720
Depreciation:				
(investment) $\times$ $(A/F, 10\%, 5)$	983	1,245	2,031	2,129
TOTAL ANNUAL EXPENSES	$8,783	$8,527	$ 8,329	$ 7,849
Increment considered		A $\rightarrow$ B	B $\rightarrow$ C	B $\rightarrow$ D
Δ Investment		$1,600	$4,800	$5,400
Δ Total annual expenses		$256	$198	$678
E.R.R.R. on Δ investment		16%	4.1%	12.5%
Is increment justified?		Yes	No	Yes

Summary Comparison of Economy Study Methods
Based on Example 6-1 Results

In Chapter 5, the section "Summary Comparison of Economic Study Methods" stated that all six methods will give a valid answer for the reinvestment assumption inherent to that method. It further stated that the equivalent present, annual, and future worth methods give the same ranking of alternatives, because all assume that funds can be reinvested at the same rate, which is normally the M.A.R.R. The equivalent worth results for Example 6-1 were as follows:

	Press				
	A	B	C	D	
Present worth–cost	$35,568	$35,205	$36,274	$34,683	⎫
Annual cost	9,383	9,287	9,569	9,149	⎬ Select D
Future worth–cost	57,283	56,697	58,420	55,858	⎭

In general, the ratio of equivalent worth results for any two alternatives, say press A and press B is a constant:

$$\frac{\text{P.W.-C}_A}{\text{P.W.-C}_B} = \frac{\text{A.C.}_A}{\text{A.C.}_B} = \frac{\text{F.W.-C}_A}{\text{F.W.-C}_B} = 1.0103$$

Further, for any alternative,

$$\text{A.C.} = \text{P.W.-C.}(A/P, i\%, N) = \text{F.W.-C.}(A/F, i, N)$$

and

$$\text{P.W.-C.} = \text{F.W.-C.}(P/F, i\%, N)$$

The rate of return methods always give consistent results regarding which investments (or incremental investments) will earn more or less than the M.A.R.R.[†] However, the calculated rates of return will differ somewhat between the methods because of their different reinvestment assumptions. The results for Example 6-1, in which M.A.R.R. = 10%, were as follows:

	Increment Considered (%)			
Method	A → B	B → C	B → D	
I.R.R.	18.6	0.8	13.8	⎫
E.R.R.	14.6	4.6	12.0	⎬ Select D
E.R.R.R.	16.0	4.1	12.5	⎭

[†]This statement is based on the frequent assumption for the E.R.R. and E.R.R.R. methods that e = M.A.R.R.

ALTERNATIVES HAVING IDENTICAL REVENUES AND DIFFERENT LIVES

In many cases an alternative that requires a larger investment of capital than another will have not only lower out-of-pocket costs but also a longer useful life. Figure 6-1 summarizes this possibility. Unequal lives somewhat complicate the analysis of alternative investments. To make economy studies of such cases, we must adopt some procedure that will put the alternatives on a comparable basis. Two types of assumptions commonly are employed: (1) the repeatability assumption and (2) the coterminated assumption.

The *repeatability assumption* involves two main conditions:

1. The period of needed service for which the alternatives are being compared is either indefinitely long or a length of time equal to a common multiple of the lives of the alternatives.
2. What is estimated to happen in an alternative's initial life span will happen also in all succeeding life spans (replacements), if any, for each alternative.

The repeatability assumption is usually made in economic analyses by default (i.e., because there is no good basis for estimates to the contrary).

Whenever the alternatives to be compared have different lives and one or both of the above conditions is *not* appropriate, it is necessary to enumerate on a cash flow diagram what disbursements are expected for each alternative for as long as service is needed. This information can then be used to make the economy study.

The *coterminated assumption* involves the use of a finite study period for all alternatives. This time span may be the period of needed service or any arbitrarily specified length of time such as (1) the life of the shorter-lived alternative, or (2) the life of the longer-lived alternative, or (3) what is sometimes descriptively called the "organization's planning horizon."

Figure 6-3 shows the effect of these two assumptions for structures M and N, which have lives of 10 and 25, years, respectively, and which are the subject of Example 6-2. The objective is to choose either M or N because here the "do nothing" option does not exist.

COMPARISONS USING THE REPEATABILITY ASSUMPTION

Consider the following example for which economic analyses will be made using different economy study methods and the repeatability assumption.

EXAMPLE 6-2

A selection is to be made between two structural designs. Because revenues do not exist (or can be assumed to be equal), only negative cash flows are shown on the following page.

Repeatability Assumption: (For lowest common multiple of lives, 50 years)

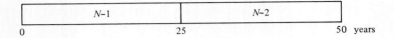

M-1	M-2	M-3	M-4	M-5

0   10   20   30   40   50 years

N-1	N-2

0   25   50 years

Coterminated Assumption

(A) At the life of the shorter–lived alternative:

M-1

0

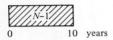

N-1

0   10 years

(B) At the life of the longer–lived alternative:

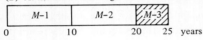

M-1	M-2	M-3

0   10   20   25 years

N-1

0   25 years

Key:

M-1   Life span 1 for structure *M*, etc.

☐   Cash flows same as for first life span.

▨   Cash flows presumably different from first (or full)
     life span because of early termination.

FIGURE 6-3   Repeatability Versus Coterminated Assumptions.

	Structure M	Structure N
First cost	$12,000	$40,000
Useful life	10 years	25 years
Salvage value at end of life	0	$10,000
Annual disbursements for operation and maintenance	$2,200	$1,000

Determine which structure is better if the M.A.R.R. is 15% using the repeatability assumption.

Solution of Example 6-2 by the P.W.-C. Method

A basic principle in comparing alternatives by the P.W.-C. method is that all alternatives should be compared over the *same length of time* (study period). By use of the repeatability assumption, this is most conveniently chosen to be the lowest common multiple of the lives for the two alternatives. In this case 50 years of service can be provided by structure M and replacing it four times— a total of five structures. The same length of service can be obtained by structure N and replacing it once—a total of two such structures.

Present Worth of Costs	Structure M
Original investment	$12,000
First replacement: $12,000($P/F$, 15%, 10)	2,966
Second replacement: $12,000($P/F$, 15%, 20)	733
Third replacement: $12,000($P/F$, 15%, 30)	181
Fourth replacement: $12,000($P/F$, 15%, 40)	44
Annual disbursements: $2,200($P/A$, 15%, 50)	14,653
TOTAL P.W.-C.	30,577

Present Worth of Costs	Structure N
Original investment	$40,000
First replacement: ($40,000 − $10,000)($P/F$, 15%, 25)	911
Annual disbursement: $1,000($P/A$, 15%, 50)	6,661
Less salvage of last replacement: $10,000($P/F$, 15%, 50)	− 9
TOTAL P.W.-C.	$47,563

Thus the P.W.-C. for structure M is lower, indicating that M is better. ■

With the repeatability assumption, the annual equivalent cash flow over each life span in the common multiple of years is identical for a particular alternative. This allows structure M and N to be compared with minimal effort as shown below.

Solution of Example 6-2 by the A.C. Method

	Structure M	Structure N
Annual costs:		
Disbursements	$2,200	$1,000
C.R. cost:		
$12,000($A/P$, 15%, 10)	2,391	
($40,000 − $10,000)($A/P$, 15%, 25)		
+ $10,000(15%)		6,141
TOTAL A.C.	$4,591	$7,141

Since $4,591 is less than $7,141, structure M is again shown to be the better economic choice. Notice that with the A.C. method, repeatability permits alternatives to be compared over their different life spans, thus reducing the amount of computational effort required. ■

Solution of Example 6-2 by the I.R.R. Method

The I.R.R. on the incremental investment of structure N over the baseline (least first cost) alternative, which here is structure M, is required. This I.R.R. on the increment can be computed by either of the following approaches:[†]

1. Finding the interest rate at which the equivalent worth (P.W., A.W., F.W.) of the net cash flow of the differences between the two alternatives is equal to zero, or
2. Finding the interest rate at which the equivalent worths (or equivalent costs) of the two alternatives are equal.

The first approach based on present worth was used in solving Example 6-1 by the I.R.R. method. We illustrate the second approach using annual worth for this example.

The A.C.'s for the two structures are equated at the unknown i' as follows:

$$\$12,000(A/P, i', 10) + \$2200 = \$40,000(A/P, i', 25)$$
$$- \$10,000(A/F, i', 25) + \$1000$$

By trial and error, it can be determined that this equality is met when $i' = 5.5\%$. Therefore, the I.R.R. on the M $\rightarrow$ N increment is 5.5%. Because 5.5% is much less than the M.A.R.R., the incremental investment required by structure N cannot be justified and the choice is, as expected, structure M. ∎

Solution of Example 6-2 by the E.R.R.R. Method with $e = 15\%$

	Structure M	Structure N
Annual expenses:		
Disbursements	$2,200	$1,000
Depreciation:		
$12,000(A/F, 15\%, 10)$	591	
$(\$40,000 - \$10,000)(A/F, 15\%, 25)$		141
TOTAL ANNUAL EXPENSES	$2,791	$1,141

Δ Annual expenses (annual savings): $2,791 − $1,141 = $1,650
Δ Investment: $40,000 − $12,000 = $28,000
E.R.R.R. on Δ Investment: $1,650/$28,000 = 5.9%

Since 5.9% < 15%, structure M is again shown to be the more desirable alternative. ∎

COMPARISONS USING THE COTERMINATED ASSUMPTION

If the period of needed service is less than a common multiple of the lives, that should be the *study period* used for purposes of an engineering economy study. Even if the period of needed service is not known, it is often thought convenient to use some arbitrarily specified study period for both alternatives. In such cases, a sometimes-important concern involves the salvage value to be assigned to any

†When the cash flow difference between alternatives changes sign more than once, the I.R.R. method can result in two or more rates of return, as demonstrated in Appendix 5-B, p. 171.

alternative that will not have reached its useful life at the end of the study period.

A cotermination point commonly used in these situations is the life of the shortest-lived alternative. Example 6-3 illustrates the coterminated assumption.

EXAMPLE 6-3

Suppose that we are faced with the same alternatives as in Example 6-2, as follows:

	Structure M	Structure N
First cost	$12,000	$40,000
Useful life	10 years	25 years
Salvage value at end of life	0	$10,000
Annual disbursements	$2,200	$1,000
M.A.R.R.	15%	15%

It is desired to compare the economics of the alternatives by terminating the study period at the end of 10 years and assuming a salvage (remaining) value for structure N at that time of, say, $25,000.

Solution of Example 6-3 by the A.C. Method

	Structure M	Structure N
Annual costs:		
Disbursements	$2,200	$1,000
C.R. cost:		
$12,000(A/P, 15%, 10)	2,391	
($40,000 − $25,000)(A/P, 15%, 10)		
+ $25,000(15%)		6,739
TOTAL A.C.	$4,591	$7,739

Structure M still has the lower annual cost (this result does not have to agree with the decision for Example 6-2 using the repeatability assumption). ■

Solution of Example 6-3 by the E.R.R. Method with e = 15%

	Structure M	Structure N
Investment	$12,000	$40,000
F.W. of annual disbursements (annual disbursements)		
× (F/A, 15%, 10)	44,668	20,304
F.W. of salvage value:		25,000
Increment considered		
Δ Investment		28,000
Δ F.W. of savings		49,364
E.R.R. on Δ investment		5.9%

Since 5.9% < 15%, structure M is again the economic choice. ∎

It should be expected that different assumptions for determining the salvage value of the longer-lived alternative *could* result in different decisions. This did not happen in Example 6-3 because the choice is insensitive to salvage value estimates of structure N. In fact, a salvage value of $40,000 for structure *N* at the end of year 10 would not reverse the preference for structure M!

Of course, in many situations it is not reasonable to expect cash flows to remain unchanged for life spans after the first. Inflation, to be discussed in Chapter 8, could cause subsequent life spans to experience different cash flows. In such situations the P.W. or F.W. method is usually easiest to use for comparison of alternatives after enumerating the magnitude and timing of expected cash flows over a common (identical) study period.

EXAMPLE 6-4

Suppose that we are faced with the same alternatives as in Example 6-2 except now we want to terminate the study period at the end of 25 years, either because the firm thinks the services of a structure will be needed only for that time, or because it arbitrarily wants to use that length of study period. Further, the firm estimates that the same first cost and annual disbursements for structure M will *not* be repeated after the first life cycle. Rather, it estimates that the first cost will be $28,000 for the first replacement and $36,000 for the second replacement; and that the annual disbursements will be $3,500 for both the first and second replacements. Since the third replacement will be terminated after only 5 years of service, it is estimated to have a salvage value then of $15,000. Estimates for structure N are the same as given for Example 6-2, and the M.A.R.R. = 15%. Figure 6-4 shows a cash flow diagram for both alternatives.

Solution of Example 6-4 by the P.W.-C. Method

Structure M:
Original investment	$12,000
First replacement: $28.000(P/F, 15%, 10)	6,921
Second replacement: $36,000(P/F, 15%, 20)	2,200
Annual disbursements:	
Years 1–10: $2,200(P/A, 15%, 10)	11,041
Years 11–25: $3,500(P/A, 15%, 15)(P/F, 15%, 10)	5,059
Less: salvage on second replacement: $15,000(P/F, 15%, 25)	− 456
TOTAL P.W.-C.	$36,765

Structure N:
Original investment	$40,000
Annual disbursements: $1,000(P/A, 15%, 25)	6,464
Less: salvage on original investment: $10,000(P/A, 15%, 25)	− 304
TOTAL P.W.-C.	$46,160

Thus, even with the changed conditions, structure M appears to be considerably more economical than structure N. With relatively high interest rates, "front-end" capital intensive alternatives such as structure N are difficult to justify

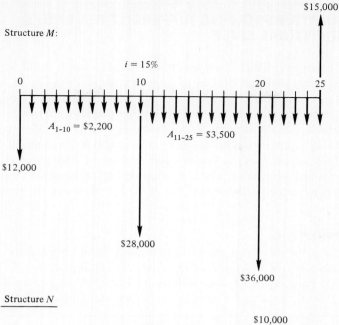

FIGURE 6-4 Cash Flow Diagram for Example 6-4.

even when the "downstream" replacement costs of *competing* alternatives are large. Thus the tendency is to choose an alternative with a low first cost. ∎

Solution of Example 6-4 by A.C. and F.W.-C. Methods

The easiest way to compare the annual costs and the future worth of costs for the alternatives in Example 6-4 is to determine these quantities directly from the equivalent present worth–cost results obtained above. Because the study period is 25 years, the annual costs are spread over that period and the future worths of costs occur at the end of that period. Thus

$$A.C. = P.W.-C. \times (A/P, 15\%, 25)$$

$$\text{For structure M: } = \$36,765(0.1547) = \$5,688$$

$$\text{For structure N: } = \$46,160(0.1547) = \$7,141$$

$$F.W.-C. = P.W.-C. \times (F/P, 15\%, 25)$$

$$\text{For structure M: } = \$36,765(32.9190) = \$1,210,265$$

$$\text{For structure N: } = \$46,160(32.9190) = \$1,519,541 \quad ∎$$

ALTERNATIVES HAVING DIFFERENT REVENUES AND IDENTICAL LIVES

It is to be expected that many investment situations will result in different revenues being produced by two or more alternatives. Any of the six basic methods of comparison presented in Chapter 5 may generally be used for such economy studies without difficulty, if their respective reinvestment assumptions are applicable. This situation is shown by the right-hand side of Figure 6-1.

In general, a firm should be willing to invest additional amounts of capital as long as each increment is justified by a sufficient return. This criterion of choice is made explicit when using a rate of return method and implicitly exists when using an equivalent worth method and is further illustrated in the following example.

EXAMPLE 6-5

Suppose that we have been requested to evaluate these six mutually exclusive alternatives with the P.W. and I.R.R. methods. The useful life of each alternative is 10 years, salvage value equals the initial investment, and the M.A.R.R. is 16%. Also, net annual revenues less expenses varies among all alternatives.

	B1	B2	C	D	E	F
Investment (first cost)	$900	$1,500	$2,500	$4,000	$5,000	$7,000
Annual revenues less expenses	150	276	400	925	1,125	1,425
Salvage value	900	1,500	2,500	4,000	5,000	7,000

Solution of Example 6-5 by the P.W. Method

These six investments have estimated salvage values that equal first costs. Such a situation may exist, for instance, when alternatives involve the purchase and rent of farmland. Table 6-7 shows the analysis of this problem with the present worth method. The alternative that maximizes net P.W. is E, and this is the recommended choice. The least desirable, though marginally acceptable, alternative is C. ∎

TABLE 6-7 Comparison of Six Alternatives by the P.W. Method

	Alternative					
	B1	B2	C	D	E	F
Present worth of: Investment	$900	$1,500	$2,500	$4,000	$5,000	$7,000
Annual revenues less expenses $\times$ (P/A, 16%, 10)	725	1,334	1,933	4,471	5,437	6,887
Salvage value $\times$ (P/F, 16%, 10)	204	340	567	907	1,133	1,587
Net P.W.	+$ 29	+$ 174	$ 0	+$1,378	+$1,570	+$1,474

Solution of Example 6-5 by the I.R.R. Method

When the first cost (P) equals the salvage value (S) as in this problem, the I.R.R. can be calculated directly by dividing the net annual revenues less expenses (A) by the investment amount. In this regard, the following must be true when an i' = I.R.R. sets the net A.W. (for example) equal to zero.

$$0 = A - [(P - S)(A/P, i', N) + Si']$$

Because $P = S$, then $i'S = A$ and $i' = A \div S = A \div P$.

The reader will recall that the symbol "Δ" is used to mean "incremental" or "change in," and that the letters on the ends of an arrow designate the projects for which the increment is considered. Table 6-8 provides the analysis of the six alternatives in view of the "shortcut" given above for determining an I.R.R. when $P = S$ and lives are equal.

From the analysis in Table 6-8, it is apparent that alternative E would be chosen because it requires the largest investment for which the last increment of investment capital is justified. That is, we desire to invest as much of the $7,000 presumably available for this purpose as long as each avoidable increment can earn 16% or better.

It is assumed in Example 6-5 (and all other examples involving mutually exclusive alternatives, unless noted to the contrary) that capital *not* committed to one of the alternatives is invested in some other opportunity where it will earn 16% per year. Therefore, the $2,000 left over by selecting E instead of F can earn 16% elsewhere, which is more than we could obtain by investing it in F. All equivalent worth methods of Chapter 5 make use of this assumption (i.e., investment opportunities exist at the M.A.R.R.), and the reader can see that the P.W. and I.R.R. methods lead to the same recommendation. Consequently, rate of return methods and equivalent worth methods produce consistent results when their underlying assumptions are identical.

Before leaving Example 6-5, it is instructional to note that alternative D has the highest I.R.R. on total investment. It *is not* the recommended choice. Furthermore, the highest I.R.R. on incremental investment also results from alternative D. The largest investment that has an I.R.R. greater than the M.A.R.R. is alternative F. It is not the choice either for reasons which are now obvious.

TABLE 6-8 Comparison of Six Alternatives by the I.R.R. Method

	Alternative					
	B1	B2	C	D	E	F
I.R.R. on initial investment = A ÷ P	16.7%	18.4%	16.0%	23.1%	22.5%	20.4%
Increment considered	B1	B1 → B2	B2 → C	B2 → D	D → E	E → F
Δ Investment (ΔP)	$900	$600	$1,000	$2,500	$1,000	$2,000
Δ Annual revenues less expenses (ΔA)	$150	$126	$124	$649	$200	$300
I.R.R. on ΔP = ΔA ÷ ΔP	16.7%	21.0%	12.4%	26.0%	20.0%	15.0%
Is increment justified?	Yes	Yes	No	Yes	Yes	No

These three common errors must be guarded against as mentioned earlier in Example 6-1. ■

ALTERNATIVES HAVING DIFFERENT REVENUES AND DIFFERENT LIVES

From the right-hand side of Figure 6-1 it is apparent that another basic type of problem involving choice among mutually exclusive alternatives arises when different revenues, costs, and lives characterize the alternatives. Either the repeatability assumption or the coterminated assumption can be utilized, depending on conditions to be met in the problem. Examples 6-6 and 6-7 illustrate these assumptions, and Example 6-8 presents a situation in which two *nonrepeating* alternatives are compared.

EXAMPLE 6-6
The following data have been estimated for two investment alternatives, A and B, for which revenues as well as costs are known and which have *different* lives. If the minimum attractive rate of return is 10%, show which project is more desirable by using the A.W., P.W., and I.R.R. methods. Use the repeatability assumption.

	A	B
Investment	$3,500	$5,000
Annual revenue	$1,900	$2,500
Annual disbursements	$645	$1,383
Useful life	4 years	8 years
Net salvage value	0	0

Solution of Example 6-6 by the A.W. Method

	A	B
Annual revenue	$1,900	$2,500
Annual expenses:		
Disbursements	645	1,383
C.R. cost:		
$3,500(A/P, 10\%, 4)$	1,104	
$5,000(A/P, 10\%, 8)$		937
TOTAL ANNUAL EQUIVALENT EXPENSES	$1,749	$2,320
Net A.W. (revenue − expenses)	$151	$180

Since alternative B has a higher net A.W., it is shown to be the better economic choice. ■

Solution of Example 6-6 by the P.W. Method

The expected lives of the alternatives are 4 and 8 years, respectively. The lowest common multiple is 8 years, which will be taken to be the length of the study period.

	A	B
Annual revenue:		
$1,900(P/A, 10\%, 8)$	$10,136	
$2,500(P/A, 10\%, 8)$		$13,337
TOTAL P.W. OF REVENUE	$10,136	$13,337
Annual disbursements:		
$645(P/A, 10\%, 8)$	3,441	
$1,383(P/A, 10\%, 8)$		7,378
Original investment	3,500	5,000
First replacement: $3,500(P/F, 10\%, 4)$	2,390	
TOTAL P.W. OF COSTS	$ 9,331	$12,378
Net P.W. (revenue − costs)	$805	$959

Since alternative B has a higher net P.W., it again is shown to be the better choice. ∎

Solution of Example 6-6 by the I.R.R. Method

Because the repeatability assumption is specified, the least common multiple of years that comprises the study period is 8 years. With this in mind, the first step is to compute the I.R.R. on the total investment in alternatives A and B. Recall that the M.A.R.R. is 10%. This determination is shown below by setting N.P.W. equal to zero over an 8-year period.

Alternative A:

$$0 = -\$3,500 + \$1,255(P/A, i_A', 8) - \$3,500(P/F, i_A', 4)$$

By trial and error, $i_A' = 16.2\%$ and alternative A is acceptable.

Alternative B:

$$0 = -\$5,000 + \$1,117(P/A, i_B', 8)$$

By trial and error, $i_B' = 15.1\%$ and alternative B is also acceptable.

If, at this point, the analyst made a choice based on maximizing the I.R.R. on total investment, alternative A would be recommended. However, from previous discussion it was emphasized that *a correct choice among alternatives often does not result from selecting the alternative having the greatest rate of return*. This is the case in Example 6-6, and the reason for such ranking errors in comparing mutually exclusive alternatives is explained subsequently.

To obtain the correct choice in this example, we must examine the incremental investment and net receipts associated with alternative B relative to alternative A. The reader will observe that alternative B has the same cash flows as A, *except for the avoidable increment* that must be justified in order to make B our choice. The I.R.R. at which the N.P.W. of incremental cash flows equals zero is determined in this manner:

alternative A → alternative B

$$0 = -\$1,500 - \$138(P/A, i', 8) + \$3,500(P/F, i', 4)$$

By trial and error, the I.R.R. on the incremental cash flow is 12.7%. Because 12.7% is greater than the M.A.R.R. of 10%, alternative B should be chosen. (It is left to the reader to show that there are no multiple rates of return in this example.)

Alternatively, the I.R.R. on the incremental cash flows could be calculated by finding the i' at which the net A.W. of the two alternatives is equal:

$$-\$3,500(A/P, i', 4) + \$1,900 - \$645$$
$$= -\$5,000(A/P, i', 8) + \$2,500 - \$1,383$$

Again, it can be determined that $i' = 12.7\%$ causes both alternatives to have equal net annual worths. Since this is greater than 10%, the avoidable increment in alternative B is justified.

The reason for ranking errors that can occur when selections among alternatives are based on maximization of rate of return (I.R.R., E.R.R., or E.R.R.R.) can be seen in Figure 6-5. When the M.A.R.R. lies to the left of I.R.R.$_{A\to B}$, an incorrect choice will be made by selecting an alternative that maximizes rate of return. This is because the I.R.R. method assumes reinvestment of cash flows at 16.2% and 15.1%, respectively, for alternatives A and B. But the N.P.W. method assumes reinvestment at 10%. In this example, reinvestment can occur at a rate as high as 12.7% without causing a reversal in preference.

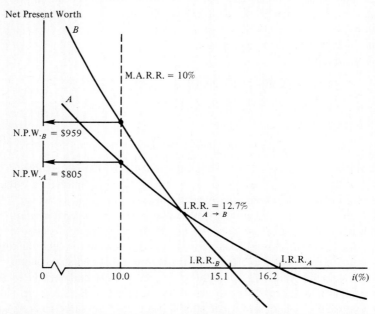

FIGURE 6-5 Illustration of the Ranking Error in Studies Using Rate of Return Methods.

From Figure 6-5 it can be seen that $N.P.W._B > N.P.W._A$ at the M.A.R.R. of 10% even though $I.R.R._A > I.R.R._B$. This ranking inconsistency is avoided by examining $I.R.R._{A \to B}$, which correctly leads to the selection of alternative B.

If the M.A.R.R. had been greater than $I.R.R._{A \to B}$, a correct choice would have been made by recommending the alternative with the greatest rate of return on total investment. ∎

EXAMPLE 6-7

Suppose that Example 6-6 is modified such that the expected period of required service from alternative A or B is only 4 years. (Perhaps our company has a firm contract to produce a manufactured good for exactly 4 years.) A choice must be made between A and B in view of the coterminated life of 4 years and a M.A.R.R. = 10%. Which project is more desirable?

Solution of Example 6-7 by the A.W. Method

As pointed out in Example 6-3, a key question here concerns the estimated salvage value of alternative B at the end of year 4 (S_4). A variety of approaches can be used to make this estimate. When there is no information to the contrary, it is convenient to assume the following for alternative B:

$$S_4 = \text{present worth of remaining capital recovery}$$

$$\text{amounts} + \text{present worth of salvage at end of year 8}$$

Because the salvage value at end of year 8 is zero, S_4 can be computed in accordance with the definition above:

$$S_4 = [\$5,000(A/P, 10\%, 8)](P/A, 10\%, 4) = \$937.22(3.1699)$$

$$= \$2,970.86$$

Calculations of net A.W. for alternative A and B are summarized as follows:

	A	B
Annual revenue	$1,900	$2,500
Annual expenses:		
Disbursements	645	1,383
C.R. cost:		
$3,500(A/P, 10%, 4)$	1,104	
$5,000(A/P, 10%, 4) - \$2,970.86(A/F, 10%, 4)$		937
TOTAL ANNUAL EQUIVALENT EXPENSES	$1,749	$2,320
Net A.W. (revenue − expenses)	$ 151	$ 180

Alternative B is the better choice. With the assumption above concerning the coterminated salvage value of B, the annual worths in Example 6-7 are identical to those in Example 6-6! This, of course, would not be true if a different salvage value had been specified for alternative B. However, the *analysis procedure* would be the same. ∎

EXAMPLE 6-8

Compare the two projects below when it is known that each will be used for a single life span, and the two life spans are not equal. That is, project J2 has a finite life of 5 years and project K4 extends over 7 years. These alternatives *cannot be repeated* over their common multiple of years (35) and their service *cannot be coterminated* at a common point in time. The M.A.R.R. equals 20%, and a comparison is desired based on the P.W., A.W., and F.W. methods.

Solution of Example 6-8

This example involves two alternatives that have nonrepeating cash flows and different useful lives. We still must adhere to the basic principle of comparing alternatives over an identical period of time, so a *reinvestment assumption* is needed for the shorter-lived alternative. Hence we presume that positive cash flows of alternative J2 can be reinvested at the M.A.R.R. during the sixth and seventh years of a common 7-year study period.

To illustrate this assumption explicitly, the net F.W. of both alternatives at the end of year 7 is first calculated, followed by determination of their net P.W. and net A.W.:

	Project	
	J2	K4
Investment	$15,000	$22,000
Annual revenues less expenses	$ 5,000	$ 6,000
Salvage value	$ 2,000	$ 1,800
Useful life	5 years	7 years

$$\text{Net F.W. of project J2} = -\$15,000(F/P, 20\%, 7)$$
$$+ \$5,000(F/A, 20\%, 5)(F/P, 20\%, 2)$$
$$+ \$2,000(F/P, 20\%, 2)$$
$$= \$2,711.81$$
$$\text{Net F.W. of project K4} = -\$22,000(F/P, 20\%, 7)$$
$$+ \$6,000(F/A, 20\%, 7) + \$1,800$$
$$= \$465.44$$

Hence Project J2 would be the recommended choice. The net P.W. can be determined directly from the data above or from net F.W.:

$$\text{Net P.W. of J2} = -\$15,000 - \$5,000(P/A, 20\%, 5)$$
$$+ \$2,000(P/F, 20\%, 5)$$
$$= \$756.82$$
$$\text{Net P.W. of K4} = -\$22,000 + \$6,000(P/A, 20\%, 7)$$

$$+ \$1,800(P/F, 20\%, 7)$$

$$= \$129.90$$

Again, the better project is seen to be J2. Notice that the N.P.W. of J2, computed over 5 years, is being directly compared with the N.P.W. of K4 computed over 7 years. This is a valid comparison, however, *only when the above reinvestment assumption is appropriate*. Finally, the net A.W. can be calculated for the 7-year study period from net F.W.:

$$\text{Net A.W. of J2} = \$2,711.81(A/F, 20\%, 7) = \$209.96$$

$$\text{Net A.W. of K4} = \$465.44(A/F, 20\%, 7) = \$36.04$$

The reader should observe that the net A.W. for each project under these study conditions is not the same as would be obtained under the repeatability assumption. Therefore, care must be taken to *identify clearly the study conditions that are appropriate* in a comparison of mutually exclusive alternatives before an economic measure of merit can be correctly applied. ∎

COMPARISON OF ALTERNATIVES BY THE CAPITALIZED WORTH METHOD

One special variation of the present worth method mentioned in Chapter 5 involves the determination of the worth of all receipts and/or disbursements over an infinitely long length of time. This is known as the *capitalized worth* (C.W.) *method*. If disbursements only are considered, results obtained by this method can be more appropriately expressed as *capitalized cost*. This is a convenient basis for comparing mutually exclusive alternatives when the period of needed service is indefinitely long or when the common multiple of the lives is very long and the repeatability assumption is applicable.

The capitalized worth of a perpetual series of end-of-period uniform payments A, with interest at $i\%$ per period, is $A(P/A, i\%, \infty)$. From the interest formulas, it can be seen that $(P/A, i\%, N) \to 1/i$ as N becomes very large. Thus capitalized worth $= A/i$ for such a series, as can also be seen from the relation

$$P = A(P/A, i\%, \infty) = A\left[\lim_{N \to \infty} \frac{(1 + i)^N - 1}{i(1 + i)^N}\right] = A\left(\frac{1}{i}\right)$$

The P is often referred to as the *capitalized worth* of A.

The annual worth of a series of payments of amount $\$X$ at the end of each kth period with interest at $i\%$ per period is $\$X(A/F, i\%, k)$. The capitalized worth of such a series can thus be calculated as $\$X(A/F, i\%, k)/i$.

EXAMPLE 6-9
Compare structures M and N given in Example 6-2 by the capitalized worth (cost) method.

Solution

	Capitalized Cost	
	Structure M	Structure N
First cost	$12,000	$40,000
Replacements:		
$12,000(A/F, 15%, 10)/0.15	3,940	
($40,000 − $10,000)(A/F, 15%, 25)/0.15		940
Annual disbursements:		
$2,200/0.15	14,667	
$1,000/0.15		6,667
TOTAL CAPITALIZED COST	$30,607	$47,607

Thus structure M is the indicated better alternative, which is, of course, consistent with the results for Example 6-2 by the other methods of comparison. As an aside, since we have previously calculated the equivalent annual costs for these two alternatives, the easiest way to have determined the capitalized costs was to use the relation

$$\text{C.C.} = \frac{\text{A.C.}}{i}$$

For structure M: $4,591/0.15 = $30,607

For structure N: $7,141/0.15 = $47,607

COMPARISON OF ALTERNATIVES BY THE PAYOUT PERIOD METHOD

As you recall from Chapter 5, the payout (payback) period is often used as a measure of project risk. Many organizations compare mutually exclusive alternatives with this method even though it can give misleading or erroneous results. Hence its use should be avoided except as a supplement to analysis by a correct method.

If the life of each alternative is the same and the risks are comparable, the payback method does correctly rank the alternatives relative to each other. Otherwise, the method is useful primarily as an indicator of relative risk. The higher the payout period, the greater the opportunity for events to not turn out as originally estimated, and thus the greater the risk. The following is a typical case involving three alternatives:

	A	B	C
Required investment	$5,000	$5,000	$6,000
Estimated useful life (N)	8 years	4 years	14 years
Available for payout (annual revenue minus out-of-pocket costs)	$2,250	$2,500	$2,800
Payout period = investment/available for payout	2.22 years	2.00 years	2.14 years

By this payout analysis, alternative B would be selected, since it has a shorter payout period than alternative A and C, and it requires less capital than C. If, however, the three alternatives are compared on the basis of their net annual worths, assuming repeatability and a M.A.R.R. = 12% with no salvage value, the results are as follows:

	A	B	C
Annual revenue minus out-of-pocket costs	$2,250	$2,500	$2,800
Less: capital recovery cost			
Investment (A/P, 12%, N)	1,007	1,646	905
Net A.W.	$1,243	$ 854	$1,895

By this analysis, alternative B is found to be the least desirable of the three, and, if the additional capital is available, alternative C is more desirable than either A or B.

Because the payout period method does not take expected life into account, its use is not recommended for comparing alternatives having different lives. Rather than indicate overall profitability or desirability, the payout method shows only the number of years required to recover investment capital, which is sometimes a useful measure of risk.

ALTERNATIVES INVOLVING INCREASING FUTURE DEMAND (DEFERRED INVESTMENT PROBLEMS)

In many projects there are clear indications that the future demands for products or services will considerably exceed those of the present. Such cases present the problem of determining whether it is more economical to provide immediately for all the foreseeable future demand or to provide only for the immediate demands and then make additional provisions at a later date. The situation may be illustrated by consideration of the installation of domestic gas mains for a new real estate subdivision. Two extreme possibilities exist. The first would be to provide a completely separate main from outside the subdivision to each house as it is completed and occupied. This would require repeated tearing up of streets and the digging of many parallel ditches. The other extreme would be to at one time install mains of sufficient size to provide service to all the houses that might ever be built in the subdivision, even though many of them might not be built for several years, if at all. Obviously, many alternatives could be provided intermediate to these two extremes.

In this case it is apparent that the first alternative would require less immediate expenditure of capital. However, it is quite possible that in the long run it might not be as economical as the second extreme. We would at once suspect that some intermediate alternative might be more economical than either of the two extremes. Thus, where provisions exist for meeting increased future demand in more than one way, it is necessary to make economy studies of two or more mutually exclusive alternatives to determine which will be the more economical.

Such studies ordinarily involve relatively long-range demands, usually from 10 to 50 or more years. They, therefore, are encountered most frequently in connection with public works—such as roads, bridges, water reservoirs, or sewers—or in connection with public utilities—such as power plants, gas and water mains, or telephone central offices and cable installations. It will be noted that all such projects involve relatively stable services for which the demand is likely to continue, and the necessity for meeting the future demands. In some cases there is a limitation of site location. For example, in locating the central office of a telephone company a difference of a block or two many mean thousands of dollars in cable costs. Unless the correct site is determined and acquired well in advance, it may not be available when the actual demand occurs.

There are two factors that make economy studies involving variable future demand different from those investment studies that have been discussed previously in this chapter. The *first factor* is that the alternatives may involve different amounts of invested capital at different times. This is illustrated in Figure 6-6. Here the demand is assumed to increase from an initial value to a larger amount at a future date. This demand could be met in two alternative ways. Alternative A would be to immediately provide capacity to meet all the anticipated future demands. Alternative B, on the other hand, would immediately provide capacity B1 sufficient only to meet the demand of the first period of years. At a later date capacity B2 would be added to enable the demand of the later years to be met. Quite clearly these two alternatives involve different investments of capital initially, and alternative B involves different amounts of capital investment at different times. It also is evident that the revenues, costs, and profits probably would not be uniform throughout the investment period, regardless of which alternative is selected.

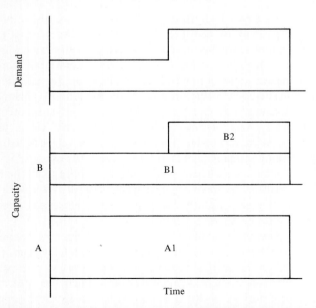

FIGURE 6-6   Two Alternatives for Providing Capacity to Meet an Increasing Demand.

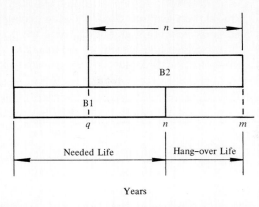

FIGURE 6-7 "Hangover Life" Resulting from a Deferred Capacity Provision which has Life Beyond that Needed.

The *second factor* that often must be dealt with in future demand studies is that illustrated in Figure 6-7. If, in the previous example, alternative B were to consist of two units, B1 and B2, each having a life of n years, it is clear that at the end of the needed life the installation B2 would have $m - n$ years of life still available. This situation, obviously, presents the same problem that was discussed earlier in this chapter in connection with alternatives that have different lives. One way of dealing with this extra life (sometimes called *hangover life*) is to assign an anticipated salvage value at the end of the needed life, as is commonly done with the coterminated assumption. However, another procedure sometimes exists in the case of future demand studies. Because it is recognized that the deferred capacity (unit B2 in this case) will be needed for only $n - q$ years (Figure 6-7), in some cases it is possible to build the second unit as an "inferior" unit that will, in fact, have a shorter life than the unit that is provided initially. This is what was portrayed for unit B2 in Figure 6-6.

The following is a typical example of an investment situation involving variable future demand, nonuniform investments and costs and coterminated asset lives. The solution is shown by the present worth–cost method. Reasons why the internal rate of return procedure cannot be applied without modification in this particular case are then given.

EXAMPLE 6-10

In 1985 a municipality is planning to build a new power plant. It believes that the plant would be needed for a number of years, but no absolute need is seen after 10 years. Two plans were being considered. Plan A is to build immediately a plant of sufficient size to meet all the needs of the next 10 years—1986 through 1995. Such a plant would be of 5,000-kilowatts capacity and cost $4,000,000. The life of this plant is estimated to be at least 10 years. Plan B would be to construct a plant of 3,000-kilowatts capacity immediately to meet the demands of the first 6 years and at the end of 6 years to enlarge this plant by an additional 2,000 kilowatts to meet the needs of the last 4 years. The 3,000-kilowatt plant would cost $3,100,000, and the additional 2,000 kilowatts would cost $1,800,000. Annual property taxes and insurance would amount to 2% of the first cost in each case. Although the 2,000-kilowatt addition would have an economic life of at least 10 years if needed, it is believed that it could be disposed of for $300,000 at the end of 1995 if it should not be needed beyond that date. Annual operating and maintenance costs for the two plans were estimated to be as follows:

	Plan A	Plan B
First and second years	$480,000	$450,000
Third and fourth years	540,000	520,000
Fifth and sixth years	610,000	600,000
Seventh and eighth years	785,000	800,000
Ninth and tenth years	960,000	970,000

The minimum attractive rate of return for the municipality is 7%.

Solution by P.W.-C. Method

For Plan A:

		P.W.-C.
Investment in 5,000-kW plant		$4,000,000
Taxes and insurance = $4,000,000 × 0.02 × $(P/A, 7\%, 10)$		562,000
Operation and maintenance costs:		
First and second years: $480,000(P/A, 7\%, 2)$	$ 868,000	
Third and fourth years: $540,000 × (P/A, 7\%, 2)(P/F, 7\%, 2)$	853,000	
Fifth and sixth years: $610,000 × (P/A, 7\%, 2)(P/F, 7\%, 4)$	841,000	
Seventh and eighth years: $785,000 × (P/A, 7\%, 2)(P/F, 7\%, 6)$	946,000	
Ninth and tenth years: $960,000 × (P/A, 7\%, 2)(P/F, 7\%, 8)$	1,010,000	
TOTAL		$4,518,000
Total P.W.-C.		$9,080,000

For Plan B:

		P.W.-C.
Investment in 3,000-kW plant		$3,100,000
Investment in 2,000-kW plant built 6 years hence: $1,800,000 × (P/F, 7\%, 6)$	$1,200,000	
Less present worth of salvage: $300,000 × (P/F, 7\%, 10)$	− 152,000	
Net present worth–cost		$1,048,000
Taxes and insurance:		
First 6 years: $3,100,000 × 0.02 × (P/A, 7\%, 6)$	296,000	
Last 4 years: $4,900,000 × 0.02 × (P/A, 7\%, 4)(P/F, 7\%, 6)$	221,000	
TOTAL		$ 517,000

Operation and maintenance costs:
First and second years:
$450,000(P/A, 7\%, 2)$ $ 814,000
Third and fourth years:
$520,000(P/A, 7\% 2)(P/F, 7\%, 2)$ 821,000
Fifth and sixth years:
$600,000(P/A, 7\%, 2)(P/F, 7\%, 4)$ 827,000
Seventh and eighth years:
$800,000(P/A, 7\%, 2)(P/F, 7\%, 6)$ 964,000
Ninth and tenth years:
$970,000(P/A, 7\%, 2)(P/F, 7\%, 8)$ 1,020,000

TOTAL	$4,446,000
Total P.W.-C.	$9,111,000

On the basis of present worth–cost, plan A is slightly more economical and would be used. However, from a practical viewpoint, this is a case where we might very well give considerable additional consideration to plan B, inasmuch as the extra present worth is not great ($31,000) and it permits a portion of the investment to be deferred. Furthermore, if the service should be required after 1995 the 2,000-kilowatt plant presumably would be usable. ■

Solution by I.R.R. Method

If we were to attempt a comparison of the two plans by the internal rate of return method, the comparison of the annual cash flows would be as follows (initial costs of plants plus taxes and insurance and operations and maintenance, less salvage value):

Year	Plan A	Plan B	Increment (B → A)
0	− $4,000,000	− $3,100,000	− $ 900,000
1	− 560,000	− 512,000	− 48,000
2	− 560,000	− 512,000	− 48,000
3	− 620,000	− 582,000	− 38,000
4	− 620,000	− 582,000	− 38,000
5	− 690,000	− 662,000	− 28,000
6	− 690,000	− 2,462,000	+ 1,772,000
7	− 865,000	− 898,000	+ 33,000
8	− 865,000	− 898,000	+ 33,000
9	− 1,040,000	− 1,068,000	+ 28,000
10	− 1,040,000	− 768,000	− 272,000

The cash flows shown in the last column in the preceding tabulation would result from the greater initial investment required for plan A. For determination of the internal rate of return, the present worth of these cash flows would be equated to zero. However, it will be noted that the final cash flow is negative, so that the cash flow reverses sign more than once—a condition that may prevent the calculation of a unique internal rate of return. A suggested methodology for solving this type of problem is described in Appendix 5-B.

An adjustment that can be made to the (B → A) cash flows in order to allow one to calculate the I.R.R. is to make equivalent moves in one or more cash flows so that there is only one reversal in sign. The interest rate at which equivalence adjustments are made is normally the M.A.R.R. For this problem, one possible adjustment is to move the $-\$272,000$ from year 10 to year 5 at $i = 7\%$. The equivalent value at year 5 is

$$P_5 = F_{10}(P/F, 7\%, 5) = -\$272,000(0.7130) = -\$194,000$$

The $-\$194,000$ added to the $-\$28,000$ already at year 5 results in a net $-\$222,000$. The cash flows, which now exhibit only one reversal in sign, are next discounted at i' such that their N.P.W. $= 0$. This is left as an exercise for the curious student. It turns out the I.R.R. of the adjusted cash flows is 7.5% ($>7\%$), again indicating that the incremental investment in plan A is justified.

∎

SELECTIONS AMONG INDEPENDENT ALTERNATIVES

All previous examples in this chapter have involved alternatives which are mutually exclusive, that is, the choice of one project excludes the choice of any other project so that at most one project under consideration will be chosen. Any of the six basic economy study methods can also be used for the comparison of *independent* projects. By independent it is meant that the choice of one project does not affect the choice of any other, and any number of projects may be chosen as long as sufficient capital is available. Independent project alternatives are often descriptively called *opportunities*.

EXAMPLE 6-11
Given the following independent projects, determine which should be chosen using the annual worth method. The minimum required rate of return is 10%, and there is no limitation on total investment funds available.

Project	Investment, P	Life, N	Salvage Value, S	Net Annual Cash Flow, A
X	$10,000	5 years	$10,000	+$2,300
Y	12,000	5 years	0	+ 2,800
Z	15,000	5 years	0	+ 4,067

Solution

Project	(1) Net Annual Cash Flow, A	(2) Capital Recovery Cost: $(P - S)$ × $(A/P, 10\%, 5) + S(10\%)$	(3) = (1) − (2) Net Annual Worth
X	+$2,300	$1,000	+$1,300
Y	+ 2,800	3,166	− 366
Z	+ 4,067	3,957	+ 110

Thus projects X and Z, having positive net annual worths, would be satisfactory for investment, but project Y would not be satisfactory. The same indication of satisfactory projects and the unsatisfactory project would be obtained using other economy study methods. ∎

In many problems involving selections among independent alternatives, different revenues (or savings) and useful lives are present. Because these alternatives are typically nonrepeating as discussed in Example 6-8, it is usually assumed that positive cash flows of shorter-lived projects are reinvested at the M.A.R.R. over a period of time corresponding to the life of the longest-lived alternative. The following example illustrates this assumption as well as a constraint on funds available for investment.

EXAMPLE 6-12

A large corporation is considering the funding of several independent, nonrepeating proposals for enlargening freshwater harbors in three areas of the country. Its available capital this year for such projects is $200 million, and the firm's M.A.R.R. is 10%. In view of the data below, which project(s), if any, should be funded?

Proposal	Investment, P	Net Annual Benefits, A	Useful Life, N
A	$93,000,000	$13,000,000	15 years
B	55,000,000	9,500,000	10 years
C	71,000,000	10,400,000	30 years

Solution

With assumed reinvestment of cash inflows at 10% interest, the net present worth of each proposal can be computed over its useful life and utilized in the selection of independent alternatives. If each proposal is acceptable, we must next determine which combination of proposals maximizes net present worth without exceeding $200 million.

Proposal	(1) Investment	(2) P.W. of Benefits $A(P/A, 10\%, N)$	(3) = (2) − (1) Net Present Worth
A	$93,000,000	$98,879,034	$ 5,879,034
B	55,000,000	58,373,388	3,373,388
C	71,000,000	98,039,910	27,039,910

All three proposals cannot be chosen without violating the limitation on available funds. Proposals A and C should be recommended because this combination maximizes net P.W. (= $32,918,944) with a total investment of $164 million. The leftover funds ($200 − $164 = $36 million) would presumably be invested elsewhere at the M.A.R.R. of 10%. ∎

CONSIDERATION OF SETS OF MUTUALLY EXCLUSIVE, INDEPENDENT, AND CONTINGENT PROJECTS

It is helpful to think of sets of projects in three major groups as follows:

1. *Mutually exclusive:* which means that at most one project out of the group can be chosen.
2. *Independent:* which means that the choice of a project is independent of the choice of any other project in the group.
3. *Contingent:* which means that the choice of a project is conditional on the choice of one or more other projects.

It is common for decision makers to be faced with sets of mutually exclusive, independent, and/or contingent investment projects. For example, a contractor might be considering investing in a dump truck, and/or a power shovel, and/or an office building. For each of these types of investments, there may be two or more mutually exclusive alternatives (i.e., brands of dump trucks, types of power shovels, and designs of office buildings). While the choice of an office building is probably independent of that of either dump trucks or power shovels, the choice of any type of power shovel may be contingent (conditional) on the decision to purchase a dump truck.

To provide a simple method of handling these types of projects as well as to provide some insight into mathematical programming formulations of this type of decision problem, a general approach is recommended in this section. This approach requires that all investment projects be listed and that all the feasible combinations of projects be enumerated. *Such combinations will then be mutually exclusive.* Each combination of projects is mutually exclusive since each is unique and the acceptance of one combination of investment projects precludes the acceptance of any of the other combinations. The cash flow of each combination is determined simply by adding, period by period, the cash flows of each project included in the mutually exclusive combination being considered.

For example, suppose that we have three projects, A, B, and C. If the projects themselves are all mutually exclusive, then the four possible mutually exclusive combinations are shown in binary form in Table 6-9. If, by chance, the firm

TABLE 6-9 Combinations of Three Mutually Exclusive Projects[a]

Mutually Exclusive Combination	Project			Explanation
	X_A	X_B	X_C	
1	0	0	0	Accept none
2	1	0	0	Accept A
3	0	1	0	Accept B
4	0	0	1	Accept C

[a]For each investment project there is a binary variable X_j that will have the value 0 or 1 indicating that project j is rejected (0), or accepted (1). Each row of binary numbers represents an investment alternative (mutually exclusive combination). This convention is used throughout this book.

TABLE 6-10 Mutually Exclusive Combinations of Three Independent Projects

Mutually Exclusive Combination	Project			Explanation
	X_A	X_B	X_C	
1	0	0	0	Accept none
2	1	0	0	Accept A
3	0	1	0	Accept B
4	0	0	1	Accept C
5	1	1	0	Accept A and B
6	1	0	1	Accept A and C
7	0	1	1	Accept B and C
8	1	1	1	Accept A, B, and C

felt that one of the projects must be chosen (i.e., it is not permissible to turn down all alternatives) then mutually exclusive combination 1 would be eliminated from consideration.

If the three projects were independent, there are eight mutually exclusive combinations, as shown in Table 6-10.

To illustrate one of many possible instances of contingent projects, suppose that A is contingent on the acceptance of both B and C, and that C is contingent on the acceptance of B. Now there are four mutually exclusive combinations, as shown in Table 6-11.

Suppose that one is considering two independent sets of mutually exclusive projects. That is, projects A1 and A2 are mutually exclusive while projects B1 and B2 are mutually exclusive. However, the selection of any proposal from the set of proposals A1 and A2 is independent of the selection of any proposal from the set of proposals B1 and B2. For example, the decision problem may be to select at most one dump truck out of two brands being considered, and to select at most one office building out of two designs being considered. Table 6-12 shows all mutually exclusive combinations for this situation.

Example 6-13 provides an illustration involving enumeration of mutually exclusive combinations to select an optimal set of projects under capital constraints.

TABLE 6-11 Mutually Exclusive Combinations of Three Projects with Contingencies

Mutually Exclusive Combination	Project		
	X_A	X_B	X_C
1	0	0	0
2	0	1	0
3	0	1	1
4	1	1	1

TABLE 6-12   Mutually Exclusive Combinations for Two Independent Sets of Mutually Exclusive Projects

Mutually Exclusive Combination	Project			
	X_{A1}	X_{A2}	X_{B1}	X_{B2}
1	0	0	0	0
2	1	0	0	0
3	0	1	0	0
4	0	0	1	0
5	0	0	0	1
6	1	0	1	0
7	1	0	0	1
8	0	1	1	0
9	0	1	0	1

EXAMPLE 6-13

The following are prospective projects, their interrelationships, and respective cash flows for the coming budgeting period. Using the net P.W. method and M.A.R.R. = 10%, determine what combination of projects is best if the capital to be invested is (a) unlimited, and (b) limited to $48,000.

Project B1 ⎱ mutually exclusive
Project B2 ⎰

Project C1 ⎱ mutually exclusive and
Project C2 ⎰ dependent on the acceptance of B2
Project D   contingent on the acceptance of C1

Project	Cash Flow ($000s) for End of Year					Net P.W. ($000s) at
	0	1	2	3	4	M.A.R.R. = 10%
B1	−50	20	20	20	20	+13.4
B2	−30	12	12	12	12	+ 8.0
C1	−14	4	4	4	4	− 1.3
C2	−15	5	5	5	5	+ 0.9
D	−10	6	6	6	6	+ 9.0

Solution

The net P.W. for each project by itself is shown on the right-hand column of the table above. As a sample calculation, the net P.W. for project B1 is

$$-\$50,000 + \$20,000(P/A,\ 10\%,\ 4) = +\$13,400$$

The feasible mutually exclusive combinations are as follows:

Mutually Exclusive Combination	Project				
	B1	B2	C1	C2	D
1	0	0	0	0	0
2	1	0	0	0	0
3	0	1	0	0	0
4	0	1	1	0	0
5	0	1	0	1	0
6	0	1	1	0	1

The combined cash flows and the net P.W. for each mutually exclusive combination are:

Mutually Exclusive Combination	Cash Flow ($000s) for End of Year					Invested Capital ($000s)	Net P.W. ($000s) at M.A.R.R. = 10%
	0	1	2	3	4		
1	0	0	0	0	0	0	0
2	−50	20	20	20	20	50	13.4
3	−30	12	12	12	12	30	8.0
4	−44	16	16	16	16	44	6.7
5	−45	17	17	17	17	45	9.0
6	−54	22	22	22	22	54	16.0

Examination of the right-hand column reveals that mutually exclusive combination 6 has the highest net P.W. if capital available (in year 0) is unlimited, as specified in part (a). If, however, capital available is limited to $48,000, as specified in part (b), both mutually exclusive combinations 2 and 6 are not feasible. Of the remaining mutually exclusive combinations, 5 is best, which means that projects B2 and C2 would be selected for a net P.W. = $9,000.∎

For problems that involve a relatively small number of projects, the general technique just presented for arranging various types of projects into mutually exclusive combinations is computationally practical. However, for larger numbers of projects the number of mutually exclusive combinations becomes quite large, and therefore this approach becomes computationally cumbersome. Linear (integer) programming provides a practical means to accomplish the same result and will be described in Chapter 17.

PROBLEMS

6-1 The Consolidated Oil Company must install antipollution equipment in a new refinery to meet federal clean air legislation. Four types of equipment are being considered, which will have investment and annual operating costs as follows:

	Equipment			
	A	B	C	D
Investment	$600,000	$760,000	$1,240,000	$1,600,000
Power	68,000	68,000	120,000	126,000
Labor	40,000	45,000	65,000	50,000
Maintenance	660,000	600,000	420,000	370,000
Taxes and insurance	12,000	15,000	25,000	28,000

Assuming a useful life of 10 years for each type of equipment, no salvage value, and that the company wants a before-tax minimum return of 15% on its capital, calculate the present worth of each alternative to determine which one should be purchased.

6-2 Work Problem 6-1 with the I.R.R. and E.R.R. methods when the reinvestment rate (e) equals 15%.

6-3 A large sailboat is being built in a coastal town. In specifying air conditioning equipment and insulation for the living quarters of the vessel, engineers have found various combinations of compressors and insulation thickness to be feasible. A small air-conditioner compressor causes the need for more insulation but has slightly lower operating and maintenance costs compared to a larger compressor. Three combinations are being considered that have the following investment and annual costs.

	Combination		
	A	B	C
First cost of compressor	$7,000	$ 4,200	$ 5,500
First cost of insulation	7,000	14,000	10,000
Out-of-pocket costs/year	950	300	600

Each system is expected to have a 20-year life, with no salvage value. The before-tax minimum attractive rate of return is 20%. Compare the alternatives with the annual worth method and the present worth method.

6-4 Work Problem 6-3 by using the future worth method and the I.R.R. method. Which alternative should be recommended?

6-5 Machines A and B are to be compared on the basis of annual cost when the minimum attractive rate of return is 12%. Assume repeatability. Pertinent cost data are as follows:

	Machine A	Machine B
First cost	$10,000	$13,000
Useful life	10 years	15 years
Salvage value	$ 1,000	$ 3,000
Annual operating costs	$ 500	$ 100
Overhaul—end of fifth year	$ 300	$ 200
Overhaul—end of tenth year	None	$ 550

6-6 A small branch office of a major retailing firm is planning to purchase a minicomputer. Three companies have supplied cost, service life, and salvage data (shown below). The minimum attractive rate of return is 12%.

	Machine		
	A	B	C
Initial cost	$800	$1,400	$900
Annual service contract cost	$180	$ 150	$170
Life	5 years	8 years	5 years
Salvage value	$200	$ 600	$100

If the expected annual saving in labor is $500 regardless of which machine is purchased, what recommendation would you make to the boss? Assume that the coterminated life assumption is valid and that after 5 years the market value of machine B is expected to be $825.

6-7 A construction company is going to purchase several heavy-duty trucks. Its M.A.R.R. before taxes is 18%. It is considering two makes, and the following relevant data are available.

	Wiltsbilt	Big Mack
Cost	$10,000	$15,000
Life (estimated by manufacturer)	3 years	5 years
Salvage value at end of life	$ 2,000	$ 3,000
Annual out-of-pocket costs	$ 4,000	$ 3,000

(a) Which type of truck should be selected when the repeatability assumption is appropriate?

(b) Which type of truck would you recommend if the study period is limited to 3 years (coterminated assumption) and it is estimated that a Big Mack truck will have a salvage value of $5,600 at that time?

6-8 A real estate operator has a 30-year lease on a plot of land. He gets estimates on the costs and income of various types of structures on the piece of land as follows.

	Cost of Structure	Receipts Less Expenses[a]
Apartment house	$300,000	$69,000/year
Theater	200,000	40,000/year
Department store	250,000	55,000/year
Office building	400,000	76,000/year

[a]Expenses do not include depreciation, of course.

Each structure is expected to have a salvage value equal to 20% of its initial cost. If the investor requires a minimum attractive rate of return of at least 12% before taxes on all his investments, which structure (if any) should he build?

(a) Use the annual worth method to determine which type of structure (if any) should be chosen.

(b) Use the I.R.R. method to investigate incremental differences between alternatives and to recommend which structure should be selected.

6-9 Work Problem 6-8 by using the E.R.R. and E.R.R.R. methods when the reinvestment rate is 12%. Calculate the present worth of each alternative at 12%. Why do all methods lead to the same choice?

6-10 A manufacturing company is considering purchasing a 10-horsepower (hp) electric motor which it estimates will run an average of 6 hours per day for 250 days per year. Past experience indicates that (1) its annual cost for taxes and insurance averages 2.5% of first cost, (2) it must make 10% on invested capital before income tax considerations, and (3) it must recover capital invested in machinery within 5 years. Two motors are offered to the company. Motor A costs $340 and has a guaranteed efficiency of 85% at the indicated operating load. Motor B costs $290 and has a guaranteed efficiency of 80% at the same operating load. Electric energy costs the company 2.3 cents per kilowatthour (kWh), and 1 hp = 0.746 kW.

Use the internal rate of return method to choose the better electric motor. Then compare the two motors with the present worth method.

6-11 A certain service can be performed satisfactorily either by process R or process S. Process R has a first cost of $8,000, an estimated service life of 10 years, no salvage value, and annual net receipts (revenues − expenses) of $2,400. The corresponding figures for process S are $18,000, 20 years, salvage value equal to 20% of first cost, and $4,000. Assuming a minimum attractive rate of return of 15% before income taxes, find the future worth of each process and specify which you would recommend. Use the repeatability assumption.

6-12 Compare the equivalent uniform annual costs of the following two electric pumps if the minimum acceptable rate of return (M.A.R.R.) is 10%. The *service requirement* for the pump will be exactly 5 years, because the government contract for which the pump is needed will continue only for the next 5 years. What assumptions did you have to make in your analysis?

	Circle D Pump	Qwik Sump Pump
First cost	$4,000	$7,000
Useful life	3 years	7 years
Annual expenses	$ 800	$ 900
Salvage value	$ 0	$1,000

6-13 In the design of certain industrial facilities, the following alternatives are under consideration:

	Alternative 1	Alternative 2	Alternative 3
First cost	− $28,000	− $18,000	− $24,000
Net cash flow/year	+ $ 5,500	+ $ 2,400	+ $ 4,800
Salvage value	+ $ 1,500	0	+ $ 500
Service life	10 years	10 years	10 years

Assume that the interest rate (M.A.R.R.) is 15% and use the present worth method to choose the best of these three alternatives.

6-14 In the Rawhide Co., Inc. (cosmetics manufacturers), plant decisions regarding approval of proposals for plant investment are based upon a stipulated minimum attractive rate of return of 20% before income taxes. The following five packaging devices were compared assuming a 10-year life and zero salvage value for each. Which one (if any) should be selected? Make any additional calculations you think are needed.

	Packaging Equipment				
	A	B	C	D	E
Investment	$38,000	$50,000	$55,000	$60,000	$70,000
Net annual return	$11,000	$14,100	$16,300	$16,800	$19,200
Rate of return (I.R.R.)	26.1%	25.2%	26.9%	25.0%	24.3%

6-15 You have been asked to evaluate the economic implications of various methods for cooling condenser effluents from a 350-megawatt steam-electric plant. In this regard, cooling ponds and once-through cooling systems have been eliminated from consideration because of their adverse ecological effects. It has been decided to use cooling towers to dissipate waste heat to the atmosphere. There are two basic types of cooling towers: wet and dry. Furthermore, heat may be removed from condenser water by (1) forcing (mechanically) air through the tower or (2) allowing heat transfer to occur by making use of natural draft. Consequently, there are four basic cooling tower designs that could be considered. Assuming that the cost of capital to the utility company is 12%, your job is to recommend the best alternative (i.e., the least expensive during the service life) in view of the data below. Further assume that each alternative is capable of satisfactorily removing waste heat from the condensers of a 350-megawatt power plant. What *non-economic* factors can you identify that might also play a role in the decision-making process? Refer to the tabled data on page 214.

6-16 (a) Compare the following two piping systems using the present worth method if M.A.R.R. = 12%. Use the repeatability assumption.

	Plastic	Copper
First cost	$2,000	$5,000
Life	2 years	5 years
Salvage value	0	$1,000
Annual expenses	$ 600	$ 300

(b) The salvage value of the copper piping system at the end of year 2 is estimated to be $1,200. With the coterminated assumption, which system would be chosen for a 2-year study period?

6-17 The following estimates of *cost* apply to equipment alternatives A and B. The minimum attractive rate of return required is 20%. These are *nonrepeating* alternatives. Assume that savings attributable to the extra investment in A can be reinvested at 20% per year.

	A	B
First cost	$12,000	$25,000
Operating savings	$5,000 at end of year 1 and increasing by $600 per year thereafter	$10,000 at end of year 1 and increasing by $1,000 per year thereafter
Overhaul costs	$5,000 every 5 years	None required
Life	20 years	15 years
Salvage value at end of life	$15,000 *if just* overhauled	Negligible

Alternative Types of Cooling Towers for a 350-Megawatt Fossil-Fired Power Plant Operating at Full Capacity[a]

	Alternative				
	Wet Tower Mech. Draft	Wet Tower Natural Draft	Dry Tower Mech. Draft	Dry Tower Natural Draft
Initial cost	$3 million	$8.7 million	$5.1 million	$9.0 million
Power for I.D. fans	40 200-hp induced-draft fans	None	20 200-hp I.D. fans	None
Power for pumps	20 150-hp pumps	20 150-hp pumps	40 100-hp pumps	40 100-hp pumps
Mechanical maintenance/ year	$0.15 million	$0.10 million	$0.17 million	$0.12 million
Service life	30 years	30 years	30 years	30 years
Salvage value	0	0	0	0

[a]100 hp = 74.6 kW; cost of power to plant is 2.2 cents per kWh or kilowatt-hour; induced-draft fans and pumps operate around the clock for 365 days/year (continuously). Assume that electric motors for pumps and fans are 90% efficient.

(a) Compare these alternatives by use of the present worth method.

(b) Compare the alternatives by use of the annual worth method.

6-18 A house is for sale at $70,000, to be financed as follows: 10% down payment ($7,000), $2,000 closing costs (paid by the buyer upon occupancy), and a loan on the balance of $63,000 at 12% to be repaid over the next 25 years. Assume that repayments of the loan are payable once each year. Each equal end-of-year payment includes interest on the unpaid balance plus some equity. The same house can be purchased under an alternative plan which calls for a $30,000 down payment, $500 closing costs, and a loan on the balance of $40,000 at 10% to be repaid in 25 years. The resale value of the house and annual property taxes *do not* vary with the financing plan. If you had $32,000 cash, which method of financing would you select? Consider your minimum attractive rate of return to be 12% (a certificate of deposit in a savings and loan association) and ignore the effects of income taxes.

6-19 A comparison is being made between alternatives M2 and N2 in view of these estimated cash flows:

	Alternative	
End of Year	M2	N2
0	− $1,500	− $12,000
1	575	4,400
2	575	4,400
3	575	4,400
4	575	4,400
5	0	4,400

A coterminated life of 4 years is used in the analysis, and the M.A.R.R. is 15%. What recommendation should be made with (a) the present worth method, and (b) the internal rate of return method? Should the recommendations be the same?

6-20 Consider the three mutually exclusive projects below and use the *external rate of return* method to make a selection. The project chosen must provide service for a 10-year period. The M.A.R.R. is 12% and *e* is also 12%. State all assumptions you make.

	Project A	Project B	Project C
Initial investment	$2,000	$8,000	$20,000
Receipts less expenses	$600/year	$2,220/year	$3,600/year
Salvage value	0	0	0
Project life (years)	5	5	10

6-21 Two high-speed backhoes are being considered by the Apex Construction Company to replace a present piece of equipment. The equipment will only be needed three years. Cost data for the proposed backhoes are as follows:

	Backhoe M	Backhoe N
Purchase price	− $50,000	− $100,000
Annual savings	+ $25,000	+ $ 60,000
Life of backhoe	3 years	5 years
Estimated salvage value at end of year 3	+ $20,000	+ $ 10,000

Over what *range* of values of the M.A.R.R. is alternative N preferred to alternative M?

6-22   A company is planning to purchase a small executive airplane. It has about decided to buy an "Airbird" model, costing $300,000, and has based its decision on an assumed economic life of 6 years. A company vice-president favors an "Eaglejet" plane, and has presented data that shows rather convincingly that its annual out-of-pocket operating costs would be $42,000 per year less, and it would have an equally long useful life. This plane, however, would cost considerably more to purchase. It is believed that either plane would have a salvage value of about 30% of first cost at the end of 6 years. If the company's capital is worth 15% and income tax effects are neglected, how much can it afford to pay for the more expensive plane?

6-23   Which type of air-conditioning equipment would you select if you knew the need for the equipment exists for 10 years only? State any assumptions you make.

	Alternatives	
	Apex Inc.	Zandir Corp.
Initial investment	− $250,000	− $210,000
Annual maintenance	− $ 8,000	− $ 7,000
Annual fuel savings	0	+ $ 3,000
Partial replacement cost at end of year 8	$ 50.000	0
Life expectancy of equipment	20 years	12 years
Salvage value	0	0
M.A.R.R.	20%	20%

6-24   A recent engineering graduate has encountered a puzzling situation in her analysis of three mutually exclusive projects. Her cash flow data and results with three different study methods are shown below when the M.A.R.R. is 8%. Comment on her results and rework the analysis as you would have done it. Which economic measure of merit is correct in this situation?

		Cash Flows for:		
End of Year		Project I	Project II	Project III
0		− $100,000	− $100,000	− $100,000
1		0	$110,000	0
2		$155,000	0	0
3		0	0	$120,000
4		0	0	0
5		− $ 21,000	$ 10,000	$ 50,000
Measure of merit:	N.P.W.	$18,595	$8,658	$29,289
	I.R.R.	19.5%	15.6%	16.2%
	Simple payback	2 years	1 year	3 years

6-25   In building the landing strip at a small municipal airport, which is not used by commercial planes, one method of construction will cost $500,000 and have an estimated life with proper maintenance of 40 years. The annual cost of such maintenance is estimated to be $10,000 per year. An alternative type of construction would cost only $275,000, but at the end of 10 years it is estimated that another $250,000 will have to be spent for resurfacing and other major repair work. It would then last another 10 years. This resurfacing would involve closing the field for a month and would result in a revenue loss of $20,000 and considerable inconvenience. If this alternative is used, it is estimated that the annual maintenance will cost $5,000. If capital costs the municipality 12%, which alternative would you recommend with the capitalized worth method?

6-26   Estimates for a proposed development are as follows. Plan A has a first cost of $50,000, a life of 25 years, a $5,000 salvage value, and annual maintenance of $1,200. Plan B has a first cost of $90,000, a life of 50 years, no salvage value, and annual maintenance of $6,000 for the first 15 years and $1,000 per year for years 16–50. Assuming interest at 10%, compare the two plans by use of the capitalized worth (cost) method.

6-27   In the design of a certain system, two alternatives are under consideration. These alternatives are as follows.

	Plan A	Plan B
First cost	$50,000	$120,000
Life	20 years	50 years
Salvage value	$10,000	$ 20,000
Annual expenses	$ 9,000	$ 5,000

If *perpetual service life* is assumed, which of these alternatives do you recommend? The M.A.R.R. is 10%.

6-28   Use the *capitalized worth method* to determine which bridge design to recommend. The minimum attractive rate of return is 15%.

	Bridge Design A	Bridge Design B
Initial cost	$274,000	$326,000
Annual costs of upkeep	$ 10,000	$ 8,000
Interim replacement costs	$ 50,000 (every sixth year)	$ 42,000 (every seventh year)
Useful life	83 years	92 years
Salvage value	$ 0	$ 0

6-29   The demand for a certain product is expected to be constant for 4 years and then to increase sharply and continue at the higher level for the foreseeable future. Two alternative methods are available for providing the equipment required to produce the product. Method A is to provide one installation, A1, at a cost of $50,000, which will have a life of 10 years and will have a capacity sufficient to meet the requirements during the first 4 years. At the end of 4 years a second unit, A2, costing $40,000, would be added, which would have a life of 6 years and would supply sufficient output to meet the increased demand for the remainder of the 10-year period. With this alternative the relevant annual out-of-pocket costs would be $2,000 during years 1–4, and $4,000 during years 5–10. Alternative B is to provide a larger installation at a cost of $75,000 which would have a 10-year life and would have sufficient capacity to provide all the needs for the 10-year period. With this installation the relevant out-of-pocket costs would be $3,000 per year during years 1–4, and $3,500 per year during years 5–10. Assuming that neither installation would have any salvage value at the end of life, and that capital is worth 15%, which installation should be made?

6-30 Given: The Highridge Water District needs an additional supply of water from Steep Creek. The engineer has selected two plans for comparison.

(a) *Gravity plan:* Divert water at a point 10 miles up Steep Creek and carry it through a pipeline by gravity to the district.

(b) *Pumping plan:* Divert water at a point near the district and pump it through 2 miles of pipeline to the district. The pumping plant can be built in two stages, with one-half capacity installed initially and the other one-half 10 years later.

All costs are to be repaid within 40 years, with interest at 8%. Salvage values can be ignored. During the first 10 years, the average use of water will be less than the average during the remaining 30 years. Costs are as follows:

	Gravity	Pumping
Initial investment	$2,800,000	$1,400,000
Additional investment in tenth year	None	$ 200,000
Operation, maintenance, and replacements	$10,000/year	$25,000/year
Power cost:		
Average first 10 years	None	$ 50,000/year
Average next 30 years	None	$100,000/year

Required; Select the more economical plan for a 40-year period on the basis of present worth.

6-31 Three independent proposals are being considered:

	Proposal		
	X	Y	Z
Initial investment	$100	$150	$200
Uniform annual savings	$ 16.28	$ 22.02	$ 40.26
Useful life	10 years	15 years	8 years
Computed I.R.R. over the useful life	10%	12%	12%

The *before-tax M.A.R.R. is 10%* so all alternatives appear to be acceptable. At the end of their useful lives, proposals X and Z will be replaced with other proposals that have a 10% internal rate of return. Which proposals should be chosen if investment funds are limited to $250?

6-32 Which of the following independent alternatives would you recommend if no more than $30,000 is available for investment? The M.A.R.R. is 20%.

End of Year	Alternative		
	1	2	3
0	− $12,000	− $10,000	− $15,000
1	5,000	5,000	6,000
2	5,000	5,000	6,000
3	5,000	3,000	6,000
4	5,000	4,000	6,000
Rate of return	24.1%	27.2%	21.9%

6-33 The Upstart Corporation is trying to decide between two industrial cranes. Crane A and crane B below are mutually exclusive, and one of them must be chosen immediately.

	Crane	
	A	B
First cost	$250,000	$370,000
Operating cost/year	$ 22,000	$ 8,000
Salvage value (end of useful life)	$100,000	$125,000
Useful life	15 years	18 years
P.W. over useful life	− $366,353	− $408,923

Crane A has an extension boom that is optional. It would cost an extra $25,000 but would *save* an estimated $5,000 per year in operating cost. Crane B comes equipped with an extension boom, but an optional auger attachment can be purchased for $10,000. The auger would save $8,000 per year in drilling setup costs. The before-tax M.A.R.R. is 15% at Upstart. Carefully state your assumptions and make a recommendation regarding which alternative to select. The availability of funds is not a limiting factor in this situation.

6-34 Four proposals are under consideration by your company. Proposals A and C are mutually exclusive; proposals B and D are mutually exclusive and cannot be implemented unless proposal A *or* C has been selected. No more than $140,000 can be spent at time 0. The before-tax M.A.R.R. is 15%. The estimated cash flows are as follows:

	Proposal			
End of Year	A	B	C	D
0	− $100,000	− $20,000	− $120,000	− $30,000
1	+ 40,000	+ 6,000	+ 25,000	+ 6,000
2	+ 40,000	+ 10,000	+ 50,000	+ 10,000
3	+ 60,000	+ 10,000	+ 85,000	+ 19,000

Form all mutually exclusive combinations in view of the specified contingencies and determine which one should be selected.

6-35 Your company has $20,000 in "surplus" funds which it wishes to invest in new revenue-producing projects. There have been *three* independent sets of mutually exclusive proposals developed. The service life of each is 5 years and all salvage values are zero. You have been asked to perform an internal rate of return analysis to select the best combination of proposals. If the cost of capital is 12%, which combination of proposals would you recommend?

	Proposal	First Cost	Net Annual Benefits
Mutually exclusive	A1	− $ 5,000	+ $1,500
	A2	− 7,000	+ 1,800
Mutually exclusive	B1	− 12,000	+ 2,000
	B2	− 18,000	+ 4,000
Mutually exclusive	C1	− 14,000	+ 4,000
	C2	− 18,000	+ 4,500

6-36 **(a)** List all mutually exclusive alternatives when these conditions exist among the proposals below:

Proposal B1
Proposal B2 } independent (one or both can be chosen)

Proposal C1
Proposal C2 } mutually exclusive and dependent on the acceptance of B2

Proposal D   dependent (contingent) on the acceptance of B1

(b) In view of the cash flow pattern of each proposal, which alternative would you select if M.A.R.R. = 12% (before taxes)?

Proposal	End-of-Year Cash Flow ($000s), Year:				
	0	1	2	3	4
B1	− 50	20	20	20	20
B2	− 30	12	12	12	12
C1	− 14	4	4	4	4
C2	− 15	5	5	5	5
D	− 10	6	6	6	6

6-37 A firm is considering the development of several new products. The products under consideration are listed below and products in each group are mutually exclusive.

Group	Product	Development Cost	Annual Net Cash Income
A	A1	$ 500,000	$ 90,000
	A2	650,000	110,000
	A3	700,000	115,000
B	B1	600,000	105,000
	B2	675,000	112,000
C	C1	800,000	150,000
	C2	1,000,000	175,000

At most one product from each group will be selected. The firm has a minimum attractive rate of return of 10% and a budget limitation on development costs of $2,100,000. The life of all products is assumed to be 10 years, with no salvage value.
(a) List all mutually exclusive combinations.
(b) Using the net present worth criterion, which alternative should be selected?

CHAPTER 7

Uncertainty, Sensitivity and Nonmonetary Attributes

In the problems discussed in Chapters 5 and 6, specific assumptions were stated concerning applicable revenues and costs and other quantities important to the economic analysis. It was assumed that a high degree of confidence could be placed in all estimated values, which is sometimes called *assumed certainty*. Decisions made solely on the basis of this kind of analysis sometimes are called *decisions under certainty*. This is a rather misleading term, in that there rarely is a case in which estimated quantities can be assumed to be certain. In virtually all situations there is doubt as to the ultimate results that will be obtained from an investment. It is the purpose of this chapter to present and discuss several methods that are helpful in analyzing investment situations where such uncertainties exist. In addition, this chapter demonstrates practical methods of dealing with nonmonetary attributes if they are to be considered in engineering economy studies.

THE MEANING OF RISK, UNCERTAINTY, AND SENSITIVITY

Risk frequently is defined as the variations of actual values from estimated or expected values† that are due to chance (random) causes. In economic analyses,

†The statistical interpretation of *expected values* will be discussed later in this chapter.

the actual values or outcomes are known only after a project is undertaken or completed. *Uncertainty* frequently refers to variations in actual values that are due to errors in estimating, the inability to make accurate estimates because of insufficient information about the factor or the future, or the failure to consider all the factors. Although we may make a technical distinction between risk and uncertainty, both can cause study results to vary from predictions, and there seldom is anything significant to be gained by attempting to treat them separately. *Therefore, in the remainder of this book "risk" and "uncertainty" are used interchangeably.*

In dealing with uncertainty, it often is very helpful to determine to what degree changes in an estimate would affect an investment decision; that is, how *sensitive* a given investment situation is to changes in a particular factor which is not known with certainty. If a particular factor can be varied over a wide range without causing much effect on the investment decision, the decision under consideration is said not to be sensitive to that particular factor. Conversely, if a small change in the relative magnitude of a factor will reverse an investment decision, the decision is highly sensitive to that factor.

EVALUATION OF UNCERTAINTY

It is useful to consider the factors that may affect the uncertainty involved in an investment, so that they may be related to the measure of merit that must be met or exceeded for an investment to be justified. Similarly, they may be included as nonmonetary (irreducible) considerations in reaching a final decision.

The factors that affect uncertainty are many and varied. It would be almost impossible to list and discuss all of them. There are four major sources of uncertainty, however, which nearly always are present in economy studies.

The first factor, which is always present, is the *possible inaccuracy of the estimates used in the study*. If exact information is available regarding the items of income and expense, the resulting accuracy should be good. If, on the other hand, little factual information is available, and nearly all the values have to be estimated, the accuracy may be high or low, depending upon the manner in which the estimated values are obtained. Are they sound scientific estimates or merely guesses?

The accuracy of the income figures is difficult to determine. If they are based upon a considerable amount of past experience or have been determined by adequate market surveys, a fair degree of reliance may be placed on them. On the other hand, if they are merely the result of guesswork, with a considerable element of hope thrown in, they must of course be considered to contain a sizable element of uncertainty.

When dealing with a saving in existing operating expenditures, there should be less uncertainty involved. It is usually easier to determine what the saving will be since we have considerable experience and past history on which to base the estimates.

In most cases the income figures will contain more error than any other element of a study, with the possible exception of estimated operating expenditures. Frequently, annual income and expenditures are discovered to be the most sensitive elements in the study. There should be no large error in estimates of capital required. Uncertainty in investment capital requirements is often reflected as a ''contingency'' above the actual cost of plant and equipment. If we feel confident that the amount allowed for this purpose is on the high side, the resulting study is apt to be conservative.

The second key factor affecting uncertainty is the *type of business involved and the future health of the economy*. Some lines of business are notoriously less stable than others. For example, most mining enterprises are more risky than large retail food stores. However, we cannot arbitrarily say that an investment in any retail food store always involves less uncertainty than investment in mining property. Whenever capital is to be invested in an enterprise, the nature and history of the business as well as expectations of future economic conditions (e.g., interest rates) should be considered in deciding what risk is present. In this connection it becomes apparent that investment in an enterprise that is just being organized, and thus has no past history, is usually rather uncertain. This is especially true when the economy is dramatically changing because of business cycles.

A third factor affecting uncertainty is the *type of physical plant and equipment involved*. Some types of structures have rather definite economic lives and secondhand values. Little is known of the physical or economic lives of others, and they have almost no resale value. A good engine lathe generally can be used for many purposes in nearly any fabrication shop. Quite different would be a special type of lathe that was built to do only one unusual job. Its value would be dependent almost entirely upon the demand for the special task that it can perform. Thus the type of physical property involved will have a direct bearing upon the accuracy of the estimated income and expenditure patterns. Where money is to be invested in specialized plant and equipment, this factor should be considered carefully.

The fourth, and very important, factor that must always be considered in evaluating uncertainty is the *length of the assumed study period*. The conditions that have been assumed in regard to income and expense must exist throughout the study period in order for us to obtain a satisfactory return on the investment. A long study period naturally decreases the probability of all the factors turning out as estimated. Therefore, a long study period, all else being equal, always increases the uncertainty in an investment.

COMMON METHODS FOR DEALING WITH UNCERTAINTY

There are numerous methods for taking uncertainty, resulting from the four major sources described above, into account. This chapter discusses and illustrates each of the following popular methods.

1. *Breakeven analysis* is commonly utilized when the selection among alternatives is heavily dependent on a single factor, such as capacity utilization, that is uncertain. A breakeven point for the factor is determined such that

two alternatives are equally desirable from an economic standpoint. It is then possible to choose between the alternatives by estimating the most likely value of the uncertain factor and comparing this estimate to the breakeven value.

2. *Sensitivity analysis* is often employed when one or more factors are subject to uncertainty. The basic questions that sensitivity analysis attempts to resolve are: (a) What is the behavior of the measure of merit (e.g., net present worth) to $\pm x\%$ changes in each individual factor? (b) What is the amount of change in a particular factor that will cause a reversal in preference for an alternative?

3. *Optimistic–pessimistic estimation* of factors included in an engineering economy study has been used to establish a range of extreme values for the economic measure of merit. If the range falls *entirely* in the acceptable region (e.g., net annual worth ≥ 0), the alternative is desirable. Often, though, all factors estimated under conservative (pessimistic) conditions result in an unacceptable alternative. This method directs attention to the best and worst outcomes of going ahead with an alternative and requires managerial judgment to make the final go–no go determination.

4. *Risk-adjusted minimum attractive rates of return* are sometimes utilized to deal with estimation uncertainties. This method involves the use of higher M.A.R.R.'s for alternatives that are classified as ''highly uncertain'' and lower M.A.R.R.'s for projects for which there appear to be fewer uncertainties.

5. *Reduction of the useful life* of an alternative is another means for attempting to include explicitly the effects of uncertainty. Here the estimated project life is reduced by a fixed percentage, for instance 50%, and each alternative is evaluated regarding its acceptability over only this reduced life span.

6. *Probability functions* for uncertain elements are estimated and directly incorporated into the analysis of alternatives. This approach involves various statistical concepts and makes use of selected descriptive measures for summarizing uncertainty in the analysis. Two procedures based on formal probabilistic concepts are presented: (a) closed-form analysis of the measure of merit, and (b) Monte Carlo simulation. Any of the basic measures of economic merit presented in Chapter 5 could be used in connection with these two procedures.

BREAKEVEN ANALYSIS

Essentially all data employed in an economy study are uncertain simply because they represent estimates of the future. Breakeven analyses are useful when one must make a decision between alternatives which are highly sensitive to a variable or parameter and when that variable is difficult to estimate. Through breakeven analysis, one can solve for the value of that variable or parameter at which the conclusion is a standoff. That value of the variable is known as the *breakeven point*. (The use of breakeven points with respect to production and sales volumes was briefly discussed in Chapter 2.) If one can then estimate whether the actual outcome of that variable will be higher or lower than the breakeven point, the best alternative becomes apparent.

The following are examples of common variables for which breakeven analyses might provide useful insights into the decision problem:

1. *Revenue and annual cost:* Solve for the annual revenue required to equal (breakeven with) annual costs. Breakeven annual costs of an alternative can also be determined in a pairwise comparison when revenues are identical for both alternatives being considered.
2. *Rate of return:* Solve for the rate of return at which a given pair of alternatives are equally desirable.
3. *Salvage value:* Solve for future equipment resale value that would result in indifference as to preference for an alternative.
4. *Equipment life:* Solve for the useful life required for an alternative to be justified.
5. *Capacity utilization:* Solve for the hours of utilization per year, for example, at which an alternative is justified or at which two alternatives are equally desirable.

The usual breakeven problem involving two alternatives can be most easily approached mathematically by equating the annual worths or present worths of the two alternatives expressed as a function of a variable. In breakeven studies, project lives may or may not be equal and care should be taken to determine whether the coterminated or repeatability assumption best fits the situation. Examples below illustrate both mathematical and graphical solutions to typical breakeven problems.

EXAMPLE 7-1

Suppose that there are two alternative electric motors that provide 100-hp output. An Alpha motor can be purchased for $1,250 and has an efficiency of 74%, an estimated life of 10 years, and estimated maintenance costs of $50 per year. A Beta motor will cost $1,600 and has an efficiency of 92%, a life of 10 years, and annual maintenance costs of $25. Taxes and insurance costs on either motor will be $1\frac{1}{2}\%$ of the investment per year. If the minimum attractive rate of return is 15%, how many hours per year would the motors have to be operated at full load for the annual costs to be equal? Assume negligible salvage values for both and that electricity costs $0.05 per kilowatthour.

Solution by Mathematics

Note: 1 hp = 0.746 kW and input = output/efficiency. If N = number of hours of operation per year, components of the annual cost, A.C.$_\alpha$, for the Alpha motor would be as follows:

Capital recovery cost (depreciation and minimum profit):

$$\$1,250 \times (A/P, 15\%, 10) = \$1,250 \times 0.1993 = \$249$$

Operating cost for power:

$$(100 \times 0.746 \times N \times \$0.05)/0.74 = \$5.04N$$

Maintenance cost:

$$\$50$$

Taxes and insurance:

$$\$1,250 \times 0.015 = \$18.75$$

Similarly, for the Beta motor components of the annual cost, A.C.$_\beta$, in the same order as above would be as follows:

Capital recovery cost (depreciation and minimum profit):

$$\$1,600 \times (A/P, 15\%, 10) = \$1,600 \times 0.1993 = \$319$$

Operating cost for power:

$$(100 \times 0.746 \times N \times \$0.05)/0.92 = \$4.05N$$

Maintenance cost:

$$\$25$$

Taxes and insurance:

$$\$1,600 \times 0.015 = \$24$$

At the breakeven point, A.C.$_\alpha$ = A.C.$_\beta$. Thus

$$\$249 + \$5.04N + \$50.00 + \$18.75 = \$319 + \$4.05N$$

$$+ \$25.00 + \$24.00$$

$$\$5.04N + \$317.75 = \$4.05N + \$368.00$$

$$N \cong 51 \text{ hours/year} \qquad \blacksquare$$

Solution by Graphics

Figure 7-1 shows a curve for the total annual costs of each motor as a function of the number of hours of operation per year. The constant annual costs (Y-intercepts) are \$317.75 and \$368.00 for Alpha and Beta, respectively, and the costs that vary directly with hours of operation per year (slope of lines) are

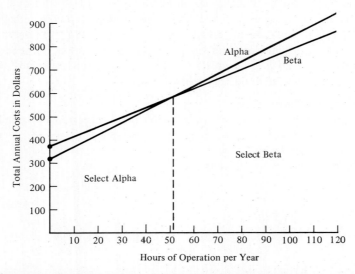

FIGURE 7-1 Graphical Solution of Breakeven Point for Example 7-1.

$5.04 and $4.05 for Alpha and Beta, respectively. Of course, the breakeven point is the value of the independent variable at which the annual cost curves for the two alternatives intersect. ■

As was pointed out in Chapter 6, there is always some uncertainty as to what future requirements will be and when they will occur. Provision of a certain amount of capacity will establish a limit to the service that can be provided in the future. It often is helpful to know at what future date a deferred investment will be needed so that an alternative permitting deferred investment will break even with one that provides immediately for all future demands. Where only the costs of acquiring the assets by the two alternatives need to be considered, or where the annual costs through the entire life are not affected by the date of acquisition of the deferred asset, the breakeven point may be determined very easily and may be helpful in arriving at a decision between alternatives. Example 7-2 illustrates this type of breakeven study.

EXAMPLE 7-2

In the planning of a two-story municipal office building the architect has submitted two designs. The first provides foundation and structural details so that two additional stories can be added at a later date without modifications to the original structure. This building would cost $1,400,000. The second design, without such provisions, would cost only $1,250,000. If the first plan is adopted, it is estimated that an additional two stories could be added at a later date at a cost of $850,000. If the second plan is adopted, however, considerable strengthening and reconstruction would be required, which would add $300,000 to the cost of a two-story addition. Assuming that the building is expected to be needed for 75 years, by what time would the additional two stories have to be built to make the adoption of the first design justified? The M.A.R.R. is 10%.

Solution

The breakeven deferment period is determined as follows:

	Provide Now	No Provision
Present worth–cost:		
First unit	$1,400,000	$1,250,000
Second unit	$850,000 × $(P/F, 10\%, N)$	$1,150,000 × $(P/F, 10\%, N)$
Equating total present worth–costs:		
$1,400,000 + $850,000 × $(P/F, 10\%, N)$ = $1,250,000 + $1,150,000 × $(P/F, 10\%, N)$		

Solving, we have

$$(P/F, 10\%, N) = 0.5$$

From the 10% table in Appendix E, $N = 7$ years (approximately). Thus, if the additional space will be required in less than 7 years, it would be more economical to make immediate provision in the foundation and structural details. If the addition would not be likely to be needed until after 7 years, greater economy would be achieved by making no such provisions in the first structure. ■

A graphical solution as shown in Figure 7-1 is particularly useful to give the analyst or decision maker a visual portrayal of the *difference* in the economic desirability for alternatives over a wide range of the variable under consideration. There are many situations in which the relationship between the dependent and independent variables is not continuous and cannot therefore be expressed readily in mathematical terms. In other cases the relationship may be complex enough that the time required to develop a mathematical formula would be so great that it would be uneconomical to solve the problem in this manner. In such cases a graphical solution may be used to determine the breakeven point.

SENSITIVITY ANALYSIS

In a great many cases a simple breakeven analysis is not feasible because several factors may vary simultaneously as the single variable under study is varied. In such instances, and in fact in most cases, it is helpful to determine how sensitive the situation is to the several variables so that proper weight and consideration may be assigned to them. Sensitivity, in general, means the relative magnitude of change in the measure of merit (such as rate of return) caused by one or more changes in estimated study parameters. Sometimes sensitivity is more specifically defined to mean the relative magnitude of the change in one or more factors that will reverse a decision among alternatives. Example 7-3 extends sensitivity explorations from one variable to several different variables through the use of tabular displays. Following that, the use of optimistic–most likely–pessimistic estimates and graphical means for describing the results of sensitivity studies are shown.

EXAMPLE 7-3

A small group of investors is considering starting a premixed-concrete plant in a rapidly developing suburban area about 5 miles from a large city. The group believes that there will be a good market for premixed concrete in this area for at least the next 10 years, and that if they establish such a local plant it would be unlikely that another local plant would be established. Existing plants in the adjacent city would, of course, continue to serve this new area. The investors believe that the plant could operate at about 75% of capacity 250 days per year since it is located in an area where the weather is mild throughout the year.

The plant will cost $100,000 and it would have a capacity of 72 cubic yards of concrete per day. Its salvage value at the end of 10 years is estimated to be $20,000, which is the value of the land. To deliver the concrete, four trucks would be required, costing $8,000 each, having an estimated life of 5 years and a trade-in value of $500 each at the end of that time. In addition to the four truck drivers, who would be paid $50.00 per day each, four people would be required to operate the plant and office, at a cost of $175.00 per day. Annual operating and maintenance costs for the plant and office are estimated at $7,000 and for each truck at $2,250, both in view of 75% capacity utilization. Raw-material costs are estimated to be $27.00 per cubic yard of concrete. Payroll taxes, vacations, and other fringe benefits would amount to 25% of the annual payroll. Taxes and insurance on each truck would be $500, and taxes and insurance on the plant would be $1,000 per year. The investors would not contribute any labor to the business, but a manager would be employed at an annual salary of $20,000.

Delivered, premixed concrete currently is selling for an average of $45 per cubic yard. A useful plant life of 10 years is expected, and capital invested elsewhere by these investors is earning about 15% before income taxes. It is desired to find the *annual worth* for the expected conditions above and to perform sensitivity studies for certain variables.

Solution by A.W. Method

Annual revenue:

$$72 \times 250 \times \$45 \times 0.75 = \$607,500$$

Annual costs:

1. Capital recovery
   Plant: $100,000(A/P, 15\%, 10)$
   $- \$20,000(A/F, 15\%, 10)$ $= \quad \$18,940$
   Trucks: $4[\$8,000(A/P, 15\%, 5)$
   $- \$500(A/F, 15\%, 5)]$ $= \quad 9,250$
              $\$ 28,190$

2. Labor:
   Plant and office: $\$175 \times 250$ $= \quad 43,750$
   Truck drivers: $4 \times \$50 \times 250$ $= \quad 50,000$
   Manager $= \quad 20,000$
              $\$113,750$

3. Payroll taxes, etc.: $\$113,750 \times 0.25$ $28,438$

4. Taxes and insurance:
   Plant $= \quad 1,000$
   Trucks: $\$500 \times 4$ $= \quad 2,000$
              $\$ 3,000$

5. Operation and maintenance at 75% capacity:
   Plant and office $= \quad 7,000$
   Trucks: $\$2,250 \times 4$ $= \quad 9,000$
              $\$ 16,000$

6. Materials: $72 \times 0.75 \times 250 \times \$27.00$ $364,500$
   TOTAL $\$553,878$

The net A.W. for these "most likely" estimates is $607,500 - \$553,878 = \$53,622$. Apparently, the project is a highly attractive investment opportunity. ■

In Example 7-3 there are three variable factors that are of great importance and that must be estimated: *capacity utilization,* the *selling price of the product,* and the *economic life of the plant.* A fourth factor—raw-material costs—is important, but any significant change in this factor also would be equally effective to competitors and probably would be reflected in a corresponding change in the selling price of mixed concrete. The other cost elements should be determinable with considerable accuracy. Therefore, we would like to investigate the effect of variations in the plant utilization, selling price, and economic life. Sensitivity analysis presents a good method for doing this.

SENSITIVITY TO CAPACITY UTILIZATION

As a first step we must determine how cost factors would vary, if at all, as capacity utilization is varied. In this case, it is probable that the cost items listed under groups 1, 2, 3, and 4 in the previous tabulation would be virtually un-affected if capacity utilization should vary over a quite wide range—from 50 to 90%, for example. To meet peak demands, the same amount of plant, trucks, and personnel probably would be required. Group 5 costs for operation and maintenance would be affected somewhat. For this type of factor we must try to determine what the variation would be or make a reasonable assumption as to the probable variation. For this case it will be assumed that one half of these costs would be fixed and the other half would vary with capacity utilization by a straight-line relationship. Certain other factors, such as the cost of materials in this case, will vary in direct proportion to the capacity utilization.

Using these assumptions, Table 7-1 shows how the revenue, costs and net annual worth would change with different capacity utilizations. It will be noted that the annual worth is moderately sensitive to capacity utilization. The plant could be operated at a little less than 65% of capacity, instead of the assumed 75%, and still produce a net annual worth greater than 0. Also, quite clearly, if they should be able to operate above the assumed 75% of capacity, the annual worth would be very good. This type of analysis provides those who must make the decision with a good idea as to how much leeway they would have in capacity utilization and still have an acceptable venture.

SENSITIVITY TO SELLING PRICE

Examination of the sensitivity of the project to the selling price of the concrete reveals the situation shown in Table 7-2. The values in this table assume that the plant would operate at 75% of capacity; thus the costs would remain constant,

TABLE 7-1 Net Annual Worth at $i = 15\%$ for Premixed-Concrete Plant for Various Capacity Utilizations[a] (Average Selling Price Equals $45 per Cubic Yard)

	50% Capacity	65% Capacity	90% Capacity
Annual revenue	$405,000	$526,500	$729,000
Annual costs:			
Capital recovery	28,190	28,190	28,190
Labor	113,750	113,750	113,750
Payroll taxes and similar items	28,438	28,438	28,438
Taxes and insurance	3,000	3,000	3,000
Operations and maintenance	13,715	15,086	17,372
Materials	243,000	315,900	437,400
TOTAL COSTS	$430,093	$504,364	$628,150
Net A.W.	−$25,093	+$22,136	+$110,850

[a]At 75% capacity utilization, $x/2 + (x/2)(0.75) = \$16,000$, so that $x = \$18,286$ at 100% capacity utilization. Therefore, at 50% utilization, the operations and maintenence cost would be $\$9,143 + 0.5(\$9,143) = \$13,715$.

TABLE 7-2   Effect of Various Selling Prices on the Net Annual Worth for Premixed-Concrete Plant Operating at 75% of Capacity

	Selling Price			
	$45.00	$43.65(3%)ᵃ	$42.75(5%)ᵃ	$40.50(10%)ᵃ
Annual revenue	$607,500	$589,275	$577,125	$546,750
Annual costs	553,878	553,878	553,878	553,878
	$ 53,622	$ 35,397	$ 23,247	$ 7,128

ᵃPercentage values shown in parentheses are reductions in price below $45.

with only the selling price varying. Here it will be noted that the project is quite sensitive to price. A decrease in price of 10% would drop the rate of return to less than 15% (i.e., the net A.W. < 0). Since a decrease of 10% is not very large, the investors would want to make a thorough study of the price structure of conrete in the area of the proposed plant, particularly with respect to the possible effect the increased competition that the new plant would create.

SENSITIVITY TO USEFUL LIFE

The effect of the third factor, assumed useful life of the plant, can be investigated readily. If a life of 5 years were assumed for the plant, instead of the assumed value of 10 years, the only factor in the study that would be changed would be the cost of capital recovery. If the salvage value is assumed to remain constant, the capital recovery cost over a 5-year period is

$100,000 (A/P, 15\%, 5) - \$20,000 (A/F, 15\%, 5) = \$26,866$ per year

which is an increase of $7,926. In this case the net annual worth would be reduced to $45,696—a decline of 14.8%. Hence, a 50% reduction in useful life causes only a 14.8% reduction in net annual worth. Clearly, the venture is quite insensitive to the assumed useful life.

With the added information supplied by the sensitivity analyses that have just been described, those who would have to make the investment decision concerning the proposed concrete plant would be in a much better position to make a decision than if they had only the initial study results, based on an assumed utilization of 75% of capacity, available to them. They would know which factors were critical and thus could seek more information about these particular items if desired.

OPTIMISTIC–PESSIMISTIC ESTIMATES AND GRAPHICAL WAYS TO DESCRIBE SENSITIVITY†

A useful method for exploring sensitivity is to change one or more factors to be estimated in a favorable (optimistic) direction and in an unfavorable (pessi-

†Adapted from J. R. Canada and W. G. Sullivan, "To understand the decision problem, say it with pictures," Proceedings of Spring National Conference, American Institute of Industrial Engineers, 1977.

TABLE 7-3 Optimistic–Most Likely–Pessimistic Estimates and Net Annual Worths for Proposed Ultrasound Device

	Estimation Condition		
	Optimistic (O)	Most Likely (M)	Pessimistic (P)
Investment	$150,000	$150,000	$150,000
Life	18 years	10 years	8 years
Salvage value	0	0	0
Annual savings	$110,000	$ 70,000	$ 50,000
Annual disbursements	$ 20,000	$ 43,000	$ 57,000
Minimum attractive rate of return	8%	8%	8%
Net annual worth (net A.W.)	+$ 73,995	+$ 4,650	−$ 33,100

mistic) direction to investigate the effect of these changes on the economy study result. This represents a simple method for including uncertainty in the analysis.

In applications of this method, the optimistic condition for a factor is often specified as a value that will be exceeded by the actual outcome with a probability of 5%. Similarly, the pessimistic condition will be exceeded by the actual outcome with a probability of 95%.

As an example, consider a proposed ultrasound inspection device for which the optimistic, pessimistic and most likely (or best) estimates are given in Table 7-3. Also shown at the end of Table 7-3 are the net annual worths (A.W.s) for all three estimation conditions. Note that the net A.W. for optimistic conditions is highly favorable (+$73,995), while for pessimistic conditions it is quite unfavorable (−$33,100). After obtaining this information, the decision maker may be willing to make a go–no go decision on the proposed device. However, he should recognize that these are extreme outcomes—the optimistic net A.W. assumes that *all* estimated factors turn out according to the optimistic estimates, and the pessimistic net A.W. assumes that *all* estimated factors turn out per the pessimistic estimates. It is reasonable to assume that this will not happen, but instead that different factors may turn out to have a mixture of optimistic, most likely and pessimistic outcomes. One good way to reflect such results is shown in Table 7-4, which shows net annual worths for all combinations of estimated

TABLE 7-4 Net Annual Worths ($) for All Combinations of Estimated Outcomes[a] for Annual Savings, Annual Disbursements, and Life—Proposed Ultrasound Device

	Annual Disbursements								
	O			M			P		
Annual Savings	Life			Life			Life		
	O	M	P	O	M	P	O	M	P
O	73,995	67,650	63,900	50,995	44,650	40,900	36,995	30,650	26,900
M	34,000	27,650	23,900	10,995	4,650	900	−3,005	−9,350	−13,100
P	14,000	7,650	3,900	−9,005	−15,350	−19,100	−23,005	−29,350	−33,100

[a]Estimates: O, optimistic; M, most likely; P, pessimistic.

outcomes for three of the key elements being estimated. This also could have been done for four or more factors if this many had been subject to significant variation, but the size of the matrix display would grow enormously. Even the 3 × 3 × 3 matrix shown in Table 7-4 quickly becomes cumbersome because of the proliferation of numbers. There are ways around this, as we shall soon see.

MAKING MATRICES EASIER TO INTERPRET

It should be recognized that the net A.W. numbers in Table 7-4 are the results of estimates subject to varying degrees of uncertainty. Hence little information of value would be lost if the numbers were rounded to the nearest thousand dollars. Further, suppose that management is most interested in the number of combinations of conditions in which the net A.W. is, say (a) more than $50,000 and (b) less than $0. Table 7-5 shows how Table 7-4 might be changed to make it easier to interpret and use in communicating with management.

From Table 7-5 it is apparent that four combinations result in a net A.W. > $50,000 while nine produce a net A.W. < $0. Each combination of conditions is not necessarily equally likely. Therefore, statements such as "There are 9 chances out of 27 that we will lose money on this project" are not appropriate in this example.

GRAPHICAL SENSITIVITY DISPLAYS

An effective way of displaying and examining sensitivity is to graph the measure of merit for independent variation of all factors of interest by expressing variation for each on a common abscissa in terms of percent deviation from its most likely value. This is shown in Figure 7-2 for the proposed ultrasound inspection device,

TABLE 7-5 Results in Table 7-4 Made Easier to Interpret (Net Annual Worths in $000s)[a,b]

Annual Savings	Annual Disbursements								
	O			M			P		
	Life			Life			Life		
	O	M	P	O	M	P	O	M	P
O	(74)	(68)	(64)	(51)	45	41	37	31	27
M	34	28	24	11	5	1	− 3	− 9	− 13
P	14	8	4	− 9	− 15	− 19	− 23	− 29	− 33

[a]Estimates: O, optimistic; M, most likely; P, pessimistic.
[b]Circled entries, net annual worth > $50,000 (4 out of 27 combinations); Underscored entries, net annual worth < $0 (9 out of 27 combinations).

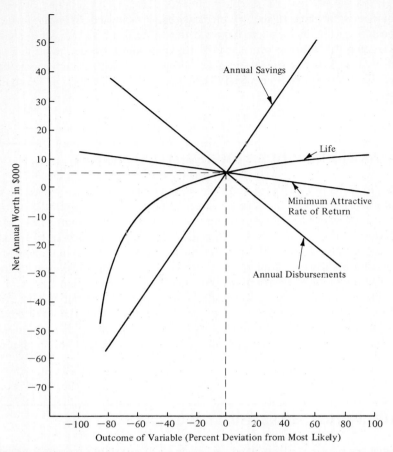

FIGURE 7-2 Example Graph of Sensitivity to Multiple Variables, Each Independently Deviating From the Most Likely Estimate.

for which the most likely estimates were given in Table 7-3. Figure 7-2 shows, among other things, that net annual worth is relatively sensitive to changes in annual savings, annual disbursements, and reductions in project life. It also shows that net annual worth is relatively insensitive to changes in the M.A.R.R. and to increases in the project life.

Another type of sensitivity test that is often quite valuable is to determine the relative (or absolute) change in one or more factors which will just reverse the decision. Applied to the example in Table 7-3, this means determining the relative change in each factor that will decrease the net A.W. by $4,650 so that it reaches $0. Figure 7-3 shows this by using a table and histogram bars of varying lengths to emphasize that the indicated economic merit of the device is (1) most sensitive to changes in the estimated annual receipts, and (2) least sensitive to changes in the salvage value.

It is clear that even with a few factors the number of possible combinations of conditions in a sensitivity analysis can become quite large, and the task of

FIGURE 7-3 Sensitivity of Decision Reversal to Changes in Selected Estimates.

	Most Likely Estimate	Required Outcome†	Amount of Change	Change Amount as Percent of Most Likely
Investment	$150,000	$181,200	+$31,200	+20.8% ———
Life	10 years	7.3 years	−2.7 years	−27.0% ———
Salvage Value	0	−67,890	−67,890	∞ ——————→
Annual Savings	70,000	65,350	− 4,650	− 6.6% —
Annual Disbursements	43,000	47,650	+ 4,650	+10.8% ——
Minimum Attractive Rate of Return (M.A.R.R.)	8%	12.5%	+ 4.5%	+56% —————

†To reverse decision (decrease A.W. to $0)

investigating all of them might be quite time consuming and costly. Ordinarily, a sensitivity analysis involves eliminating from detailed consideration those factors for which the measure of merit is quite insensitive and highlighting the conditions for other factors to be studied further in accordance with the degree of sensitivity of each. Thus the number of combinations of conditions included in the analysis hopefully can be kept to a manageable size. Digital computers are quite useful for sensitivity analyses, particularly when evaluating the effect of simultaneous variations in several elements. Creative displays of relevant information in tabular and graphical form can be invaluable in promoting insight into, and communicating about, decision problems.

DECISION RULES FOR INTERPRETING OPTIMISTIC–PESSIMISTIC ESTIMATES

The estimates of Table 7-3 provided a variety of tabular and graphical results for the net annual worth of a proposed ultrasound detection device. However, the absence of objective or subjective degrees of belief that each set of estimates (O, M, P) would come true caused some difficulty in interpreting the results of this particular method. One way to remedy the shortcoming is to utilize selected decision rules that allow the analyst to make a recommendation when several *mutually exclusive* alternatives are under consideration.

Example 7-4
The following is a description of several decision rules that can be applied to two or more investment alternatives. Alternative A is the equipment described in Table 7-3. Two other alternatives have been estimated in the same fashion, and all resultant net annual worths for optimistic–most likely–pessimistic conditions for these alternatives are presented in Table 7-6

TABLE 7-6 Net Annual Worths at M.A.R.R. = 8% for Three Investment Alternatives

Alternative	Estimation Conditions		
	Optimistic	Most Likely	Pessimistic
A	$73,995	$4,650	− $33,100
B	51,450	8,940	1,475
C	18,230	9,600	2,180

Solution by Maximin (or Minimax) Rule

The maximin rule is quite conservative, for it involves determining the minimum net annual worth associated with each alternative and then selecting the alternative that maximizes (is the highest of) these minimum net annual worths. Similarly, in the case of costs, the minimax rule suggests that the decision maker determine the maximum cost associated with each alternative and then select the alternative that minimizes (is the lowest of) those maximum costs.

For the problem depicted in Table 7-6, the minimum possible profits are: − $33,100 if alternative A is chosen; $1,475 if alternative B is chosen; and $2,180 if alternative C is chosen. The $2,180 for alternative C is the maximum of these minimum net annual worths, so C would be chosen by this conservative rule. ∎

Solution of Maximax (or Minimin) Rule

The maximax rule is the most optimistic, or nonconservative, of the decision rules for complete uncertainty. It involves determining the maximum net annual worth associated with each alternative and then selecting the alternative that maximizes these maximum annual worths. Similarly, in the case of costs, the minimin rule suggests that the decision maker determine the minimum cost outcome for each alternative and then select the alternative that minimizes (is the lowest of) those minimum costs.

For the example in Table 7-6, the maximum possible net annual worth is $73,995 if alternative A is chosen; $51,450 if alternative B is chosen; and $18,230 if alternative C is selected. Consequently, alternative A would be the indicated choice by this highly optimistic decision rule. ∎

Solution by Laplace Rule

This rule simply assumes that all possible estimation conditions are equally likely so that the choice is based on the average outcomes as calculated using equal probabilities for all outcomes. There is a common tendency to use this assumption implicitly in situations where there is no evidence to the contrary, but the assumption (and hence the rule) is of highly questionable merit.

For the example in Table 7-6, the average outcome, E, for each method can be calculated as:

$$E_A = \$73,995(1/3) + \$4,650(1/3) - \$33,100(1/3) = \$15,182$$

$$E_B = \$51,450(1/3) + \$8,940(1/3) + \$1,475(1/3) = \$20,622$$

$$E_C = \$18,230(1/3) + \$9,600(1/3) + \$2,180(1/3) = \$10,003$$

Thus alternative B has the highest average net annual worth and would be the choice by this rule. ∎

Solution by Hurwicz Principle or Rule

This rule is intended to reflect any degree of moderation between extreme optimism (as in maximax rule) and extreme conservatism (as in maximin rule) which the decision maker may wish to choose. The rule may be stated explicitly as follows: Select an index of optimism, H, such that $0 \leq H \leq 1$. For each alternative, compute the weighted outcome: $H \times$ (value of profit or cost if more favorable outcome occurs) $+ (1 - H) \times$ (value of profit or cost if least favorable outcome occurs). Choose the alternative that maximizes the weighted outcome (i.e., select the largest positive A.W. or P.W. or the smallest negative A.W. or P.W.).

When this rule is applied to the example in Table 7-6, assuming an index of optimism, H, of 0.2, the weighted outcome, W, of each alternative is

$$W_A = \$73,995(0.2) - \$33,100(0.8) = -\$11,681$$

$$W_B = \$51,450(0.2) + \$ 1,475(0.8) = \$11,470$$

$$W_C = \$18,230(0.2) + \$ 2,180(0.8) = \$ 5,390$$

Thus alternative B is seen to be the best choice using the Hurwicz rule and an index of optimism of 0.2. ∎

Actually, for the problem in Example 7-4, many decision makers would eliminate alternative A from consideration because of the possibility of a significant loss being incurred. Rather few business executives want to undertake a venture when there is considerable likelihood of a loss being incurred, even though the possible gain may be quite great. Instead, they prefer to select between alternatives that offer varying amounts of gain that are relatively certain. Consequently, our unwillingness to gamble, with loss as a possible outcome, often is an important factor in investment decisions, and the extent of such unwillingness will vary from person to person, and even for a given person at different times. Therefore, it is doubtful that a completely logical basis for decision making can be achieved. Probably the best that can be expected is that decision makers can recognize and justify the bases on which they make decisions and achieve a reasonable degree of consistency in their decisions. Perhaps life is sometimes more interesting because decisions cannot always be predicted on a "logical" basis!

RISK-ADJUSTED MINIMUM ATTRACTIVE RATES OF RETURN

Uncertainty causes factors integral to economy studies, such as cash flows and project life, to become random variables in the analysis. (Simply stated, a random variable is a function that assigns a unique numerical value to each possible outcome of a probabilistic quantity.) A widely used industrial practice for explicitly including uncertainty is to increase the M.A.R.R. when a project is thought to be relatively uncertain. Most likely estimates of other factors are then utilized in the study. Hence a procedure has emerged that employs "risk-adjusted" interest rates. It should be noted, however, that many pitfalls of performing studies of financial profitability with risk-adjusted M.A.R.R.'s have been identified.†

In general, the preferred practice to account for uncertainty in estimates (cash flows, project life, etc.) is to directly deal with their suspected variations in terms of probability assessments rather than to manipulate the M.A.R.R. as a means of reflecting the "virtually certain" versus "highly uncertain" status of a capital investment project. Intuitively, the risk-adjusted interest rate procedure makes good sense because much more uncertainty regarding the overall profitability of a project exists in the early years compared to, say, the last 2 years of its life. By increasing the M.A.R.R., emphasis is placed on early cash flows rather than on longer-term benefits, and this would appear to compensate for time-related project uncertainties. But the question of uncertainty in cash flow amounts is not directly addressed. The following example illustrates how this method of dealing with uncertainty can lead to an illogical recommendation.

EXAMPLE 7-5

The Atlas Corporation is considering two alternatives, both affected by uncertainty to different degrees, for increasing the recovery of a precious metal from its smelting process. Data are shown below concerning capital investment requirements and estimated annual savings of both alternatives. The firm's M.A.R.R. for its "risk-free" investments is 10%.

End-of-Year Cash Flow	Alternative	
	P	Q
0	− $160,000	− $160,000
1	+ 120,000	+ 20,827
2	+ 60,000	+ 60,000
3	0	+ 120,000
4	+ 60,000	+ 60,000

Because of technical considerations involved, alternative P is thought to be *more uncertain* than Q. Therefore, according to the Atlas Corporations's engineering economy handbook, the risk-adjusted M.A.R.R. applied to P will be 20% and the risk-adjusted M.A.R.R. for Q has been set at 17%. Which alternative should be recommended?

Solution

At the risk-free M.A.R.R. of 10%, both alternatives have the same net present worth of $39,659. All else being equal, alternative Q would be chosen because

†A. A. Robichek, and S. C. Myers, "Conceptual problems in the use of risk-adjusted discount rates," *Journal of Finance*, Vol. 21 (December 1966), pp. 727–730.

it is less uncertain than alternative P. Now a net present worth analysis is performed for the Atlas Corporation using their prescribed risk-adjusted M.A.R.R.'s for the two options.

$$\text{N.P.W.}_P(20\%) = -\$160,000$$
$$+ \$120,000(P/F, 20\%, 1) + \$60,000(P/F, 20\%, 2)$$
$$+ \$60,000(P/F, 20\%, 4) = \$10,602$$
$$\text{N.P.W.}_Q(17\%) = -\$160,000 + \$20,827(P/F, 17\%, 1)$$
$$+ \$60,000(P/F, 17\%, 2)$$
$$+ \$120,000(P/F, 17\%, 3)$$
$$+ \$60,000(P/F, 17\%, 4) = \$8,575$$

Without considering uncertainty (i.e., M.A.R.R. = 10%), the selection was seen to be alternative Q. But when the more uncertain alternative P is "penalized" by applying a higher risk-adjusted M.A.R.R. to compute its net P.W., the comparison of alternatives favors alternative P! One would expect to see alternative Q recommended with this procedure. This illogical result can be seen clearly in Figure 7-4, which demonstrates the general situation where contradictory results might be expected. ∎

Even though the intent of the risk-adjusted M.A.R.R. is to make more uncertain projects appear less economically attractive, the opposite was shown to be true in Example 7-5. Furthermore, a related shortcoming of the risk-adjusted M.A.R.R. procedure is that cost-only projects are made to appear more desirable (less negative net P.W., for example) as the interest rate is adjusted upward to account for uncertainty. At extremely high interest rates, the alternative having

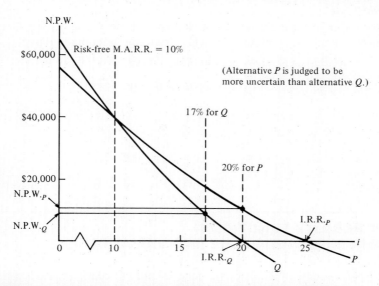

FIGURE 7-4 Graphical Portrayal of Risk-Adjusted Interest Rates.

the lowest investment requirement would be favored regardless of subsequent cost cash flows. Because of such difficulties as illustrated above, this procedure is not generally recommended as an acceptable means of dealing with uncertainty.

REDUCTION OF USEFUL LIFE

Many of the methods for dealing with uncertainty that have been discussed to this point have attempted to compensate for potential losses that could be incurred if conservative decision-making practices are not followed. Thus, dealing with uncertainty in an engineering economy study tends to lead to the adoption of conservative (pessimistic) estimates of factors so as to reduce "downside risks" of making a wrong decision.

The method considered in this section makes use of a truncated project life that is often considerably less than the estimated useful life. By dropping from consideration those revenues (savings) and costs that may occur after the reduced study period, heavy emphasis is placed on rapid recovery of investment capital in the early years of a project's life. Consequently, this method is closely related to the discounted payback technique discussed in Chapter 5, and it suffers from most of the same deficiencies that beset the payback method.

EXAMPLE 7-6

Suppose that the Atlas Corporation referred to in Example 7-5 decided not to utilize risk-adjusted interest rates as a means of recognizing uncertainty in their economy studies. Instead, they have decided to truncate the study period at 75% of the most likely estimate of useful life. Hence all cash flows past the third year would be ignored in the analysis of alternatives. By using this method, should alternative P or Q be selected when M.A.R.R. = 10%?

Solution

Based on the net present worth criterion, it is apparent that neither alternative would be the choice with this procedure for recognizing uncertainty.

$$\text{N.P.W.}_\text{P}(10\%) = -\$160{,}000 + \$120{,}000(P/F, 10\%, 1)$$

$$+ \$60{,}000(P/F, 10\%, 2)$$

$$= -\$1{,}322$$

$$\text{N.P.W.}_\text{Q}(10\%) = -\$160{,}000 + \$20{,}827(P/F, 10\%, 1)$$

$$+ \$60{,}000(P/F, 10\%, 2)$$

$$+ \$120{,}000(P/F, 10\%, 3) = -\$1{,}322 \quad\blacksquare$$

PROBABILITY FUNCTIONS—
THEIR EXPECTED VALUE AND VARIANCE

Expected values and *variances* are quite helpful in making decisions when uncertainty is involved. Expected values and variances are associated with the concept of probability, which is usually considered to be the long-run relative frequency with which an event occurs, or the subjectively judged likelihood that it will occur.

The expected value is the *product* of the probability that a revenue, cost, or other variable will occur and its value (numerical outcome) if it does occur, summed over all possible outcomes of the variable. Thus, if the useful life of a project is estimated to be 10 years with 0.4 probability and 15 years with 0.6 probability, the expected project life is $10 \times 0.4 + 15 \times 0.6 = 4 + 8 = 12$ years. Factors having probabilistic outcomes are often called *random variables*.

In general, the expected value of a random variable whose discrete outcomes, x_i, occur in accordance with a known probability function can be computed as

$$E(X) \doteq \sum_{i=1}^{n} x_i \, Pr \, (x_i) \qquad (7\text{-}1)$$

where $E(X)$ = expected value of the random variable X
$Pr \, (x_i)$ = probability of x_i occurring

$$\sum_{i=1}^{n} Pr \, (x_i) = 1; \qquad i = 1, 2, \ldots, n \text{ outcomes}$$

The variance of a random variable is a measure of how much dispersion in possible outcomes exists about the expected value and tends to indicate the degree of uncertainty associated with a random variable. Specifically, the variance of a discrete random variable, X, is given by the following equation:

$$V(X) = \sum_{i=1}^{n} x_i^2 \, Pr \, (x_i) - [E(X)]^2 \qquad (7\text{-}2)$$

The standard deviation of X, SD (X), is merely the square root of $V(X)$.

Expected value and variance concepts apply theoretically to long-run conditions in which it is assumed the event is going to occur repeatedly. However, application of these principles is often useful even when investments are not going to be made repeatedly over the long run.

EXAMPLE 7-7

We now apply the expected value and variance principles to the premixed-concrete plant discussed earlier in Example 7-3. Suppose that the probabilities of attaining various capacity utilizations are

Capacity (%)	Probability
50	0.10
65	0.30
75	0.50
90	0.10

It is desired to determine the expected value and variance of *annual revenue*.

Subsequently, the expected value and variance of net A.W. for the project can be computed. By evaluating both $E(A.W.)$ and $V(A.W.)$ for the concrete plant, an indication of the venture's average profitability and its uncertainty is obtained.

Solution for Annual Revenue:

i	Capacity (%)	(A) Probability, Pr (x_i)	(B) Revenue,[a] x_i	(A) × (B): Expected Revenue	(C) = (B)^2 x_i^2	(A) × (C)
1	50	0.10	$405,000	$ 40,500	1.64×10^{11}	0.164×10^{11}
2	65	0.30	526,500	157,950	2.77×10^{11}	0.831×10^{11}
3	75	0.50	607,500	303,750	3.69×10^{11}	1.845×10^{11}
4	90	0.10	729,000	72,900	5.31×10^{11}	0.531×10^{11}
				$575,100		3.371×10^{11}

[a]From Table 7-1.

Expected value of annual revenue: $\Sigma (A \times B) = \$575,100$

Variance of annual revenue: $\Sigma (A \times C) - (575,100)^2 = 6,400 \times 10^6$

Solution for Net Annual Worth:

i	Capacity (%)	(A) Pr (x_i)	(B) Annual Worth, x_i	(A) × (B): Expected A.W.	(C) = (B)^2 (A.W.)^2	(A) × (C)
1	50	0.10	− $ 25,093	− $ 2,509	0.63×10^9	0.063×10^9
2	65	0.30	+ 22,136	+ 6,641	0.49×10^9	0.147×10^9
3	75	0.50	+ 53,622	+ 26,811	2.88×10^9	1.440×10^9
4	90	0.10	+ 100,850[a]	+ 10,085	10.17×10^9	1.017×10^9

[a]See Table 7-1.

Expected value of net A.W.: $\Sigma (A \times B) = \$41,028$

Variance of net A.W.: $\Sigma (A \times C) - (41,028)^2 = 984 \times 10^6$

Standard deviation of net A.W.: $31,370

The standard deviation of net A.W. is less than the expected net A.W., and only the 50% capacity utilization situation results in a negative net A.W. Consequently, the investors in this undertaking may well judge the venture to be an acceptable one. ∎

EVALUATION OF ALTERNATIVES CONSIDERING PROBABILISTIC LOSSES

There are situations, such as flood control projects, in which future losses due to natural or human-made risks can be decreased by increasing the amount of capital that is invested. Drainage ditches or dams, built to control floodwaters, may be constructed in different sizes, costing different amounts. If correctly designed and used, the larger the size, the smaller will be the resulting damage loss when a flood occurs. As we might expect, the most economical size would provide satisfactory protection against most floods, although it could be anticipated that some overloading and damage may occur at infrequent periods.

EXAMPLE 7-8

A drainage ditch in a mountain community in the West where flash floods are experienced has a capacity sufficient to carry 700 cubic feet per second. Engineering studies produce the following data regarding the probability of water flow in any one year versus the probability the flow will be exceeded and the cost of enlarging the ditch:

Water Flow (ft³/sec)	Probability of a Greater Flow Occurring in Any One Year	Investment to Enlarge Ditch to Carry This Flow
700	0.20	—
1,000	0.10	$20,000
1,300	0.05	30,000
1,600	0.02	44,000
1,900	0.01	60,000

Records indicate that the average property damage when serious overflow occurred will amount to $20,000. It is believed that this would be the average damage whenever the storm flow *was greater* than the capacity of the ditch. Reconstruction of the ditch would be financed by 40-year bonds bearing 8% interest. It is thus computed that the annual cost for debt repayment (principal of the bond plus interest) would be 8.39% of the initial cost, for $(A/P, 8\%, 40) = 0.0839$. It is desired to determine the most economic ditch size (water flow capacity).

Solution

The total expected annual cost for the structure and property damage for all alternative ditch sizes would be as follows:

Water Flow (ft³/sec)	Annual Investment Cost	Expected Annual Property Damage[a]	Total Expected Annual Cost
700		$20,000 × 0.20 = $4,000	$4,000
1,000	$20,000 × 0.0839 = $1,678	20,000 × 0.10 = 2,000	3,678
1,300	30,000 × 0.0839 = 2,517	20,000 × 0.05 = 1,000	3,517
1,600	44,000 × 0.0839 = 3,692	20,000 × 0.02 = 400	4,092
1,900	60,000 × 0.0839 = 5,034	20,000 × 0.01 = 200	5,234

[a]These amounts are obtained by multiplying $20,000 by the probability of greater water flow occurring.

From these calculations it may be seen that the minimum expected annual cost would be achieved by enlarging the ditch so that it would carry 1,300 cubic feet per second, with the expectation that a greater flood might occur 1 year out of 20 on the average and cause property damage of $20,000. ■

It should be noted that when loss of life or limb might result such as in Example 7-8, there usually is considerable pressure to disregard pure economy and build such projects in recognition of the nonmonetary values associated with human safety.

The following example illustrates the same principles in Example 7-8 except that it applies to safety alternatives involving electrical circuits.

EXAMPLE 7-9

There are three alternatives being evaluated for the protection of electrical circuits, with the following required investments and probabilities of failure:

Alternative	Investment	Probability of Loss in Any Year
A	$ 90,000	0.40
B	100,000	0.10
C	160,000	0.01

If a loss does occur, it will cost $80,000 with probability 0.65, and $120,000 with probability 0.35. The probabilities of loss in any year are independent of the probabilities associated with the resultant cost of a loss if one does occur. Each alternative has a useful life of 8 years and no salvage value at that time. The M.A.R.R. is 12%, and annual maintenance is expected to be 10% of the capital investment. It is desired to determine which alternative is best based on expected total annual costs.

Solution

The expected cost of a loss, if it occurs, can be calculated as

$$\$80,000(0.65) + \$120,000(0.35) = \$94,000$$

Alternative	Capital Recovery Cost = Investment × $(A/P, 12\%, 8)$	Annual Maintenance = Investment (0.10)
A	$ 90,000(0.2013) = 18,117	$ 9,000
B	100,000(0.2013) = 20,130	10,000
C	160,000(0.2013) = 32,208	16,000

Alternative	Expected Annual Cost of Failure	Total Expected Annual Cost
A	$94,000(0.40) = $37,600	$64,717
B	94,000(0.10) = 9,400	39,530
C	94,000(0.01) = 940	49,148

Thus alternative B is shown to be the best based on total expected annual cost, which is a long-run average cost. However, one might rationally choose alternative C so as to reduce significantly the chance of an $80,000 or $120,000 loss occurring in any year in return for a 24.3% increase in the total expected annual cost. ■

One of the prime problems when expected values are to be computed is the determination of the probabilities. In most situations there is no history of previous cases for the particular venture being considered. Therefore probabilities seldom can be based on historical data and rigorous statistical procedures. In most cases it is necessary that the analyst, or person making the decision, rely on judgment or even intuition in estimating the probabilities. This fact makes some persons hesitate to use the expected-value concept, inasmuch as they cannot see the merit in applying such a technique to improve the evaluation of uncertainty when so much apparent subjectivity is present. Although this argument has merit, the fact is that economy studies always deal with future events and there must be an extensive amount of estimating. Furthermore, even if the probabilities could be based accurately on past history, there rarely is any assurance that the future will repeat the past. In such situations structured methods for assessing "subjective" probabilities are used often in practice.† Also, even if we must estimate the probabilities, the very process of doing so requires us to give some thought to the uncertainty that is inherent in all estimates going into the analysis. Such forced thinking is likely to produce better results than little or no thinking about such matters.

EVALUATIONS THAT CONSIDER UNCERTAINTY, USING VARIANCES AND EXPECTED VALUES

In addition to using expected values of random variables to deal with uncertainty, some situations require the variance of the variable to be also taken into account in adequately portraying the dispersion in study results. Therefore, the uncertainty associated with an alternative can be represented more realistically in terms of its variability in addition to its expected value.

The first procedure for dealing with probabilistic factors is to compute mathematically their expected values and variances. This is referred to as "closed-form" analysis. A second general procedure for treating probabilistic information is to utilize Monte Carlo simulation.

One general type of problem to which both procedures can be applied involves a cash flow that varies according to some known probability function and a project life which is certain. In this type of problem, the expected value and variance of the equivalent worth of the cash flow can be determined. A second type of problem involves cash flows that are certain and a probabilistic project life, and the project life does have a known probability function. These first two

†For further reading, see W. G. Sullivan and W. W. Claycombe, *Fundamentals of Forecasting* (Reston, Va.: Reston Publishing Company, Inc., 1977), Chap. 6.

types of problems are addressed in this section. A third type of problem involves probabilistic cash flows and probabilistic project lives. This is a more complicated type of problem and is analyzed with Monte Carlo simulation methods in a later section.

Three examples are presented that demonstrate closed-form analysis for the first two types of problems mentioned above. In these examples, all variables are assumed to be statistically independent.

EXAMPLE 7-10

Assume that net annual benefits for a project during each year of its life are discretely estimated and have the following probabilities.

Net Annual Benefits (NAB)	Pr (NAB)
$2,000	0.40
3,000	0.50
4,000	0.10

The life is 3 years for certain and the initial investment is $7,000, with negligible salvage value. The M.A.R.R. is 15%. Determine E(N.P.W.) and the probability that net present worth is greater than zero [i.e., Pr (N.P.W. $\geq$ 0)].

Solution

The net present worth for each value of NAB can be determined in this manner:

$$\text{N.P.W.} = -7{,}000 + \text{NAB} \,(P/A, 15\%, 3)$$

$$\text{N.P.W.} = -7{,}000 + \text{NAB} \,(2.2832)$$

If the NAB is:	Then N.P.W. Equals:	Which Occurs with Probability:
$2,000	− $2,434	0.40
3,000	− 150	0.50
4,000	2,133	0.10

$$E(\text{N.P.W.}) = (0.40)\,(-\$2{,}434) + (0.50)\,(-\$150) + (0.10)\,(\$2{,}133)$$

$$= -\$835$$

Another method of working this problem is

$$E(\text{NAB}) = \$2{,}000(0.40) + \$3{,}000(0.50) + \$4{,}000(0.10) = \$2{,}700$$

$$E(\text{N.P.W.}) = -7{,}000 + E(\text{NAB}) \times (P/A, 15\%, 3)$$

$$= -\$7{,}000 + \$2{,}700\,(2.2832) = -\$835$$

The probability of each value of net present worth is illustrated in the figure.

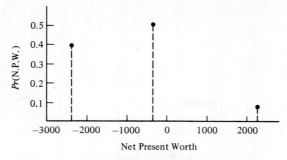

EXAMPLE 7-10

As can be seen from the diagram, Pr (N.P.W. ≥ 0) = 0.10. ∎

Of perhaps more interest is the situation in which annual cash flows are random variables having estimated expected values and variances. In this case, the analyst frequently assumes that annual cash flows are distributed according to a normal distribution.† The example below illustrates normally distributed and independent cash flows during each year of a project that has a certain life of 3 years.

EXAMPLE 7-11

For the following cash flow estimates, find E(N.P.W.), V(N.P.W.), and Pr (I.R.R. ≤ M.A.R.R.). Assume that the cash flows are normally distributed and that M.A.R.R. = 15%.

Year, k	Expected value of Cash Flow, A_k	Standard Deviation (SD) of Cash Flow, A_k
0	− $7,000	0
1	3,500	$600
2	3,000	500
3	2,800	400

A graphical portrayal of these normally distributed cash flows is shown in Figure 7-5 (on following page).

Solution

The expected net present worth is calculated as follows:

$$E\text{(N.P.W.)} = -\$7{,}000 + \$3{,}500(P/F, 12\%, 1) + \$3{,}000(P/F, 15\%, 2)$$
$$+ \$2{,}800(P/F, 15\%, 3)$$

$$E\text{(N.P.W.)} = \$153$$

---

†This frequently encountered probability function is discussed in any good statistics book, such as that by R. L. Scheaffer and J. T. McClare, *Statistics for Engineers* (Boston: Duxbury Press, 1982).

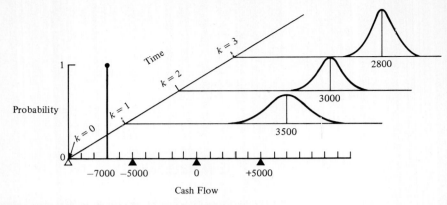

FIGURE 7-5 Probabilistic Cash Flows Over Time.

To determine V(N.P.W.), we use this relationship:

$$V(C_kA_k) = C_k^2V(A_k)$$

where A_k is the net cash flow in year k $(0 \leq k \leq N)$ and C_k is the single-payment present worth factor, $(P/F, 15\%, k)$. Note that $V(A_k)$ = (standard deviation of $A_k)^2$. Furthermore,

$$\sum_{k=1}^{N} V(C_kA_k) = \sum_{k=1}^{N} C_k^2V(A_k)$$

Thus

$$V(\text{N.P.W.}) = 0^21^2 + 600^2(P/F, 15\%, 1)^2 + 500^2(P/F, 15\%, 2)^2$$
$$+ 400^2(P/F, 15\%, 3)^2$$
$$= 484{,}324$$

and

$$\text{SD(N.P.W.)} = \sqrt{V(\text{N.P.W.})} = \$696$$

where SD(N.P.W.) is the standard deviation of N.P.W. For a decreasing N.P.W.(i) function having a unique internal rate of return, the probability that the I.R.R. is less than the M.A.R.R. is the same as the probability that N.P.W. is less than 0.

Consequently, by using the standardized normal distribution in Appendix G, we can determine the probability that N.P.W. is less than or equal to zero by utilizing the fact that the sum of independently distributed normal random variables is also normally distributed.†

$$Z = \frac{\text{N.P.W.} - E(\text{N.P.W.})}{\text{SD (N.P.W.)}} = \frac{0 - 153}{696} = -0.22$$

---

†For a random variable, X, that is normally distributed with mean μ and standard deviation σ, the transformed (standardized) variable is defined as

$$Z = \frac{X - \mu}{\sigma}$$

$$\text{Pr (N.P.W.} \leq 0) = \text{Pr } (Z \leq -0.22)$$

From Appendix G we find that Pr $(Z \leq -0.22) = 0.4129$. ■

The approach used in Example 7-11 can be generalized with two formulas. For independent random variables, the expected value of the net present worth is given by

$$E(\text{N.P.W.}) = \sum_{k=0}^{N} (1 + i)^{-k} E(A_k) \qquad (7\text{-}3)$$

where A_k is the net cash flow in year k, $E(A_k)$ is the expected value of A_k, and N is the certain project life. Similarly, the variance is given by

$$V(\text{N.P.W.}) = \sum_{k=0}^{N} (1 + i)^{-2k} V(A_k) \qquad (7\text{-}4)$$

Note that the $(1 + i)^{-2k}$ term equals either $(P/F, i, 2k)$ or $(P/F, i, k)^2$.

A different approach must be used for the second general type of problem in which the project life is a random variable. In general, the expected value of net present worth can be obtained by determining the expected value of the series present worth factor and *not* by determining the expected life of the project. The following example illustrates this principle for a project having a useful life that is a random variable and uniform net annual benefits that are known with certainty.

EXAMPLE 7-12
Suppose that the project life of a proposed venture is discretely distributed as follows:

Life, N	Pr (N)
2 years	0.40
3 years	0.50
4 years	0.10

The net annual benefits equal $4,000 over the life of the project and the initial investment is $8,000 with no salvage value. Determine the expected value and variance of net present worth, assuming that M.A.R.R. equals 15%.

Solution
The expected value is computed as follows:

$$E(\text{N.P.W.}) = -\$8,000 + \$4,000 \sum_{N=2}^{4} (P/A, 15\%, N) \text{ Pr } (N)$$

$$= -\$8,000 + \$4,000[1.626(0.40) + 2.283(0.50)$$

$$+ 2.855(0.10)]$$

$$= \$310$$

Hence for a 2-year project life, the net present worth is $-\$8,000 + \$4,000(1.626)$ $= -\$1,496$ with probability 0.40. For a 3-year life, the N.P.W. is $+\$1,132$ with probability of 0.50 that life will equal 3 years. Similarly, the N.P.W. is $+\$3,420$ when the project life is 4 years, with probability 0.10. The expected N.P.W. is identical to that obtained above:

$$E(\text{N.P.W.}) = -\$1,496(0.4) + \$1,132(0.50) + \$3,420(0.10)$$

$$= \$310$$

By inspection it is apparent that $\Pr(\text{N.P.W.} \leq 0) = 0.40$.

From Equation 7-2, the variance of the net present worth is determined in this manner:

$$V(\text{N.P.W.}) = E(\text{N.P.W.}^2) - [E(\text{N.P.W.})]^2$$

where

$$E(\text{N.P.W.})^2 = \sum_{N=2}^{4} [-\$8,000 + \$4,000(P/A, 15\%, N)]^2 \Pr(N)$$

$$= (-\$1,496)^2(0.40) + (\$1,132)^2(0.50)$$

$$+ (\$3,420)^2(0.10)$$

$$= 2,705,558$$

Thus

$$V(\text{N.P.W.}) = 2,705,558 - (310)^2 = 2,609,458$$

and

$$\text{SD}(\text{N.P.W.}) = \$1,615 \qquad \blacksquare$$

The reader should realize that the examples presented in this section have been very simple and have involved straightforward cash flows. Also, all variables have been assumed to be independent. It should be apparent that additional complications in a problem formulation could result in a very complex and time-consuming solution. Monte Carlo simulation techniques are typically used in such situations.

EVALUATION OF UNCERTAINTY, USING MONTE CARLO SIMULATION†

The development of computers has resulted in the increased use of Monte Carlo simulation as an important tool for analysis of project uncertainties. For complicated problems, Monte Carlo simulation generates random outcomes for prob-

†Adapted from W. G. Sullivan and R. Gordon Orr, "Monte Carlo simulation analyzes alternatives in uncertain economy," *Industrial Engineering*, Vol. 14, No. 11 (November 1982). Reprinted with permission from *Industrial Engineering* magazine. Copyright © Institute of Industrial Engineers, Inc., 25 Technology Park/Atlanta, Norcross, GA.

TABLE 7-7 Probability Distribution for Useful Life

Number of Years, N		Pr (N)	
3	possible values	0.20	$\Sigma = 1.00$
5		0.40	
7		0.25	
10		0.15	

abilistic factors so as to imitate the randomness inherent in the original problem. In this manner a solution to a rather complex problem can be inferred from the behavior of these random outcomes.

To perform a simulation analysis, the first step is to construct an analytical model that represents the actual investment opportunity. This may be as simple as developing an equation for the net present worth of a proposed industrial robot in an assembly line or as complex as examining the effects of various U.S.-imposed trade barriers on sales of a proposed new product in international markets. The second step is development of a probability distribution from subjective or historical data for each uncertain factor in the model. Sample outcomes are randomly generated by using the probability distribution for each uncertain quantity and then utilized to determine a *trial* outcome for the model. By repeating this sampling process a large number of times, a frequency distribution of trial outcomes for a desired measure of merit, such as net present worth or annual worth, is created. The resulting frequency distribution can then be used to make probabilistic statements about the original problem.

To illustrate the Monte Carlo simulation procedure, suppose that the probability distribution for the useful life of a piece of machinery has been estimated as shown in Table 7-7.

The useful life can be simulated by assigning random numbers to each value such that they are proportional to the respective probabilities. (A random number is selected in a manner such that each number has an equal probability of occurrence.) Because two-digit probabilities are given in Table 7-7, random numbers can be assigned to each outcome as shown in Table 7-8. Next, a single outcome is simulated by choosing a number at random from a table of random numbers.† For example, if any random number between and including 00 and

TABLE 7-8 Assignment of Random Numbers

Number of Years, N	Random Numbers
3	00–19
5	20–59
7	60–84
10	85–99

†The last two digits of randomly chosen telephone numbers in a telephone directory are usually quite close to being random numbers.

TABLE 7-9 Random Normal Deviates (RNDs)

-1.565	0.690	-1.724	0.705	0.090
0.062	-0.072	0.778	-1.431	0.240
0.183	-1.012	-0.844	-0.227	-0.448
-0.506	2.105	0.983	0.008	0.295
1.613	-0.225	0.111	-0.642	-0.292

19 is selected, the useful life is 3 years. As a further example, the random number 74 corresponds to a life of 7 years.

If the probability distribution that describes a random variable is normal, a slightly different approach is followed.† Here the simulated outcome is based on the mean and standard deviation of the probability distribution and on a random normal deviate, which is a random number of standard deviations above or below the mean of a standardized normal distribution. An abbreviated listing of typical random normal deviates is shown in Table 7-9. For normally distributed random variables, the simulated outcome is based on Equation 7-5:

$$\text{outcome value} = \text{mean} + [\text{random normal deviate} \times \text{standard deviation}] \quad (7\text{-}5)$$

For example, suppose that net *annual* cash flow is assumed to be normally distributed with a mean of $50,000 and a standard deviation of $10,000, as shown below.

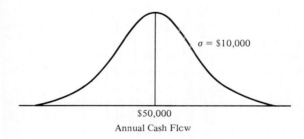

$\sigma = \$10,000$

$50,000
Annual Cash Flow

Simulated cash flows for a period of 5 years are listed in Table 7-10.

Notice that the average net annual cash flow is $248,850/5, which equals $49,770. This approximates the known mean of $50,000 with an error of 0.46%.

If the probability distribution that describes a random event is uniform and continuous with a minimum value of A and a maximum value of B, another procedure should be followed to determine the simulated outcome. Here the outcome can be computed with Equation 7-6:

†A random variable, X, is normally distributed with mean μ and standard deviation σ in accordance with the following equation:

$$f(X) = \frac{1}{\sigma\sqrt{2\pi}} \exp\left\{-\left[\frac{(X - \mu)^2}{2\sigma^2}\right]\right\}$$

The standardized normal distribution, $f(Z)$, of the variable $Z = (X - \mu)/\sigma$ has a mean of 0 and a standard deviation of 1.

TABLE 7-10 Example of the Use of Random Normal Deviates

Year	Random Normal Deviate (RND)	Net Annual Cash Flow [$50,000 + RND ($10,000)]
1	0.090	$50,900
2	0.240	52,400
3	−0.448	45,520
4	0.295	52,950
5	−0.292	47,080

$$\text{simulation outcome} = A + \frac{R.N.}{R.N._m}[B - A] \qquad (7\text{-}6)$$

where $R.N._m$ is the maximum possible random number (9 if one digit is used, 99 if two, etc.) and R.N. is the random number actually selected. This equation should be used when the minimum outcome, A, and the maximum outcome, B, are known.

For example, suppose that the salvage value in year N is uniformly and continuously distributed between $8,000 and $12,000. A value of this random variable would be generated as follows with a random number of 74:

$$\text{simulation outcome} = \$8,000 + \frac{74}{99}(\$12,000 - 8,000) = \$10,990$$

Proper use of these procedures, coupled with an accurate model, will result in an approximation to the actual outcome. But how many simulation trials are necessary for an *accurate* approximation? In general, the greater the number of trials, the more accurate the approximation will be. One method to determine if a sufficient number of trials has been conducted is to keep a running average of results. At first, this average will vary considerably from trial to trial. The amount of change between successive averages should decrease as the number of simulation trials increases. Eventually, this running average should level off at an accurate approximation.

EXAMPLE 7-13

This example illustrates how Monte Carlo simulation can simplify the analysis of a relatively complex problem. The estimates provided below relate to a capital investment opportunity being considered by a large manufacturer of air-conditioning equipment. Subjective probability functions have been estimated for the four independent uncertain factors as follows.

Investment
Normally distributed with a mean of −$50,000 and a standard deviation of $1,000.

Useful life
Uniformly distributed with a minimum life of 10 years and a maximum of 14 years.

Annual revenue

$35,000 with a probability of 0.4
$40,000 with a probability of 0.5
$45,000 with a probability of 0.1

Annual expense

Normally distributed with a mean of $-\$30,000$ and a standard deviation of $2,000.

Management of this company wishes to determine the probability that the investment will be a profitable one using an interest rate of 10%. In order to answer this question, the net present worth (N.P.W.) of the venture will be simulated.

Solution

For purposes of illustrating the Monte Carlo simulation procedure, five trial outcomes are computed manually in Table 7-11. The estimate of the average net present worth based on this very small sample is $\$19,004/5 = \$3,801$. For more accurate results, hundreds or even thousands of repetitions would be required. ■

The applications of Monte Carlo simulation for investigating uncertainty are many and varied. However, it must be remembered that the results can be no more accurate than the model and the probability estimates used. In all cases the procedure and rules are the same: careful study and development of the model; accurate assessment of the probabilities involved; true randomization of outcomes as required by the Monte Carlo procedure; and calculation and analysis of the results. Furthermore, a sufficiently large number of Monte Carlo trials should always be used to reduce the range of error to an acceptable level.

PERFORMING MONTE CARLO SIMULATION WITH A COMPUTER

It is apparent from the preceding section that a Monte Carlo simulation of a complex project requiring several thousand trials could be accomplished only with the help of a computer. Indeed, there are numerous simulation programs that can be obtained from software companies and universities. To illustrate the computational features and output of a typical simulation program, Example 7-13 has been evaluated with a computer program developed by R. Gordon Orr, an industrial engineering graduate student at the University of Tennessee. The computer queries and user responses (shown in boxes) are indicated on pages 257–258. Simulation results for 3,160 trials are shown in Figure 7-6. (This many trials were needed for the cumulative average present worth to stabilize to a variation of $\pm 0.5\%$.)

TABLE 7-11 Monte Carlo Simulation of N.P.W. Involving Four Independent Factors

Trial Number	Random Normal Deviate (RND_1)	Investment, P [\$50,000 + RND_1 (\$1,000)]	Three Random Numbers (RN)	Project Life, N [$10 + \frac{RN}{999}(14-10)$]	Project Life, N (Nearest Integer)
1	−1.003	\$48,997	807	13.23	13
2	−0.358	49,642	657	12.63	13
3	1.294	51,294	488	11.95	12
4	−0.019	49,981	282	11.13	11
5	0.147	50,147	504	12.02	12

One Random Number	Annual Revenue, R \$35,000 for 0–3 40,000 for 4–8 45,000 for 9	Random Normal Deviate (RND_2)	Annual Expense, E [\$30,000 + RND_2 (\$2,000)]	$N.P.W. = -P + (R - E)(P/A, 10\%, N)$
2	\$35,000	−0.036	\$29,928	−\$12,970
0	35,000	0.605	31,210	− 22,724
4	40,000	1.470	32,940	− 3,189
9	45,000	1.864	33,728	+ 23,233
8	40,000	−1.223	27,554	+ 34,654
				+\$19,004

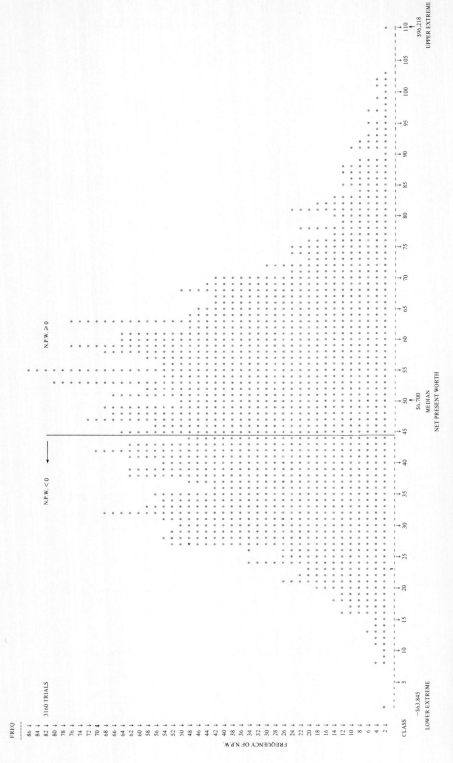

FIGURE 7-6 Histogram of Net Present Worth for Example 7-13.

THE FOLLOWING PROGRAM USES MONTE CARLO SIMULATION
TECHNIQUES AS APPLIED TO RISK ANALYSIS PROBLEMS OF
ENGINEERING ECONOMY.

WILL YOU BE USING A REMOTE PRINTER FOR OUTPUT ? (Y OR
N) ⊡Y

INPUT A RANDOM NUMBER BETWEEN 1 AND 1000. ⊡199

MAXIMUM NUMBER OF ITERATIONS YOU WISH TO RUN ? ⊡1000

WHAT INTEREST RATE (PERCENT) IS TO BE USED ? ⊡10

THE DATA FOR EACH RANDOM VARIABLE INVOLVED MAY BE
FORMULATED AS FOLLOWS:

1.   SINGLE VALUE OR ANNUITY
2.   SINGLE VALUE WITH ARITHMETIC GRADIENT
3.   SINGLE VALUE WITH GEOMETRIC GRADIENT
4.   DISCRETE DISTRIBUTION
5.   UNIFORM DISTRIBUTION
6.   NORMAL DISTRIBUTION
7.   A SERIES OF YEARLY CASH FLOWS
8.   SALVAGE VALUE DEPENDENT ON PROJECT LIFE
9.   TRIANGULAR DISTRIBUTION

INFORMATION FOR INITIAL CASH FLOW:

    DISTRIBUTION IDENTIFICATION NUMBER =   ⊡6

    MEAN VALUE =   ⊡-50000

    STANDARD DEVIATION =   ⊡1000

INFORMATION FOR YEARLY CASH FLOW:

    THIS CASH FLOW MAY CONSIST OF A NUMBER OF
    DIFFERENT ELEMENTS WHICH MAY FOLLOW DIFFERENT
    DISTRIBUTIONS.
    PLEASE INPUT THE DATA ONE ELEMENT AT A TIME AND
    YOU WILL BE PROMPTED FOR ADDITIONAL INFORMATION.

    DISTRIBUTION IDENTIFICATION NUMBER =   ⊡4

    NUMBER OF VALUES = ⊡3

    INPUT VALUES IN ASCENDING ORDER:

    VALUE 1 = ⊡35000

                WITH PROBABILITY ⊡0.4

    VALUE 2 = ⊡40000

                WITH PROBABILITY ⊡0.5

    VALUE 3 = ⊡45000

                WITH PROBABILITY ⊡0.1

```
IS THERE ADDITIONAL ANNUAL CASH FLOW DATA? (Y OR
N)   [Y]

DISTRIBUTION IDENTIFICATION NUMBER =   [6]

MEAN VALUE =   [-30000]

STANDARD DEVIATION =   [2000]

IS THERE ADDITIONAL ANNUAL CASH FLOW DATA? (Y OR
N)   [N]


INFORMATION FOR SALVAGE VALUE:

DISTRIBUTION IDENTIFICATION NUMBER =   [1]

CASH VALUE =   [0]


INFORMATION FOR PROJECT LIFE:

DISTRIBUTION INDENTIFICATION NUMBER =   [5]

MINIMUM VALUE =   [10]

MAXIMUM VALUE =   [14]

EXPECTED VALUE OF PRESENT WORTH =            7759.60
VARIANCE OF PRESENT WORTH =           680623960.00
STANDARD DEVIATION OF PRESENT WORTH =       26088.77
PROBABILITY THAT PRESENT WORTH IS GREATER THAN
ZERO =                                        0.595

EXPECTED VALUE OF ANNUAL WORTH =             1114.15
VARIANCE OF ANNUAL WORTH =             14611587.00
STANDARD DEVIATION OF ANNUAL WORTH =         3822.51
PROBABILITY THAT ANNUAL WORTH IS GREATER THAN
ZERO =                                        0.595
```

The average N.P.W. is $7,759.60, which is larger than the $3,801 obtained from Table 7-11. This underscores the importance of having a sufficient number of simulation trials to ensure reasonable accuracy in Monte Carlo analyses.

The histogram indicates that the *median net present worth* of this investment is $6,700 and that the dispersion of present worth trial outcomes is considerable. The standard deviation of simulated trial outcomes is one way to measure this dispersion. Based on Figure 7-6, there are 59.5% of all simulation outcomes with a net present worth of $0 or greater. Consequently, this project may be too risky for the cautious company to undertake because the "down-side" risk of failing to realize at least a 10% return on the investment is about 4 chances out of 10. Perhaps another investment should be considered.

A typical application of simulation involves the analysis of several mutually exclusive projects. In such studies how can one compare projects that have

TABLE 7-12 Simulation Results for Three Mutually Exclusive Alternatives

Project	E(P.W.)	SD(P.W.)	E(P.W.) ÷ SD(P.W.)
A	$37,382	$1,999	18.70
B	49,117	2,842	17.28
C	21,816	4,784	4.56

different expected values and standard deviations of, for instance, net present worth? One approach is to select the alternative that *minimizes* the probability of attaining a net present worth that is less than zero. Another popular response to this question utilizes a graph of expected value (a measure of the ''reward'') plotted against standard deviation (an indicator of uncertainty) for each alternative. An attempt is then made to assess subjectively the tradeoffs that result from choosing one alternative over another in pairwise comparisons.

To illustrate the latter concept, suppose three projects having varying degrees of uncertainty have been analyzed with Monte Carlo computer simulation, and the results shown in Table 7-12 have been obtained. These results are plotted in Figure 7-7, where it is apparent that alternative C is inferior to alternative A because of its lower E(P.W.) and larger standard deviation. Therefore, C offers a smaller present worth that has a greater amount of uncertainty associated with it! Unfortunately, the choice of B versus A is not as clear because the increased expected present worth of B has to be balanced against the increased uncertainty of B. This trade-off *may or may not* favor B, depending on management's perception of accepting the additional uncertainty associated with a larger expected reward. The comparison also presumes that alternative A is acceptable to the decision maker. One simple procedure for selecting between A and B is

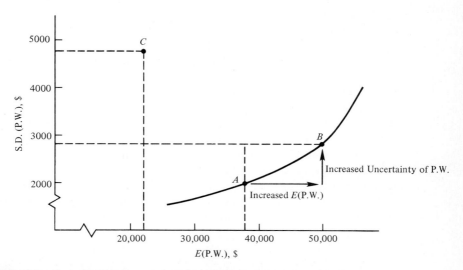

FIGURE 7-7 Graphical Summary of Computer Simulation Results.

to rank projects based on the ratio of E(P.W.) to SD(P.W.). In this case, project A would be chosen because it has the more favorable (larger) ratio.

SELECTED METHODS FOR INCLUDING NONMONETARY ATTRIBUTES

Real-world decision making among alternatives normally involves numerous nonmonetary as well as monetary attributes. These attributes, also known as *criteria*, can be anticipated whenever a decision must be made in consideration of nonmonetary and monetary objectives.

For instance, would you purchase a new car or select a job based strictly on monetary criteria? Probably not. Similarly, engineering and business decisions cannot (typically) avoid criteria related to employee morale and safety, image of the firm, legal and environmental issues, and so on.

Once the decision problem has been clearly identified and objectives to be met have been established, it is essential to select suitable attributes by which the best alternative can be identified. In this regard several guidelines can be suggested for making sure that the appropriate attributes have been included in the analysis. First, a sufficient number of attributes should be chosen to reflect adequately the objectives to be satisfied by selecting the best alternative. Second, each attribute should distinguish between two or more alternatives (i.e., the values of each attribute should not be identical for all alternatives). Therefore, each attribute should take on a unique value that is determined by differences among alternatives. Third, an attribute should not be completely redundant with others included in the analysis. This means that the value of a given attribute cannot be implied or predicted by one or more other attributes.

The remainder of this section presents several methods that may be used separately, or in combination, for explicitly incorporating nonmonetary attributes into the analysis of complex decision problems. These methods are classified into two categories.

The *first* group of methods analyzes available information in its total (full) number of dimensions.† For example, if r attributes have been chosen to characterize the alternatives, the predicted values for all r attributes are considered in the choice. Full dimension methods, even though recognizing the problem in its entirety, may not reduce the complexity of the decision problem. However, they are frequently useful in eliminating, or screening, inferior alternatives from the analysis.

A *second* category of methods reduces all nonmonetary and monetary attributes to a single dimension or value (usually numerical). Because these methods do not attempt to analyze data in terms of their initial units of measurement but rather to reduce the information to a single unit of measurement, the decision problem's multiple attributes have been compressed to one dimension.

---

†The dimension corresponds to the number of scales of measurement for the various nonmonetary and monetary attributes which are necessary to discriminate among alternatives.

**TABLE 7-13 Methods for Dealing
with Nonmonetary Attributes in
Economy Studies**

Dimension	Name of Method
Full	1. Dominance
	2. Feasible ranges
Single	3. Lexicography
	4. Effectiveness index

A summary of the four methods to be considered is given in Table 7-13. Each is described and then illustrated with an example problem.

DOMINANCE

When one alternative is better than another with respect to all evaluation criteria, there is no problem in deciding between them. In this case the first alternative dominates the second one. By comparing each possible pair of alternatives to determine whether the monetary and nonmonetary attribute values for one are no worse than those for the other, it may be possible to eliminate several candidates from further consideration or to select the single alternative clearly superior to the others. In many cases this approach does not result in the determination of a preferred alternative because of its inability to discriminate among alternatives so that a unique choice is possible.

The effects of estimation uncertainties can be included in the dominance procedure when approximations for the value of each criterion are represented as a range of uncertainty or as a random variable. A weak form of dominance would involve an alternative's being dominated if, in pairwise comparison, the upper and lower values of its range of uncertainty are not better than the corresponding values for another alternative over all attributes being used to evaluate the alternatives.

To illustrate the full-dimensioned dominance principle, consider the problem of deciding among four alternative radiology department designs in a proposed hospital expansion. Suitable monetary and nonmonetary attributes were chosen to distinguish among alternatives after much deliberation between hospital administrators and radiologists. For each criterion selected, a performance measure was specified by which the various nonmonetary and monetary considerations could be assessed. Data developed for two monetary attributes and five nonmonetary attributes are shown in Table 7-14 for each alternative. Ranges of estimates have been included to indicate the presence of uncertainty.

In reviewing the four alternatives, it can be observed that alternative IV is dominated by alternatives I and II. That is, IV is no better than I or II in each attribute. For some attributes it is *as good as* I or II, but in no instance is it *better than* I or II. Comparing these alternatives in a pairwise fashion results in no other instances of dominance. Consequently, with this method alternative IV would be dropped from further consideration. No further discrimination among

TABLE 7-14 Illustrative Data for Evaluation of Radiology Department Designs

Evaluation Attributes (and Performance Measures)	Alternatives			
	I	II	III	IV
Monetary				
i. Construction cost (thousands of dollars)	1,000 ± 10	1,400 ± 10	1,100 ± 10	1,400 ± 10
ii. Annual operating cost (thousands of dollars)	350 ± 10	500 ± 10	450 ± 10	500 ± 15
Nonmonetary				
a. Patient travel distance (thousands of feet/year)	2,200 ± 400	900 ± 200	1,500 ± 300	2,400 ± 400
b. Physician travel distance (thousands of feet/year)	1,000 ± 300	400 ± 100	800 ± 200	1,200 ± 300
c. Information retrieval time (time to locate old records and film)	2–3 hours	15–30 minutes	5–6 hours	3–4 hours
d. Accuracy of information (excellent, very good, good, fair, poor)	E-VG	F-P	VG-G	F-P
e. Remunerative rewards (excellent, very good, good, fair, poor)	E-VG	VG-G	VG-G	VG-G

alternatives is possible and other methods, either separately applied or in combination with dominance, are required to arrive at the selection of a single alternative.

FEASIBLE RANGES

As the name of this method implies, feasible ranges for each attribute must be established. By carefully reviewing maximum performance expectations and minimum performance levels which are practical and attainable for the attributes, upper and lower bounds are placed on the acceptability of predicted criteria measurements. These ranges establish two fictitious alternatives against which maximum and minimum performance expectations of feasible alternatives can be defined. By bounding the permissible values of attributes from two sides (or from one), a choice among decisions may be substantially simplified.

When the decision maker is concerned with meeting or exceeding *minimum* performance requirements, for example, the acceptability of an alternative can be altered by "tightening" or "loosening" these minimum values. By carefully reviewing the goals to be achieved and changing minimum performance requirements, the feasible set of alternatives can be reduced so that the decision problem is simplified. After several iterations, there may remain only one alternative that satisfies minimal conditions. Clearly, this method forces the decision maker to develop a thorough understanding of goals to be fulfilled.

TABLE 7-15 Feasible Ranges for Evaluation Attributes

Attribute (from Table 7-14)	Minimum Acceptable (Worst Value Permissible)	Maximum Attainable (Best Realizable Value)
i	$1,800,000	$800,000
ii	$600,000	$300,000
a	3,000,000 ft/yr	700,000 ft/yr
b	1,800,000 ft/yr	200,000 ft/yr
c	4 hours	10 minutes
d	Poor	Excellent
e	Good	Excellent

This method can be applied to the data of Table 7-14 only after feasible ranges for the various attributes have been established as shown in Table 7-15. Low values for some of the attributes are desirable (e.g., total construction cost, patient travel distance, and information retrieval time), while high values of certain other attributes are preferable (e.g., accuracy of information and remunerative rewards). To interpret the feasible ranges shown in Table 7-15, the "minimum" end of the range represents the worst value for an attribute which is still acceptable. On the other hand, the "maximum" end of the range represents the best possible value which is conceivably attainable for an attribute.

As the minimum predicted value of each attribute in Table 7-14 is compared with the least acceptable value in its feasible range (from Table 7-15), it is apparent that alternative III does not provide satisfactory performance in one attribute. The information retrieval time was predicted to be 5 to 6 hours (worse than the allowable 4 hours). Therefore, alternative III cannnot be regarded as a feasible alternative and is eliminated at this point. The other alternatives are feasible with respect to the minimum performance requirements shown in Table 7-15.

LEXICOGRAPHY

This method is particularly suitable for decisions in which a single attribute is judged to be much more important than all other attributes. For example, in choosing among the four radiology department designs, the safety of the patient as indicated by travel distances to and from the department may be an *overriding* consideration in the final selection. Thus a final choice of alternative II for this problem might be based solely on the most acceptable value for the patient travel distance criterion.

By comparing alternatives over one attribute, the decision problem has been reduced to a single dimension (i.e., the measurement scale of the predominant attribute). The alternative having the best value for the most important attribute is chosen. However, when two or more alternatives have identical values for the predominant attribute, the second most important attribute must be specified and used to break the deadlock. If ties continue to occur, the decision maker successively names the next most important factor until a single alternative is chosen.

Clearly, lexicography requires that the decision maker rank the importance of each attribute. If a selection is made by using one, or a few, of the attributes, this procedure fails to take into account all the available data. Lexicography does not require comparability across attributes, but it does process information in its original form of measurement. Therefore, the method is suitable for the analysis of quantitative or qualitative data. Because of the simplicity of lexicography, it has intuitive appeal in many situations where monetary and nonmonetary attributes must be considered.

EFFECTIVENESS INDEX

An effectiveness index is a composite measure (single dimensioned) of the "worth" of important monetary and nonmonetary attributes as satisfied by each alternative. Numerous methods for constructing an effectiveness index are available. The method discussed here is called an "additive weighting" procedure, and it involves these basic steps:

1. Rank attributes in order of decreasing importance.
2. Assign weights to the attributes such that their relative importance is quantified on a dimensionless scale from, for instance, 0 to 1 inclusive.
3. Scale the performance (effectiveness) of each alternative in attaining the maximum worth attainable by each attribute.
4. Multiply the scaled performance for each attribute in step 3 by the weight of the attribute from step 2 and sum over all attributes for each alternative. The resultant "score" is an indication of the total worth of an alternative.
5. Select the alternative that has the maximum score.

To illustrate the additive weighting method of dealing with monetary and nonmonetary attributes, assume that alternatives III and IV of Table 7-14 have already been eliminated so that alternatives I and II remain to be evaluated. The seven attributes have been ranked in importance from high to low with these results (step 1):

Rank[a]	Attribute
1	Patient travel distance (a)
2	Accuracy of information (d)
3	Construction cost (i)
4	Information retrieval time (c)
5	Physician travel distance (b)
6	Annual operating cost (ii)
7	Remunerative rewards (e)

[a]To obtain ordinal rankings of attributes, the reader interested in more detail should refer to W. T. Morris, *Engineering Economic Analysis* (Reston, Va.: Reston Publishing Company, Inc. 1976), Chap. 7.

Weights are next assigned to each attribute in view of their rankings, with 1.00 initially being given to "patient travel distance." The second most important factor, "accuracy of information," would be assigned a value of 0.80, for example, if it is about 80% as important as "patient travel distance." When all attributes are considered in this manner, suppose that the following results are obtained:

Weight[a]	Attribute	Normalized Weight
1.00	Patient travel distance (a)	0.26
0.80	Accuracy of information (d)	0.21
0.65	Construction cost (i)	0.17
0.40	Information retrieval time (c)	0.10
0.40	Physician travel distance (b)	0.10
0.35	Annual operating cost (ii)	0.09
0.30	Remunerative rewards (e)	0.07
Σ = 3.90		Σ = 1.00

[a]For additional information, refer to J. R. Canada and J. A. White, *Capital Investment Decision Analysis for Management and Engineering* (Englewood Cliffs, N.J.; Prentice-Hall, Inc., 1980), Chap. 19.

"Normalized" weights are often computed by dividing each weight above by the sum of all weights.

The performance of each alternative is next calculated (scaled) as a simple percentage of the feasible range for each attribute presented in Table 7-15. A linear relationship between the "worth" of each attribute's performance and its average outcome as shown in Table 7-14 is assumed here, but this relationship could be very nonlinear in other problems.†

An example of how average performance can be transformed to a (0, 1) scale in this step of the additive weighting procedure is shown in Figure 7-8 for attributes a and d. Alternative I has a value of 0.35 for attribute a and a value

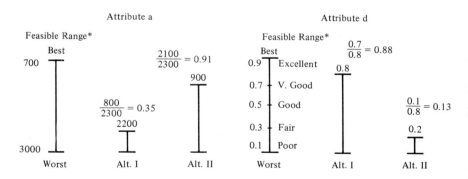

*From Table 7-15.

FIGURE 7-8 Performance Scaling of Two Attributes.

†For example, see J. R. Miller, *Professional Decision Making* (New York: Praeger Publishers, Inc., 1970).

TABLE 7-16 Calculation of an Effectiveness Index

Attribute	Normalized Weight	Alternative I Effectiveness	Product[a]	Alternative II Effectiveness	Product[a]
a	0.26	0.35	0.091	0.91	0.237
d	0.21	0.88	0.185	0.13	0.027
i	0.17	0.80	0.136	0.40	0.068
c	0.10	0.39	0.039	0.94	0.094
b	0.10	0.50	0.050	0.88	0.088
ii	0.09	0.83	0.075	0.33	0.030
e	0.07	0.75	0.053	0.25	0.018
		Score $= \Sigma$	0.629	Score $= \Sigma$	0.562

[a]Product = normalized weight $\times$ effectiveness.

of 0.88 for attribute d. The corresponding values for alternative II are 0.91 and 0.13. These values are developed from the following relationship for any attribute and alternative:

$$\text{effectiveness measure} = \frac{\text{expected outcome} - \text{worst outcome}}{\text{best outcome} - \text{worst outcome}}$$

An overall summary of all five steps of the additive weighting procedure is presented in Table 7-16. Based on this method, alternative I would be the recommended choice because its score is higher than that of alternative II.

PROBLEMS

7-1 Why should the effects of uncertainty be considered in engineering economy studies? What are some likely sources of uncertainty in these studies?

7-2 Consider these two alternatives:

	Alternative 1	Alternative 2
First cost	$4,500	$6,000
Annual receipts	$1,600	$1,850
Annual disbursements	$ 400	$ 500
Estimated salvage value	$ 800	$1,200
Useful life	8 years	10 years

Suppose that the salvage value of alternative 1 is known with certainty. By how much would the estimate of salvage value for alternative 2 have to vary so that the initial decision based on the data above would be reversed? The annual minimum attractive rate of return is 15%.

7-3 Referring to Problem 7-2, suppose that there is concern over the accuracy of estimated annual receipts for alternative 2. Is this concern justified in terms of the sensitivity of the initial decision to possible changes in this quantity?

7-4 (a) In Problem 7-2, determine the life of alternative 1 for which the net annual worths are equal.
(b) If the period of required service from either alternative 1 or 2 is exactly 7 years, which alternative should be selected? Assume that the salvage value remains unchanged.

7-5 Hodnett County is planning to build a three-story building. It is expected that some years later three more stories will have to be added to the building. Two alternative plans have been proposed, as follows: Design A is a conventional design for a three-story building. Its estimated first cost is $750,000. It is estimated that the three additional stories will cost $1,000,000 whenever they are added. Design B has an initial first cost of $900,000. The addition of three more stories is estimated to cost $800,000 whenever they are added. The total life of the building (design A or B) will be 60 years, and its market value will be zero at that time. Maintenance and energy costs will be $5,347 per year *less* with design B for each of the 60 years. If the addition to the building with either design is made, how soon must the additional stories be constructed to justify the selection of design B? The M.A.R.R. is 15% per year.

7-6 Two electric motors are being considered to power an industrial hoist. Each is capable of providing 90 horsepower. Pertinent data for each motor are as follows:

	Motor	
	D-R	Westhouse
Investment	$2,500	$3,200
Electrical efficiency	0.74	0.89
Maintenance/year	$40	$60
Life	10 years	10 years
M.A.R.R.	12%/year	12%/year

If the expected usage of the hoist is 500 hours per year, what would the cost of electrical energy have to be (in cents/kWh) before the D-R motor is favored over the Westhouse motor? [*Note:* 1 horsepower (hp) = 0.746 kilowatt.]

7-7 Suppose that for a capital investment project the optimistic–most likely–pessimistic estimates are as follows:

	Optimistic	Most Likely	Pessimistic
Investment	$80,000	$95,000	$120,000
Useful life	12 years	10 years	6 years
Salvage value	$30,000	$20,000	0
Net annual cash flow	$35,000	$30,000	$20,000
Minimum required rate of return	12%	12%	12%

(a) What is the net annual worth for each of the three estimation conditions?
(b) It is thought that the most critical elements are useful life and net annual cash inflow. Develop a table showing the net annual worth for all combinations of estimates for these two factors assuming all other factors remain at their most likely values.

7-8 A new steam flow monitoring device must be purchased immediately by a local municipality. These "most likely" estimates have been developed by a group of engineers:

Investment	$140,000
Annual savings	$ 25,000
Useful life	12 years
Residual value (end of year 12)	$ 40,000
M.A.R.R.	10%/year

Because considerable uncertainty surrounds these estimates, it is desired to evaluate the sensitivity of net present worth to ±50% changes in the most likely estimates of (a) annual savings, (b) life, and (c) residual value. Graph the results and determine to which factor the decision is most sensitive.

7-9 Two gasoline-fueled pumps are being considered for a certain agricultural application. Both pumps operate at a rated output of 5 hp but differ in their first cost and thermal efficiency. Gasoline costs $1.50 per gallon, and the M.A.R.R. is 15%. Following is a summary of "most likely" data:

	Pump M	Pump N
First cost	$1,700	$2,220
Maintenance cost per 100 hours of operation	$ 40	$ 25
Fuel consumption per hour of operation	$2\frac{1}{2}$ gallons	2 gallons
Useful life	10 years	10 years
Salvage value	$ 0	$ 0

Because of erratic weather conditions, the most difficult element to estimate is hours of operation per year. Optimistic, most likely, and pessimistic estimates of annual operation are 3,000 hours, 5,000 hours, and 6,500 hours, respectively.

(a) Plot net annual worth of each pump as a function of hours of operation.

(b) Based on results in part (a), which pump would you recommend?

7-10 A bridge is to be constructed now as part of a new road. Engineers have determined that traffic density on the new road will justify a two-lane road and bridge at the present time. Because of uncertainty regarding future use of the road, the time at which an extra two lanes will be required is currently being studied. The estimated probabilities of having to widen the bridge to four lanes at various times in the future are as follows:

Widen bridge in:	Probability
3 years	0.1
4 years	0.2
5 years	0.3
6 years	0.4

The two-lane bridge will cost $200,000 and the four-lane bridge, if built at one time, will cost, $350,000. The cost of widening a two-lane bridge will be an extra $200,000 plus $25,000 for every year that construction is delayed. If money can earn 12% interest, what would you recommend?

7-11 In Problem 7-10, perform an analysis to determine how sensitive the choice of a four-lane bridge built at one time versus a four-lane bridge that is constructed in two stages is to the interest rate. Will an interest rate of 15% per year reverse the initial decision?

7-12 In reference to Problem 7-10, suppose that instead of specifying probabilities for times at which the four-lane bridge will be required, these estimates have been made:

Pessimistic estimate	4 years
Most likely estimate	5 years
Optimistic estimate	7 years

In view of these estimates, what would you recommend? What difficulty do you have in interpreting your results? List some advantages and disadvantages of this method of preparing estimates.

7-13 Consider the matrix of profits below for five mutually exclusive alternatives (only one of the five will be chosen).

Alternative	Degree of Tax Reform		
	Pessimistic	Most Likely	Optimistic
A	12	18	18
B	14	14	15
C	14	26	15
D	10	22	14
E	18	12	10

(a) Which alternative is best when using the maximin rule?
(b) Which is best by the Laplace rule?
(c) Determine the best alternative by the Hurwicz principle when $H = 0.7$.

7-14 For the following matrix of costs, select the best course of action with each of these decision rules: (a) minimax; (b) Laplace; (c) minimin; (d) Hurwicz principle when $H = 0.3$.

Mutually Exclusive Alternative	Future Condition			
	θ_1	θ_2	θ_3	θ_4
S	18	18	10	14
T	14	14	14	14
U	5	26	10	14
V	14	22	10	10
W	10	12	12	10

7-15 In Problem 7-2, suppose that alternative 2 is believed to be more uncertain than alternative 1. A risk-adjusted M.A.R.R. of 18% will therefore be used to determine the net annual worth of alternative 2. Which alternative would be recommended with this method?

7-16 Pump M in Problem 7-9 is manufactured in a foreign country and is believed to be less reliable than pump N. To cope with this uncertainty, a risk-adjusted M.A.R.R. of 20% is utilized to calculate its net annual worth. When hours of operation per year total 5,000 hours, which pump would be selected? What difficulty is encountered with this method?

7-17 To deal with estimation uncertainties in Problem 7-2, the useful lives of alternatives 1 and 2 have been reduced to 6 years and 8 years, respectively. Does this affect the choice that should be made?

7-18 In a certain building project, the amount of concrete to be poured during the next week is uncertain. The foreman has estimated the following probabilities of concrete poured.

Amount (cubic yards)	Probability
1,000	0.1
1,200	0.3
1,300	0.3
1,500	0.2
2,000	0.1

Determine the expected value (amount) of concrete to be poured next week. Also compute the variance of concrete to be poured.

7-19   Consider the following two random variables, p and q.

Price, p	Pr (p)	Quantity Sold, q	Pr (q)
$6	$\frac{1}{3}$	10	$\frac{1}{3}$
5	$\frac{1}{3}$	15	$\frac{1}{3}$
4	$\frac{1}{3}$	20	$\frac{1}{3}$

Assume that p and q are independent. What is the mean and variance of the probability distribution for revenue $(p \times q)$?

7-20   Suppose that a random variable (e.g., salvage value for a piece of equipment) is normally distributed with mean = 175 and variance = 25. What is the probability that the actual salvage value is *at least* 171?

7-21   The net annual worth of project R-2 is normally distributed with a mean of $1,500 and a variance of 810,000. Determine the probability that this project's net A.W. is less than $1,700.

7-22   A dam is being planned for a river that is subject to frequent flooding. From past experience, the probabilities that water flow will exceed the design capacity of the dam, plus relevant cost information, are as follows:

Design of Dam	Probability of Greater Flow	Required Investment
A	0.100	$180,000
B	0.050	195,000
C	0.025	208,000
D	0.015	214,000
E	0.006	224,000

Estimated damages that occur if water flows exceed design capacity are $150,000, $160,000, $175,000, $190,000, and $210,000 for design A, B, C, D, and E, respectively. The life of the dam is expected to be 50 years, with negligible salvage value. For an interest rate of 8%, determine which design should be implemented. What nonmonetary considerations might be important to the best selection?

7-23   A diesel generator is needed to provide auxiliary power in the event that the primary source of power is interrupted. At any given time, there is a 0.1% probability that the generator will be needed. Various generator designs are available, and more expensive generators tend to have higher reliabilities should they be called on to produce power. Estimates of reliabilities, investment costs, maintenance costs, and damages resulting from a complete power failure (i.e., the standby generator fails to operate) are given below for three alternatives.

Alternative	First Cost	Operating and Maintenance Costs/Year	Reliability	Cost of Power Failure	Salvage Value
R	$200,000	$5,000	0.96	$400,000	$40,000
S	170,000	7,000	0.95	400,000	25,000
T	214,000	4,000	0.98	400,000	38,000

If the life of each generator is 10 years and M.A.R.R. = 10%, which generator should be chosen?

7-24 The owner of a ski resort is considering installing a new ski lift, which will cost $90,000. Costs, other than depreciation, for operating and maintaining the lift are estimated to be $150 per day when operating. The Weather Service estimates that there is a 60% probability of 80 days of skiing weather per year, 30% probability of 100 days, and 10% probability of 120 days per year. The operators of the resort estimate that during the first 80 days of adequate snow in a season, an average of 500 people will use the lift each day, at a fee of $1 each. If 20 additional days are available, the lift will be used by only 400 people per day during the extra period; and if 20 more days of skiing are available, only 300 people per day will use the lift during those days. The owners desire to recover any invested capital within 5 years, and want at least a 25% rate of return. Should the lift be installed?

7-25 Every time an automatic welding machine fails, the cost in idle labor and repairs is $1,000. The outage time averages 4 working hours and the plant works 2,000 hours a year. Suppose that the machine could fail a maximum of 250 times in a given year. The probabilities of failure during a year are assessed from a certain trade association to be: no failures, 0.050; 1, 0.113; 2, 0.209; 3, 0.333; 4, 0.201; 5, 0.080; 6, 0.009; 7, 0.003; 8, 0.001; 9, 0.0008; 10, 0.00008. A standby machine with an expected life of 10 years can be procured for $10,000 with $2,000 salvage value. The annual disbursements to keep it ready to run are $500. The probability of its breaking down during a standby run are nil. The minimum required rate of return is 15%. Make a recommendation regarding whether to purchase the standby machine.

7-26 For the following cash flow estimates, determine the E(NPW) and V(NPW). Also find the probability that N.P.W. will exceed $0. The cash flows are normally distributed and the M.A.R.R. is 12% per year.

End of Year	Expected Value of Cash Flow	Standard Deviation of Cash Flow
0	- $14,000	0
1	6,000	$ 800
2	4,000	400
3	4,000	400
4	8,000	1,000

7-27 The useful life of a certain machine is 5 years for certain. Its investment cost is $6,000 and its net annual savings are as follows:

Net Annual Benefits	Probability
$1,000	0.3
1,800	0.5
2,500	0.2

Determine (a) $E(N.P.W.)$, (b) $V(N.P.W)$, and (c) Pr $(N.P.W. \geq 0)$. The M.A.R.R. is 10% per year.

7-28 A. proposed venture has an initial investment of $80,000, annual receipts of $30,000, and an uncertain useful life, N, as follows:

N	Probability of N
1	0.05
2	0.15
3	0.20
4	0.30
5	0.20
6	0.05
7	0.05

Determine the expected present worth of this investment when the M.A.R.R. is 20% per year.

7-29 Consider the following project, which has a minimum life of 1 year and a maximum life of 5 years.

End of Year, N	Certain Cash Flow	Probability of N
0	− $7,000	—
1	+ 5,000	0.20
2	+ 4,000	0.10
3	+ 2,500	0.40
4	+ 2,500	0.25
5	+ 4,000	0.05

(a) What is the probability that this project has a present worth greater than $1,000 if the interest rate is 20% compounded annually?
(b) Determine the expected value of the project's present worth.

7-30 Consider Problem 7-24 when, in addition to uncertainty regarding number of skiing days per year, the useful life of the venture is *also* uncertain as follows:

Life, N	Pr (N)
4	0.2
5	0.6
6	0.2

Finally, the salvage value (S) of the ski lift is a function of the venture's life:

$$S = \$10,000(7 - N)$$

(a) Set up a table and use Monte Carlo simulation to determine five trial outcomes of the venture's net annual worth. Recall that the M.A.R.R. is 25% per year.
(b) Based on your simulation outcomes, should the lift be installed? State any assumptions you make.

7-31 Consider the following estimates for a new piece of production equipment:

Factor	Expected Value	Type of Probability Distributions
Investment cost	$150,000	Known with certainty
Useful life, N	13 years	Uniform in [8, 18]
Salvage value	$ 2,000 (13 − N)	Normal, $\sigma = \$500$
Annual savings	$ 70,000	Normal, $\sigma = \$4,000$
Annual disbursements	$ 43,000	Normal, $\sigma = \$2,000$
Minimum attractive rate of return (M.A.R.R.)	8%/year	

(a) Set up a table and simulate five trials of the equipment's net present worth.
(b) Compute and mean of the five trials and recommend whether the equipment should be purchased.

7-32 Simulation results are available for two mutually exclusive alternatives. A large number of trials have been run with a computer, with the results shown in the figure.

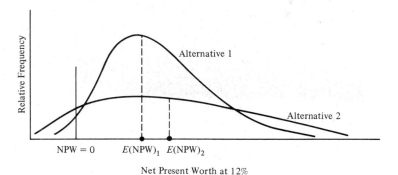

PROBLEM 7-32

Discuss the issues that may arise when attempting to decide between these two alternatives.

7-33 Three large industrial centrifuge designs are being considered for a new chemical plant. By using the following data, recommend a preferred design with each method that was discussed in this chapter for dealing with nonmonetary attributes.

Attribute	Weight	Design A	B	C	Feasible Range
Initial cost	0.25	$140,000	$180,000	$100,000	$80,000–$180,000
Maintenance	0.10	Good	Excellent	Fair	Fair–excellent
Safety	0.15	Not known	Good	Excellent	Good–excellent
Reliability	0.20	98%	99%	94%	94–99%
Product quality	0.30	Good	Excellent	Good	Fair–excellent

7-34 Use all the nonmonetary methods of this chapter to decide which job offer you should accept. Attempt to develop the data where "?'s" appear for your particular situation in view of the attributes shown.

Attribute	Job Offer 1	2	3	Weight	Feasible Range
Location	Phoenix	Buffalo	Raleigh	?	—
Annual salary	$26,000	$27,500	$23,000	?	?
Proximity to relatives	?	?	?	?	?
Quality of leisure time	?	?	?	?	?
Promotion potential	Fair	Excellent	Excellent	?	?
Commuting time/day	1 hour	$1\frac{1}{2}$ hours	$\frac{1}{2}$ hour	?	?
Fringe benefits	Excellent	Very good	Good	?	?
Type of work	Factory	Hospital	Government	?	?

7-35 You have volunteered to serve as a judge in a beauty contest—to select Sunshine, the most beautiful pig in the world. The four finalists are shown together with your assessments for the various attributes used to distinguish among contestants:

Attribute	Contestant I	II	III	IV
Facial beauty	Cute, but plump	Sad eyes, great snout	Beautiful lips	A real knockout
Poise[a]	10	8	8	3
Body tone[a]	5	10	7	8
Weight (lb)	400	325	300	380
Coloring	Brown	Spotted, black and white	Gray	Brown and white
Disposition	Friendly	Tranquil	Easily excited	Sour

[a]Scaled from 1 to 10, with 10 being the highest rating possible.

(a) Use dominance, feasible ranges, lexicography, and additive weighting to select your winner. Develop your own feasible ranges and weights for the attributes.

(b) If there were two other judges, discuss how the final selection of this year's "Sunshine" might be made.

7-36   You have decided to buy a new automobile and are willing to spend a maximum of $10,000 from your savings account. (Money not spent will remain in the account, where it earns effective interest of 12% per year.) The choice has been narrowed to three cars having these attribute values:

Attribute	Alternative		
	Domestic 1	Domestic 2	Foreign
Negotiated price	$8,400	$10,000	$9,300
Gas mileage (average)	25 mpg	30 mpg	35 mpg
Type of fuel	Gasoline	Gasoline	Diesel
Comfort	Very good	Excellent	Excellent
Aesthetic appeal	5 out of 10	7 out of 10	9 out of 10
Number of passengers	4	6	4
Ease of servicing	Excellent	Very good	Good
Performance on road	Fair	Very good	Very good

Use four methods for dealing with nonmonetary attributes (dominance, feasible ranges, lexicography, and additive weighting) and determine whether a selection can be made with each. You will need to develop additional data that reflects your preferences.

Cost Estimating and Consideration of Inflation

Because engineering economy studies deal with outcomes of decisions made at the present time that extend into the future, estimates of future cash flows, project lives, market conditions, and so on, are a vital ingredient of any study. This chapter addresses the matters of how to prepare estimates of future cash flows and other factors required in an analysis and how to deal correctly with inflation after this key consideration has been estimated. Consequently, the two topics in Chapter 8 can be viewed as important extensions of the general subject of uncertainty, which was discussed in Chapter 7.

COST ESTIMATING

The purpose of the estimating function of a firm is not to produce exact data but to obtain numbers having a high probability of falling within an acceptable range. Neither a preliminary estimate nor a final estimate is expected to be exact; rather, it should adequately suit the need at a reasonable cost. Preparation of precise estimates would be excessively time consuming and expensive, and even if they were possible to obtain they would still be subject to estimation errors.

Deviations between estimated and actual outcomes result from innumerable factors. Two of the most important are human errors in making estimates and unpredictable changes in circumstances. There are so many possible approaches to estimating a given quantity that the determination of which approach(es) should be used in what level of detail is a significant problem in itself.

It is useful to think of future events as resulting in part from factors that have determined these events in the past, and resulting in part from factors that are new and different. To the extent that, say, future production costs are the product of the same factors that have determined past production costs, the analyst may use explicit prediction techniques to extend past experience into forecasts of the future. However, to the extent that future production costs depend on factors that were not active in the past, the past data fail and one must rely on managerial experience and judgment. Forecasting, perhaps more than any other aspect of decision making, involves the combined skill and experience of both the analyst and management.

Types of Cost Estimation

Cost estimates can be classified according to detail, accuracy, and their intended use as follows:

1. *Order of magnitude estimates:* used in the planning and initial evaluation stage of a project.
2. *Semidetailed or budget estimates:* used in the design stage of a project.
3. *Definitive estimates:* used in the detailed engineering/construction stage of a project.

Order of magnitude estimates are used to screen alternatives. They typically provide accuracy in the range of ± 30 to 50% and are developed through semi-formal means such as conferences, questionnaires, and generalized equations.

Budget (semidetailed) estimates are compiled after a project has been given the "go-ahead." Their accuracy usually lies in the range of $\pm 15\%$. These estimates are obtained by many of the same methods used to screen alternatives, with the main difference being the fineness of cost breakdowns and the amount of effort spent on the estimate.

Detailed estimates are used as the basis for bids. Their accuracy is about $\pm 5\%$. They are made from specifications, drawings, site surveys, vendor quotations, and "in-house" historical records. Often detailed estimates are prepared with the help of computer software packages such as the Pricing and Cost Estimating (PACE) code developed by N.A.S.A.

Thus it is apparent that a cost estimate can vary from an instant, top-of-the-head guess to a very detailed and hopefully accurate prognostication of the future. The level of detail and accuracy of an estimate should depend on:

1. Difficulty of estimating the item in question.
2. Methods or techniques employed.
3. Qualifications of the estimator(s).
4. Time and effort available and justified by the importance of the study.
5. Sensitivity of study results to the particular estimate.

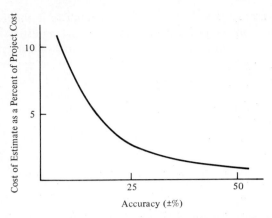

FIGURE 8-1   Accuracy of a Cost Estimate Versus the Cost of Making It.

As cost estimates become more detailed, the accuracy of the estimate typically improves but the cost of estimating increases dramatically. This general relationship is shown in Figure 8-1 and illustrates the idea that cost estimates should be prepared in full recognition of how accurate a particular study requires them to be.

Regardless of how estimates are made, individuals who use them should have specific recognition that the estimate will be in error to some extent. Even the use of sophisticated estimation techniques will not, in itself, eliminate error. However, it will hopefully minimize estimation errors and will at least provide better recognition of the anticipated degree of error.

Sources of Data for Estimating

The variety of sources from which information useful in estimating can be obtained is too great for complete enumeration. The following four major sources of information, ordered roughly according to decreasing importance, are described in subsequent sections.

1. Accounting records.
2. Other sources within the firm.
3. Sources outside the firm.
4. Research and development.

1. *Accounting records.* It should be emphasized that although data available from the records of the accounting function are a prime source of information for economic analyses, such data are very often not suitable for direct, unadjusted use.

A very brief and simplified description of the accounting process was given in Chapter 1. In its most basic sense accounting consists of a series of procedures for keeping a detailed record of monetary transactions between established categories of assets, each of which has an accepted interpretation useful for its own purposes. The data generated by the accounting function are often inherently misleading for economic analyses, not only because they are based on past results, but also because of the following limitations:

(a) The accounting system is rigidly categorized. Categories of various types of assets, liabilities, net worth, income, and expenses for a given firm may be perfectly appropriate for operating decisions and financial summaries, but rarely are they fully appropriate to the needs of economic analyses and decision making involving long-term considerations.

(b) Standard accounting conventions cause misstatements of some types of financial information to be "built into" the system. These misstatements tend to be based on the philosophy that management should avoid overstating the value of its assets or understating the value of its liabilities and should therefore assess them very conservatively. This leads to such practices as (1) not changing the stated value of one's resources as they appreciate due to rising market prices, and (2) depreciating assets over a much shorter life than actually expected. As a result of such accounting practices, the analyst should always be careful about treating such resources as cheaply (or, sometimes, as expensively!) as they might be represented in the accounting records.

(c) Accounting data have illusory precision and implied authoritativeness. Although it is customary to present data to the nearest dollar or the nearest cent, the records are not nearly that accurate in general.

In summary, accounting records are a good source of historical data, but have some limitations when used in making estimates for economic analyses. Moreover, accounting records rarely contain direct statements of incremental costs or opportunity costs, both of which are essential in most economic analyses.

2. *Other sources within the firm.* The usual firm has a large number of people and records that may be excellent sources of estimates or information from which estimates can be made. Examples of functions within firms that keep records useful to economic analyses are sales, production, inventory, quality, purchasing, industrial engineering, and personnel. Professional colleagues, supervisors, and workers in production/service areas can provide insights or suggest sources that can be obtained readily.

3. *Sources outside the firm.* There are numerous sources outside the firm that provide information helpful for estimating. The main problem is in determining those that are most beneficial for particular needs. The following is a listing of commonly utilized outside sources:

(a) *Published information,* including technical directories, buyer indexes, U.S. government publications, reference books, and trade journals offer a wealth of information. For instance, *Standard and Poor's Industry Surveys* gives monthly information regarding key industries. *The Statistical Abstract of the United States* is a remarkably comprehensive source of cost indexes and data. The Bureau of Labor Statistics publishes many periodicals which are good sources of labor costs, such as the *Monthly Labor Review, Employment and Earnings, Current Wage Developments, Handbook of Labor Statistics,* and the *Chartbook on Wages, Prices and Productivity.* A buyer's index of manufacturers is *Thomas' Register of American Manufacturers,* which can be used to obtain addresses of vendors to inquire for prices. An annual construction cost handbook, *Building Construction Cost Data*, is published by the R. S. Means Company, Kingston, Massachusetts. It includes standard crew sizes, unit prices, and prevailing wage rates for various regions of the country. The R. S. Means Company also publishes other volumes of cost data on mechanical and electrical work, repair and remodeling, and selected cost indexes.

(b) *Personal contacts* are excellent potential sources. Vendors, salespeople, professional acquaintances, customers, banks, government agencies, chambers of commerce, and even competitors are often willing to furnish needed information on the basis of a serious and tactful request.

4. *Research and development (R&D)*. If the information is not published and cannot be obtained by consulting someone who knows, the only alternative may be to undertake R&D to generate it. Classic examples are developing a pilot plant and undertaking a test market program. These activities are usually expensive and may not always be successful; thus this final step is taken only in connection with very important decisions, and when the sources mentioned above are known to be inadequate.

Cost Indexes

A cost index is a dimensionless number that indicates how costs and prices change (typically inflate) with time. Such indexes provide a convenient means of developing certain types of order-of-magnitude cost estimates. For example one simple index, I_n, is given by Equation 8-1:

$$I_n = \left(\frac{C_{n1}}{C_{k1}} + \frac{C_{n2}}{C_{k2}} + \cdots + \frac{C_{nm}}{C_{km}} + \cdots + \frac{C_{nM}}{C_{kM}} \right) \Big/ M \qquad (8\text{-}1)$$

where k = reference year (e.g., 1974)
n = year for which an index is to be determined $(n > k)$
M = total number of items in the index $(1 \leq m \leq M)$
C_{nm} = unit cost (or price) of the mth item in year n
C_{km} = unit cost (or price) of the mth item in reference year k

Notice that all M items are given equal weight in Equation 8-1.

A more popular form of Equation 8-1 is the general weighted index which places more emphasis on some items and less on others:

$$\bar{I}_n = \frac{W_1(C_{n1}/C_{k1}) + W_2(C_{n2}/C_{k2}) + \cdots + W_M(C_{nM}/C_{kM})}{W_1 + W_2 + \cdots + W_M} \qquad (8\text{-}2)$$

where $W_1, W_2, \ldots, W_M$ are the weights on items $1, 2, \ldots, M$. These weights can sum to any positive number, but 1.00 or 100 are often used for practical purposes.

EXAMPLE 8-1

Based on the data below, develop a weighted index for the price of a gallon of gasoline in 1980, when 1974 is the reference year having an index value of

100. The weight placed on *unleaded regular* gasoline is three times that of either leaded regular or premium because roughly three times as much is sold compared to leaded regular or premium.

	Price (cents/gal) in year						
	1974	1975	1976	1977	1978	1979	1980
Regular leaded	53	57	59	62	63	86	119
Premium	57	61	64	67	69	92	128
Regular unleaded	55	60	61	66	67	90	125

Solution

In Example 8-1, k is 1974 and n is 1980. From Equation 8-2, the value of $\bar{I}_{1980}$ is

$$0.20\left(\frac{119}{53}\right) + 0.20\left(\frac{128}{57}\right) + 0.60\left(\frac{125}{55}\right) = 2.262$$

Because $\bar{I}_{1974}$ equals 100, the weighted index in 1980 would have a value of 226.2. If the index in 1984, for example, is estimated to be 284.7, it is a simple matter to detemine the corresponding prices of gasoline from $\bar{I}_{1980} = 226.2$.

$$\textit{Regular leaded:} \quad 119 \text{ cents/gal}\left(\frac{284.7}{226.2}\right) = 150 \text{ cents/gal}$$

$$\textit{Premium:} \quad 128 \text{ cents/gal}\left(\frac{284.7}{226.2}\right) = 161 \text{ cents/gal}$$

$$\textit{Regular unleaded:} \quad 125 \text{ cents/gal}\left(\frac{284.7}{226.2}\right) = 157 \text{ cents/gal} \quad \blacksquare$$

An estimate of the present cost of an item in year n can be obtained by multiplying the cost of the item at an earlier point in time (year k) by the ratio of the index value in year n to the index value in year k. This is sometimes referred to as the "ratio technique" of updating costs. Thus when using this technique, the cost of a piece of equipment, plant, or component part can be taken from historical data with a specified base year and updated with a cost index.

EXAMPLE 8-2

A certain index for the cost of purchasing and installing utility boilers is keyed to 1964, where its baseline value was arbitrarily set at 100. Company XYZ installed a 50,000-lb/hr boiler in 1972 for $200,000 when the index had a value of 178. This same company must install another boiler of the same size in 1984. The index in 1984 is 312. What is the approximate cost of the new boiler?

Solution

Relative to 1972 when the initial boiler was installed, the cost index has risen $(312/178)100 = 175\%$. Therefore, a rough estimate of its cost in 1984 is $200,000(1.75) = $350,000. $\blacksquare$

Many indexes are periodically published, including the *Engineering News-Record* Construction Index which incorporates labor and material costs, and the Marshall and Stevens cost index. The *Statistical Abstract of the United States* publishes government indexes on yearly materials, labor, and construction costs. The Bureau of Labor Statistics publishes the *Wholesale Prices and Price Indexes* and the *Consumer Price Index Detailed Report*. Indexes of price changes are frequently used in engineering economy studies.

Estimates Needed for Typical Economic Analyses

Some judgment is required in determining how far to investigate individual variables to be estimated. This judgment should weigh which variables are dominant and deserve more study and which variables, even if drastically misjudged, will not produce significant changes in the overall estimate.

Perhaps the most serious sources of error in estimating result from overlooking important types of costs. A tabular form or checklist is a good means of preventing such oversights, but it is no better than the completeness of the checklist for the particular situation being estimated. Technical familiarity with the project is essential in assuring completeness as well as reasonable accuracy of project estimates.

The following is a brief listing of the types of estimates that are typically needed in economic analyses, together with some discussion of how those estimates might be obtained.

1. *First (investment) costs*, which may consist of two types:

(a) *Fixed capital investment*, such as for design and engineering, land purchase and improvement, buildings, equipment, installation, promotional and legal fees, and startup costs.

(b) *Working capital*, such as for inventories, accounts receivable, cash for wages, materials, and other accounts payable. Working capital is a revolving fund needed to get a project started and to meet subsequent obligations. Normally, it is assumed that working capital can be recovered in total (i.e., 100% salvage value) by the end of the life of a project.

2. *Labor costs* are a function of skill level, labor supply, and time required. Standards for the normal amount of output per labor hour have been developed for many classes of work. Standard times combined with expected wage rates, adjusted for productivity patterns, if any, provide a reasonable estimate of labor costs for repetitive jobs. Labor costs for specialized work can be predicted from bid estimates or quotes by agencies offering the service. It should be remembered that labor costs should consider fringe benefits as well as direct wages.

3. *Maintenance costs* are the ordinary costs required for the upkeep of property and minor changes required for more efficient use. Maintenance costs tend to increase with the age of an asset because more upkeep is required later in life.

4. *Property taxes and insurance* are usually expressed as an annual percentage of first cost in economic comparisons.

5. *Quality and scrap costs* depend upon the types of products and associated quality standards as well as upon abilities of the work force, learning time, and rework possibilities.

6. *Overhead costs* are by definition those costs which cannot be conveniently and practicably charged to particular products or services, and thus are normally prorated among the products or centers on some arbitrary basis. One should guard against using these arbitrary allocations in economic analyses, for the differences in overhead costs brought about by various alternatives being considered are rarely described by these rates. In general, one should consider each individual cost element included in the overhead and estimate how much, if any, each cost element is affected by each alternative.

7. *Revenues* are normally very difficult to predict for all except the most stable markets because of the buying whims of the public.

8. *Useful lives* are often very difficult to estimate and crucial to the analysis result.

9. *Salvage values* are typically a function of the useful life and, in the case of long lives, are relatively unimportant to the analysis result.

10. *Inflation rates* are also difficult to predict, and often long-range estimates for a particular industry are purchased from a consulting firm specializing in economic forecasts.

How Estimates Are Accomplished

Estimates can be prepared in limitless ways, some of which can be categorized as follows:

1. A *conference* of various people who are thought to have good information or bases for estimating the quantity in question. A special version of this is the *Delphi method*, which involves cycles of questioning and feedback in which the opinions of individual participants are kept anonymous.

2. *Comparison* with similar situations or designs about which we have more information and from which we can extrapolate estimates for alternatives under consideration. The comparison method may be used to approximate the cost of a design or product that is new. This is done by taking the cost of a more complex design for a similar item as an upper bound, and the cost of a less complex item of similar design as a lower bound. The resulting approximation is usually not very accurate, but the comparison method does have the virtue of setting bounds that might be useful for decision making.

3. *Quantitative techniques*, which do not always have standardized names. Some examples are given here with the names used being generally suggestive of the approaches.

(a) *Unit technique* involves utilizing an assumed or estimated ''per unit factor'' that can be estimated effectively. Examples are:

Capital cost of plant per kilowatt of capacity.

Fuel cost per kilowatthour generated.

Capital cost per installed telephone.

Revenue per long-distance call.

Temperature loss per 1,000 feet of steam pipe.

Operating cost per mile.

Maintenance cost per hour.

Such factors, when multiplied by the appropriate unit, give a total estimate of cost or savings.

There are limitless possibilities for breaking quantities into units that can be estimated readily. Examples are:

> In different units, such as dollars per week, to convert to dollars per year.
>
> A proportion, rather than a number, such as % defective, to convert to number of defects.
>
> A number, rather than a proportion, such as number defective and number produced, to convert to % defective.
>
> A rate, rather than a number, such as miles per gallon, to convert to gallons consumed.
>
> A number, rather than a rate, such as miles and hours traveled, to convert to average speed.
>
> Using an adjustment factor to increase or decrease a known or estimated number, such as defectives reported, to convert to total defectives.

As a simple example, suppose that we need a preliminary estimate of the cost of a particular house. Using a unit factor of, say, $35 per square foot and knowing that the house is approximately 2,000 square feet, the estimated cost would be $35 \times 2,000 = $70,000.

While the unit technique is very useful for preliminary estimating purposes, one can be dangerously misled by such average values. In general, more detailed methods can be expected to result in greater estimation accuracy.

(b) *Segmenting technique* involves decomposing an uncertain quantity into parts that can be separately estimated and then added together. As an example, suppose that we desire to estimate sales of a product X in the Dakotas. The simplest possible segmenting would be to estimate sales separately in North Dakota and South Dakota and then to add the two together.

(c) *Factor technique* is an extension of the unit method and the segmenting method in which one sums the product of several quantities or components and adds these to any components estimated directly. That is,

$$C = \sum_d C_d + \sum_m f_m \times U_m \qquad (8\text{-}3)$$

where C = value (cost, price, etc.) being estimated

C_d = cost of selected component d that is estimated directly

f_m = cost per unit of component m

U_m = number of units of component m

As a simple example, suppose that we need a slightly refined estimate of the cost of a house consisting of 2,000 square feet, two porches, and a garage. Using unit factors of $30 per square feet, $3,000 per porch, and $5,000 per garage, we can calculate the estimate as

$$\$30 \times 2,000 + \$3,000 \times 2 + \$5,000 = \$71,000$$

(d) *Power-sizing technique* is a sophistication of the unit method frequently used for costing industrial plant and equipment in which it is recognized that cost varies as some power of the change in capacity or size. That is,

$$(C_A \div C_B) = (S_A \div S_B)^X \qquad (8\text{-}4)$$

where C_A = cost for plant A ⎫ (both in \$ as of point in time for
$ C_B$ = cost for plant B ⎭ which estimate is desired)
$ S_A$ = size of plant A ⎫
$ S_B$ = size of plant B ⎬ (both in same physical units)
$ X$ = cost-capacity factor to reflect economies of scale†

As an example, suppose that it is desired to make a preliminary estimate of the cost of a 600-MW fossil power plant if built now. It is known that a 200-MW plant cost \$50 million in 1960 when the appropriate cost index was 400, and that cost index is now 1,200. The power-sizing model estimate, with $X = 0.79$, is

Cost now of 200-MW plant: \$50 million $\times$ (1,200 $\div$ 400) = \$150 million
(call it C_B)

Cost now of 600-MW plant: $C_A \div$ \$150 million = $(600 \div 200)^{0.79}$
(call it C_A)

$$C_A = \$150 \text{ million} \times 2.38 = \$357.3 \text{ million}$$

(e) *Miscellaneous statistical and mathematical modeling techniques* can be used for estimating or forecasting the future. Typical of the numerous mathematical modeling techniques that allow one to break down difficult problems to make more and/or reliable estimates, but which will not be described herein, are econometric models, demographic (population characteristic) models, network models, stochastic process models, mathematical programming models, input–output tables, regression models, and exponential smoothing models.

A Manufacturing Example

Manufacturers are faced with the problem of making a product that can be sold at a competitive price so that they can make a reasonable profit. The price of their product is based on the overall cost of making the item plus a built-in profit. Some companies that make a variety of products do not have a precise idea of exactly what each product costs—to find out might be prohibitively expensive—but they still need rough estimates to help them make decisions about what to produce and how to price their products.

As discussed in Chapter 1, product costs are classified as direct or indirect. Direct costs are easily assignable to a specific product, while indirect costs are not easily allocated to a certain product. For instance, direct labor would be the wages of a machine operator; indirect labor would be supervision.

Manufacturing costs have a distinct relationship with production volume in

†May be calculated/estimated from experience. See p. 137 of W. R. Park, *Cost Engineering Analysis* (New York: John Wiley & Sons, Inc., 1973), for typical factors. For example, $X = 0.68$ for nuclear generating plants and 0.79 for fossil-fuel-generating plants.

that they may be fixed, variable, or step-variable. Generally, administrative costs are fixed regardless of volume, material costs vary directly with volume, and equipment cost is a step function of production level.

The general cost categories of manufacturing expense include engineering and design, development costs, tooling, manufacturing labor, materials, supervision, quality control, reliability and testing, packaging, plant overhead, general and administrative, distribution and marketing, financing, taxes and insurance. Where do we start?

When estimating the cost of a manufactured product, we need drawings, specifications, production schedules, historical records of the company's labor cost, a bill of materials, and the process plan. The process plan tells us all operations that must be done to a product and the labor hours involved.

Engineering and design costs consist of design, analysis, and drafting, together with miscellaneous charges, such as reproductions. The engineering cost may be allocated to a product on the basis of how many engineering labor hours are involved. Other major types of costs that must be estimated are:

Tooling costs, which consist of repair and maintenance, plus the cost of any new equipment.

Manufacturing labor costs, which are determined from standard data, historical records, or the accounting department.

Materials costs, which can be obtained from historical records, vendor quotations, and the bill of materials. Scrap allowances must be included.

Supervision is a fixed cost that is based on the salaries of supervisory personnel.

Plant overhead, which includes utilities, maintenance, and repairs. There are various methods used to apply overhead, such as in proportion to direct labor dollars, or direct labor hours, or machine hours. If we use direct labor hours to allocate overhead, then

$$\text{overhead rate} = \frac{\text{total factory overhead}}{\text{total direct labor hours}}$$

Administrative costs are often included with the factory overhead (or burden).

The following simple example shows the general procedure for making a "per unit" product estimate and illustrates the use of a typical costing worksheet.

The worksheet on page 287 shows the determination of the cost of a throttle assembly. The 36.48 direct labor hours are multiplied by the composite labor rate, $4.54 per hour, to yield $165.62. Planning labor and quality control are expressed as 12% and 11% of direct labor cost, respectively. This gives a total labor cost of $203.71. Factory burden and general and administrative expense are applied as percentages of the total labor cost. The cost of production material and parts from outside vendors are also entered on the worksheet. Packing costs are added at a rate of 5% of all previous costs, giving a total direct charge of $576.81. Other direct charges are figured in to give a total manufacturing cost of $582.58. A 10% profit is added in giving a total selling price of $640.84. Since there are 50 parts in the production run, the unit selling price is $12.82.

WORKSHEET

Customer _____Aqua Boat Company_____

Model	CDX75Y	Estimator	Sam Steward
Part Name	Throttle Assembly		
Part No.	00681	Date	July 31, 1982
Parts Req'd.	50	Page	1 of 1

MANUFACTURING COST	CHARGE		
	Hours	Rate	Dollars
Factory Labor	36.48	$4.54	$165.62
Planning & Liaison Labor		12%	19.87
Quality Control		11%	18.22
TOTAL LABOR			$203.71
Factory Burden		105%	213.90
General & Admin. Expense		15%	30.56
Production Material			87.17
Outside Manufacture			14.00
SUBTOTAL			$549.34
Packing Costs		5%	27.47
Premium Pay			
Total Direct Charge			576.81
Other Direct Charge		1%	5.77
Facility Rental			
Total Manufacturing Cost			582.58
Profit/Fee @ 10%			58.26
Total Selling Price			640.84
Quantity—(Unit)			50
Ship Set (Unit) Selling Price			$ 12.82

Source: T. F. McNeill, and D. S. Clark, *Cost Estimating and Contract Pricing* (New York: American Elsevier Publishing Co., Inc., 1966).

CONSIDERATION OF INFLATION

Prior to this chapter we have assumed that prices for goods and services in the marketplace are relatively unchanged over substantial periods of time, or that the effect of such changes is constant among all cash flows for the alternatives under consideration. Unfortunately, these are not generally realistic assumptions. *Inflation*, which is the phenomenon of rising prices bringing about a reduction in the purchasing power of a given unit of money, is a fact of life and can significantly affect the economic comparison of alternatives.

Until the mid-1950s, money was universally accepted as a fixed measure of the worth of resources. Because the purchasing power of a given sum of money now is not constant, individuals and companies alike realize that investment opportunities must be evaluated with money treated as a variable measure of the worth of a resource.

If all cash flows in an economic comparison of alternatives are inflating at the same rate, inflation can be disregarded in before-tax studies. In cases where all incomes and all expenses are not inflating at the same rate, inflation gives rise to differences in economic attractiveness among alternatives that must be taken into account. Thus, when the effects of inflation are not included in an engineering economy study, an erroneous choice among competing alternatives can result. That is, reversals in preference may occur by assuming that inflation affects all investment opportunities to the same extent. Consequently, the objective of maximizing shareholders' (owners') wealth is inadvertently compromised.

A typical case is when a project is financed by means of borrowing at a fixed rate of interest as in the four plans of Table 4-1. The effect of inflation is to reduce the cost of borrowing in "real" terms. Conversely, in the same situation the lender fails to realize his desired return on the capital loaned out.

In this section the effects of inflation on before-tax studies are discussed. In Chapter 10 the after-tax effects of inflation will be considered.

Annual rates of inflation (often referred to as escalation) vary widely for different types of goods and services and over different periods of time. For example, the U.S. government-prepared Consumer Price Index rose less than 2% per year during the 1950s, but increased to approximately 7% per year during the 1970s. It is expected that inflation averaging 5 to 6% per year will continue to be a major concern in our economy throughout the 1980s.† Although it appears that such inflation will extend into the long-term future, it is possible that its opposite, deflation, can occur as was true during the depression of the 1930s.

Actual and Real Dollars

We now offer definitions of terms and describe general methods for considering inflation in engineering economy studies. First, two distinct types of cash flow estimates are defined, either of which can be correctly used in engineering economy studies.

1. *Actual dollars (A$)*: the actual number of dollars associated with a cash flow (or depreciation amount) as of the point in time it occurs. For example, persons typically anticipate their salaries 2 years hence in terms of actual

---

†The U.S. government does not publish forecasts of inflation rates. However, many companies prepare their own forecasts or purchase them from consulting firms such as Wharton Econometric Associates, Inc. and Data Resources, Inc.

dollars. Sometimes A$ are referred to as *current* dollars, and they include an allowance for inflation.

2. *Real dollars (R$)*: dollars expressed in terms of the same purchasing power relative to a particular point in time, regardless of when the corresponding actual dollar amounts occur. For instance, the future costs of various forms of energy are often estimated in real dollars (relative to some base year) to provide a consistent means of comparison. Often R$ are termed *constant* dollars.

Actual dollars as of any time, n, can be converted into real dollars of purchasing power as of any base point in time, k, by the relation

$$R\$ = A\$ \times \left(\frac{1}{1+f}\right)^{n-k} = A\$ \times (P/F, f\%, n - k) \qquad (8\text{-}5)$$

where f is a constant inflation rate per period over the n periods. It is most common to express real dollars as dollars of purchasing power at the time estimates are made or at the beginning of the study period under consideration. Thus, for purposes of this and following chapters, a value of k equal to 0 is usually used in Equation 8-5 to correspond with "end-of-year 0" on a cash flow diagram.

EXAMPLE 8-3

Suppose that your salary will increase at 6% per year for each of the next 4 years and is expressed in A$ as follows.

Year, n	Salary (A$)
1	$25,000
2	26,500
3	28,090
4	29,775

If the inflation rate (f) is expected to average 8% per year, what is the R$ equivalent of these A$ salary amounts? Assume that the base point in time is year 1.

Solution

By using Equation 8-5, the R$ salary equivalents are readily calculated for the base point in time, $k = 1$:

Year	Salary (R$ in Year 1)
1	$25,000(P/F, 8\%, 0) = \$25,000$
2	$26,500(P/F, 8\%, 1) = 24,537$
3	$28,090(P/F, 8\%, 2) = 24,083$
4	$29,775(P/F, 8\%, 3) = 23,636$

These numbers illustrate a situation that reflects the experience of many people in recent years; that is, even though salaries or wages have been increasing, the purchase power (R$) of those salaries has not increased correspondingly. Indeed, many people have experienced declining purchasing power because the rate of salary increases has not been as 'great as the rate of inflation. ∎

EXAMPLE 8-4

Referring to Example 8-3, determine the R$ salaries expressed in year 3 spending power.

Solution

With $k = 3$ and Equation 8-5, the following R$ salaries are obtained.

Year	Salary (R$ in Year 3)
1	$25,000(F/P, 8\%, 2)$ = $29,160
2	$26,500(F/P, 8\%, 1)$ = 28,620
3	$28,090(P/F, 8\%, 0)$ = 28,090
4	$29,775(P/F, 8\%, 1)$ = 27,569

∎

Inflation Versus Escalation

It is important in before-tax economy studies to distinguish between general inflation, differential escalation, and effective annual escalation. These definitions are used throughout the remainder of this book:

1. *General inflation rate (f)*: a general measure of the annual decrease in purchasing power of a dollar. The annual general inflation rate is defined by a selected, and broadly based, index of price changes. We define f to be the projected annual compound general inflation rate based on the Consumer Price Index (CPI), or the Implicit Price Index (IPI) for the Gross National Product, for a specified future interval of time. Many organizations have their own index of general inflation that reflects the particular business environment in which they work.

2. *Differential annual escalation rate* (e_j'): the annual change in the price of a specific commodity or service j which can be more or less than the general inflation rate. (The subscript j is used to label different commodity, service, or revenue cash flows.) Differential escalation results from technological breakthroughs, increased demand for a commodity or service with restricted supply, and so on, and can be positive or negative in sign.

3. *Effective annual escalation rate* (e_j): the total annual rate of increase in the price of commodity or service j. The effective annual escalation rate includes the effects of general inflation (f) and the applicable differential price change (e_j') above or below the general inflation rate. It is possible for e_j to be less than f. The following equation expresses the relationship between the rates defined above:

$$1 + e_j = (1 + f)(1 + e_j') \quad \text{or} \quad e_j' = \frac{e_j - f}{1 + f} \quad (8\text{-}6)$$

TABLE 8-1 A Comparison of Effective Annual Escalation Rates and the Consumer Price Index

Year	Wheat		Imported Crude Oil		Consumer Price Index	
	Cost per Bushel[a]	Escalation Rate (%)	Cost per Barrel[b]	Escalation Rate (%)	Index Value[c]	General Inflation Rate (%)
1966	$1.63	—	$ 1.80	—	97.2	—
1967	1.39	− 14.7	1.80	0.0	100.0	2.9
1968	1.24	− 10.8	1.80	0.0	104.2	4.2
1969	1.25	0.8	1.80	0.0	109.8	5.4
1970	1.33	6.4	1.80	0.0	116.3	5.9
1971	1.34	0.8	2.18	21.1	121.3	4.3
1972	1.76	31.3	2.48	13.8	125.3	3.3
1973	3.95	124.4	3.98	60.5	133.1	6.2
1974	4.09	3.5	10.83	172.1	147.7	11.0
1975	3.56	− 13.0	10.99	1.5	161.2	9.1
1976	2.73	− 23.3	11.51	4.7	170.5	5.8
1977	2.33	− 14.7	12.40	7.7	181.5	6.5
1978	2.94	26.2	12.70	2.4	195.4	7.7
Annual compound price change rate[d]		5.0		17.7	Annual compound general inflation rate[d]	6.0

[a]From U.S. Department of Agriculture, U.S. Agricultural Marketing Service.
[b]From *Petroleum Institute Weekly,* Arab light-benchmark crude (cheapest crude).
[c]From U.S. Department of Labor, Bureau of Labor Statistics.
[d]The annual compound rates are computed as follows: [(price or index value in final year ÷ price or index value in initial year)$^{1/N}$ − 1]100, where N equals 12 years in this case.
Source: W. G. Sullivan and J. A. Bontadelli, "The Industrial Engineer and Inflation," *Industrial Engineering* March 1980, pp. 24–33.

The data in Table 8-1 provide a representative comparison of the different effective annual escalation rates (e_j) for two important world commodities and constrasts them with the CPI. In addition, the data illustrate that the escalation rate may vary significantly for a commodity or service but can be represented with an annual compound rate over a longer time period.

EXAMPLE 8-5

Suppose that the general rate of inflation (f) is expected to average 8% during the next 4 years. For a certain commodity, the differential annual escalation rate has been projected to be $+2\%$. What will the marketplace prices be for this commodity over the next 4 years if it is now selling for $6.30 per unit?

Solution

By utilizing Equations 8-6 and 8-5, respectively, we can determine the A$ selling price of this commodity in view of its effective annual escalation rate:

$$1 + e = (1.08)(1.02) \quad \text{or} \quad e = (1.08)(1.02) - 1 = 0.1016$$

End of Year	Selling Price per Unit (A$)
1	$6.30(F/P, 10.16\%, 1) = \$6.94$
2	$6.30(F/P, 10.16\%, 2) = 7.65$
3	$6.30(F/P, 10.16\%, 3) = 8.42$
4	$6.30(F/P, 10.16\%, 4) = 9.28$

■

EXAMPLE 8-6

A recent engineering graduate has received annual salaries shown below over the past 4 years. During this time, the CPI has performed as indicated. Determine the engineer's annual salaries in *year 0 dollars* ($k = 0$) if the CPI is the appropriate indicator of general inflation for this person.

End of Year	Salary (A$)	CPI
1	$16,000	7.1%
2	18,000	5.4%
3	20,500	8.9%
4	23,200	11.2%

Solution

The engineer's salary has increased by 12.5%, 13.9%, and 13.2% in years 2, 3, and 4, respectively. These are effective annual inflation rates. By using Equation 8-5 with each year's inflation taken into account separately, the R$ equivalents in year 0 dollars are calculated as follows.

End of Year	Salary (R$ in Year 0)	
1	$16,000(P/F, 7.1\%, 1)$	$= \$14,939$
2	$18,000(P/F, 7.1\%, 1)(P/F, 5.4\%, 1)$	$= 15,946$
3	$20,500(P/F, 7.1\%, 1)(P/F, 5.4\%, 1)(P/F, 8.9\%, 1)$	$= 16,676$
4	$23,200(P/F, 7.1\%, 1)(P/F, 5.4\%, 1)(P/F, 8.9\%, 1)(P/F, 11.2\%, 1)$	$= 16,972$

■

Real and Combined Interest Rates

There are two fundamental types of interest rates employed for discounting/compounding cash flows. After these rates are defined, we shall demonstrate their proper use with numerous examples. The two interest rates are:

1. *Real interest rate* (i_r): the marginal cost of capital that does not include a marketplace adjustment for the expected general inflation rate. It is therefore the increase in real purchasing power expressed as a percent per period. The real discount rate reflects potential earning power of money in the absence of general inflation and is utilized when moving real dollar amounts forward or backward in time. It is generally accepted that real interest rates are roughly constant when plotted against time.

2. *Combined interest rate* (i_c): the marginal cost of capital that includes a marketplace adjustment for the anticipated general inflation rate in the economy. Thus it represents the increase in future cash flows to cover real interest and general inflation expectations. This annual rate takes into account both inflation and the potential earning power of money, and is used when moving actual dollar amounts forward or backward in time. Equation 8-7 shows the relationship between the real and combined interest rates.

$$1 + i_c = (1 + f)(1 + i_r) \qquad \text{or} \qquad i_c = i_r + f + i_r(f) \qquad \text{(8-7)}$$

EXAMPLE 8-7

An investor lends $10,000 today to be repaid in a lump sum at the end of 10 years with interest at 10% $(= i_c)$ compounded annually. What is his real rate of return, assuming that inflation is 8% $(= f)$ compounded annually?

Solution

In 10 years, the investor will receive his original $10,000 plus interest that has accumulated:

$$F = \$10,000(F/P, 10\%, 10) = \$25,937$$

The purchasing power of a dollar bill, however, has been reduced (or eroded) at 8% annual inflation using Equation 8-5 to

$$P = \$1(P/F, 8\%, 10) = \$0.4632$$

Thus the $25,937 is only worth, in today's purchasing power,

$$\$25,937 \times 0.4632 = \$12,014$$

The $12,014 of today's purchasing power that is returned for the use of $10,000 represents a real rate of return that may be calculated by finding the $i'\%$, at which

$$\$10,000 = \$12,014 \times (P/F, i'\%, 10) \qquad \text{or} \qquad i'\% = 1.85\%$$

The real interest rate can also be directly calculated from Equation 8-7 to be

$$i_r = \frac{i_c - f}{1 + f} \qquad \text{or} \qquad i_r = 0.0185$$

The value of i_r can be approximated much more expeditiously by simply subtracting the inflation rate, f, from the $i_c = 10\%$ being charged. Thus the real rate is roughly equal to $10\% - 8\% = 2\%$. This approximation is fairly close to the correct real interest rate at low rates of inflation but becomes increasingly less accurate as the inflation rate increases. ■

EXAMPLE 8-8

An investor established an Individual Retirement Account (I.R.A.) in 1981 that involves a *series* of 20 deposits (rather the one lump-sum deposit in Example 8-7) as shown in the figure. The account is expected to compound at an average interest rate of 12% per year through the year 2015. Inflation is expected to average 6% per year during this time.

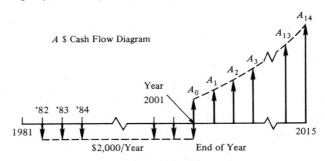

EXAMPLE 8-8

(a) What is the future worth of the I.R.A. at the end of year 2001 if inflation is expected to average 6% per year?
(b) What is the future worth of the I.R.A. in 1981 spending power?
(c) A withdrawal plan is desired so that once the first amount, A_0, has been withdrawn, each subsequent withdrawal will be 10% greater than that of the previous year. What is the value of A_0? Refer to the section "Interest Formulas Relating a Geometric Gradient Series to Its Present and Annual Worth" in Chapter 4.

Solution

(a) F (in A$) $= \$2,000(F/A, i_c = 12\%, 20) = \$144.105$.
(b) F (in R$) $= F$ (in A$)$(P/F, f, 20)$
$$= \$144,105(P/F, 6\%, 20)$$
$$= \$44,933$$
(c) $\$144,105 = A_0 + \dfrac{A_1}{1.10}\left(P/A, \dfrac{12\% - 10\%}{1.10}, 14\right)$
$$= A_0 + A_0(P/A, 1.82\%, 14)$$
$$= A_0 + A_0(12.261); \qquad A_0 = \$10,867$$

What Interest Rate to Use in Engineering Economy Studies

In general, the interest rate that is appropriate for equivalence calculations in engineering economy studies depends on the type of cash flow estimates:

Method	If Cash Flows Are in Terms of:	Then the Interest Rate to Use Is:
A	Real dollars (R$)	Real interest rate, i_r
B	Actual dollars (A$)	Combined interest rate, i_c

This table should make intuitive sense as follows. If one is estimating cash flows in terms of real dollars, the real (uninflated) interest rate is used. Similarly, if one is estimating cash flows in terms of actual (inflated) dollars, the combined (inflated) interest rate is used. Thus one can make economic analyses using either R$ or A$ with equal validity provided that the appropriate interest rate is used for equivalence calculations.

Many government agencies require that economy studies be conducted by using R$ cash flow estimates with differential escalation (if any) directly taken into account. They would utilize a real interest rate for equivalence calculations. Most companies in the private sector of this country perform engineering economy studies for which A$ cash flow estimates have been prepared. *Consequently, most M.A.R.R.'s used by private industry include anticipated inflation, and thus they are combined interest rates.* Furthermore, the A$ cash flow estimates include an effective annual escalation rate for commodity j. The effects of escalation must always be explicitly dealt with in A$ studies unless all cash flows escalate at a single rate that equals the general inflation rate.

EXAMPLE 8-9

To illustrate the relationship of the two methods described above, consider a project requiring an investment of $20,000 which is expected to return, in terms of actual dollars, $6,000 at the end of the first year, $8,000 at the end of the second year, and $12,000 at the end of the third year. The rate of inflation is 5% per year, and the real interest rate is 10% per year. Compare the N.P.W. of this project using methods A and B.

Solution

From Equation 8-7, the combined interest rate is 15.5%. Table 8-2 shows how the net present worth of the project would be calculated to be $-$1,060$ using method B.

If method A for considering inflation is used, the cash flows should be estimated in terms of real dollars. For the data given and an inflation rate of 5%, if the estimator is precisely consistent, his projections in terms of real dollars (as of time of investment) should be $20,000, $5,720, $7,240, and $10,360 for each year, respectively. Table 8-3 shows how, using Equation 8-5 and a real interest rate of 10%, the net present worth of the project would be calculated to be the same $-$1,060$ as when using method A. ■

TABLE 8-2 Calculation of Net Present Worth with Estimates in Actual Dollars

Year, n	Outcome (A$)	Discount Factor for Real Interest and Inflation $(P/F, 15.5\%, n)$	Present Worth (Rounded)
0	$-$20,000	1.000	$-$20,000
1	6,000	0.867	5,200
2	8,000	0.745	5,980
3	12,000	0.647	7,760
			$\Sigma \cong -\$ 1,060$

TABLE 8-3 Calculation of Net Present Worth Estimates in Real Dollars

Year n	Outcome (R$)	Discount Factor for Real Interest Only $(P/F, 10\%, n)$	Present Worth (Rounded)
0	− $20,000	1.000	− $20,000
1	5,720	0.909	5,200
2	7,240	0.826	5,980
3	10,360	0.751	7,760
			$\Sigma \cong$ − $ 1,060

In Example 8-9 it is worthy of note that had the outcomes in actual dollars been discounted by the real interest rate of $10%, the net present worth would have been calculated to be $1,080, indicating a favorable project. This error is regretably quite common in economy studies. The resulting N.P.W. = $1,080 is in contrast with the − $1,060 net present worth calculated when the correct interest rates that correspond to the types of dollars under consideration are utilized.

EXAMPLE 8-10

A $5,000 corporate bond that matures in 8 years can be purchased now for $3,900. The nominal rate of interest ($= i_c$) on the bond, payable annually, is 10%. Thus interest of $500 per year is paid on the bond. What is the combined interest rate and the real interest rate being earned if $f = 6\%$?

Solution

All the cash flows in this problem are A$ amounts. When the unknown interest rate is determined by conventional means, it will therefore be a combined interest rate:

$$\$3,900 = \$500(P/A, i'_c, 8) + \$5,000(P/F, i'_c, 8)$$

By trial and error $i'_c = 14.9\%$. The real interest rate from Equation 8-7 is

$$i_r = \frac{i_c - f}{1 + f} = \frac{0.149 - 0.06}{1.06} = 0.084$$

or $i_r \cong 8.4\%$. ∎

Fixed and Responsive Annuities

Whenever future investment receipts are predetermined by contract, as in the case of a bond or a fixed annuity, these receipts do not respond to inflation. In cases where the future receipts are not predetermined, however, they may respond to inflation. The degree of response varies from case to case. To illustrate the nature of inflation, let us consider two annuities. The first annuity is fixed

TABLE 8-4 Illustration of Fixed and Responsive Annuity with Inflation Rate of 6% per Year

Year	Fixed Annuity In Actual Dollars	Fixed Annuity In Equivalent Real Dollars[a]	Responsive Annuity In Actual Dollars	Responsive Annuity In Equivalent Real Dollars[a]
1	$2,000	$1,887	$2,120	$2,000
2	2,000	1,780	2,247	2,000
3	2,000	1,679	2,382	2,000
4	2,000	1,584	2,525	2,000
5	2,000	1,495	2,676	2,000
6	2,000	1,410	2,837	2,000
7	2,000	1,330	3,007	2,000
8	2,000	1,255	3,188	2,000
9	2,000	1,184	3,379	2,000
10	2,000	1,117	3,582	2,000

[a]See Equation 8-5 when the base point in time is assumed to be year 0.

(unresponsive to inflation) and yields $2,000 per year for 10 years. The second annuity is of the same duration and yields enough future dollars to be equivalent to $2,000 per year in real purchasing power. Assuming an inflation of 6% per annum, pertinent values for the two annuities over a 10-year period are shown in Table 8-4.

Thus, when the receipts are constant in actual dollars (unresponsive to inflation), their equivalent value in real dollars of investment declines over the 10-year interval to $1,117 in the final year. When receipts are fixed in value of real dollars of investment (responsive to inflation), their equivalent in actual dollars rises to $3,582 by year 10.

Included in actual dollar studies are certain quantities unresponsive to inflation, such as depreciation, lease fees, and interest charges based on some existing contract or loan agreement. For instance, depreciation write-offs, once determined, do not increase (with present accounting practices) to keep pace with inflation; lease fees and interest charges typically are contractually fixed for a given period of time.

A real dollar analysis requires that all cash flow estimates be made in terms of equivalent purchasing power at some reference point in time. (In this chapter we let $k = 0$ for convenience.) Often this is not as easy as it would first appear. Any actual dollar quantities that are unresponsive to inflation will decrease in terms of their real dollar values. Furthermore, the argument that a real dollar analysis involves fewer details than an actual dollar analysis is not necessarily valid when several commodities or services that increase (or decrease) in price at different rates are involved. In a real dollar analysis these price differentials must be estimated just as the total increase or decrease in price for each cash flow element is projected in an actual dollar analysis.

EXAMPLE 8-11

A person is to receive a fixed annuity of $2,000 a year for 17 years, the first payment being made on April 1, 1993. He offers to sell the annuity on April 1, 1986, for an amount, P, in dollars equivalent to the annuity's above-described stream of payments. Assuming that he desires a real return on his money of 4%, determine the value of P when the average inflation rate is 6% per year.

Solution

The combined interest rate to be applied to the constant annuity in A$ is $i_c = 0.04 + 0.06 + 0.0024 = 0.1024$. The lump-sum equivalent value of the annuity on April 1, 1992, is $2,000(P/A, 10.24\%, 17) = \$15,807.65$. Therefore, $P = \$15,807.65 \times (P/F, 10.24\%, 6)$, or $P = \$8,807$ on April 1, 1986.

∎

EXAMPLE 8-12

Consider a situation where an annuity of $3,000 in year 0 spending power is increasing at a constant inflation rate of 5% for 15 years. The first cash flow occurs at the end of year 1. A combined interest rate of 15.5% is computed from $i_r = 10\%$ and $f = 5\%$. Calculate the present and annual worths of the annuity receipts.

Solution

$$P.W. = \$3,000\left(P/A, \frac{i_c - f}{1 + f}, 15\right)^{\dagger}$$

$$= \$3,000\left(P/A, \frac{15.5\% - 5\%}{1.05}, 15\right)$$

$$= \$3,000(P/A, 10\%, 15)$$

$$= \$22,818.24$$

The equivalent uniform annual worth in A$ is determined with this relationship:

$$A.W. = P.W.(A/P, i_c, 15)$$

$$= \$22,818(A/P, 15.5\%, 15)$$

$$= \$3,997.10$$

If the equivalent annual worth $(A.W._0)$ is desired in R$ at the present time such that it inflates at $f = 5\%$ per year, this calculation would be made:

$$A.W._0 = P.W.(A/P, i_r, 15)$$

$$= \$22,818(A/P, 10\%, 15)$$

$$= \$3,000.$$

∎

†Refer to earlier discussion of geometric gradients in Chapter 4.

Graphically, the results of Example 8-12 can be summarized as shown in the figure.

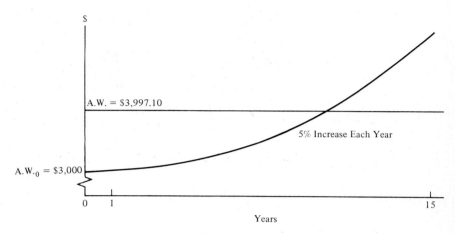

$ (y-axis)

A.W. = $3,997.10

5% Increase Each Year

A.W.$_0$ = $3,000

0   1   15

Years

EXAMPLE 8-12

EXAMPLE 8-13

As an individual homeowner, suppose that you are interested in purchasing a heat pump in early 1985 to replace your present heating and air-conditioning system. You calculate that on the average 24,000 kWh of electricity will be required to heat and cool your home with this heat pump. In early 1985 the cost per kilowatthour to you is 4.5 cents, and this cost is expected to increase by 15% per year (e) over the next 15 years. During the same period, the general inflation rate is expected to average 6%. If your present system averages 30,000 kWh of electrical consumption each year and your opportunity cost of capital is 10% ($= i_c$), how much could you afford to spend now for a heat pump?

Solution
Savings per year in early 1985 dollars equal

$$\frac{6000 \text{ kWh}}{\text{year}} (\$0.045/\text{kWh}) = \$270/\text{year}$$

Because these savings will escalate at 15% per year for 15 years, we need to calculate the present worth of this geometric gradient (refer to Chapter 4) to determine how much we can afford to spend now. A solution based on R$ savings is the following.

$$P = \sum_{k=1}^{15} \frac{270(1.15)^k}{(1.10)^k}$$

$$= \$270(P/A, -4.35\%, 15)$$

$$= \$5,888$$

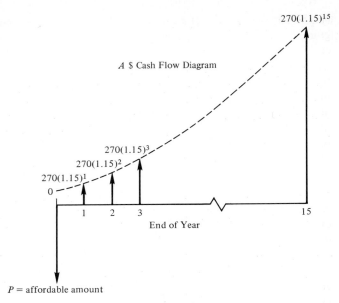

A \$ Cash Flow Diagram

$270(1.15)^{15}$

$270(1.15)^3$
$270(1.15)^2$
$270(1.15)^1$
0

1   2   3   15

End of Year

P = affordable amount

EXAMPLE 8-13

Thus when the rate of annual escalation of a commodity such as electricity exceeds the annual opportunity cost of capital, a negative real interest rate $[i_r = (0.10 - 0.15)/1.15]$ will result as demonstrated in this example. ■

Thus far we have concentrated on the treatment of inflation because it is the dominant condition that we have experienced for the past 50 years or so. As noted earlier, *deflation* is also possible where there is a decrease in the monetary price of goods and services. This translates into an increase over time in the purchasing power of a fixed sum of money. When deflation is present, the same procedures described for inflation apply except that the value of f is negative rather than positive. Hence cash flow estimates may be made in R\$ or A\$ and discounting/compounding takes place at i_r or i_c, respectively. The combined interest rate would be

$$i_c = (1 + i_r)(1 - f) - 1$$

$$= i_r - f - i_r f \qquad (8\text{-}8)$$

where f is the annual rate of deflation.

An Example That Considers Escalation

In before-tax engineering economy studies, the effects of inflation on prices are often ignored since all cash flows are *presumably* influenced to the same extent. If costs and benefits of competing alternatives are initially estimated in real (year 0) dollars and if all cash flows are assumed to increase each year at the general rate of inflation, an economy study can be conducted in terms of R$ cash flow estimates, and discounting occurs at the firm's real M.A.R.R.

When differential escalation is estimated for various types of cash flows comprising one or more alternatives, it is no longer possible to assume that all cash flows constitute constant annuities in real dollars. In such cases the effects of differential escalation must be considered in before-tax economy studies, and ignoring it may lead to an incorrect selection among alternatives. This possible source of error in reversing an alternative's attractiveness is demonstrated by the following example.

EXAMPLE 8-14

A company that routinely estimates cash flows in real dollars is attempting to evaluate a new venture. Revenues are expected to inflate at the general rate of inflation (f), which is expected to average 6% per year. The two basic cost components are labor and energy, with annual escalation rates of 4% and 9%, respectively. The capital investment required for the new venture is $10,000, and the company's real M.A.R.R. is 4%. Should the proposed venture be undertaken in view of the data developed by the company (shown below)?

| End of Year | Annual Cash Flows (R$) | | | |
	Revenue	Labor	Energy	Investment
0	0	0	0	− $10,000
1	+ $5,000	− $1,000	− $1,700	0
2	+ 5,000	− 1,000	− 1,700	0
3	+ 5,000	− 1,000	− 1,700	0
4	+ 5,000	− 1,000	− 1,700	0
5	+ 5,000	− 1,000	− 1,700	0

Company Solution

Because the *average* annual escalation rate on costs was close to 6%, analysts in this company decided to ignore differential escalation. Their analysis of net present worth is the following.

$$\text{N.P.W.} = -\$10,000 + (5,000 - 2,700)(P/A, 4\%, 5)$$

$$= +\$239.19$$

Thus it is concluded that the new venture is marginally attractive. ∎

Solution Including Differential Escalation

The differential escalation on revenues is 0% since they are expected to increase by 6% per year. Moreover, the differential escalation on labor (e_L') and energy (e_E') are approximately −2% and +3%, respectively. Consequently the R$ annual cash flow estimates for this venture would be as shown below.

End of Year	Annual Cash Flows (R$)		
	Revenue	Labor ($e_L' = -2\%$)	Energy ($e_E' = 3\%$)
0	0	0	0
1	+$5,000	$-\$1,000(1 - 0.02)^1 = -980$	$-\$1,700(1 + 0.03)^1 = -1,751$
2	+ 5,000	$- 1,000(0.98)^2 = -960$	$- 1,700(1.03)^2 = -1,804$
3	+ 5,000	$- 1,000(0.98)^3 = -941$	$- 1,700(1.03)^3 = -1,858$
4	+ 5,000	$- 1,000(0.98)^4 = -922$	$- 1,700(1.03)^4 = -1,913$
5	+ 5,000	$- 1,000(0.98)^5 = -904$	$- 1,700(1.03)^5 = -1,971$

The net before-tax cash flow (BTCF) is

End of Year	BTCF	
0	$-\$10,000$	
1	+ 2,269	
2	+ 2,236	
3	+ 2,201	N.P.W.(4%) = $-\$196$
4	+ 2,165	
5	+ 2,125	

With differential escalation taken into consideration, the venture changes from marginally attractive (N.P.W. = $239) to unattractive (N.P.W. = $-\$196$).

The N.P.W. of R$ negative cash flows above could have been easily determined by using a single, weighted differential escalation rate:

$$\bar{e}' = -2\%\left(\frac{1,000}{2,700}\right) + 3\%\left(\frac{1,700}{2,700}\right) = +1.15\%/\text{year}$$

This applies when the annuities expressed in year 0 dollars do not change during the study period. Thus

$$\text{N.P.W.}(4\%) \cong -\$10,000 - \$2,700\left(P/A, \frac{4\% - 1.15\%}{1.0115}, 5\right)$$

$$+ \$5,000(P/A, 4\%, 5) \cong -\$170$$

A closer agreement to the correct answer of $-\$196$ would have been possible if Equation 8-6 had been used to obtain more accurate approximations of e_L' and e_E'. ∎

For large projects having many types of cash flows that escalate at differential rates, it is advisable to include the effects of inflation/escalation in before-tax engineering economy studies.

Based on solutions for Examples 8-11 to 8-14, it is worthwhile to summarize the exact interest rate to be applied to actual dollar versus real dollar cash flow estimates:

Cash Flows	Interest Rate to Use	
	When $e = f$	When $e \neq f$
A$	i_c (Example 8-11)	i_c (In Example 8-13 cash flows were shown to escalate at e)
R$	i_r (Example 8-12)	i_r (In Example 8-14 cash flows were estimated in terms of e')

A Public Utility Case Study
Involving Variable Capacity Factors and Inflation

An electric utility has obtained estimates for the investment and recurring expenses associated with a 300-MW coal-fired station. The investment amount is $300 million, which corresponds to $1,000 per kilowatt installed. Regarding the recurring annual expenses, these estimates were developed in real dollars (relative to year 0) for a capacity factor of 100%, a heat rate of 9000 Btu/kWh, and coal costing $1.60 per million Btu.

Fuel: $\left(\dfrac{\$1.60}{10^6 \text{ Btu}}\right)\left(\dfrac{9,000 \text{ Btu}}{\text{kWh}}\right)\left(\dfrac{8,760 \text{ hr}}{\text{yr}}\right)(300,000 \text{ kW}) = \$37.84$ million

Other: $\begin{cases} \text{fixed operating and maintenance (O\&M)} = \$5.0 \text{ million/year} \\ \text{variable operating and maintenance} = \$10.5 \text{ million/year at } 100\% \text{ capacity} \end{cases}$

Capacity factors are expected to vary over the 30-year plant life as follows:

Years	Capacity Factor[a]
1–5	0.80
6–10	0.65
11–30	0.50

[a]This is the ratio of kWh per year actually generated to the maximum possible kilowatt-hour production during the year.

The utility has made the assumption that all recurring annual expenses inflate at 8% per year. Their weighted average cost of capital is 12%, which is a combined interest rate based on general inflation averaging 6% per year.

System planners have been asked to determine the net present worth of this generating station as well as the *uniform* annual cost of energy (\$/kWh) over the plant life, expressed in actual dollars. Their calculation of net present worth and "levelized" cost of electricity is shown below:

Step 1: The real interest rate equals

$$\left(\frac{0.12 - 0.06}{1.06}\right)100 = 5.66\%$$

The differential escalation rate on fuel and O&M is

$$\left(\frac{0.08 - 0.06}{1.06}\right)100 = 1.89\%$$

Step 2: Because recurring annual expenses are assumed to vary linearly with the capacity factor, this setup is appropriate for determining the net present worth of the plant (millions of dollars):

$$\text{N.P.W.} = \$300 + \$48.34(0.8)\left(P/A, \frac{5.66\% - 1.89\%}{1.0189}, 5\right)^{\dagger}$$

$$+ \$48.34(0.65)(P/A, 3.8\%, 5)(P/F, 3.8\%, 5)$$

$$+ \$48.34(0.5)(P/A, 3.8\%, 20)(P/F, 3.8\%, 10)$$

$$+ \$5.00(P/A, 3.8\%, 30)$$

$$= \$880.17$$

Step 3: The total uniform annual cost (A) of the plant in actual dollars is

$$A = \text{N.P.W.}(A/P, 12\%, 30)$$

$$= \$880.17(0.1241)$$

$$= \$109.27$$

Unfortunately, the annual amount of electricity produced is not uniform over the life of the plant:

Years	Energy Produced, E_k
1–5	300,000 kW(8,760 hr/yr)(0.8) = 2.1×10^9 kWh/yr
6–10	300,000 kW(8,760 hr/yr)(0.65) = 1.7×10^9 kWh/yr
11–30	300,000 kW(8,760 hr/yr)(0.5) = 1.3×10^9 kWh/yr

To obtain a uniform annual cost per kilowatthour over a 30-year period, this expression is useful:

$$\text{cost/kWh} = \sum_{k=0}^{30}\left[\frac{C_k}{E_k}\left(P/F, \frac{5.66\% - 1.89\%}{1.0189}, k\right)\right]$$

$\dagger$

$\dfrac{i_r - e'}{1 + e'} = 0.038$, or 3.8% per year.

where C_k represents the total recurring annual cost in year k, and $C_0 = \$300 \times 10^6$. Notice that

$$\sum_{k=0}^{30} C_k\left(P/F, \frac{5.66\% - 1.89\%}{1.0189}, k\right)$$

has previously been computed in step 2 and equals $\$880.17 \times 10^6$. Therefore, the expression for cost per kilowatthour above simplifies to this:

$$\text{cost/kWh} = \frac{\text{N.P.W.}}{E_{1-5}(P/A, 3.8\%, 5) + E_{6-10}(P/A, 3.8\%, 5)(P/F, 3.8\%, 5)}$$
$$+ E_{11-30}(P/A, 3.8\%, 20)(P/F, 3.8\%, 10)$$

$$= \frac{\$880.17 \times 10^6}{28.42 \times 10^9 \text{ kWh}}$$

$$= \$0.031$$

or

$$\text{cost/kWh} = \frac{\overline{A}}{\overline{E}} = \frac{\$109.27 \times 10^6}{(28.42 \times 10^9 \text{ kWh})(A/P, 12\%, 30)}$$

$$= \$0.031$$

A Process Industry Case Study Including Inflation

The R Square Corporation is considering two plans, A and B, for expanding the capacity of its urea manufacturing plant. The relevant data and analysis for both plans are given below.

PLAN A
R Square Corporation's office of agricultural and chemical development operates a urea plant that was built in 1968 at a cost of \$4.3 million and capacity of 300,000 lb/yr. Due to increasing use of urea-based fertilizers and increasing maintenance costs on the existing unit, the feasibility of constructing a new plant with 750,000 lb/yr capacity is being studied. Time does not permit an in-depth cost estimation for the new plant, so engineers decide to obtain the cost of the new plant by scaling up the cost of the old plant. The cost-capacity factor is known to be 0.65 and the construction-cost index has increased an average of 10.5% for the past 15 years. What cost should the engineers report to the project review committee?

Urea plant estimates:

$$C_0 = \$4.3 \text{ million: cost of plant in 1968}$$

$$S_0 = 300,000 \text{ lb/yr: capacity of existing plant}$$

$$C_n = ? : \text{cost scaled up (but based on 1968 pricing)}$$

$$S_N = 750,000 \text{ lb/yr: capacity of plant being studied}$$

$$C_{NE} = \text{cost reported to committee in 1983 dollars}$$

$$X = \text{cost-capacity factor} = 0.65$$

Solution

$$\frac{C_N}{C_0} = \left(\frac{S_N}{S_0}\right)^x$$

$$C_N = C_0\left(\frac{S_N}{S_0}\right)^x$$

$$C_N = \$4.3 \text{ million} \left(\frac{750,000 \text{ lb/yr}}{300,000 \text{ lb/yr}}\right)^{0.65}$$

$$C_N = \$7.8 \text{ million (cost basis in 1968)}$$

$$C_{NE} = C_N(F/P, 10.5\%, 15) = \$7.8 \text{ million}(1.105)^{15},$$

or

$$C_{NE} = \$34.88 \text{ million (estimated cost of new urea unit in 1983 dollars)}$$

PLAN B

As an alternative plan, it is learned that a manufacturer of these units can prefabricate and install a unit on-site for a total cost of $22 million. The R Square Corporation's constructed unit has an estimated life of 15 years and the prefabricated unit has a life of 12 years. The real dollar operating costs for the company unit are $40,000 per year for years 1–10 and $30,000 per year for years 10–15. The real dollar operating costs for the prefabricated unit are $50,000 per year. The real interest rate is assumed to be 10%. Which alternative should they select? What assumptions are involved?

Solution

Note: Assume negligible salvage value for both alternatives, and compare the alternatives using the annual cost method.

	Prefabricated Unit	Company Unit
Initial cost	$22,000,000	$34,880,000
Annual costs	$50,000/yr	$40,000/yr, years 1–10
		$30,000/yr, years 10–15
Life	12 years	15 years

Prefabricated unit:

$$\overset{0.1468}{\text{A.C.} = \$50,000 + \$22,000,000(A/P, 10\%, 12)}$$

$$= \$3,279,600$$

Company unit:

$$\text{A.C.} = \$30,000 + \$10,000 \overset{6.1446}{(P/A, 10\%, 10)} \overset{0.1315}{(A/P, 10\%, 15)}$$
$$+ \$34,880,000(A/P, 10\%, 15)$$
$$= \$4,893,884$$

By a wide margin, the R Square Corporation should let the outside manufacturer construct and install the urea unit. However, numerous nonmonetary considerations could shift the decision to plan A. The monetary risks associated with expanding in an uncertain and highly competitive market may well cause neither plan to be acceptable. An indefinitely long study period has been assumed in the analysis above.

SUMMARY

Accurate cost estimating is a vital component of any engineering economy study. Several sources of data and methods for utilizing these data in preparing cash flow estimates have been discussed. Since correctly taking inflation into account is an important aspect of estimating costs and in evaluating alternatives, much of this chapter has dealt with incorporating price and cost changes into before-tax economy studies. In this regard, it must be ascertained whether cash flows have been estimated in actual dollars or real dollars. The appropriate interest rate to use when discounting or compounding actual dollar amounts is a combined or "marketplace" rate, while the corresponding rate to apply in real dollar studies is the firm's real interest rate. After-tax studies involving inflation (or deflation) offer some difficulties caused by depreciation and other fixed actual dollar annuities such as interest charges and lease fees. These problems are discussed in detail in Chapter 10.

PROBLEMS

8-1 Give as many reasons as you can for closely scrutinizing accounting data before using them in an engineering economy study.

8-2 Why might a company's purchasing department be a good source of estimates for equipment required 2 years from now for a modernized production line? Which department(s) could provide good estimates of labor costs and maintenance associated with operating the production line? What information might the accounting department provide that would bear upon the incremental costs of the modernized line?

8-3 Prepare an index for housing construction costs in 1984 using these data:

Type of Housing	Percent	Reference Year ($I = 100$)		1984	
Single units	70	$28		$38	
Duplex units	5	23	$/ft²	34	$/ft²
Multiple units	25	20		31	

8-4 Manufacturing equipment that was purchased in 1970 for $200,000 must be replaced in

1986. What is the estimated cost of the replacement based on the equipment cost index below?

Year	Index	Year	Index
1960	100	1983	708
1970	223	1984	779
1980	600	1985	841
1982	681	1986	972

8-5   Suppose that your brother-in-law has decided to start a company that produces synthetic lawns for lazy homeowners. He anticipates starting production in 18 months. In estimating future cash flows of the company, which of the following would be relatively easy versus relatively difficult to obtain? Also, suggest how each might be estimated with "reasonable" accuracy.
(a) Cost of land for a 10,000-square-foot building.
(b) Cost of the building (cinder block construction).
(c) Fixed capital investment.
(d) Initial working capital.
(e) First year's labor and material costs.
(f) First year's sales revenues.

8-6   In a new industrial park, telephone poles and lines must be installed. Altogether it has been estimated that 10 miles of telephone lines will be needed and each mile of line costs $14,000 (includes labor). In addition, a pole must be placed every 40 yards on the average to support the lines, and the cost of the pole and its installation is $210. What is the estimated cost of the entire job?

8-7   If an ammonia plant that produces 500,000 pounds per year cost $2,500,000 to construct 8 years ago, what would a 1,500,000-pound-per-year plant cost now? Suppose that the construction cost index has increased an average rate of 12% per year for the past 8 years and that the cost-capacity factor to reflect economy of scale, x, is 0.65.

8-8   A reactor vessel cost $75,000 10 years ago. The reactor had the capacity of producing 500 pounds of product per hour. Today, it is desired to build a vessel of 1,000 pounds per hour capacity. With an inflation rate of 5% per year, and assuming a cost-capacity factor to reflect economies of scale, x, to be 0.75, what will the approximate future cost of the new reactor be in 5 years?

8-9   Refer to the graph and estimate the *total* production costs for a certain chemical compound when the capacity of the plant is 2,500,000 pounds per year.

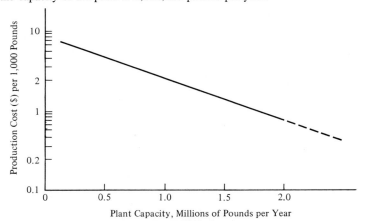

PROBLEM 8-9

8-10 Using the costing work sheet provided in this chapter, estimate the unit cost of manufacturing metal wire cutters in lots of 100 when these data have been obtained:

$$\text{Factory labor} = 4.2 \text{ hours at } \$5.20/\text{hour}$$

$$\text{Factory burden} = 150\% \text{ of labor}$$

$$\text{Outside manufacture} = \$74.87$$

$$\text{Production material} = \$14.22$$

$$\text{Packing costs} = 7\% \text{ of factory labor}$$

$$\text{Profit} = 12\%$$

8-11 Your rich aunt is going to give you end-of-year gifts of $1,000 for each of the next 10 years.
(a) If inflation is expected to average 6% per year during the next 10 years, what is the total value of these gifts at the present time? The real interest rate is considered to be 4% per year.
(b) Suppose that your aunt specified that the annual gifts of $1,000 are to be increased by 6% each year to keep pace with inflation. With a real interest rate of 4%, what now is the present value of the gifts?

8-12 Because of inflation in our economy, the purchasing power of the dollar shrinks with the passage of time. If the average inflation rate is expected to be 8% per year into the foreseeable future, how many years will it take for the dollar's purchasing power to be one-half of what it is now? (That is, there is a future point in time when it takes 2 dollars to buy what can be purchased today for 1 dollar.)

8-13 Which of these situations would you prefer?
(a) You invest $2,500 in a certificate of deposit that earns an effective rate of 8% per year. You plan to leave the money alone for 5 years, and the inflation rate is expected to average 5% per year. Taxes are ignored.
(b) You spend $2,500 on an antique piece of furniture. In 5 years you believe the furniture can be sold for $4,000. Assume that the average inflation rate is 5% per year. Again taxes are ignored.

8-14 Operating and maintenance costs for two alternatives have been estimated on different bases as follows.

End of Year	Alternative A—Costs Estimated in Actual (Inflated) Dollars	Alternative B—Costs Estimated in Real (Constant) Dollars with $k = 0$
1	$120,000	$100,000
2	132,000	110,000
3	148,000	120,000
4	160,000	130,000

If the average inflation rate is expected to be 6% per year and money can earn a real rate of 9% per year, show which alternative has the least negative present worth at time 0.

8-15 Suppose that you deposit $1,000 in a Swiss bank account that earns an effective rate of 18% per year, and you withdraw the principal after 6 years. You receive interest each year and spend it on your favorite hobby. What is the real annual rate of return on your investment if the general rate of inflation is 10%/yr.? Be exact!

8-16 If you buy a lathe now, it costs $100,000. If you wait 2 years to purchase the lathe, it will cost $135,000. Suppose you decide to purchase the lathe now, reasoning that you can earn 18% per year on your $100,000 if you do not purchase the lathe. If the inflation rate (f) in the economy during the next 2 years is expected to average 12% per year, did you make the right decision?

8-17 Your company has just issued bonds with a face value of $1,000. They mature in 10 years and pay annual dividends of $100. At present they are being sold for $887. If the average annual inflation rate over the next 10 years is expected to be 6%, what is the real rate of return per year on this investment?

8-18 A man desires to have $30,000 in a savings account when he retires in 20 years. This amount is to be equivalent to $30,000 in today's purchasing power. If the expected average inflation rate is 7% per year and the savings account earns 5% interest, what lump sum of money should the man deposit now in his savings account?

8-19 Determine the present worth of the following inflating cash flows that start in year 0 and continue for the next 100 years. The combined interest rate is 15% per year and the rate of inflation is 8% per year.

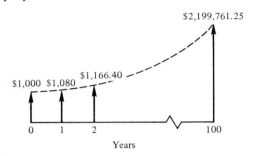

PROBLEM 8-19

8-20 The AZROC Corporation needs to acquire a small computer system for one of its regional sales offices. The purchase price of the system has been quoted at $50,000, and the system will reduce manual office expenses by $18,000 per year in real dollars. Historically, these manual expenses have inflated at an average rate of 8% per year and this is expected to continue into the future. A maintenance agreement will also be contracted for and its cost per year in actual dollars is constant at $3,000.

What is the minimum (integer-valued) life of the system such that the new computer can be economically justified? Assume that the computer's salvage value is zero at all times. The firm's M.A.R.R. is 25% and includes an adjustment for anticipated inflation in the economy. Show all calculations.

8-21 Addition of a new assembler to a certain company's computer can cut down computing time by 40%. Computing time costs $0.08 per kilobyte-hr. The assembler will be used for 5 years and has a first cost of $1,800. The older assembler is being used 17,500 kilobyte-hr/yr. Should the assembler be bought? An alternative is to invest the money at 15% per year interest during the 5 years. The average rate of inflation is 6% per year.

8-22 The incremental design and installation costs of a total solar system (heating, air conditioning, hot water) in a certain Tennessee home were $14,000 in 1984. The annual savings in electricity (in 1984 dollars) have been estimated at $2,500. Assume that the life of the system is 15 years.
(a) What is the internal rate of return on this investment if electricity prices do not escalate during the system's life?
(b) What average annual escalation rate on electricity would have to be experienced over the system's life to provide a rate of return of 25% for this investment?

8-23 A gas-fired heating unit is expected to meet an annual demand for thermal energy of 500 million Btu, and the unit is 80% efficient. Assume that each thousand cubic feet of natural gas, if burned at 100% efficiency, can deliver 1 million Btu. Suppose further that natural gas is now selling for $2.50 per thousand cubic feet. What is the present worth of fuel cost for this heating unit over a 12-year period if natural gas prices are expected to escalate at an average rate of 10% per year? The firm's M.A.R.R. is 18%.

8-24   A large corporation's electricity bill now amounts to $400 million. During the next 10 years, electricity usage is expected to increase by 75% and the estimated electricity bill 10 years hence has been projected to be $920 million. Assuming electricity usage and rates increase at uniform annual rates over the next 10 years, what is the annual rate of escalation of electricity prices expected by this corporation?

8-25   A 30-year home mortgage loan of $60,000 is obtained by Mr. and Mrs. Smith at 12% nominal interest. Monthly payments will be made to the bank for the next 360 months in this amount:

$$\$60,000(A/P,\ 1\%/\text{month},\ 360) = \$617.17$$

Because these payments are so high, the bank has agreed to permit the couple to make their first payment in the amount of $358.56(1 + e)$, and payments will then escalate at this constant rate for each subsequent payment. Graphically, the situation is as shown in the figure.

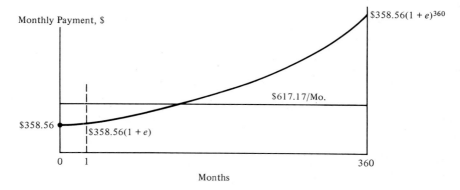

PROBLEM 8-25

What is the constant escalation rate that makes these two loan repayment plans equivalent (ignoring income taxes)?

8-26   A liberal arts graduate has decided to invest 5% of her first-year's salary in a money market fund. This amounts to $1,000 at the end of the first year. She also makes payments of $120 per month on a blue BMW. She has been told that her savings should keep up with expected salary increases of 8% per year. Thus she plans to invest an extra 8% each year starting at the end of year 2. At the end of year 1 she invests $1,000; in year 2, $1,080, in year 3, $1,166.40; and so on, up through year 10. If the average rate of inflation is expected to be 5% over the next 10 years and if she will earn an average of 10% per year in this account, what is the future worth of the money market fund at the end of the tenth year?

8-27   Suppose that you have just graduated with a B.S. degree in engineering and the Omega Corporation offers you $30,000 per year as a starting salary. The company is located in New York City, and your guaranteed raises over the next 5 years will be 15% per year starting in the second year of employment. Another company offers you a position in Jonesville but says that it is willing to negotiate a starting salary with you. Guaranteed raises with the Jonesville-based company will be 8% per year over a 5-year period. What starting salary should you request of the Jonesville company if you are indifferent about where you would like to work? Ignore the effects of income taxes. Other data are the following:
  (1) Your opportunity cost of capital (i_c) is 15% per year.
  (2) The average annual cost of living index for the next 5 years is 108 in New York City, and it is 100 in Jonesville (based on A$ spending power differences).
  (3) The average rate of inflation (f) in the economy is projected to be 8% over the next 5 years.

8-28 A small heat pump, including the duct system, now costs $2,500 to purchase and install. It has a useful life of 15 years and incurs annual maintenance of $100 per year in real (year 0) dollars over its useful life. A compressor replacement is required at the end of the eighth year at a cost of $500 in real dollars. The yearly cost of electricity for the heat pump is $680 based on prices at the beginning of the investor's time horizon. Electricity prices are projected to escalate at an annual rate of 10%. All other costs are expected to escalate at 6%, which is the projected general inflation rate. The firm's interest rate, which includes an allowance for general inflation, is 15%. No salvage value is expected from the heat pump at the end of 15 years.

(a) What is the annual equivalent cost, expressed in actual dollars, of owning and operating this heat pump?

(b) What is the annual cost in year 0 dollars of owning and operating the heat pump?

8-29 Suppose that you are faced with the problem of deciding between the following two automobiles. All cost estimates are based on 15,000 driving miles per year and are expressed in year 0 dollars.

	Alternative		Escalation Rate
	Domestic	Import	
Negotiated purchase price (including tax)	$8,500	$9,200	—
Fuel/year	1,050	700	8%/yr
Maintenance/year	200	400	5%/yr
Insurance/year	400	440	5%/yr
Miscellaneous/year	100	125	7%/yr
Ownership period, N	5 years	5 years	—
Trade-in value at end of year N	$1,000	$2,000	6%/yr

Your personal M.A.R.R. will average 12% per year over the next 5 years, and this is a "marketplace" rate of interest that includes an allowance for general inflation ($f = 7\%$). Which automobile would you select based solely on monetary considerations?

8-30 An electric utility company in the Northeast is trying to decide whether to switch from oil to coal at one of its generating stations. After much investigation, the problem has been reduced to these trade-offs:

	Oil	Coal
Cost of retrofitting boilers to burn coal	—	?
Annual fuel cost (year 0 dollars)	25×10^6	17×10^6
Escalation rate	10%/yr	6%/yr
Life of plant	25 years	25 years

Determine the cost of retrofitting the boilers (to burn coal) that could be justified at this generating station. The utility's real annual cost of capital is 3% and the general inflation rate in the economy will average 6% over the next 25 years.

CHAPTER 9

Depreciation

Depreciation is the decrease in value of physical properties with the passage of time.† Although the fact that depreciation does occur is easily ascertained and recognized, the determination of its magnitude in advance is not easy. In fact, the actual amount of depreciation can never be determined until the asset is retired from service. Because depreciation is a noncash cost that must be considered properly in after-tax economy studies in Chapter 10, the analyst often encounters problems in dealing with it.

It is clear that depreciation will be an estimate, and it most likely will not be entirely accurate. However, the analyst possibly may find some solace in the fact that accountants and business managers face equally perplexing problems in dealing with depreciation.

Basically, from a business viewpoint, a physical asset has value because one expects to receive future monetary benefits through the possession and use of it. These benefits are in the form of future cash flows resulting from (1) the use of the asset to produce salable goods or services, or (2) the ultimate sale of the asset. It is because of these anticipated cash flows that the asset has commercial value. Depreciation, then, represents a decrease in value because the ability of

†The term *property* is used in this chapter in a general sense. It includes buildings, machines, goods, and so on.

the asset to produce these future cash flows decreases, as the result of one or more of several causes, with the passage of time.

DEFINITIONS OF VALUE

Because depreciation is defined as decrease in *value*, it is necessary to give some consideration to the meaning of that term. Unfortunately, we discover that there are several meanings attached to it. Probably the best definition of value, in a commercial sense, is that it is the present worth of all the future profits that are to be received through ownership of a particular property. This undoubtedly excellent definition is, however, difficult to apply in actual practice, inasmuch as we can seldom determine profits far in advance. Thus several other measures of value are commonly used, some of which are approximations of the foregoing definition.

The most commonly encountered measure of value is *market value*. This is what will be paid by a willing buyer to a willing seller for a property where each has equal advantage and is under no compulsion to buy or sell. The buyer is willing to pay the market price because he believes it approximates the present value of what he will receive through ownership *with some rate of interest or profit included*. In most matters relating to depreciation, it is market value that is used. For new properties the cost on the open market is used as the original value.

Next to market value, probably the most important kind of value is *use value*. This is what the property is worth to the owner as an operating unit. A property may be worth more to the person who possesses it and has it in operation than it would be to someone else who, if he purchased it, might have to spend additional funds to move it and get it into operation. Use value is, of course, very closely akin to the original definition of value that was stated previously. It is difficult to determine for the same reasons.

A third type of value is known as *fair value*. This usually is determined by a disinterested party in order to establish a price that is fair to both seller and buyer.

Book value is the worth of a property as shown on the accounting records of a company. It is ordinarily taken to mean the original cost of the property less the amounts that have been charged as depreciation expense. It thus represents the amount of capital that remains invested in the property and must be recovered in the future through the depreciation accounting process. It should be remembered, however, that because companies may use various depreciation accounting methods that produce different results, book value may have little or no relationship to the actual or market value of the property involved.

Salvage, or *resale, value* is the price that can be obtained from the sale of the property after it has been used. Salvage value implies that the property has further utility. It is affected by several factors. The reason of the present owner for selling may influence the salvage value. If the owner is selling because there is very little commercial need for the property, this will affect the resale value; change of ownership will probably not increase the commercial utility of the article. Salvage value will also be affected by the present cost of reproducing the property; price levels may either increase or decrease the resale value. A third factor that may affect salvage value is the location of the property. This is particularly true in the case of structures that must be moved in order to be of further use.

It may be seen that the various definitions of value vary considerably. Although a person normally possesses property so that he may receive benefits from it, some of the benefits frequently are not in the form of money. This fact further complicates the setting of value in monetary terms in order to place an ordinary commercial value on property.

PURPOSES OF DEPRECIATION

Because property decreases in value, it is desirable to consider the effect that this depreciation has on engineering projects. Primarily, it is necessary to consider depreciation for two reasons:

1. To provide for the recovery of capital that has been invested in physical property.
2. To enable the cost of depreciation to be charged to the cost of producing products or services that result from the use of the property. *Depreciation cost is deductible in computing profits on which income taxes are paid.*

To understand these purposes, consider the following example.

EXAMPLE 9-1

Mr. Doe invested $3,000 in a machine for making a special type of concrete building tile as an avocation. He found that with his own labor in operating the machine he could produce 500 tiles per day. Working 300 days per year he could make 150,000 tiles. He was able to sell the tiles for $50 per thousand. The necessary materials and power cost $20 per thousand tiles.

At the end of the first year he had sold 150,000 tiles and computed his total profit, at the rate of $30 per thousand, to be $4,500. This continued for 2 more years, at which time the machine was worn out and would not operate longer. To continue in business, he would have to purchase a new machine.

During the 3-year period, believing he was actually making a profit of $4,500 per year, he had spent the entire amount on other interests. He suddenly found that he no longer had his original $3,000 of capital, his machine was worn out, and he had no money with which to purchase a new one. What error had Mr. Doe made in his reasoning and accounting?

Solution

Analysis of the situation described in Example 9-1 reveals that Mr. Doe had not recognized that depreciation was occurring, and he had made no provision for recovering the capital invested in the tile machine. The machine, which was valued at $3,000 when purchased, had decreased in value until it was worthless.

Through this depreciation, $3,000 of capital had been used in making tiles. Depreciation was just as much a cost of producing the tiles as was the cost of the material and power. However, depreciation differs from these other costs in that *it always is paid or committed in advance. Thus it is essential that depreciation be considered so that the capital that is used to prepay this cost may be recovered.* Failure to do this will always result ultimately in the depletion of capital.

Because capital must be maintained, it is necessary that the recovery be made by charging the depreciation that has taken place to the cost of producing whatever has been produced. Thus, in the case of the tile machine, production of 450,000 tiles "consumed" the machine. We might say that each thousand tiles produced decreased the value of the machine $3,000/450 = $6.67. Therefore, $6.67 should be charged as the cost of depreciation for making each thousand tiles. Adding this cost of depreciation to $20, the cost of materials and power, gives the true cost of producing 1,000 tiles. With the true cost known, the actual profit can then be determined. At the same time, with depreciation charged as a cost, a means for recovery of capital is provided. ■

Thus depreciation accounting has a twofold purpose. First, it provides for the maintenance of capital. Second, it enables the proper amounts to be charged as the cost of depreciation in determining production costs, and ultimately in determining profits. It is this second purpose that is of primary importance to the engineer in making economy studies.

ACTUAL DEPRECIATION REVEALED BY TIME

Depreciation differs from other costs in several respects. *First*, although its actual magnitude cannot be determined until the asset is retired from service, it always is paid or committed in advance. Thus, when we purchase an asset, we are prepaying all the future depreciation cost. *Second*, throughout the life of the asset we can only estimate what the annual or periodic depreciation cost is. Consequently, we must estimate the depreciation cost in economy studies. Obviously, it follows that such estimates will not be entirely accurate, but this should not be too disturbing inasmuch as the same is true of virtually all other cost items in an economy study.

A *third* difference is the fact that while much usually can be done to control the ordinary out-of-pocket costs, such as labor and material costs, relatively little can be done to control depreciation cost once an asset has been acquired, except, perhaps, through maintenance expenditures. Further, many of the factors that affect depreciation costs are external to the person or organization that owns the asset. If future conditions change and the demand for a product decreases, there may be a decline in the amount of material used, and probably a decrease in the profits. However, the depreciation cost, having been prepaid, may continue as before, and the result may be a loss of capital through failure to recover what has been prepaid.

TYPES OF DEPRECIATION

Depreciation, or the decrease in value of an asset, has several causes, some of which are very difficult to predict or anticipate. Decreases in value with the passage of time may be classified as follows:

1. Normal depreciation: (a) physical, (b) functional.
2. Depreciation due to changes in price level.
3. Depletion.

Physical depreciation is due to the lessening of the physical ability of a property to produce results. Its common causes are wear and deterioration. These cause operation and maintenance costs to increase and output to decrease. As a result, the profits may decrease. Physical depreciation is mainly a function of time and use.

Functional depreciation, often called *obsolescence*, is more difficult to determine than physical depreciation. It is the decrease in value that is due to the lessening in the demand for the function that the property was designed to render. This lessening may be brought about in many ways. Styles change, population centers shift, more efficient machines are produced, or markets are saturated. Increased demand may mean that an existing machine is no longer able to produce the required volume. Thus *inadequacy* is a cause of functional depreciation.

Depreciation due to changes in price levels is almost impossible to predict and is seldom accounted for in economy studies. When price levels rise during inflationary periods, even if all the capital invested at the time of original purchase has been recovered, this recovered capital will not be sufficient to provide an identical replacement. Although there has been a recovery of the invested capital, the capital has decreased in value. Thus it is the capital, not the property, that has depreciated. Inflating annual depreciation to compensate for this phenomenon is not permitted in determining profits for income tax purposes.

REQUIREMENTS OF A DEPRECIATION METHOD

From the standpoint of management, a depreciation method should:

1. Provide for the recovery of invested capital as rapidly as is consistent with the economic facts involved; known and computed salvage should agree, if possible.
2. Not be too complex.
3. Assure that the book value will be reasonably close to the market value at any time.
4. Be accepted by the Internal Revenue Service, if the method is also to be used for determining federal income taxes.

These requirements are somewhat contradictory and are not easily met. As a result, numerous methods for computing depreciation have been devised. Each is based on some hypothesis regarding loss of an asset's value versus time and is an attempt to solve the complex depreciation problem in a reasonably simple and satisfactory manner. Because there are conflicting factors involved, and because future, unknown factors exist, it can be expected that perfection will not be achieved through the use of any depreciation method.

For economy study purposes the requirements of a depreciation method are somewhat different. Obviously, it should provide for the recovery of capital and the proper assignment of depreciation cost *over the estimated life of the asset*. Property used in connection with the production of income is depreciated when its estimated life is greater than 1 year. The amount of depreciation claimed in a given year is influenced by an asset's value, estimated life, salvage value and date in service, and the method of calculating depreciation. But, equally important, a depreciation method should account properly for the flow of capital funds that are recovered, and which thereby reduce the amount of capital remaining invested in a project. These recovered funds thus are available to the firm for other use or investment as seen earlier in Figure 1-2. Finally, the method used must permit the proper evaluation of the *profitability* of an investment being considered in an engineering economy study.

THE ECONOMIC RECOVERY TAX ACT OF 1981

In August 1981 Congress passed the Economic Recovery Tax Act of 1981 (ERTA). This Act resulted in many significant changes in permissible depreciation ("cost recovery") practices, among other things, in the United States. In this regard new rules introduced by ERTA for computing depreciation are referred to as the Accelerated Cost Recovery System (A.C.R.S.). An important consequence of A.C.R.S. is improved capital recovery of business investments through accelerated depreciation schedules. A.C.R.S. is but one incentive offered by ERTA to stimulate technological progress by making capital investments more attractive. Other incentives, such as lower federal income tax rates, are discussed in Chapter 10.

For purposes of this book, the most important aspect of A.C.R.S. is the greatly reduced use of several traditional depreciation methods which had been in effect for decades. However, ERTA does not retroactively affect depreciation practices for assets placed in service prior to January 1, 1981, so many of the traditional methods will continue to be in force for many years to come. ACRS applies to most assets used to produce income that were placed in service after December 31, 1980. For example, in Chapter 11 the subject of replacement analysis is discussed, and it is imperative that traditional methods of calculating depreciation be taken into account for "old" assets in making such studies. Furthermore, the A.C.R.S. still allows some of these methods to be used in special circumstances.

Consequently, depreciation methods discussed in this chapter are subdivided into *two distinct parts*. The first deals with several widely used traditional (pre-A.C.R.S.) methods of computing depreciation, while the second discusses the determination of depreciation, or "cost recovery" allowances, with the A.C.R.S. In this manner a wide assortment of engineering economy problems, in which federal income taxes represent a significant cost of doing business, can be realistically dealt with in Chapters 10, 11, and 13.

PRE-A.C.R.S. METHODS OF DEPRECIATION

In 1971 the Internal Revenue Service (IRS) instituted guideline procedures, entitled the Class Life Asset Depreciation Range (CLADR) system, for determining depreciable lives and estimated salvage values of numerous classes of property. Obviously, one of the most important matters in computing depreciation is the life over which the value of the depreciable property can be spread. In this regard, sample CLADR guideline periods, plus their lower and upper limits, are shown in Table 9-1.

TABLE 9-1 Range of Useful Life Allowed for the Depreciation of Selected Classes of Assets Under the CLADR System of the Internal Revenue Service

Description of Depreciable Assets	Asset Depreciation Range (years)		
	Lower Limit	Guideline Period	Upper Limit
Transportation			
Automobiles, taxis	2.5	3	3.5
Buses	7	9	11
General-purpose trucks:			
Light	3	4	5
Heavy	5	6	7
Air transport (commercial)	9.5	12	14.5
Petroleum			
Exploration and drilling assets	11	14	17
Refining and marketing assets	13	16	19
Manufacturing			
Sugar and sugar products	14.5	18	21.5
Tobacco and tobacco products	12	15	18
Carpets and apparel	7	9	11
Lumber, wood products, and furniture	8	10	12
Chemicals and allied products	7.5	9.5	11.5
Cement	16	20	24
Fabricated metal products	9.5	12	14.5
Electronic components	5	6	7
Rubber products	11	14	17
Communication			
Telephone			
Central-office buildings	36	45	54
Distribution poles, cables, etc.	28	34	42
Radio and television broadcasting	5	6	7
Electric utility			
Hydraulic plant	40	50	60
Nuclear	16	20	24
Transmission and distribution	24	30	36
Services			
Office furniture and equipment	8	10	12
Computers and peripheral equipment	5	6	7
Recreation—bowling alleys, theaters, etc.	8	10	12

Source: Depreciation, U.S. Internal Revenue Service Publication 534, U.S. Department of the Treasury, December 1981 (rev.).

Depreciation methods used in conjunction with the CLADR system and earlier IRS systems are discussed and illustrated in this section and include the following:†

1. Straight line method.
2. Declining balance method.
3. Declining balance with switchover.
4. Sum-of-the-years'-digits method.
5. Sinking fund method.
6. Service output method.

Even though the CLADR system was repealed by the Economic Recovery Tax Act of 1981, the methods listed above continue to apply to assets placed in service before January 1, 1981, and are essential in several types of economy studies treated later in this book. Furthermore, in some instances the A.C.R.S. guidelines do not apply to an item and the analyst may elect to use any of the six methods listed above.

To illustrate the proper use and calculation of depreciation for purposes of engineering economy studies, the methods listed above are applied to Example 9-2.

EXAMPLE 9-2

A new electric saw for cutting lumber in a furniture manufacturing plant has an installed cost, or *basis*, of $4,000 and a 10-year estimated life (see the CLADR guideline period in Table 9-1). It was placed in service on January 1, 1979. The permissible salvage value of the saw is zero at the end of 10 years. What will be the depreciation cost, or "write-off," during the sixth year, the book value at the end of the sixth year, and the cumulative depreciation cost through the sixth year? An asset's book value is its original cost less the accumulated depreciation through a particular year.

The Straight Line Method

The straight line method of computing depreciation assumes that the loss in value is directly proportional to the age of the asset. This straight line relationship gives rise to the name of the method. Given

N = depreciable life of the asset in years

P = original cost

d_k = annual cost of depreciation in the kth year ($1 \le k \le N$)

BV_k = book value at the end of k years

S = salvage value at the end of the life of the asset

---

†The straight line and declining balance methods may still be used in connection with A.C.R.S.

$$D_k^* = \text{total depreciation through the } k\text{th year}$$

then

$$d_k = \frac{P - S}{N} \tag{9-1}$$

$$D_k^* = \frac{k(P - S)}{N} \tag{9-2}$$

$$BV_k = P - \frac{k(P - S)}{N} \tag{9-3}$$

The term $(P - S)$ is often referred to as the *depreciable value* of an asset.

Solution by the Straight Line Method

Applying these equations to Example 9-2, we obtain

$$d_6 = \frac{\$4,000 - \$0}{10} = \$400 \text{ per year}$$

$$D_6^* = \frac{6(\$4,000 - \$0)}{10} = \$2,400$$

$$BV_6 = \$4,000 - \frac{6(\$4,000 - \$0)}{10} = \$1,600 \qquad \blacksquare$$

This method of computing depreciation is widely used. It is simple and gives a uniform annual depreciation charge. Its proponents hold that inasmuch as other costs, as well as the depreciable life, must be estimated, there is little reason for attempting to use a more complex formula.

The Declining Balance Method

In the declining balance method, sometimes called the *constant percentage method* or the *Matheson formula*, it is assumed that the annual cost of depreciation is a fixed percentage of the book value at the beginning of the year. The ratio of the depreciation in any one year to the book value at the beginning of the year is constant throughout the life of the asset and is designated by R ($0 \leq R \leq 1$). Thus

Depreciation during the first year:

$$d_1 = P \times R \tag{9-4}$$

Depreciation for the kth year:

$$d_k = P(1 - R)^{k-1} \cdot R \tag{9-5}$$

Cumulative depreciation through the kth year:

$$D_k^* = P[1 - (1 - R)^k] \tag{9-6}$$

Book value at the end of k years:

$$BV_k = P(1 - R)^k \qquad (9\text{-}7)$$

Book value (salvage value) at the end of N years:

$$BV_N = P(1 - R)^N \qquad (9\text{-}8)$$

The declining balance procedure is rather simple to apply. However, it has two weaknesses. The annual cost of depreciation is different each year and, from a calculation viewpoint, this is inconvenient. Also, with this formula an asset can never depreciate to zero value because S is not utilized in Equations 9-4 to 9-8. For the same reason, the book value with the declining balance method may "undercut" the desired salvage value when S is greater than zero. These are not serious difficulties since a switch from the declining balance method to any slower method of depreciation (such as straight line) can be made so that a zero (or some other) book value in year N is reached. Furthermore, it is a simple matter to switch to zero depreciation when BV_k becomes equal to S as shown in Figure 9-1.

The declining balance depreciation rate allowed by the IRS depends on the type of depreciable property and is stated in terms of the straight line depreciation rate, which is $1.0/N$ (for 0 salvage value). Because these rates are all greater than the straight line rate, the declining balance method is often referred to as an *accelerated* depreciation method. When double the straight line rate is used, the method is called *double declining balance depreciation*.

Type of Property	Allowable Declining Balance Rate, R
All new depreciable property except real estate	Double straight line, $2/N$
All used depreciable property and new real estate property	$1\frac{1}{2}$ straight line, $1.5/N$
Used rental residential property	$1\frac{1}{4}$ straight line, $1.25/N$

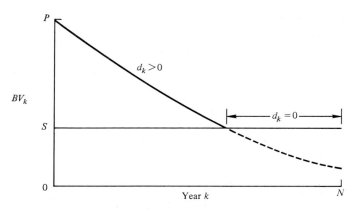

FIGURE 9-1 Illustration of the Declining Balance Depreciation Method Undercutting $S > 0$.

Solution by the Double Declining Balance Method

Applying the declining balance relationships in Equations 9-5, 9-6, and 9-7 to Example 9-2, we obtain

$$R = 2/N = 0.2$$

$$d_6 = \$4,000(1 - 0.2)^5 (0.2) = \$262.14$$

$$D_6^* = \$4,000[1 - (1 - 0.2)^6] = \$2,951.42$$

$$BV_6 = \$4,000(1 - 0.2)^6 = \$1,048.58 \qquad \blacksquare$$

Proponents of the declining balance method assert that its results more nearly parallel the actual market value of an asset than do those obtained by the straight line method. This undoubtedly is true in the case of such things as automobiles, where new models and style changes are large factors in the establishment of the market value. However, it is not true of many industrial and commercial structures and some equipment.

Declining Balance with Switchover to Straight Line

Because the declining balance method never reaches a book value of zero, it is permissible to switch from this method to the straight line method so that an asset's BV_N will be zero or some other desired salvage amount. In fact, a switchover can be made only once from an accelerated method of depreciation such as double declining balance to any slower method to enable a legitimate salvage value to be realized. It is *not* permissible to switch from a slower method to a faster method, however.

Table 9-2 illustrates a switchover from double declining balance depreciation to straight line depreciation for Example 9-2. The switchover occurs in the year where a larger depreciation amount is obtained from the straight line method.

From Table 9-2, we obtain as before:

$$d_6 = \$262.14$$

$$D_6^* = \$2,951.42$$

$$BV_6 = \$1,048.58 \qquad \blacksquare$$

The reader will observe that BV_{10} is $\$4,000 - \$3,570.50 = \$429.50$ without switchover to the straight line method in Table 9-2. With switchover, BV_{10} equals 0. It is clear that this asset's d_k, D_k^*, and BV_k in years 7 through 10 are different with the switchover.

The Sum-of-the-Years'-Digits Method

In order to obtain the depreciation charge in any year of life by the *sum-of-the-years'-digits method* (commonly designated as SYD), the digits corresponding

TABLE 9-2 Switchover from the Double Declining Balance Method to the Straight Line Method

		Depreciation Method		
Year, k	(1) Beginning-of-Year Book Value[a]	(2) Double Declining Balance Method[b]	(3) Straight Line Method[c]	(4) Depreciation Amount Selected[d]
1	$4,000.00	$ 800.00	> $400.00	$ 800.00
2	3,200.00	640.00	> 355.56	640.00
3	2,560.00	512.00	> 320.00	512.00
4	2,048.00	409.60	> 292.57	409.60
5	1,638.40	327.68	> 273.07	327.68
6	1,310.72	262.14	= 262.14	262.14* (switch)
7	1,048.58	209.72	< 262.14	262.14
8	786.44	167.77	< 262.14	262.14
9	524.30	134.22	< 262.14	262.14
10	262.16	107.37	< 262.14	262.14
		$3,570.50		$4,000.00

[a]Column 1 for year k less column 4 for year k equals the entry in column 1 for year $k + 1$.
[b]20% ($= 2/N$) of column 1.
[c]Column 1 divided by remaining years from beginning of year through the tenth year.
[d]Select the larger amount in column 2 and column 3.

to the number of each permissible year of life are first listed in reverse order. The sum of these digits is then determined.† The depreciation factor for any year is the number from the reverse-ordered listing for that year divided by the sum of the digits. For example, for a property having a life of 5 years, SYD depreciation factors are as follows:

Year	Number of the Year in Reverse Order (digits)	SYD Depreciation Factor
1	5	5/15
2	4	4/15
3	3	3/15
4	2	2/15
5	1	1/15
Sum of the digits =	15	

The depreciation for any year is the product of the SYD depreciation factor for that year and the depreciable value, $P - S$. The general expression for the annual cost of depreciation for any year k, when the total life is N, is

$$d_k = (P - S)\left[\frac{2(N - k + 1)}{N(N + 1)}\right]$$

(9-9)

The book value at the end of year k is

$$BV_k = P - \left[\frac{2(P - S)}{N}\right]k + \left[\frac{(P - S)}{N(N + 1)}\right]k(k + 1)$$

(9-10)

†The sum of digits for a life N equals $1 + 2 + \cdots + N = N(N + 1)/2$.

and the cumulative depreciation through the kth year is simply

$$D_k^* = P - BV_k \qquad (9\text{-}11)$$

Solution by the SYD Method

When Equations 9-9 to 9-11 are applied to the data of Example 9-2, the results are:

Sum of the years' digits $= 10(11)/2 = 55$

Depreciation factor for the sixth year $= 5/55$

$$d_6 = \$4,000 \left[\frac{2(10 - 6 + 1)}{10(11)} \right] = \$363.64$$

$$BV_6 = \$4,000 - \left[\frac{2(4,000)}{10} \right] 6 + \left[\frac{4,000}{10(11)} \right] 6(7) = \$727.27$$

$$D_6^* = \$4,000 - \$727.27 = \$3,272.73 \qquad \blacksquare$$

By setting up a table similar to Table 9-2, it could be determined whether a switchover to the straight line method would be advantageous in this situation. (It is left to the reader as an exercise to show that this switchover should *not* be made.)

Book values for the four methods discussed to this point have been plotted in Figure 9-2. In Figure 9-2 it will be noted that the SYD method, like the declining balance method, provides for very rapid (accelerated) depreciation during the early years of life. Further, the SYD method enables properties to be depreciated to zero value and is easier to use than the declining balance

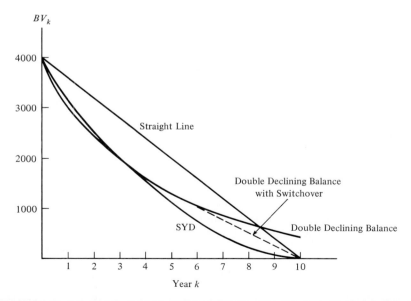

FIGURE 9-2 Comparison of Book Values Obtained by Various Depreciation Methods in Example 9-2.

method. This method also tends to reduce chances that the book value of an asset will exceed actual, or resale, value at any time. However, use of the SYD, or any other accelerated depreciation method, in effect reduces the computed profits of a corporation during the early years of asset life and thus reduces income taxes in those early years. This is discussed further in Chapter 10.

The Sinking Fund Method

The *sinking fund method* assumes that a sinking fund is established in which funds will accumulate for replacement purposes. The total depreciation that has taken place up to any given time is assumed to be equal to the accumulated value of the sinking fund (including interest earned) at that time.

With this method, if the estimated life, salvage value, and *interest rate* on the sinking fund are known, a uniform yearly deposit can be computed. The cost of depreciation for any year is the sum of this deposit and accumulated interest for that year. For an interest rate of $i\%$, these equations are used in connection with the sinking fund method of depreciation:

$$d = (P - S)(A/F, i\%, N) \tag{9-12}$$

$$d_k = d(F/P, i\%, k - 1) \tag{9-13}$$

$$D_k^* = (P - S)(A/F, i\%, N)(F/A, i\%, k) \tag{9-14}$$

$$BV_k = P - [(P - S)(A/F, i\%, N)(F/A, i\%, k)] \tag{9-15}$$

$$D_k^* = P - BV_k \tag{9-16}$$

Solution by the Sinking Fund Method

Applying these equations to the data of Example 9-2 produces the following results when $i = 8\%$:

$$d = (\$4,000 - \$0)(A/F, 8\%, 10)$$

$$= (\$4,000)(0.0690) = \$276.12$$

$$d_6 = (\$276.12)(F/P, 8\%, 5) = \$405.71$$

$$BV_6 = \$4,000 - \$4,000(A/F, 8\%, 10)(F/A, 8\%, 6)$$

$$= \$4,000 - \$2,025.60 = \$1,974.40$$

$$D_6^* = \$4,000 - \$1,974.40 = \$2,025.60 \qquad \blacksquare$$

Although the sinking fund method almost never is used for accounting purposes because of its low depreciation charges in the early years of asset life, it has importance when using the E.R.R.R. method.

The Service Output Method

Some companies attempt to compute the depreciation of equipment on the basis of its output. When equipment is purchased, an estimate is made of the amount of service it will render during its useful life. Depreciation for any period is then charged on the basis of the service that has been rendered during that period. The *service output method* has the advantages of making the unit cost of depreciation constant and giving low depreciation expense during periods of low production. That it is difficult to apply may be understood by realizing that not only the depreciable life, but also the total amount of service that the equipment will render during this period, must be estimated.

The *machine-hour method* of depreciation is a variation of the service output method. It is applied to Example 9-2.

Solution by the Machine-Hour Method

Suppose it is expected that the electric saw will be used a total of 10,000 hours over a period of 10 years and then will have no salvage value. In the sixth year of operation, the estimated usage is 800 hours and the cumulative usage by the end of year 6 should be about 6,400 hours. If depreciation is based on hours of use, determine d_6, D_6^*, and BV_6 for the electric saw.

$$\text{depreciation/hour} = \frac{\$4,000}{10,000 \text{ hours}} = \$0.40/\text{hour}$$

$$d_6 = 800 \text{ hours}(\$0.40/\text{hr}) = \$320$$

$$D_6^* = 6,400 \text{ hours}(\$0.40/\text{hr}) = \$2,560$$

$$BV_6 = \$4,000 - \$2,560 = \$1,440 \qquad \blacksquare$$

ACCELERATED COST RECOVERY SYSTEM (A.C.R.S.)

Depreciation determined by A.C.R.S. provisions is mandatory under the Economic Recovery Tax Act of 1981 (ERTA) for most tangible depreciable assets placed in service after December 31, 1980. Under the A.C.R.S. many *simplifications* to depreciation methods presented in the preceding section have been instituted. The most important ones concern the determination of salvage values and depreciable lives and the subsequent calculation of annual depreciation writeoffs.

For instance, with the A.C.R.S. an asset's salvage value is permitted to be zero in depreciation calculations and its depreciable life is categorized into one of four class lives. Prior to ERTA the IRS had published depreciation guideline procedures under the name Class Life Asset Depreciation Range (CLADR), which utilized approximately 100 different categories to determine the asset's useful life (a sample of these was seen in Table 9-1). In contrast, the A.C.R.S. recovers the value of tangible depreciable property over a period of either 3, 5, 10, or 15 years. Most personal property† is depreciated in 3 or 5 years, and real property is recovered over 10 or 15 years.

---

†A personal asset is property, either tangible (e.g., equipment) or intangible (e.g., franchises), owned by an individual or a corporation that is not real estate or a permanent structure situated on real estate.

TABLE 9-3   A.C.R.S. Recovery Percentages for Property Placed in Service in (a) 1981–1984, (b) 1985, and (c) After 1985

	The Applicable Percentage for the Class of Property is:			
If the Recovery Year is:	3-Year	5-Year	10-Year	15-Year Public Utility
(a) FOR PROPERTY PLACED IN SERVICE, 1981–1984				
1	25	15	8	5
2	38	22	14	10
3	37	21	12	9
4		21	10	8
5		21	10	7
6			10	7
7			9	6
8			9	6
9			9	6
10			9	6
11				6
12				6
13				6
14				6
15				6
(b) FOR PROPERTY PLACED IN SERVICE IN 1985				
1	29	18	9	6
2	47	33	19	12
3	24	25	16	12
4		16	14	11
5		8	12	10
6			10	9
7			8	8
8			6	7
9			4	6
10			2	5
11				4
12				4
13				3
14				2
15				1

APPLICATIONS OF ENGINEERING ECONOMY

TABLE 9-3 (Continued)

	The Applicable Percentage for the Class of Property is:			
If the Recovery Year is:	3-Year	5-Year	10-Year	15-Year Public Utility
(c) FOR PROPERTY PLACED IN SERVICE AFTER 1985				
1	33	20	10	7
2	45	32	18	12
3	22	24	16	12
4		16	14	11
5		8	12	10
6			10	9
7			8	8
8			6	7
9			4	6
10			2	5
11				4
12				3
13				3
14				2
15				1

Three separate schedules for calculating A.C.R.S. depreciation are to be used over the period 1981–1986. The allowable recovery periods for all three schedules are given in Table 9-3, and the recovery percentage assigned to each year in the three schedules is also indicated.† Note that in all schedules the first year percentage is low because half-year ownership is assumed regardless of when the property is purchased during the year.

The permissible depreciation in an ownership year is simply figured by multiplying the value (basis) of the property by the applicable percentage from Table 9-3. *To illustrate A.C.R.S. provisions, only the 1986 schedule is utilized in this and subsequent chapters in accordance with ERTA as originally approved by Congress in 1981.*

Applicable percentages from Table 9-3, or others that may be approved by Congress during the 1980s, can be substituted in examples and problems of this and following chapters by the instructor if he or she so desires. Once the applicable percentages have been specified, reworking the examples will serve as an ideal source of new material for classroom discussion and homework exercises.

†Periodic changes to depreciation regulations in the United States will be made during the 1980s. For example, The Tax Equity and Fiscal Responsibility Act of 1982 (TEFRA) specified that the 1981–1984 A.C.R.S. schedule is to apply after 1984 or until repealed. Because of uncertainty in future regulations regarding depreciation, the *original* provisions of ERTA are utilized throughout this book.

Under ERTA, property that is eligible to be written off over 3 years includes current CLADR property with a class of life in 4 years or less (primarily cars and light-duty trucks), research and experimental equipment, and certain other short-lived property. Most other personal tangible property is in the 5-year class, including CLADR personal property having a guideline life of over 4 years, single-purpose agricultural or horticultural structures, and petroleum storage facilities. The 10-year property is primarily real property† with a CLADR class life of 12.5 years or less and includes "theme" park structures, most public utility property, manufactured homes, and in addition, railroad tank cars. The 15-year property is real property with a CLADR class life greater than 12.5 years and certain long-lived public utility property. It is also noted that components of a building (storm windows, air conditioners, etc.) are depreciated in the same way as the building itself.

EXAMPLE 9-3

Calculate yearly A.C.R.S. deductions for the electric saw considered earlier in Example 9-2 in accordance with ERTA as originally approved by Congress in 1981. Suppose that the same saw is purchased and placed in service on November 21, 1986. The installed cost of the saw will be $4,000.

Solution

Under A.C.R.S. this property has a class life of 5 years and the depreciation for this recovery period based on 1986 procurement will be:

Year	Installed Cost	A.C.R.S. Percentages	Depreciation
1986	$4,000	20	$ 800
1987	4,000	32	1,280
1988	4,000	24	960
1989	4,000	16	640
1990	4,000	8	320
		100	$4,000

Notice that the company can claim depreciation for an entire year in 1986 even though the equipment will be placed in service near the end of the year. With pre-A.C.R.S. depreciation methods, depreciation taken in the first year of ownership normally is prorated based on when the equipment was actually placed in service. In this regard the electric saw in Example 9-2 was placed in service on January 1, 1979, so that a full year of depreciation could be claimed in that year. This difference in procedure is another major simplification and advantage of the A.C.R.S. ■

To summarize many differences effected by the A.C.R.S. of ERTA, depreciation write-offs resulting from the four major methods discussed thus far are shown in Table 9-4. For comparison purposes only, the various methods have

†Real property is land and generally anything tangible that is erected on, growing on, or attached to land.

TABLE 9-4 Comparison of Annual Depreciation Amounts with Four Methods

Year	A.C.R.S.	Straight Line	Sum-of-the-Years' Digits[a]	Double Declining Balance[b]
1986	$ 800	$ 400	$ 727.27	$ 800.00
1987	1,280	400	654.55	640.00
1988	960	400	581.82	512.00
1989	640	400	509.09	409.60
1990	320	400	436.36	327.68
1991		400	363.64	262.14
1992		400	290.91	262.14
1993		400	218.18	262.14
1994		400	145.45	262.14
1995		400	72.73	262.14
TOTAL	$4,000	$4,000	$4,000.00	$4,000.00

[a]It is not advantageous to switch to straight line depreciation.
[b]Converts to straight line depreciation in year 6.

been tabulated from 1986 through 1995. The net present worth as of January 1, 1986, of the *difference* between A.C.R.S. depreciation and that of each of the other methods is indicated below, assuming an interest rate of 12%:

Depreciation Method	N.P.W. of A.C.R.S. Depreciation over Method Shown
Straight line	$746.22
Sum-of-the-years' digits	320.08
Double declining balance	435.28

It can be seen that the A.C.R.S. offers substantial savings in terms of net present worth of depreciation charges relative to methods in common use prior to 1981 when the CLADR guideline life is appreciably longer than the A.C.R.S. class life. If the date that the equipment is placed in service had been identical, these savings would have been even *greater*.

If desired, an alternative recovery period and the straight line depreciation method can be used in conjunction with A.C.R.S. according to the following schedule:

In the Case of:	This Alternative Recovery Period May Be Elected (years)
3-year property	3, 5, or 12
5-year property	5, 12, or 25
10-year property	10, 25, or 35
15-year real or public utility property	15, 35, or 45

As in the normal A.C.R.S. calculations, a salvage value of zero is used in all depreciation calculations.

If an alternative recovery percentage is elected (the straight line percentage is 100%/N), all A.C.R.S. property put in service that tax year must be depreciated by the same method, and the method applies throughout the recovery period.

In general, these alternative methods would only be used by capital intensive or marginally profitable companies that would not have sufficient taxable income to take advantage of accelerated depreciation write-offs in the early years of an asset's recovery period.

In addition, if property can be depreciated under a method that is not based on a term of years, the property can be excluded from the A.C.R.S. provisions. For example, if an asset's life can be defined in terms of the production of a certain number of units, depreciation of the asset would be based on the service output method rather than the A.C.R.S.

EXAMPLE 9-4

In this example, the AIR-O Company operates a small travel agency and calculates its depreciation on the basis of a calendar year. It has been in business since early 1979. Several types of personal tangible assets have been purchased during the firm's 6-year existence. Compute this year's (1984) total allowable depreciation assuming for the sake of simplicity that the A.C.R.S. recovery percentages in Table 9-3 are applicable. The assets purchased by the firm and data relevant to this example are as follows:

Item	Description	Date Placed in Service	Cost	Salvage Value	Life or Recovery Period (years)	Method	Previous Depreciation (through 1983)
1	Typewriter	4/1/79	$1,000	$100	10	SYD	$614
2	Desk 1	4/1/79	600	50	5	SL	523
3	Desk 2	10/3/81	700	—	5	ACRS	532
4	Automobile	5/31/83	8,000	—	5	SL/ACRS	800

Solution

Item 1 could be depreciated with the sum-of-the-years'-digits method a maximum of $10/55(\$1,000 - \$100) = \$163.64$ in 1979, but this must be prorated over the 9 months that the typewriter was in service. Its d_1 is therefore $\$163.64(3/4) = \$123$, and its d_k in 1980–1984 are also listed below to the nearest dollar.

Year	1979	1980	1981	1982	1983	1984
SYD Depreciation	$123	$147	$131	$115	$98	$82

Item 2 is treated similarly to item 1 except that straight line depreciation is utilized. Its depreciation schedule (d_k) for years 1979–1984 is thus

Year	1979	1980	1981	1982	1983	1984
SL Depreciation	$83	$110	$110	$110	$110	$27

Item 3 qualifies for A.C.R.S. depreciation, and the 1981 cost recovery percentages of Table 9-3 apply to the full cost of the desk because $S = 0$. The entire first year amount can be taken in 1981, and the depreciation schedule for 1981–1984 is

Year	1981	1982	1983	1984
A.C.R.S. Depreciation[a]	$105	$154	$147	$147

[a]1985 entry not shown

Finally, item 4 also comes under the A.C.R.S. and the firm has elected to switch from the 3-year recovery percentages of Table 9-3, which would normally apply to this asset, to a constant straight line percentage of 20% over 5 years. According to the "half-year convention" of the IRS, half of the first year straight line depreciation amount can be claimed since the automobile was purchased in late May of 1983.† This is shown below.

Year	1983	1984
SL/A.C.R.S. Depreciation[a]	$800	$1600

[a]1985 and beyond not shown

In view of these depreciation schedules, the total depreciation taken in 1984 would be $82 + $27 + $147 + $1,600 = $1,856$. This highly simplified example illustrates how pre-A.C.R.S. methods and A.C.R.S. methods are often utilized side by side in computing the total annual depreciation write-off for a company. ■

In before-tax economic analyses all methods of depreciation yield an identical capital recovery cost. Recall that if the retirement of an asset takes place at the age predicted and at the salvage value predicted, the equivalent annual sum of depreciation and interest on the undepreciated balance for any method of depreciation can be shown to be equal to the *capital recovery cost*. This is true for *any* method of depreciation as long as P, S, and N are the same, and it was illustrated in Table 5-1 for the straight line method. However, in *after-tax* analyses, the method of depreciation does affect how much income taxes are paid. Thus different methods of depreciation are not equivalent in after-tax economic analyses. This will be demonstrated in Chapter 10.

MULTIPLE ASSET DEPRECIATION ACCOUNTING

To this point we have considered depreciation on individual assets. Very frequently, for accounting purposes, identical or even somewhat dissimilar assets

†Refer to *Depreciation*, U.S. Internal Revenue Service Publication 534, U.S. Department of the Treasury, December 1981 (rev.).

may be considered as a group in regard to depreciation. This group procedure has some advantages.

Although it may be known that the *average* life of assets in a group will be N years, it is recognized that some items will last less than N years, and some will last longer than N years. If N years is assumed to be the depreciable life of an individual asset, and if it should be retired from service in less than N years at a salvage value less than the book value, there would be a loss of capital. On the other hand, if it should be used for more than N years and then sold for any amount, a gain of capital would be realized. By using group, or *multiple asset*, accounting, adjustments for such losses or gains are avoided, since such a "dispersion effect" in the lives is recognized when the average life for the group is adopted.

It is fairly common practice, for tax purposes, to use item accounts for buildings, structures, and high-value equipment and to use group accounts for general equipment. In the following portions of this book we shall assume item depreciation, since this will provide some degree of simplicity and will highlight the effects of income taxes relative to gains and losses that may occur.

DEPLETION

When natural resources are being consumed in producing products or services, the term *depletion* is used to indicate the decrease in value of the resource base that has occurred. The term is commonly used in connection with mining properties, oil and gas wells, timber lands, and so on. In any given parcel of mineral property, for example, there is a definite quantity of ore, oil, or gas available. As some of the mineral is mined and sold, the reserve decreases and the value of the property normally diminishes.

However, there is a difference in the manner in which the amounts recovered through depletion and depreciation must be handled. In the case of depreciation, the property involved usually may be replaced with similar property when it has become fully depreciated. In the case of depletion of mineral or other natural resources, such replacement usually is not possible. Once the gold has been removed from a mine, or the oil from an oil well, it cannot be replaced. Thus, in a manufacturing or other business where depreciation occurs, the principle of maintenance of capital is practiced, and the amounts charged for depreciation expense are reinvested in new equipment so that the business may continue in operation indefinitely. On the other hand, in the case of a mining or other mineral industry, the amounts charged as depletion cannot be used to replace the sold natural resource, and the company, in effect, may sell itself out of business, bit by bit, as it carries out its normal operations. Such companies frequently pay out to the owners each year the amounts recovered as depletion. Thus the annual payment to the owners is made up of two parts: (1) the profit that has been earned, and (2) a portion of the owner's capital that is being returned, marked as depletion. In such cases, if the natural resource were eventually completely consumed, the company would be out of business, and the stockholder would hold stock that was theoretically worthless but would have received back all his or her invested capital.

In the actual operation of many natural resource businesses, the depletion funds may be used to acquire new properties, such as new mines and oil-producing properties, and thus give continuity to the enterprise.

Although the theoretical depletion for a year would be

$$\frac{\text{cost of property}}{\text{number of units in the property}} \times \text{units sold during year}$$

in actual practice the depletion is based upon a percentage of the year's income as permitted by the IRS. Depletion allowances on oil, gas, and mineral properties may be computed as a percentage of the gross income, provided that the amount charged for depletion does not exceed 50% of the taxable income before deduction of the depletion allowance. Typical percentage depletion allowances as of 1981 were:

Oil and gas wells	20%
Most ores and minerals	22%
Most precious metals	15%
Coal, sodium chloride	10%
Brick and tile clay, sand, gravel	5%

It is apparent that the total amount that can be charged for depletion over the life of a property under this procedure may be far more than the original cost.

High depletion allowances often are defended as being necessary for the encouragement of the discovery and development of mineral resources, on the basis that such ventures involve a high degree of risk and uncertainty. However, the high mortality rate of new businesses, and even some older ones, in the manufacturing and service industries is rather strong proof that they also involve considerable risk and uncertainty. There thus remains considerable evidence that much political pressure is involved in the special tax consideration given to depletion.

ACCOUNTING FOR DEPRECIATION FUNDS

The procedure by which accounting is made of the flow of depreciation funds and their subsequent reinvestment in the business often somewhat mystifies engineers who have not had any training in accounting. Quite probably, some of the terminology that is applied to some of the accounts involved has added to the confusion.

The basic procedure can be illustrated by the simplified balance sheets shown in Figure 9-3. These relate to a man, John Doe, who starts out with an investment of $3,000 in cash; this condition is portrayed in balance sheet (a) of Figure 9-3. Mr. Doe then uses his $3,000 to purchase a truck. The relationship of his assets, liabilities, and ownership for this condition is shown in balance sheet (b). At the end of a year he computes the depreciation on his truck to be $1,000, assuming a 3-year life and zero salvage value and using straight line deprecia-

FIGURE 9-3 Balance Sheets, Showing the Manner in Which Cash is Converted into Depreciable Assets and Recovered through Depreciation Accounting.

Balance Sheet (a)

Assets			Liabilities and Ownership		
Cash		$3,000	John Doe, Ownership		$3,000
	TOTAL	$3,000		TOTAL	$3,000

Balance Sheet (b)

Assets			Liabilities and Ownership		
Cash		$ 0	John Doe, Ownership		$3,000
Truck		3,000			
	TOTAL	$3,00		TOTAL	$3,000

Balance Sheet (c)

Assets			Liabilities and Ownership		
Cash and other assets		$1,000	John Doe, Ownership		$3,000
Truck	$3,000				
Less amount charged for depreciation to date	1,000				
		2,000			
	TOTAL	$3,000		TOTAL	$3,000

tion. He therefore sets aside, out of revenue, this $1,000 and reinvests it in the business, perhaps keeping some of it in the form of cash but using some of it for the purchase of other needed assets. This new financial state is portrayed in balance sheet (c), which at the same time shows the original value of the truck and also its depreciated value. Thus, in this general manner, the flow of depreciation money into the company, and the fact that it is retained and used, either in the form of cash or other assets, are recorded. Unfortunately, the item labeled "Less amount charged for depreciation to date" in Figure 9-3 is called by a variety of names, including *reserve for depreciation*, a term that is very often confusing.

VALUATION

Frequently, an engineer is called upon to decide the value of engineering properties. An adequate discussion of the methods used to arrive at the correct value of any property would require at least a good-sized volume; it is obviously beyond the scope of this book. However, a few of the principles involved will be considered, inasmuch as they are so intimately connected with the subject of depreciation.

The reasons for determining the value of property vary. Similarly, a valuation that is correct for one need not be correct for another. The need for determining the value of property often occurs when a private buyer is purchasing the property from a private owner. Again, the value of property may need to be known to serve as a tax base. When a municipality wishes to purchase a privately owned utility plant, the value must be decided upon. In establishing utility rates, the regulating bodies must arrive at a fair value of the property that is used to render the service. If a company examines the book value of its property at periodic intervals to determine whether the established depreciation rates are adequate, it must have some method of estimating the correct value of the property.

Obviously, both physical and functional depreciation affect the value of a property. Additions to and deletions from the property also will affect its value. But another, and very troublesome, factor is the matter of price changes due to inflation and deflation.

Theoretically, the value of a property should be a measure of the present worth of the future net profits that can be derived through ownership. Such a determination would, of course, necessitate the ability to predict accurately the future profits, and some interest rate would have to be agreed upon and used. Thus such a procedure is difficult to apply. Consequently, several generalized methods are used, each being based upon a certain hypothesis and producing results that are dependent upon the conditions contained in the hypothesis. It thus is well to keep in mind the dictum of the United States Supreme Court that stated that there is no single way to determine value. The following methods are in widespread use today.

The *first* method that might be used is sometimes called *historical cost less depreciation*. In arriving at value by this method, we consider the actual costs that have been incurred in obtaining the property being valued. This historical cost is then reduced by the amount of depreciation that appears to have occurred. This depreciation must be considered in terms of the ability of the present property to render service. The depreciated historical cost is then taken as the true value.

It is apparent that the historical-cost method neglects many factors. For example, no consideration is given to advances in technological methods that have occurred since the property was acquired. It is obvious that if an equivalent plant could be built, using modern methods and equipment, that would render the future service at much less unit cost, the existing plant could never be worth more than the more modern plant, regardless of how much might have been expended in acquiring the old plant.

So that the weaknesses of the historical-cost method can be overcome, a fictitious equivalent plant or property is usually assumed to be obtained by building it by the most modern methods, according to the most efficient design. This theoretical plant is then depreciated until it would be capable of rendering only the same amount of future service as the existing plant. Consequently, a *second* method of valuation is often called *reproduction-cost-new less depreciation*. By its use, technological progress and changes in price level are given consideration.

The results of this method are of obvious advantage to the seller of industrial property when price levels have risen. The method, however, does not consider whether the existing property is of the correct size for actual future demand or whether it is the result of unwise investment policy. The question also remains as to whether or not the owners should be given full advantage of increases in price levels that may have occurred.

DETERMINATION OF PROPERTY LIFE FROM MORTALITY DATA

As has been pointed out in this chapter, one of the most difficult problems in connection with estimating depreciation costs is the determination of the life that a property may be expected to have. With the Accelerated Cost Recovery System, the IRS has simplified the question of selecting a depreciable life by limiting the number of class lives to four categories. However, there are situations in which the *probable life* of an asset is desired—independent of IRS guidelines. For certain properties and under certain conditions, well-established statistical techniques can be used. However, it must be pointed out that the number of cases to which these techniques can be applied is limited for two basic reasons. First, it is necessary to wait until substantially all of a group of identical properties have been retired from service, or have in service a sub-

TABLE 9-5 Mortality Data for Electric Lamps

Life (hours)	Lamps in Service (Number Surviving)	Remaining Life Expectancy	Probable Life
1,999.5	0	0	1,999.5
1,899.5	755	50.0	1,949.5
1,799.5	1,897	89.8	1,889.3
1,699.5	4,237	112.5	1,812.0
1,599.5	7,559	141.1	1,740.6
1,499.5	11,334	177.2	1,676.7
1,399.5	15,637	214.9	1,614.4
1,299.5	20,620	250.5	1,550.0
1,199.5	26,131	287.4	1,486.9
1,099.5	31,969	325.8	1,425.3
999.5	37,857	367.5	1,367.0
899.5	43,745	411.2	1,310.7
799.5	49,483	457.7	1,257.2
699.5	54,994	506.8	1,206.3
599.5	59,977	560.6	1,160.1
499.5	64,280	619.7	1,119.2
399.5	68,055	682.5	1,082.0
299.5	71,377	748.5	1,048.0
199.5	73,717	823.1	1,022.6
99.5	74,859	909.8	1,009.3
−0.5	75,614	1,000.2	999.7

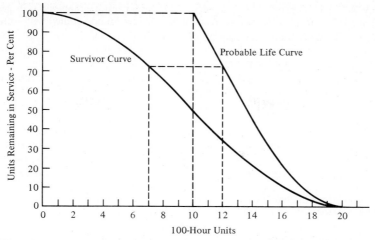

FIGURE 9-4 **Survivor and Probable Life Curves for Electric Lamps.**

stantial number of identical properties of all possible ages. Second, where rapid technological progress is occurring, or functional depreciation is a primary, determining factor, these techniques will be of little help. Thus these procedures are of definite, but limited, assistance in solving the problem of economic life in economy studies.

Just as human beings are born, live, grow old, and die, physical properties are produced, put into service, render service, and are removed from use. The same procedures that are used to determine the probable mortality, average life, and life expectancy for human beings may also be used for determining and estimating the life that may be expected from physical property, *provided that certain conditions exist*. These basic conditions are as follows:

1. A sufficient number of basically identical units must be involved so that averages may be used.
2. The service conditions must be the same for future units as for those for which the mortality data were obtained.
3. The study period must be long enough to assure valid data.

Table 9-5, for example, gives mortality data for a certain type of electric lamp, and Figure 9-4 shows the survivor and probable life curves for these lamps. Note that the average life of the entire group is shown (approximately 1,000 hours). Also shown, for example, is that lamps which have attained an age of 700 hours have a probable life of 1,200 hours.

PROBLEMS

9-1   It has been said that depreciation is probably one of the least understood but most important aspects of engineering economic analyses. Explain why this is likely to be true.

9-2   The accounting need for depreciation arises for several reasons. Describe what these reasons are. How are they related to definitions of the *value* of an asset?

9-3   How is the cost of depreciation different from other production or service expenses such as labor, material, and overhead?

9-4   Why would a company want to depreciate a capital investment rapidly in its early years of productive life? Which depreciation methods allow for accelerated write-offs?

9-5   Explain the difference between depreciation and depletion. Does an uncertain future favor a company that depreciates its physical assets or one that depletes its nonrenewable resources? Explain your reasoning.

9-6   What is the difference between individual asset (item) accounting for depreciation purposes and multiple asset (group) accounting?

9-7   On accounting balance sheets, ''reserves for depreciation'' are often confusing. Because taxable income can be legally reduced each year by the total amount of depreciation that the IRS allows, depreciation reserves represent cash that a firm supposedly has on hand for use in replacing wornout or obsolescent assets. What actually happens to these ''cash reserves'' in practice, and how does a firm acquire new assets when they are needed?

9-8   An asset for drilling that was purchased by a petroleum refinery has a first cost of $60,000 and an estimated salvage value of $12,000. The CLADR guideline period for useful life is taken from *IRS Publication 534* (Table 9-1), and the A.C.R.S. recovery period is 10 years. Compute the depreciation amount in the third year and the book value at the end of the fifth year of life by each of these methods:
(a) The straight line method.
(b) The sum-of-the-years'-digits method.
(c) The double declining balance method (with $R = 2/N$).
(d) The A.C.R.S. method.

9-9   Rework Example 9-2 when the salvage value is $1,000. Use all depreciation methods that have been discussed.

9-10   By each of the methods indicated below, calculate the book value of a highpost binding machine at the end of 4 years if the item originally cost $1,800 and had an estimated salvage value of $400. The CLADR guideline period is 8 years and the A.C.R.S. recovery period is 5 years.
(a) Straight line depreciation.
(b) 6% sinking fund depreciation.
(c) The double declining balance method ($R = 2/N$).
(d) Sum-of-the-years'-digits depreciation.

9-11   An asset costs $24,000 and has a class life of 10 years and a salvage value of $4,000 at that time. Find the depreciation charge for the third year and the book value at the end of the third year using the following depreciation methods: (a) straight line, (b) sum-of-the-years' digits, (c) declining balance with $R = 2/N$, and (d) A.C.R.S.

9-12   A piece of equipment that cost $5,000 was found to have a trade-in value of $4,000 at the end of the first year, $3,200 at the end of the second year, $2,560 at the end of the third year, $2,048 at the end of the fourth year, and $1,638 at the end of the fifth year.
(a) Determine the depreciation that occurred during each year.
(b) Using a 10% interest rate, compute the interest on the remaining investment each year as was done in Table 5-1.
(c) Determine the sum of depreciation and interest on the unrecovered investment for each year.
(d) Compare the results from (c) with the uniform year-end amount that represents capital recovery with interest.

9-13   A company purchased a machine for $30,000 and depreciated it using the 1986 A.C.R.S. recovery percentages for a 5-year class life. Show that the annual costs of depreciation plus interest at 12% per year on the beginning of year book value equal the uniform year-end capital recovery cost (refer to Table 5-1).

9-14   An asset was purchased 6 years ago for $6,400. At that time its life and salvage value were estimated to be 10 years and $1,000, respectively. If the asset is sold *now* for $1,500, what is the difference between its market value of $1,500 and its present book value if depreciation has been by:
(a) The straight line method?
(b) The sum-of-the-years'-digits method?
(c) The A.C.R.S. method?

9-15 A mining venture involved investment of $2,000,000 in the purchase and development of a mineral property. At the end of the following 5 years, the stockholders were paid $780,000, $708,000, $628,000, $552,000, and $476,000 as payment for depletion and profits. The property was then sold for $100,000, and this amount was also paid to the stockholders. The property had been depleted at a uniform rate throughout the period. What uniform rate of profit did the stockholders receive on the capital they had invested in the property at the end of each year?

9-16 A company purchased a machine for $15,000. It paid shipping costs of $1,000 and nonrecurring installation costs amounting to $1,200. At the end of 3 years it had no further use for the machine, so it spent $500 for having it dismantled and was able to sell it for $1,500.
(a) What are the total investment and depreciation costs for this machine?
(b) The company had depreciated the machine on a straight line basis, using an estimated life of 5 years and $1,000 salvage value. By what amount did the recovered depreciation fail to cover the actual depreciation?

9-17 The Ajax Plumbing Corporation uses the sum-of-the-years'-digits method of depreciation. It purchases an automatic soldering machine for $18,000. The Corporation plans to transfer the machine to a subsidiary after 6 years of service for $11,000. Determine the specified life, N, of the machine so that the book value after 6 years will be $11,000 if the salvage value after N years should be $4,000 according to IRS guidelines.

9-18 A machine costing $4,000 is estimated to be usable for 400 units and then have no salvage value.
(a) What would be the depreciation charge for a year in which 150 units were produced?
(b) What would be the total depreciation charged after 600 units were produced?

9-19 A building was constructed 25 years ago for $110,000 and was expected to last 50 years and then be worth $10,000 salvage value. Estimated rental income is $12,000 per year and estimated annual operating expenses are $3,000. Straight line depreciation is used. It is estimated that if rebuilt, the replacement cost of the building would be greater than the original cost by 1% per year. Determine the value of the building by each of the following methods: (a) historial-cost less depreciation, and (b) reproduction-cost-new less depreciation.

9-20 The mortality statistics on a certain telephone cable are as follows:

Years in Use	Percent Displaced Each Year	Years in Use	Percent Displaced Each Year
1	0.4	15	7.3
2	0.8	16	7.1
3	1.5	17	6.5
4	1.9	18	5.4
5	2.1	19	5.0
6	2.9	20	4.1
7	3.4	21	3.4
8	4.0	22	2.6
9	4.5	23	2.4
10	5.1	24	1.9
11	5.3	25	1.4
12	5.9	26	1.1
13	6.5	27	0.5
14	7.0		

Assume a given installation consists of 100 miles of cable that costs $10,000 per mile.
(a) If 12% of the installation is 15 years old, find the expectancy and present value of this portion of the installation, using sinking fund depreciation with interest at 10%.
(b) If the probable life of all of the cable is taken arbitrarily to be 18 years, what is the present value of the 12% that is 15 years old? (Use sinking fund depreciation.)

The Effects of Income Taxes

Before discussing the consequences of income taxes in engineering economy studies, we should make a clear distinction between income taxes and several other types of taxes:

1. *Income taxes* are assessed as a function of gross revenues minus certain allowable deductions and exemptions. They are levied by the federal, most state, and occasionally municipal, governments.
2. *Property taxes* are assessed as a function of the "value" of real estate, business, and personal property. Hence they are independent of the income or profit of an individual or a business. They are levied by municipal, county, and/or state governments.
3. *Sales taxes* are assessed on the basis of purchases of goods and/or services, and are thus independent of gross income or profits. They are normally levied by state, municipal, and/or county governments.
4. *Excise taxes* are federal taxes assessed as a function of the sale of certain goods or services often considered "nonnecessities," and hence are independent of the income or profit of an individual or a business. Although they are usually charged to the manufacturer or original provider of the goods or services, the cost is passed on to the consumer.

In this chapter we will be concerned only with income tax considerations. Income taxes resulting from the profitable operation of a firm normally are taken into account in evaluating engineering and business ventures. The reason is quite simple; income taxes associated with a proposed project represent a major cash outflow that should be considered together with other cash inflows and outflows in assessing the overall economic attractiveness of that project. There are many other taxes not directly associated with the income-producing capability of a new project (e.g., property taxes and excise taxes), but they are usually negligible when compared with state and federal income taxes.

Federal (and most state) income tax regulations are fairly complex and subject to rather frequent changes. For instance, numerous provisions and regulations set forth in the Economic Recovery Tax Act of 1981 (ERTA) have already been changed by the Tax Equity and Fiscal Responsibility Act approved by Congress in August 1982. Additional revisions to federal tax law are expected throughout the 1980s. As a result, it is practically impossible to keep federal tax regulations up to date in a textbook. Consequently, *the original provisions of the Economic Recovery Tax Act of 1981 are utilized throughout this and subsequent chapters.*† The practitioner who is making engineering economy studies should refer to applicable IRS publications for current income tax rates, depreciation provisions, and other regulations. Fortunately, the computational *procedure* for performing after-tax engineering economy studies during the 1980s will not vary from that demonstrated in Chapter 10.

GENERAL RELATIONSHIP BETWEEN BEFORE-TAX AND AFTER-TAX STUDIES

Thus far in this book there has been no consideration of income taxes in economy studies. That is, only *before-tax* studies have been made for two reasons. First, in the types of studies considered thus far, income taxes often do not have any major effect on the decisions that would be made. Therefore, the accuracy of the study would not be improved by their inclusion. Second, by not complicating the studies with income tax effects, we could place primary emphasis on basic economy study principles. However, there is a wide variety of individual and corporate capital investment problems in which income taxes do affect the choice among alternatives and consequently after-tax studies must be made.

An approximation of the before-tax M.A.R.R. requirement that includes the effect of income taxes in studies involving only before-tax cash flows can be determined from the following relationship:

(before-tax M.A.R.R.)[1 − (effective income tax rate] ≅ after-tax M.A.R.R.

Thus

$$\text{before-tax M.A.R.R.} \cong \frac{\text{after-tax M.A.R.R.}}{1 - (\text{effective income tax rate})} \qquad (10\text{-}1)$$

An expression for determining the effective income tax rate is given in the next section.

---

†In particular, the A.C.R.S. percentages for 1986 are utilized. Updated percentages can be substituted by the instructor and the problems then reworked as class exercises.

If the investment under study is nondepreciable and there is no capital gain or loss or investment tax credit, the relationship above is exact and not an approximation. If salvage value is less than 100% of first cost and if life of the property is finite, the depreciation method selected for income tax purposes affects the timing of income tax payments and, therefore, error can be introduced by use of Equation 10-1. Indeed, we will show through the results of Examples 10-7 to 10-10 that the earlier (sooner) in an asset's life that depreciation charges are made, the earlier income taxes are deferred, and the higher the after-tax rate of return will be. In practice, it is usually desirable to make after-tax analyses for any income tax-paying enterprise unless one knows that income taxes would have no effect on the alternatives being considered.

After-tax analyses can be performed by exactly the same methods (P.W., I.R.R., etc.) as before-tax analyses. The only difference is that after-tax cash flows should be used in place of before-tax cash flows, and the calculation of a measure of merit would be based on an after-tax minimum attractive rate of return. The adjustments needed to make before-tax cash flows into after-tax cash flows are the subject of this chapter.

The mystery behind the sometimes-complex computation of income taxes is reduced when one recognizes that income taxes *paid* are just another type of disbursement, while income taxes *saved* (due to business deductions or expenses or direct tax credits) are the same as any other kind of disbursement savings.

The basic concepts of federal and state income tax regulations that apply to most economic analyses of capital investments generally can be understood and applied without great difficulty. This chapter is not intended to be a comprehensive treatment of federal tax law. Rather, herein are described some of the more important provisions of the Economic Recovery Tax Act of 1981, followed by illustrations of general procedures for computing after-tax cash flows and making after-tax economic analyses.

BASIC INCOME TAX PRINCIPLES AND CALCULATION OF EFFECTIVE INCOME TAX RATES

Income taxes are levied on both personal and corporate incomes. Although the regulations of most of the states with income taxes have the same basic features as the federal regulations, there is great variation in the tax rates. State income taxes are in most cases much less than federal taxes and often can be closely approximated as a constant percentage of federal taxes. Therefore, no attempt will be made to discuss state income taxes. An understanding of the applicable federal income tax regulations usually will enable the analyst to apply the proper procedures if state income taxes must also be considered.

As an example of an effective (combined federal and state) income tax rate for a corporation, suppose that the federal income tax rate is 46% and the state income tax rate is 7.5%. Further assume the common case in which taxable income is computed the same way for both taxes except that state taxes are deductible from taxable income for federal tax purposes but federal taxes are not deductible from taxable income for state tax purposes. The general expression for the effective income tax rate (t) is

$$t = \text{federal rate} + \text{state rate} - (\text{federal rate})(\text{state rate}) \qquad (10\text{-}2)$$

In this example the effective income tax rate would be

$$t = 0.46 + 0.075 - 0.46(0.075) = 0.5005 \quad \text{or} \quad \text{about } 50\%$$

It is the effective income tax rate on increments of *taxable income* that is of importance in engineering economy studies. The definition of taxable income for individuals and corporations is discussed next.

TAXABLE INCOME OF INDIVIDUALS

An individual's total earned income is essentially what is called *adjusted gross income* for federal tax purposes. From this amount individuals may subtract personal exemptions and allowable deductions to determine the *taxable income*. Personal exemptions are provided at the rate of $1,000 for the taxpayer and each dependent, with extra exemptions permissible for blind persons and persons 65 years of age or over. Allowable deductions are provided for such items as excessive medical costs, state and local taxes, interest on borrowed money, charitable contributions, and casualty losses, within some limits. A standard deduction may be taken by people who do not itemize their allowable deductions. As of 1982 the amount of the standard deduction was $2,300 for a single taxpayer, $1,700 for married taxpayers filing separate returns, and $3,400 for married taxpayers filing joint returns. Thus, for individual taxpayers,

$$\text{Taxable income} = \text{adjusted gross income}$$
$$- \text{personal exemption deductions}$$
$$- \text{other allowable deductions} \qquad (10\text{-}3)$$

TAXABLE INCOME OF BUSINESS FIRMS

At the end of each tax year a corporation must calculate its net (or taxable) income or loss. Several steps are involved in this process, as shown previously in Figure 1-2, beginning with the calculation of *gross income*. Gross income represents the gross profits from operations (sales minus the cost-of-goods-sold) plus income from dividends, interest, rent, royalties, and gains (or losses) on the exchange of capital assets. From gross income the corporation may deduct all ordinary and necessary expenses to conduct the business except for capital expenditures. As discussed in Chapter 9, capital expenditures can be deducted as an expense for each tax period only to the extent of allowable depreciation charges on those capital expenditures. Thus for business firms,

Taxable income = gross income

　　　　　　　　− all deductions except capital expenditures

　　　　　　　　− depreciation charges　　　　　　　　　　　(10-4)

This taxable income is often referred to as net income before taxes and when income taxes are subtracted from it, the remainder is often called the *net income after taxes*. There are two types of income for tax computation purposes: *ordinary income* (and losses) and *capital gains* (and losses). Each of these is explained below.

ORDINARY INCOME (AND LOSSES)

Ordinary income is the net income that results from the regular business operations (such as the sale of products or services) performed by a corporation or individual. For federal income tax purposes, virtually all ordinary income adds to taxable income and is subject to a graduated rate scale with provision for higher rates with higher income. The recent and planned federal income tax rates for corporations resulting from the Economic Recovery Tax Act of 1981 are given in Table 10-1. For example, suppose that a firm in 1984 has a gross income of $5,270,000, expenditures (excluding capital) of $2,927,500, and depreciation of $1,874,300. Its taxable income and federal income tax would be determined with Equation 10-4 and Table 10-1 as follows:

taxable income
  = gross income − expenditures − depreciation
  = $5,270,000 − $2,927,500 − $1,874,300
  = $468,200

income tax =

15% of first $25,000 = 0.15 (25,000)	= $ 3,750
+ 18% of next $25,000 = 0.18 (25,000)	= $ 4,500
+ 30% of next $25,000 = 0.30 (25,000)	= $ 7,500
+ 40% of next $25,000 = 0.40 (25,000)	= $ 10,000
plus 46% of remaining $368,200 = 0.46 (368,200)	= $169,372
	$195,122

In this case the *average* income tax rate is $195,122 ÷ $468,200 = 0.417, or 41.7%. However, the income tax on the last increment of taxable income is

TABLE 10-1　Corporate Income Tax Rates in Effect in 1981 and Planned for 1983 and Beyond

Taxable Income	1981	1983 and Beyond
Less than $ 25,000	17%	15%
$ 25,000–$ 50,000	20%	18%
$ 50,000–$ 75,000	30%	30%
$ 75,000–$100,000	40%	40%
Greater than $100,000	46%	46%

clearly 46%. Because engineering economy studies are concerned with incremental differences among alternatives, we shall be using a 46% *incremental federal income tax rate*, as shown in Figure 10-1, for most studies in corporations having large taxable incomes.

EXAMPLE 10-1

A corporation now is expecting an annual taxable income of $45,000. It is considering investing an additional $100,000, which is expected to create an added annual net cash flow (receipts minus disbursements) of $35,000 and an added annual depreciation charge of $20,000. What is the corporation's federal tax liability based on rates in effect in 1981 and in 1983: (a) without the added investment and (b) with the added investment?

Solution

(a)

Income Taxes/Year	1981 Rate	1981 Amount	1983 and Beyond Rate	1983 and Beyond Amount	Change in Tax Liability (from 1981 to 1983)
On first $25,000	17%	$4,250	15%	$3,750	− $500
On next $45,000 − 25,000					
$20,000	20%	4,000	18%	3,600	− 400
TOTAL		$8,250		$7,350	− $900

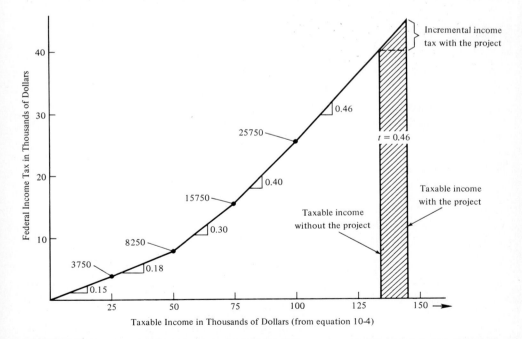

FIGURE 10-1   The 1983 Federal Income Tax Rates for Corporations (As Planned by the Economic Recovery Tax Act of 1981).

(b) Taxable income/year

Before added investment	$45,000
+ added net cash flow	35,000
− depreciation	− 20,000
TOTAL	$60,000

Income Taxes/Year		1981		1983 and Beyond		Change in Tax Liability (from 1981 to 1983)
		Rate	Amount	Rate	Amount	
On first	$25,000	17%	$ 4,250	15%	$ 3,570	− $ 500
On next	$25,000	20%	5,000	18%	4,500	− 500
On next	$60,000					
	− 50,000					
	$10,000	30%	3,000	30%	3,000	− 0
TOTAL			$12,250		$11,250	− $1,000
Increased liability from investment			$ 4,000		$ 3,900	

The maximum impact of the 1983 income tax schedule is $1,000 (17% − 15% = 2% of the first $25,000 and 20% − 18% = 2% of the next $25,000), regardless of the size of the anticipated investment. For taxable income over $50,000 the difference in income taxes is 0 as seen from Table 10-1.

As an added note, the determination of change in tax liability can usually be determined more readily by an incremental approach. For example, this problem involved changing the taxable income from $45,000 to $60,000 as a result of the new investment. Thus the change in income taxes per year (for 1983 and beyond) could be calculated as:

First $50,000 − $45,000 = $ 5,000 at 18%	=	$ 900
Next $60,000 − $50,000 = $10,000 at 30%	=	3,000
TOTAL		$3,900

The average federal income tax rate on the additional $15,000 ($35,000 − $20,000) is calculated as $3,900/$15,000 = 0.26 or 26%. ■

For individual taxpayers, two important changes were instituted by the Economic Recovery Tax Act of 1981. First, taxes were reduced by nearly 25% over a 3-year period according to the following schedule:

5% reduction effective October 1, 1981.

10% reduction effective July 1, 1982.

10% reduction effective July 1, 1983.

Second, the graduated rate scale for individual taxpayers was reduced from a range of 14 to 70% to a range of 11 to 50%. The maximum impact of both of these changes can be seen by comparing 1981 rates to the rates planned to be in effect in 1984 as shown in Table 10-2.

The rate structure varies according to filing status (e.g., single versus married) and it is also different in 1982 and 1983 as the result of progressive tax cuts being phased in.

APPLICATIONS OF ENGINEERING ECONOMY

TABLE 10-2 Federal Income Tax Rate for Persons Categorized as "Married Filing Joint Return"

If Taxable Income Exceeds This Base Amount	But Does Not Exceed This Amount	1981 Income Tax Rates		1984 Income Tax Rates	
		You Pay This Amount	Plus This % of the Excess Over the Base	You Pay This Amount	Plus This % of the Excess over the Base
0	$ 3,400	0		0	
$ 3,400	5,500	0	14	0	11
5,500	7,600	$ 294	16	$ 231	12
7,600	11,900	630	18	483	14
11,900	16,000	1,404	21	1,085	16
16,000	20,200	2,265	24	1,741	18
20,200	24,600	3,273	28	2,497	22
24,600	29,900	4,505	32	3,465	25
29,900	35,200	6,201	37	4,790	28
35,200	45,800	8,162	43	6,274	33
45,800	60,000	12,720	49	9,772	38
60,000	85,600	19,678	54	15,168	42
85,600	109,400	33,502	59	25,920	45
109,400	162,400	47,544	64	36,630	49
162,400	215,400	81,464	68	62,600	50
215,400		117,504	70		

One additional benefit of ERTA is an *indexing* provision beginning in 1985 to avoid so-called "bracket creep" caused by inflation. Bracket creep is the phenomenon whereby a taxpayer is pushed into higher tax brackets through incremental salary increases caused, in part, by inflation. For example, if your *taxable income* in 1981 is $24,000, you would be in the 28% incremental tax bracket (see Table 10-2). A 10% raise in 1982 would place you in the next higher bracket, which is 32% of taxable income. As a result, the *actual dollar* equivalent of your raise is less than 10%! If the inflation rate is averaging 10%, your standard of living, as measured by the purchasing power of your salary, is declining. ERTA attempts to remedy this situation by annually adjusting the individual income tax brackets to reflect changes in the Consumer Price Index.

EXAMPLE 10-2

John Doakes and his wife Jane file a joint federal income tax return on an adjusted gross income of $44,750. They can claim two personal exemptions of $1,000 each and have allowable deductions of $4,000. What is their taxable income and federal income tax owed, based on the rates in effect for 1981 and after 1984?

Solution

From Equation 10-3,

$$\text{taxable income} = \$44{,}750 - 2(\$1{,}000) - \$4{,}000 = \$38{,}750$$

Income Tax:	1981	1984
On first $35,200	$8,162	$6,274
On next ($38,750 − $35,200) = $3,550	1,427 at 43%	1,172 at 33%
TOTAL	$9,686	$7,446

This is a reduction of ($9,686 − $7,446)/$9,686 = 23% from 1981 to 1984.

∎

CAPITAL GAINS (AND LOSSES)

When a capital asset† is disposed of for more (or less) than its book value, the resulting capital gain (or capital loss) often is taxed (or saves taxes) at a rate which is different from that for ordinary income. The magnitude of the tax effect depends on the amount of capital gain or loss, type of property, whether it is long term or short term, whether other capital gains or losses are involved, and the applicable rate or rates. It should be noted that the recovery of previously charged depreciation (i.e., depreciation "recapture") upon disposal of depreciable property may, in some situations, be taxed as a capital gain. Usually, it is taxed as ordinary income, however. The income tax liability arising from such a transaction is explained further in Appendix 10-A and occurs when the disposal ("market") value of an asset is greater than its book value.

In equation form the determination of capital gains and losses is straightforward:

$$\text{capital gain (loss)} = \text{net selling or disposal price}$$
$$- \text{ book value}$$

or

$$\text{capital gain (loss)} = \text{net selling or disposal price}$$
$$- \text{ original first cost}$$
$$+ \text{ accumulated depreciation charges} \quad (10\text{-}5)$$

The distinction between long-term and short-term capital gains and losses is that the holding period must be *at least* 12 months in order for a gain or loss to qualify for "long-term" tax treatment. Thus gains or losses on capital assets

---

†Capital assets include all property owned by a taxpayer *except* (a) property held mainly for sale to customers, (b) most accounts or notes receivable, (c) depreciable property utilized to carry out the production process, (d) real property used by the business, and (e) copyrights and certain types of short-term, non-interest-bearing state/federal notes. Stock owned in another company is a familiar example of a capital asset.

sold before the minimum 12-month holding period are subject to "short-term" tax treatment. The rules for taxation of capital gains and losses are quite involved; a simplified summary is provided in Table 10-3. Further discussion and examples are given in Appendix 10-A, p. 390.

For tax computation purposes, short-term gains and losses are added separately, and the total of one is subtracted from the total of the other to determine whether a net short-term gain or loss results. The long-term gains and losses are treated in the same manner to determine the net long-term gain or loss. The *total net gain or loss* is then determined by merging the net short-term capital gain or loss with the net long-term capital gain or loss. A simple numerical example for an individual is given below. Corporate examples are included in Appendix 10-A.

EXAMPLE 10-3

During the year a married individual had the following transactions:

Long-term capital gain	$5,400	
Long-term capital loss	(1,750)	
Net long-term capital gain		$3,650
Short-term capital gain	$2,600	
Short-term capital loss	(3,790)	
Net short-term capital loss		($1,190)
Excess of net long-term capital gain over net short-term capital loss		$2,460

TABLE 10-3 Tax Treatment of Capital Gains and Losses Under the Economic Recovery Tax Act of 1981

	Short Term	Long Term
For individuals:		
Capital gain	Taxed as ordinary income	40% of capital gain taxed as ordinary income
Capital loss	Subtract capital loss from any capital gains; balance may be deducted from ordinary income, but not more than $3,000[a] per year	Subtract capital losses from any capital gains; half the balance may be deducted from ordinary income, but not more than $3,000[a] per year
	Any balance may be carried over to succeeding years.	
For corporations:		
Capital gain	Taxed as ordinary income	Taxed at ordinary income tax rate or 28%, whichever is smaller
Capital Loss	Corporations may deduct capital losses only to the extent of capital gains. Any capital loss in the current year that exceeds capital gains is carried back for 3 years and, if necessary, forward for up to 5 years and applied against capital gains until it is completely absorbed.	

[a]Limitations given are for married taxpayers filing a joint return. Married persons filing separate returns and unmarried taxpayers are entitled to only one-half the limitation shown. The rules for tax treatment of capital gains and losses are quite involved and are explained in greater detail in Appendix 10-A.

Thus the $2,460 would be taxed as a *long-term capital gain*. What is the individual's incremental (added) federal tax liability due to the above if his taxable income *before* consideration of the capital gains and losses is $23,000?

Solution (using Tables 10-2 and 10-3)
With a taxable income of $23,000, this taxpayer would be taxed on 40% of the long-term gain, or 0.40($2,460) = $984. The incremental income tax rate for the individual is 22% according to the 1984 schedule, so the capital gains tax would be 0.22($984) = $216.48. After taxes, the taypayer is left with $2,243.52.

∎

Investment Credit

A special provision of the federal income tax law, which, when in force, may have direct bearing on economic analyses, is the *investment credit*, also commonly called the investment tax credit (ITC). This allows businesses to subtract from their tax liability as much as 10% of what they invest in "qualifying" property. By qualifying is meant depreciable equipment or tangible personal property used in a business. The investment credit is determined by several factors: the A.C.R.S. property class or the expected useful life, the amount of used property, and the total tax claimed. The limitation according to 1981 federal tax law is summarized below.†

A.C.R.S. Class (years)	Maximum Possible Investment Credit (as a percentage of investment) Is:
3	6%
5, 10, and 15	10%

In addition, the allowable tax credit in any year cannot exceed the tax liability. If the tax liability is more than $25,000, the tax credit cannot exceed $25,000 plus 90% of the tax liability above $25,000. However, the investment credit can be carried back for 3 years and forward as many as 15 years.

A.C.R.S. property that is disposed of *before* reaching its class life, or recovery period, will *recapture* a portion of the investment credit. The recaptured amount becomes an income tax liability in the year of disposal. Each full year the property is held before disposition reduces the recapture by 2%, as shown below:

†The Tax Equity and Fiscal Responsibility Act of 1982 changes the maximum investment credit to 8% (5-, 10-, and 15-year A.C.R.S. periods) when depreciation is based on 100% of the purchase price of qualifying investment. If depreciation is based on 95% of the purchase price (i.e., A.C.R.S. percentages are applied to 0.95P), the full 10% tax credit can be taken.

If Property Disposed Of:	Recapture Is:	
	3-Year Property	5-Year Property
Within 1 year	6%	10%
Within 2 years	4%	8%
Within 3 years	2%	6%
Within 4 years	—	4%
Within 5 years	—	2%

EXAMPLE 10-4

A firm has invested in the following assets:

Amount	Expected Life	A.C.R.S. Class
$250,000	10 years	5 years
150,000	5 years	5 years
450,000	4 years	3 years
200,000	2 years	Excluded

What is the maximum possible investment tax credit if all property is new?

Solution

$$\$250,000 \times 10\% = \$25,000$$
$$\$150,000 \times 10\% = \$15,000$$
$$\$450,000 \times 6\% = \$27,000$$
$$\$200,000 \times 0\% = \underline{\$0}$$
$$\$67,000$$

∎

EXAMPLE 10-5

Suppose that a firm in a given year has a tax liability of $65,000 before consideration of credits and a maximum possible investment credit of $67,000 (such as calculated above). What is the firm's tax liability for that year?

Solution

Maximum allowable investment credit:

$$\$25,000 + 90\%(\$65,000 - \$25,000) = \$61,000$$

Tax liability for the year:

$$\$65,000 - \$61,000 = \$4,000$$

∎

It should be noted that in the example above, the difference between the maximum possible tax credit and the allowable investment credit for the year is $6,000 ($67,000 − $61,000). This "unused" investment credit can be carried back or forward to reduce taxes paid or payable in other years.

As an additional note, even though the investment credit results in a cash savings and, therefore, is a reduction in the net investment, ETRA provides that the credit does not reduce the investment to be depreciated. Thus a firm can depreciate an item based on the original purchase price of $P even though the net cash outlay or commitment for the item was as little as $0.9P (i.e., $P - $0.1P$).†

EXAMPLE 10-6

Given a $20,000 investment which is expected to have a life of 10 years (A.C.R.S. recovery period of 5 years) and a $500 salvage value, what is the net investment? What is the depreciable investment and the A.C.R.S. depreciation charges?

Solution

Investment tax credit:

$$\$20,000 \times 10\% = \$2,000$$

Net investment:

$$\$20,000 - \$2,000 = \$18,000$$

A.C.R.S. depreciation charges:

Year	Investment		1986 A.C.R.S. Recovery Percentages		A.C.R.S. Depreciation
1	$20,000	×	20%	=	$ 4,000
2	20,000	×	32%	=	6,400
3	20,000	×	24%	=	4,800
4	20,000	×	16%	=	3,200
5	20,000	×	8%	=	1,600
					$20,000

In this example, notice that the salvage value for tax purposes is zero. ∎

An additional feature of the investment credit of common interest is the business energy investment credit. It amounts to 10%, 11%, or 15% of the qualifying property. This credit is in addition to the normal investment tax credit and is added to other investment tax credits when evaluating credit limitations. The 10% credit applies to such items as recycling equipment, alternative energy property, shale oil equipment, and so on. The 11% credit applies to qualified hydroelectric generating equipment, and the 15% credit applies to solar or wind energy property, ocean thermal equipment, and geothermal equipment. Generally, property will not qualify for the business energy investment credit after 1982 with the exception of property such as solar, wind, geothermal, and ocean thermal property, which, as of 1981, will qualify through 1985.

The discussion above provides only a minimal description of some of the main provisions of the Economic Recovery Tax Act of 1981 that are important to engineering economy studies. It is by no means complete, but it is intended to provide a basis for illustrating after-tax (i.e., after income tax) economic

†See the footnote at the bottom of page 352.

analysis. In general, the analyst should either search out specific provisions of the federal and/or state income tax law‡ affecting projects being studied or seek information from persons qualified in income tax law.

The remainder of the chapter will illustrate various after-tax economy studies using a suggested tabular form for computing after-tax cash flows.

GENERAL PROCEDURE FOR MAKING AFTER-TAX ECONOMIC ANALYSES

After-tax economic analyses can be performed by exactly the same methods as before-tax analyses. The only difference is that after-tax cash flows should be used in place of before-tax cash flows by including disbursements or savings due to income taxes and then making equivalent worth calculations using an after-tax minimum attractive rate of return. While the tax rates and governing regulations may be complex and subject to changes, once those rates and regulations have been translated into their effect on after-tax cash flows, the remainder of the after-tax analysis is straightforward.

To formalize the procedure described in previous sections for determining the net income before income taxes, net income after income taxes, and after-tax cash flow of the incremental project shown in Figure 10-1, the following notation and equations are applicable.

For any given year k of the project life, $k = 0, 1, 2, \ldots, N$, let

R_k = operating revenues from the project; this is the positive cash flow from the project during period k

E_k = cash outflows (negative cash flow) during year k for all deductible expenses

D_k = sum of all noncash, or book, costs (negatively signed) during year k, such as depreciation or depletion

t = effective ordinary income tax rate (federal, state, and other)

T_k = income taxes paid during year k

ATCF_k = after-tax cash flow from the project during year k

Because the net income before taxes (i.e., *taxable income*) is ($R_k + E_k + D_k$), the ordinary *income tax liability* when $R_k > E_k + D_k$ is computed with Equation 10-6:

$$T_k = t(R_k + E_k + D_k) \qquad (10\text{-}6)$$

‡Some applicable publications are:

Tax Guide for Small Business, U.S. Internal Revenue Service Publication 334, published annually.

Your Federal Income Tax, U.S. Internal Revenue Service Publication 17, published annually.

Handbook on the Economic Recovery Tax Act of 1981 (Englewood Cliffs, N.J.: Prentice-Hall, Inc., 1981).

J. K. Lasser, *Your Income Tax* (New York: Simon and Schuster, Publishers), published annually.

Recall that R_k is positive cash flow, while E_k and D_k are negatively signed amounts. The *net income after taxes* (NIAT) is then simply taxable income minus the amount determined by Equation 10-6:

$$\underbrace{(R_k + E_k + D_k)}_{\text{taxable income}} - \underbrace{t(R_k + E_k + D_k)}_{\text{income tax}}$$

or

$$\text{NIAT} = (R_k + E_k + D_k)(1 - t) \qquad (10\text{-}7)$$

The *after-tax cash flow* associated with a project equals the net income after taxes minus noncash items (negatively signed) such as depreciation:

$$\text{ATCF}_k = (R_k + E_k + D_k)(1 - t) - D_k \qquad (10\text{-}8)$$

In effect, depreciation is *added back* in Equation 10-8 because D_k is a negative amount.

In many economic analyses of engineering and business projects, after-tax cash flows in year k are computed in terms of year k before-tax cash flows, BTCF_k:

$$\text{BTCF}_k = R_k + E_k \qquad (10\text{-}9)$$

and then

$$\text{ATCF}_k = \text{BTCF}_k - T_k \qquad (10\text{-}10)$$

A suggested tabular format to facilitate the computation of after-tax cash flows with Equations 10-6 to 10-10 is shown below:

Year k	(A) Before-Tax Cash Flow, $R_k + E_k$	(B) Depreciation, D_k	(C) = (A) + (B) Taxable Income, $R_k + E_k + D_k$	(D) = $-$(C) $\times t$ Cash Flow for Income Taxes, $-t(R_k + E_k + D_k)$	(E) = (A) + (D) After-Tax Cash Flow, $(R_k + E_k + D_k)(1 - t) - D$

Column A consists of the same information used in "before-tax" analyses—the cash revenues or savings plus the cash direct, indirect, and overhead expenses (negatively signed). Column B contains depreciation and/or depletion that can be claimed for tax purposes. Column C is the income or amount subject to income taxes. Column D contains the income taxes paid or saved. Finally, column E shows the "after-tax" cash flows to be used directly in after-tax economic analyses just the same as the cash flows in column A are used in before-tax economic analyses. A graphical summary of the process of determining *net income after income taxes* was shown earlier in Figure 1-2. *After-tax cash flow* is simply net income after taxes minus the negative signed depreciation, which results in adding back depreciation.

The column headings indicate the arithmetic operations for computing columns C, D, and E for annual operating results when $k = 1, 2, \ldots, N$. When $k = 0$ and $k = N$, capital investment amounts are usually involved and their tax treatment (if any) is illustrated in examples that follow. It is intended that the table be used with the conventions of "+" for cash inflow or savings and "−" for cash outflow or opportunity forgone. Slight deviations from this are the conventions of assigning a "−" to a depreciation charge even though it is not a cash expense and, in column C, using "+" for increases and "−" for decreases in taxable income. When all events affecting cash flow are the same for more than 1 year, a time-saving convention is to use one line in the table to describe entries for all of the years indicated in the left-hand column.

ILLUSTRATION OF COMPUTATIONS OF AFTER-TAX CASH FLOWS

The following problem series (Examples 10-7 to 10-10) illustrates the computation of after-tax cash flows as well as various common situations that affect income taxes. All problems include the common assumption that any tax disbursement or savings occurs at the same time (year) as the income or expense that affects the taxes. For purposes of comparison of the effect of various situations, the after-tax *internal rate of return* is computed for each example. One can observe from the results of Examples 10-7 to 10-10 that the faster (earlier) the depreciation write-off, the higher the after-tax internal rate of return.

In accordance with ERTA, depreciable property in use prior to January 1, 1981, must continue to be depreciated by "pre-A.C.R.S." methods discussed in Chapter 9. To illustrate these methods *in addition to* A.C.R.S. depreciation, the dates at which assets are placed in service are carefully noted in the following examples.

EXAMPLE 10-7
Certain new machinery is estimated to cost $180,000 installed. It is expected to reduce net annual operating disbursements by $36,000 per year for 10 years and to have a $30,000 market value at the end of the tenth year. (a) Develop a year-by-year tabulation of before-tax and after-tax cash flows, and (b) calculate the before-tax and after-tax internal rate of return using A.C.R.S. depreciation with a federal income tax rate of 46% plus a state income tax rate of 7.5%. State income taxes are deductible from federal income taxes. This machinery has an A.C.R.S. recovery period of 5 years and is to be placed in service in early 1986. In this example the useful life of the machinery is 10 years but the tax life is only 5 years.

Solution
(a) A tabulation of annual before-tax and after-tax cash flows for Example 10-7 is shown in Table 10-4. In column D the *effective* income tax rate is very close to 0.50 (from Equation 10-2) based on information provided above.

(b) The *before-tax internal rate of return* is determined from column A:

$$0 = -\$180,000 + \$36,000(P/A, i'\%, 10) + \$30,000(P/F, i'\%, 10)$$

By trial and error, $i' = 16.1\%$.

The entry in the last year is shown to be $30,000 since the machinery will

TABLE 10-4 After-Tax Cash Flow Analysis of Example 10-7

Year, k	(A) Before-Tax Cash Flow	(B) Depreciation			(C) = (A) + (B) Taxable Income	(D) = -(C) × 0.50 Cash Flow for Income Taxes	(E) = (A) + (D) After-Tax Cash Flow
		Cost	× A.C.R.S. Percentage	= Amount			
0	-$180,000						-$180,000
1	36,000	-$180,000	× 20%	= -$36,000	0	0	36,000
2	36,000	- 180,000	× 32%	= - 57,600	-$21,600	+$10,800	46,800
3	36,000	- 180,000	× 24%	= - 43,200	- 7,200	+ 3,600	39,600
4	36,000	- 180,000	× 16%	= - 28,800	+ 7,200	- 3,600	32,400
5	36,000	- 180,000	× 8%	= - 14,400	+ 21,600	- 10,800	25,200
6–10	36,000				36,000	- 18,000	18,000
10	30,000				30,000^a	- 15,000^b	15,000

^{a}Depreciation recapture = selling price - book value = $30,000 - 0 = $30,000.
^{b}Tax on depreciation recapture = $30,000(0.50) = $15,000 (see Appendix 10-A).

N.P.W.$(i\%)$ = $-180,000 + 36,000(P/F, i\%, 1) + 46,800(P/F, i\%, 2) + 39,600(P/F, i\%, 3) + 32,400(P/F, i\%, 4) + 25,200(P/F, i\%, 5) + 18,000(P/A, i\%, 4)(P/F, i\%, 5) + (18,000 + 15,000)(P/F, i\%, 10)$.

have this estimated market value. However, the asset was depreciated to zero with the A.C.R.S. method. Therefore, when the machine is sold at the end of year 10, there will be $30,000 of "recaptured depreciation" that is taxed at the ordinary income tax rate of 50%. This tax entry is shown in column D. More information is given on this subject in Appendix 10-A.

By trial and error, the *after-tax internal rate of return* for Example 10-7 is found to be 11.1%. As a matter of interest, this I.R.R. can be quickly computed by entering after-tax cash flow amounts in the operational computer program given in Appendix C.

It is also of interest to note the impact that the investment tax credit (ITC) would have on the after-tax I.R.R. The full 10% ITC could be claimed for this investment and would reduce other income taxes payable by the firm, thereby reducing the effective investment by $18,000 to $162,000†. As a result, the after-tax internal rate of return is increased to 14.1% (the depreciation schedule is not affected). The affected portion of the cash flow table would look as follows:

Year, k	Before-Tax Cash Flow	Depreciation	Taxable Income	Cash Flow for Income Taxes	After-Tax Cash Flow
0	− $180,000			+ $18,000	− $162,000
1	36,000	− $36,000	0	0	36,000
2	36,000	− 57,600	− $21,600	+ 10,800	46,800
.	.	.	.	.	.
.	.	.	.	.	.
.	.	.	.	.	.

The investment credit is treated as "year 0" cash flow due to the quarterly income tax payments required of corporations. The end of the first quarter is obviously closer to the beginning than the end of the first year. ■

EXAMPLE 10-8

Suppose that the new machinery described in Example 10-7 had been placed in service on January 1, 1980. The effective income tax rate remains 50%, but the CLADR guideline life is 10 years. (a) Develop a table of before-tax and after-tax cash flows when $S = \$30,000$ is used in conjunction with straight line depreciation, and (b) calculate the before-tax and after-tax internal rate of return. Assume that the investment tax credit cannot be taken.

Solution

$$\text{(a) Annual depreciation} = \frac{\$180,000 - \$30,000}{10} = \$15,000$$

†Based on ERTA (1981).

Year, k	(A) Before-Tax Cash Flow	(B) Depreciation	(C) = (A) + (B) Taxable Income	(D) = −(C) × 0.50 Cash Flow for Income Taxes	(E) = (A) + (D) After-Tax Cash Flow
0	− $180,000	—	—	—	− $180,000
1–10	+ 36,000	− $15,000	+ $21,000	− $10,500	+ 25,500
10	+ 30,000	—	—	—	+ 30,000[a]

[a]When market value equals book value, there is no taxable gain (recapture of depreciation) or loss, and hence no income tax.

(b) The before-tax cash flows are identical in Examples 10-7 and 10-8 and hence the before-tax I.R.R. is equal to 16.1%. In this example the after-tax I.R.R. is computed by trial and error to be approximately 8.6%.

$$0 = -\$180,000 + \$25,500(P/A, i'\%, 10) + \$30,000(P/F, i'\%, 10);$$
$$i' = 8.6\%$$ ∎

EXAMPLE 10-9

Repeat the calculations of Example 10-8 when sum-of-the-years'-digits depreciation is applicable over the 10-year CLADR guideline life.

Solution

$$SYD = 1 + 2 + 3 + \cdots + 10 = 55$$

$$\text{First-year depreciation} = \frac{10}{55} (\$180,000 - \$30,000) = \$27,273$$

$$\text{Decrease in depreciation each year thereafter} = \frac{1}{55} (\$180,000 - \$30,000)$$

$$= \$2,727$$

Year, k	(A) Before-Tax Cash Flow	(B) Depreciation	(C) = (A) + (B) Taxable Income	(D) = −(C) × 0.50 Cash Flow for Income Taxes	(E) = (A) + (D) After-Tax Cash Flow
0	− $180,000				− $180,000
1	+ 36,000	− $27,273	+ $8,727	− $4,364	+ 31,636
Each year 2–10	+ 36,000	Decreases by $2,727 each year	Increases by $2,727 each year	$1,364 more negative each year	Decreases by $1,364 each year
10	+ 30,000				+ 30,000

The after-tax I.R.R. with sum-of-the-years'-digits depreciation can be determined as follows:

$$0 = -\$180,000 + \$31,636(P/A, i'\%, 10) - \$1,364(P/G, i'\%, 10)$$
$$+ \$30,000(P/F, i'\%, 10)$$

By trial and error, $i' = 9.5\%$. ∎

EXAMPLE 10-10

This example illustrates numerous "complications" that can arise in connection with Example 10-8. First, sum-of-the-years'-digits (SYD) depreciation is computed using the *lower limit* of the asset depreciation range (see Table 9-1), which is 8 years. Second, a salvage value for computing SYD depreciation is assumed to be $30,000 *even though the actual market value* at the end of year 10 is believed to be closer to $50,000. Finally, a 10% investment tax credit can be claimed. In view of these modifications, what is the after-tax internal rate of return?

Solution

$$SYD = 1 + 2 + \cdots + 8 = 36$$

$$\text{First-year depreciation} = \frac{8}{36}(\$180,000 - \$30,000) = \$33,333$$

$$\text{Decrease in depreciation over the next 7 years} = \frac{1}{36}(\$150,000) = \$4,167$$

Year, k	(A) Before-Tax Cash Flow	(B) Depreciation	(C) = (A) + (B) Taxable Income	(D) = −(C) × 0.50 Cash Flow for Income Taxes	(E) = (A) + (D) After-Tax Cash Flow
0	− $180,000			+ $18,000[a]	− $162,000
1	+ 36,000	− $33,333	+ $2,667	− 1,334	+ 34,666
Each year 2–8	+ 36,000	Decreases by $4,167 each year	Increases by $4,167 each year	$2,084 more negative each year	Decreases by $2,048 each year
9	+ 36,000	0	+ 36,000	− 18,000	+ 18,000
10	+ 36,000	0	+ 36,000	− 18,000	+ 18,000
10	+ 50,000		+ 20,000[b]	− 10,000[c]	+ 40,000

[a]Investment tax credit = 0.10($180,000).
[b]Depreciation recapture = $50,000 − $30,000 = $20,000.
[c]Income taxes on recapture (at 50%) = $10,000.

The computer program of Appendix C was utilized in this example to determine the after-tax I.R.R. of 12.8%. ∎

In Examples 10-7 to 10-10 the before-tax I.R.R. was identical, namely 16.1%. However, the after-tax I.R.R.'s varied considerably among the examples as summarized below. Notice that accelerated depreciation, coupled with investment tax credits, serve to increase after-tax I.R.R.'s and improve the economic attractiveness of capital investment opportunities.

Example	Conditions	After-Tax I.R.R.
10–7	A.C.R.S. depreciation over 5 years, no investment tax credit	11.1% (14.1% with 10% ITC)
10–8	Straight line depreciation over 10 years, no investment tax credit	8.6%
10–9	Sum-of-the-years'-digits depreciation over 10 years, no investment tax credit	9.5%
10–10	SYD depreciation over 8 years, $20,000 depreciation recapture, and a 10% investment tax credit	12.8%

ILLUSTRATION OF AFTER-TAX ECONOMIC COMPARISONS USING DIFFERENT METHODS

EXAMPLE 10-11

It is desired to compare the economics of two industrial material handling systems. The pertinent data are as follows:

	System X	System Y
First cost	$80,000	$200,000
Useful life (also depreciation life)	20 years	40 years
Annual before-tax cash disbursement	$18,000	$6,000
Salvage value at end of life	$20,000	$40,000

Use a table to compute the after-tax cash flows based on straight line depreciation using the lives and salvage values given and a 30% effective income tax rate to cover both state and federal income taxes. Assuming a 10% minimum attractive after-tax rate of return, show which alternative is best by the (a) internal rate of return (I.R.R.) method and (b) present worth–cost (P.W.-C.) method. This is a large firm that is profitable in its overall operation.

Before applying the various economy study methods, it should be recognized that the solutions shown below for each method employ the repeatability assumption for comparisons of alternatives having different economic lives as discussed in Chapter 6. That is:

1. The period of needed service over which the alternatives are being compared is either indefinitely long or a length of time equal to a common multiple of the lives.
2. What is estimated to happen in the first life span will happen in all succeeding life spans, if any.

It is also assumed that system X must be depreciated over 20 years and system Y must be depreciated over 40 years.

Table 10-5 shows the calculation of after-tax cash flows to be used in the following solutions. Notice that negative taxable income and positive income taxes (i.e., reduction in taxes owed) are permissible because the firm is profitable in its overall activity.

Solution by I.R.R. Method

The I.R.R. cannot be found for either alternative alone, but one can use this method to determine if the incremental investment is justified. Since the economic lives differ, the easiest way is to find the $i'\%$ at which the annual equivalent costs of the two alternatives are equal (or the difference in the annual costs for the two alternatives is zero). Thus

$$\text{A.C.}_X = \$80,000(A/P, i'\%, 20) - \$20,000(A/F, i'\%, 20) + \$11,700$$

$$\text{A.C.}_Y = \$200,000(A/P, i'\%, 40) - \$40,000(A/F, i'\%, 40) + \$3,000$$

The calculated results are as follows:

	$i' = 5\%$	$i' = 10\%$
A.C.$_X$:	$17,515	$20,748
A.C.$_Y$:	14,325	23,362
A.C.$_X$ - A.C.$_Y$:	$ 3,190	-$ 2,614

Thus

$$i'\% = \text{I.R.R.} = 5\% + \frac{3,190}{3,190 + 2,614}(10\% - 5\%)$$

$$= 5\% + 2.8\% = 7.8\%$$

Since $7.8\% < 10\%$ (M.A.R.R.), the incremental investment is not justified, and system X is the indicated choice. ∎

TABLE 10-5   Determination of After-Tax Cash Flows for Example 10-11

Year	(A) Before-Tax Cash Flow	(B) Depreciation	(C) = (A) + (B) Taxable Income	(D) = -(C) × 30% Cash Flow for Income Taxes	(E) = (A) + (D) After-Tax Cash Flow
System X					
0	- $ 80,000				- $ 80,000
1–20	- 18,000	- $3,000[a]	- $21,000	+ $6,300	- 11,700
20	+ 20,000				+ 20,000
System Y					
0	- 200,000				- 200,000
1–40	- 6,000	- 4,000[b]	- 10,000	+ 3,000	- 3,000
40	+ 40,000				+ 40,000

[a]Annual depreciation for system X = ($80,000 - $20,000)/20 = $3,000.
[b]Annual depreciation for system Y = ($200,000 - $40,000)/40 = $4,000.

Solution by P.W.-C. Method

Using the lowest common multiple of lives = 40 years as the study period, we obtain the following data:

	System X	System Y
Annual disbursements:		
$11,700(P/A, 10\%, 40)$	$114,415	
$3,000(P/A, 10\%, 40)$		$ 29,337
Original investment	80,000	200,000
First replacement: ($80,000 − $20,000) ×		
$(P/F, 10\%, 20)$	8,919	
Less: salvage after 40 years		
$-20,000(P/F, 10\%, 40)$	−442	
$-40,000(P/F, 10\%, 40)$		−884
TOTAL P.W.-C.	$202,892	$228,453

Since P.W.-C.$_X$ < P.W.-C.$_Y$, system X is again the indicated choice. ∎

EXAMPLE 10-12

It is desired to compare the four presses in Example 6-1 on an after-tax basis by using the A.W. method. Assume an effective income tax rate of 50%, that 1986 A.C.R.S. percentages are used for depreciation purposes with a 5-year recovery period, and that the after-tax M.A.R.R. is 10%. The essential data are shown below, together with a tabular computation of the after-tax annual cash flows for each press in Table 10-6.

	Press			
	A	B	C	D
Investment	−$6,000	−$7,600	−$12,400	−$13,000
Useful life	5 years	5 years	5 years	5 years
Salvage value	0	0	0	0
Annual amounts, years 1–5:				
before-tax cash flow	−$7,800	−$7,282	−$ 6,298	−$ 5,720

Solution

Based on the after-tax cash flows calculated in Table 10-6, the equivalent after-tax annual worth of press A at a M.A.R.R. of 10% is

$$A.W._A(10\%) = -\$6,000(A/P, 10\%, 5)$$
$$- [\$3,300(P/F, 10\%, 1) + \$2,940(P/F, 10\%, 2)$$
$$+ \$3,180(P/F, 10\%, 3) + \$3,420(P/F, 10\%, 4)$$
$$+ \$3,660(P/F, 10\%, 5)] \times (A/P, 10\%, 5)$$
$$= -\$4,861$$

TABLE 10-6 Data and After-Tax Annual Cash Flow Computations for Comparison of the Four Molding Presses in Example 10-12 (Original Data Given in Example 6-1)

Press	Year	Before-Tax Cash Flow	Depreciation (see Table 9-3)	Taxable Income	Income Taxes	After-Tax Cash Flow
A	0	− $ 6,000				− $ 6,000
	1	− 7,800	− $1,200	− $ 9,000	+ $4,500	− 3,300
	2	− 7,800	− 1,920	− 9,720	+ 4,860	− 2,940
	3	− 7,800	− 1,440	− 9,240	+ 4,620	− 3,180
	4	− 7,800	− 960	− 8,760	+ 4,380	− 3,420
	5	− 7,800	− 480	− 8,280	+ 4,140	− 3,660
B	0	− 7,600				− 7,600
	1	− 7,282	− 1,520	− 8,802	+ 4,401	− 2,881
	2	− 7,282	− 2,432	− 9,714	+ 4,857	− 2,425
	3	− 7,282	− 1,824	− 9,106	+ 4,553	− 2,729
	4	− 7,282	− 1,216	− 8,498	+ 4,249	− 3,033
	5	− 7,282	− 608	− 7,890	+ 3,945	− 3,337
C	0	− 12,400				− 12,400
	1	− 6,298	− 2,480	− 8,778	+ 4,389	− 1,909
	2	− 6,298	− 3,968	− 10,266	+ 5,133	− 1,165
	3	− 6,298	− 2,976	− 9,274	+ 4,637	− 1,661
	4	− 6,298	− 1,984	− 8,282	+ 4,141	− 2,157
	5	− 6,298	− 992	− 7,290	+ 3,645	− 2,653
D	0	− 13,000				− 13,000
	1	− 5,720	− 2,600	− 8,320	+ 4,160	− 1,560
	2	− 5,720	− 4,160	− 9,880	+ 4,940	− 780
	3	− 5,720	− 3,120	− 8,840	+ 4,420	− 1,300
	4	− 5,720	− 2,080	− 7,800	+ 3,900	− 1,820
	5	− 5,720	− 1,040	− 6,760	+ 3,380	− 2,340

Similarly, the equivalent annual worths of presses B, C, and D have been determined to be:

$$A.W._B(10\%) = -\$4,858$$

$$A.W._C(10\%) = -\$5,135$$

$$A.W._D(10\%) = -\$4,942$$

Hence press B is the recommended choice, followed closely by press A. In the before-tax study of Example 6-1, press D was the choice. This demonstrates clearly that income tax considerations can change the results of before-tax economy studies. ∎

In cost-only alternatives such as those above, the *negative taxable income* and resultant *positive income taxes* should be interpreted as offsets against positive taxable income and disbursements for income taxes, respectively, that arise in other profitable areas of activity in a large firm.

The Impact of Inflation on After-Tax Economic Analyses

As noted in Chapter 8, economy studies that include the effects of inflation present some difficulties because interest charges, depreciation write-offs and lease agreements are actual dollar amounts based on past commitments and generally are *unresponsive* to inflation. At the same time, other types of cash flows are *responsive* to inflation.

A simple evaluation of the effect of inflation is illustrated by reworking Example 10-7 when an average inflation rate of 8% per year applies to savings of operating disbursements. It is also assumed that the salvage value is estimated in terms of early 1986 (i.e., real) dollars and that it will also inflate at 8% per year. No investment tax credit is considered here. A portion of Table 10-4, which deals with Example 10-7, is reproduced below:

Year	Before-Tax Cash Flow	A.C.R.S. Depreciation (1986 percentages)
0	− $180,000	
1	36,000	− $36,000
2	36,000	− 57,600
3	36,000	− 43,200
4	36,000	− 28,800
5	36,000	− 14,400
6–10	36,000	0
10	30,000	

If these before-tax cash flow estimates are in early 1986 dollars, the impact of depreciation is overstated in an after-tax economy study that is performed in real dollars. The values must be adjusted according to Equation 8-5 when the base point in time (year 0) is early 1986:

$$R\$ \text{ in year } 0 = A\$ \times (P/F, f\%, n)$$

The required adjustments to the Example 10-7 depreciation amounts in a R$ analysis are indicated in Table 10-7.

Based on the R$ after-tax cash flows of Table 10-7, the internal real rate of return can be found by trial and error to be $i'_r = 9.3\%$. When inflation was ignored in Example 10-7, the after-tax I.R.R. was 11.1%. It can be concluded, then, that inflation reduces the real after-tax I.R.R. because of the devaluation of fixed depreciation schedules (or other types of unresponsive A$ annuities such as interest payments). Such unresponsive amounts cause taxable income to increase in an *actual dollar analysis* so that income taxes increase and after-tax cash flows decrease. This point is demonstrated in Table 10-8, where Example 10-7 is reworked in terms of actual dollars when the inflation rate is 8% per year. A comparison of Table 10-8 with the analysis of Example 10-7 in Table 10-4 shows a dramatic increase in taxable income.

The after-tax I.R.R. from Table 10-8 is 18%, which could have been calculated directly with Equation 8-7 from the R$ analysis:

TABLE 10-7 Real Dollar Analysis of Example 10-7 When the Inflation Rate is 8%

Year	Before-Tax Cash Flow (R$)	A.C.R.S. Depreciation (A$)		$\left(\dfrac{1}{1+f}\right)^{year}$		A.C.R.S. Depreciation (R$)[a]	Taxable Income	Cash Flow for Income Taxes	After-Tax Cash Flow (R$)
0	−$180,000			1.0000					− $180,000
1	36,000	−$36,000	×	0.9259	=	−$33,333	+$ 2,667	−$ 1,333	+ 34,667
2	36,000	− 57,600	×	0.8573	=	− 49,383	− 13,383	+ 6,691	+ 42,691
3	36,000	− 43,200	×	0.7938	=	− 34,294	+ 1,706	− 853	+ 35,147
4	36,000	− 28,800	×	0.7350	=	− 21,169	+ 14,831	− 7,416	+ 28,584
5	36,000	− 14,400	×	0.6806	=	− 9,800	+ 26,200	− 13,100	+ 22,900
6–10	36,000	0				0	+ 36,000	− 18,000	+ 18,000
10	30,000						+ 30,000	− 15,000	+ 15,000

[a]A$ × (P/F, f%, year) = R$

$$i_c' \text{ (from Table 10-8)} = i_r' + f + i_r'(f)$$

$$= 0.093 + 0.08 + 0.093(0.08)$$

$$= 0.18$$

The importance in economic analyses of understanding how the cash flows have been estimated and the components of the internal rate of return should be obvious from Tables 10-7 and 10-8. To reiterate, if cash flows are estimated in real dollars, equivalent worth calculations must be made with a real after-tax M.A.R.R. On the other hand, if cash flows are estimated in actual dollars, equivalent worth calculations must utilize a combined after-tax M.A.R.R.

This is an *important* point, and it is illustrated further by computing the net present worth of after-tax cash flows in Tables 10-7 and 10-8 assuming a real, after-tax M.A.R.R. of 6%. In view of an average inflation rate equal to 8% per year, the combined after-tax M.A.R.R. for discounting Table 10-8 cash flows is

$$i_c = \text{M.A.R.R. (A\$ analysis)}$$

$$= 0.06 + 0.08 + 0.06(0.08)$$

$$= 0.1448 \quad \text{or} \quad 14.48\%$$

The calculations of net present worths based on R\$ and A\$ after-tax cash flows are as follows:

Real \$ (Table 10-7)

$$\begin{aligned}
\text{N.P.W.}(6\%) = & -\$180{,}000 + 34{,}667(P/F, 6\%, 1) \\
& + 42{,}691(P/F, 6\%, 2) + 35{,}147(P/F, 6\%, 3) \\
& + 28{,}584(P/F, 6\%, 4) + 22{,}900(P/F, 6\%, 5) \\
& + 18{,}000(P/A, 6\%, 5)(P/F, 6\%, 5) \\
& + 15{,}000(P/F, 6\%, 10) = \$24{,}998
\end{aligned}$$

Actual \$ (Table 10-8)

$$\begin{aligned}
\text{N.P.W.}(14.48\%) = & -\$180{,}000 + 37{,}440(P/F, 14.48\%, 1) \\
& + 49{,}695(P/F, 14.48\%, 2) \\
& + 44{,}275(P/F, 14.48\%, 3) \\
& + 38{,}889(P/F, 14.48\%, 4) \\
& + 33{,}648(P/F, 14.48\%, 5) \\
& + 28{,}564(P/F, 14.48\%, 6) \\
& + 30{,}849(P/F, 14.48\%, 7) \\
& + 33{,}317(P/F, 14.48\%, 8) \\
& + 35{,}982(P/F, 14.48\%, 9) \\
& + (38{,}861 + 32{,}384)(P/F, 14.48\%, 10) \\
= & \ \$24{,}998
\end{aligned}$$

TABLE 10-8 Actual Dollar Analysis of Example 10-7 When the Inflation Rate Is 8%

Year	Before-Tax Cash Flow (R$)	Adjustment $(1 + f)^{year}$	Adjusted Cash Flow (A$)[a]	A.C.R.S. Depreciation (A$)	Taxable Income	Cash Flow for Income Taxes	After-Tax Cash Flow (A$)
0	-$180,000	1.0000					-$180,000
1	36,000 × 1.0800 =	$38,880	-$36,000	$ 2,880	-$ 1,440	37,440
2	36,000 × 1.1664 =	41,990	- 57,600	- 15,610	+ 7,805	49,795
3	36,000 × 1.2597 =	45,350	- 43,200	2,151	- 1,075	44,275
4	36,000 × 1.3605 =	48,978	- 28,800	20,178	- 10,089	38,889
5	36,000 × 1.4693 =	52,896	- 14,400	38,496	- 19,248	33,648
6	36,000 × 1.4869 =	57,128	0	57,128	- 28,564	28,564
7	36,000 × 1.7138 =	61,698	0	61,698	- 30,849	30,849
8	36,000 × 1.8509 =	66,634	0	66,634	- 33,317	33,317
9	36,000 × 1.9990 =	71,964	0	71,964	- 35,982	35,982
10	36,000 × 2.1589 =	77,721	0	77,721	- 38,862	38,861
10	30,000 × 2.1589 =	64,768		64,768	- 32,384	32,384

[a]R$ × $(F/P, f\%, \text{year})$ = A$

When carried out properly, after-tax studies made with R$ and A$ estimates will have identical net present worths. Internal rates of return on after-tax R$ and A$ cash flows will at first appear not to be comparable, but it was demonstrated earlier in this section that the two rates of return can be easily reconciled.

The after-tax analysis of the two mutually exclusive alternatives subject to inflation is provided in Example 10-13.

EXAMPLE 10-13

Your firm has decided to acquire a new piece of machinery that includes the latest safety features required by O.S.H.A. The machinery may either be (1) purchased for cash or (2) it may be leased from another company. Mr. Williams, the president of your firm, has requested that you perform an after-tax analysis of these two means of obtaining the machinery when the following estimates and conditions are applicable.

1. The study period is 5 years and the estimated useful life of the machinery also is 5 years. Straight line depreciation is elected under A.C.R.S. provisions, with an estimated zero net salvage value at the end of the useful life of the purchased machinery. The effective incremental income tax rate (t) is 50%. Also, a 10% investment tax credit can be taken. All recaptured depreciation (if any) is taxed at 50% of the gain.
2. The following interest rates and inflation estimates are used:
   (a) The real after-tax M.A.R.R. is 10% per year.
   (b) Annual inflation (f) is expected to average 8%.
   (c) The combined after-tax M.A.R.R. (i_c) is 18.8% [i.e., $i_c = 0.10 + 0.08 + 0.10(0.08) = 0.188$ or 18.8%].
3. Annual savings, operating costs, maintenance costs, and the terminal *market value* respond to inflation. In the case of leasing the machinery, the yearly lease payment does *not* grow with inflation. When purchasing the machinery, depreciation write-offs do not respond to inflation.
4. Annual cash flow estimates:

	Machinery	
	Purchase	Lease
Savings	$5,000	$5,000
Operating costs	2,000	2,000
Maintenance cost	1,000	(included in lease contract)
Lease fee	—	6,000 (payable at end of year)

All of the annual cash flows above have been estimated in real dollars.

5. Investment costs:

Purchase Machine	Lease Machine
Initial cost = $20,000	Deposit = $1,500 (refundable at end of 5 years)
Market value = $1,500 at end of year 5 (in real dollars)	

6. The analysis is to be performed after taxes and an inflation rate (f) of 8% is estimated to apply to *all* cash flows that respond to inflation.

Based on this information, should your firm purchase or lease the machinery?

Solution

An after-tax cash flow analysis is performed with actual dollar estimates in Table 10-9 for the "purchase machinery" option and in Table 10-10 for the "lease machinery" alternative. Because annual savings, operating costs and maintenance costs inflate each year at 8%, the real dollar before-tax cash flows are converted to the corresponding actual dollar estimates in column C by using Equation 8-5. The column C entries are then combined with depreciation writeoffs in Table 10-9 and lease payments in Table 10-10 to arrive at the taxable income associated with each alternative. Notice that your firm cannot claim depreciation on the leased machinery.

After-tax cash flow in column G of Table 10-9 is determined in view of an investment tax credit in year 0 and depreciation recapture, which is taxed at 50%, in year 5. The net present worth at $i_c = 18.8\%$ is $-\$7,601$ for the "purchase machinery" alternative. Similarly, the net present worth of column G in Table 10-10 is $-\$4,395$ for the leasing alternative. A recommendation should be made to lease the machinery so that present worth of cost is minimized. ∎

The reader should again observe that a combined interest rate is used to discount the after-tax cash flows in Table 10-9 and 10-10. In this regard, *the after-tax M.A.R.R.'s of most companies are directly stated as combined interest rates*. Furthermore, most companies make economy studies in terms of A$ estimates because decision makers lean toward a measure of financial profitability that includes the effects of inflation.

Referring to Example 10-13, if an after-tax analysis had been performed *ignoring the effects of inflation*, the recommended course of action would have been to purchase the equipment! Thus, assuming a 0% inflation rate in this particular example would have led to an incorrect selection between alternatives.

TABLE 10-9 Purchase Equipment Alternative—Actual Dollar Analysis

Year	(A) Before-Tax Cash Flow (R$)	(B) Adjustment, $(1 + f)^{year}$	(C) Before-Tax Cash Flow, (A$)	(D) Depreciation (A$)	(E) Taxable Income: C + D	(F) Cash Flow for Income Taxes: $-t \times$ (E)	(G) After-Tax Cash Flow (A$)[a]: C + F
0	-$20,000	1.000	-$20,000				-$18,000
1	2,000[c]	1.080	2,160	-$4,000	-$1,840	$2,000[b]	3,080
2	2,000	1.166	2,332	- 4,000	- 1,668	920	3,166
3	2,000	1.260	2,520	- 4,000	- 1,480	834	3,260
4	2,000	1.360	2,720	- 4,000	- 1,280	740	3,360
5	2,000	1.469	2,938	- 4,000	- 1,060	640	3,469
5	1,500	1.469	2,204		2,204	531	1,102
						- 1,102	

[a]P.W. at 18.8% = - $7,601.
[b]Investment tax credit = 0.10($20,000).
[c]$5,000 (annual savings) - $2,000 - $1,000 = $2,000.

TABLE 10-10 Lease Equipment Alternative—Actual Dollar Analysis

Year	(A) Before-Tax Cash Flow (R$)	(B) Adjustment $(1 + f)^{year}$	(C) Before-Tax Cash Flow (A$)	(D) Lease Payments (A$)	(E) Taxable Income: C + D	(F) Cash Flow for Income Taxes: $-t \times$ (E)	(G) After-Tax Cash Flow (A$)[a]: C + D + F
0	−$1,500	1.000	−$1,500				−$1,500
1	3,000[b]	1.080	3,240	−$6,000	−$2,760	$1,380	− 1,380
2	3,000	1.166	3,498	− 6,000	− 2,502	1,251	− 1,251
3	3,000	1.260	3,780	− 6,000	− 2,220	1,110	− 1,110
4	3,000	1.360	4,080	− 6,000	− 1,920	960	− 960
5	3,000	1.469	4,407	− 6,000	− 1,593	796.5	− 796.5
5	1,500[c]		1,500				1,500

[a] P.W. at 18.8% = −$4,395.
[b] $5,000 (annual savings) − $2,000 = $3,000.
[c] The deposit on the leased equipment, refunded with no interest.

A CASE STUDY—DEALING WITH ESCALATION IN AFTER-TAX STUDIES

In most engineering economy studies conducted in industry, escalation rates on numerous types of revenues and/or costs have to be considered in addition to various income tax provisions set forth by the Economic Recovery Tax Act of 1981. To illustrate this situation, a fairly comprehensive and realistic case study is presented.

EXAMPLE 10-14

A potential investment opportunity is expected in early 1986 that requires the investment of $20,000 in new production equipment to increase output of an assembly plant. As a result of this investment, the revenue obtained from the modified plant is expected to increase. The following information applies to the investment opportunity.

Analysis period	10 years
Estimated useful life of the equipment	10 years
A.C.R.S. recovery period	5 years
Federal income tax rate	46%
State income tax rate	7.5%
Effective income tax rate (t)	$50\% = [0.46 + 0.075 - (0.46)(0.075)]100$
Real after-tax M.A.R.R. (i_r)	6%
General inflation rate (f)	8%
Combined discount rate (i_c)	$14.48\% = [0.06 + 0.08 + (0.06)(0.08)]100$
Increased revenue (assume that revenue escalates at the general inflation rate of 8%)	$15,000 per year in real dollars (early 1986)
Market value in 10 years	10% of the investment cost in terms of real dollars in 1986

Annual Cost:

Category	Cost (R$ in 1986)	Category Escalation Rate
Material	$1,200	10%
Labor	2,500	5.5%
Energy	2,500	15%
Other cost	500	8%

Leased equipment is also required, which can be obtained for the first 5 years at a rate of $800 per year. The contract will be renegotiated at the beginning of the sixth year at an escalated value based on the general inflation rate.

This investment qualifies for an investment tax credit.

Perform an analysis of the net present worth of this project using the appropriate income tax regulations and including the effects of differential escalation.

Conduct the analysis with (a) real-dollar estimates and (b) actual-dollar estimates to show that identical results are obtained.

Solution

(a) *Real-dollar analysis:* The adjustments required for the information given above are first described for a real-dollar analysis of the investment opportunity. Here the reference point in time for R$ cash flow estimates is early 1986.

1. *Revenue:* As stated, the revenue in real dollars is $15,000 per year and is not adjusted since it is assumed to escalate at the general inflation rate.
2. *Material:* The material escalation rate (e) is 10% per year versus a general inflation rate (f) of 8%. The annual differential escalation rate (e') for a cost category is $(e - f)/(1 + f)$ from equation 8-6. Hence the differential escalation rate for materials (e'_M) is

$$e'_M = \frac{0.10 - 0.08}{1.08} = 0.0185$$

   Material costs are adjusted by $(1 + e'_M)^k$ so that materials cash flows in year $k = -\$1,200(1.0185)^k$.
3. *Labor:* The labor escalation rate is 5.5% per year, resulting in a differential escalation rate of

$$e'_L = \frac{0.055 - 0.08}{1.08} = -0.0231$$

   labor cost in year $k = -\$2,500(1 - 0.0231)^k = -\$2,500(0.9769)^k$
4. *Energy:* The annual energy escalation rate is 15%, resulting in a differential rate of

$$e'_E = \frac{0.15 - 0.08}{1.08} = 0.0648$$

   energy cost in year $k = -\$2,500(1.0648)^k$
5. *Other costs:* Other costs escalate at the general inflation rate and, therefore, need no adjustment for the real-dollar analysis.
6. *Leased equipment:* Since the lease rate is set by the terms of the contract, the annual cost of the lease does not escalate during years 1 to 5. This results in a differential escalation rate of

$$e'_{lease} = \frac{0 - 0.08}{1.08} = -0.0741$$

   lease cost in year $k = -\$800(1 - 0.0741)^k = -\$800(0.9259)^k$
7. *Depreciation:* This is calculated with the Accelerated Cost Recovery System method. The impact of depreciation will be overstated because it is an actual-dollar amount. Real dollar equivalents are shown below.

Year k	Initial Cost	1986 A.C.R.S. Recovery Percentages	A.C.R.S. Depreciation (A\$)	$(P/F, f = 8\%, k)$	A.C.R.S. Depreciation (R\$)
1	$20,000	20	$4,000	0.9259	$3,704
2	20,000	32	6,400	0.8573	5,487
3	20,000	24	4,800	0.7938	3,810
4	20,000	16	3,200	0.7350	2,352
5	20,000	8	1,600	0.6806	1,089

8. *Investment tax credit:* Since the useful life is greater than 5 years, the full 10% investment tax credit can be taken.

$$\text{investment credit} = 0.1(\$20,000) = \$2,000$$

No adjustment is necessary because the $2,000 occurs at the present (early 1986).

9. *Market value*: The market value is estimated to be 10% of the original purchase price, or $0.1(\$20,000) = \$2,000$. This was stated to be a real-dollar estimate, so no adjustment is required.

10. *Recapture of depreciation:* Because the asset was depreciated to a zero book value, the estimated market value of $2,000 represents a recovery of depreciation and is taxed as ordinary income at the 50% rate.

All of the items above have been converted to their real-dollar equivalents in early 1986 and are shown in Table 10-11. The net present worth at $i_r = 6\%$ is $14,811. Thus the investment opportunity appears to be financially attractive.

(b) *Actual-dollar analysis:*

1. *Revenue:* The revenue of $15,000 per year must be increased each year by the general inflation rate.

$$\text{revenue in year } k = \$15,000(1.08)^k$$

2. *Material, labor, energy, and other costs:* These costs in year k are escalated by the appropriate escalation rate:

$$\text{material} = -\$1,200(1.1)^k$$
$$\text{labor} = -\$2,500(1.055)^k$$
$$\text{energy} = -\$2,500(1.15)^k$$
$$\text{other cost} = -\$500(1.08)^k$$

3. *Leased property:* The lease will be adjusted at the end of year 5 to account for 5 years of inflation at 8% per year:

$$\text{lease (years 6–10)} = -\$800(1.08)^5 = -\$1,175$$

4. *Depreciation:* The A\$ A.C.R.S. depreciation previously shown in the real-dollar analysis is utilized here.

TABLE 10-11 Real-Dollar and Actual-Dollar Cash Flow Analyses for Example 10-14

1986

Year	Revenue	Initial Investment	Material	Labor	Energy	Other Cost	Leased Equipment	Total Cost	Before-Tax Cash Flow	Depreciation	Taxable Income	Cash Flow from Income Tax	After-Tax Cash Flow
					Real-Dollar Analysis								
0		−$20,000						−$20,000	−$20,000			+$2,000ᵃ	−$18,000
1	$15,000		−$1,222	−$2,442	−$2,662	−$500	−$741	7,567	7,433	$3,704	$3,729	−1,865	5,569
2	15,000		−1,245	−2,386	−2,834	−500	−686	7,751	7,349	5,487	1,862	−931	6,418
3	15,000		−1,268	−2,331	−3,018	−500	−635	7,752	7,248	3,810	3,438	−1,719	5,529
4	15,000		−1,291	−2,277	−3,214	−500	−588	7,870	7,130	2,352	4,778	−2,389	4,741
5	15,000		−1,315	−2,224	−3,422	−500	−544	8,005	6,995	1,089	5,906	−2,953	4,042
6	15,000		−1,340	−2,173	−3,644	−500	−741	8,397	6,603		6,603	−3,301	3,301
7	15,000		−1,364	−2,123	−3,880	−500	−686	8,553	6,447		6,447	−3,224	3,224
8	15,000		−1,390	−2,074	−4,131	−500	−635	8,729	6,271		6,271	−3,135	3,135
9	15,000		−1,415	−2,026	−4,399	−500	−588	8,928	6,072		6,072	−3,036	3,036
10	15,000		−1,441	−1,979	−4,684	−500	−544	9,345	5,852		5,852	−2,926	2,926
10	2,000ᵇ							2,000	2,000		2,000ᶜ	−1,000	1,000
											Net P.W. at 6% =		14,811
					Actual-Dollar Analysis								
0		−20,000						20,000	20,000			+ 2,000ᵃ	18,000
1	16,200		−1,320	−2,638	−2,875	−540	−800	8,173	8,028	4,000	4,028	−2,014	6,014
2	17,496		−1,452	−2,783	−3,306	−583	−800	8,924	8,572	6,400	2,172	−1,086	7,486
3	18,896		−1,597	−2,936	−3,802	−630	−800	9,765	9,131	4,800	4,331	−2,166	6,965
4	20,407		−1,757	−3,097	−4,373	−680	−800	10,707	9,700	3,200	6,500	−3,250	6,450
5	22,040		−1,933	−3,267	−5,028	−735	−800	11,763	10,277	1,600	8,677	−4,388	5,939
6	23,803		−2,126	−3,447	−5,783	−793	−1,175	13,324	10,479		10,479	−5,239	5,240
7	25,707		−2,338	−3,637	−6,650	−857	−1,175	14,657	11,050		11,050	−5,525	5,525
8	27,764		−2,572	−3,837	−7,648	−925	−1,175	16,157	11,607		11,607	−5,803	5,803
9	29,985		−2,830	−4,048	−8,795	−1,000	−1,175	17,846	12,139		12,139	−6,070	6,070
10	32,384		−3,112	−4,270	−10,114	−1,079	−1,175	19,751	12,633		12,633	−6,316	6,316
10	4,318ᵇ							4,318	4,318		4,318ᶜ	−2,159	2,159
											Net P.W. at 14.48% =		14,811

ᵃInvestment tax credit.
ᵇEstimated market value.
ᶜRecovery of depreciation—taxed as ordinary income.

5. *Market value:* The 10% market value is a real-dollar amount and must be increased to account for the 8% annual inflation rate.

$$\text{Market value} = 0.1(\$20,000)(1.08)^{10} = \$4,318$$

6. *Recaptured Depreciation:* The market value in A$ of $4,318 represents the recovery of depreciation and is taxed as ordinary income at the 50% rate.

The A$ after-tax cash flows of Table 10-11 are discounted at $i_c = 14.48\%$ to obtain the net present worth of $14,811. As expected, this N.P.W. is the same as that obtained in the real-dollar analysis. ∎

By now it is apparent that the consideration of inflation in Example 10-7 (Tables 10-7 and 10-8), Example 10-13 and particularly Example 10-14 causes solutions to be quite time-consuming when calculations are performed manually. To reduce the computational burden associated with after-tax problems having actual-dollar cash flows where differential price changes are present, a BASIC computer program is provided in Appendix H. This program can also be used to analyze after-tax problems in which the effects of inflation are ignored or are believed to be not relevant.

A sample run of the program for Example 10-14 is shown in Figure 10-2 (below and pages 379–381).The small difference in results is due to combining ''material'' and ''other'' costs into one cost category in the sample run (labeled ''material''), which inflates at an average rate of 9.4% per year.

This program allows one to conduct an actual dollar analysis of an investment alternative on an after-tax basis when inflation/escalation of revenues and costs are involved. The reference year for all cash flows, except salvage value, is the time at which the initial investment is made (i.e. time 0). These are termed 'real dollars' in the prompts below. Multiple replacements of an alternative are permitted when the study period is longer than useful life. Conversely the study period can be shorter than useful life. Salvage value at the end of the study period is specified by the analyst. A combined interest rate is required for equivalence calculations.

Note: Enter all percentages and rates in decimal form. For example, 10% should be entered in this program as 0.10.

<u>INVESTMENT</u>

```
 INPUT INITIAL COST OF INVESTMENT ?20000
 INPUT INVESTMENT TAX CREDIT PERCENTAGE IN DECIMAL
FORM ?.1
 INPUT THE USEFUL LIFE (IN YEARS) OF THE INVESTMENT
?10
 HOW LONG IS THE STUDY PERIOD ?10
 INPUT THE SALVAGE VALUE OF THE INVESTMENT IN ACTUAL
DOLLARS FOR YEAR 10 ?4318
```

FIGURE 10-2 Sample Run of BASIC Computer Program (in Appendix H) for Example 10-14.

DEPRECIATION

```
 INPUT ACRS TAX LIFE OF INVESTMENT IN YEARS ?5
 INPUT THE APPROPRIATE ACRS RECOVERY PERCENTAGE IN
DECIMAL FORM
YEAR   1    ?.20
YEAR   2    ?.32
YEAR   3    ?.24
YEAR   4    ?.16
YEAR   5    ?.08
```

REVENUES AND EXPENSES

```
INPUT THE ANNUAL REVENUES IN REAL DOLLARS ?15000
INPUT EFFECTIVE ANNUAL ESCALATION RATE FOR REVENUES IN
 DECIMAL FORM ?.08
INPUT THE ANNUAL COST OF UTILITIES IN REAL DOLLARS
 ?2500
INPUT EFFECTIVE ANNUAL ESCALATION RATE FOR UTILITIES
 IN DECIMAL FORM ?.15
INPUT THE ANNUAL COST OF LABOR IN REAL DOLLARS ?2500
INPUT EFFECTIVE ANNUAL ESCALATION RATE FOR LABOR IN
 DECIMAL FORM ?.055
INPUT THE ANNUAL COST OF MATERIALS IN REAL DOLLARS
 ?1700
INPUT EFFECTIVE ANNUAL ESCALATION RATE FOR MATERIALS
 IN DECIMAL FORM ?.094
INPUT THE ANNUAL LEASE COST IN ACTUAL DOLLARS ?800
INPUT NUMBER OF YEARS BEFORE THE LEASE HAS TO BE
 RENEGOTIATED ?5
```

OTHER INFORMATION

```
INPUT INCREMENTAL INCOME TAX RATE IN DECIMAL FORM ?.50
INPUT TAX RATE ON CAPITAL GAIN OR LOSS IN DECIMAL FORM
 ?.28
INPUT INFLATION RATE FOR REPLACEMENT INVESTMENTS IN
 DECIMAL FORM ?.08
INPUT THE M.A.R.R. FOR AFTER-TAX ANALYSIS IN DECIMAL
 FORM ?.1448
```

FIGURE 10-2 (continued)

ACTUAL DOLLAR ANALYSIS

YR	UTILITY -COST-	LABOR COST-	MATERIAL --COST--	LEASE COST-	REVENUE	BTCF
0	0.00	0.00	0.00	0.00	-20000.00	-20000.00
1	-2875.00	-2637.50	-1859.80	-800.00	16200.00	8027.70
2	-3306.25	-2782.56	-2034.62	-800.00	17496.00	8572.57
3	-3802.19	-2935.60	-2225.88	-800.00	18895.68	9132.01
4	-4372.52	-3097.06	-2435.11	-800.00	20407.33	9702.65
5	-5028.39	-3267.40	-2664.01	-800.00	22039.92	10280.12
6	-5782.65	-3447.11	-2914.42	-1175.46	23803.11	10483.47
7	-6650.05	-3636.70	-3188.38	-1175.46	25707.36	11056.77
8	-7647.56	-3836.72	-3488.09	-1175.46	27763.95	11616.13
9	-8794.69	-4047.74	-3815.97	-1175.46	29985.07	12151.21
10	-10113.89	-4270.36	-4174.67	-1175.46	36701.87	16967.49

Yr	BTCF	DEPR	TAXABLE INCOME	INCOME TAXES	ATCF	PRESENT VALUE
0	-20000.00	0.00	0.00	2000.00	-18000.00	-18000.00
1	8027.70	-4000.00	4027.70	-2013.85	6013.85	5253.19
2	8572.57	-6400.00	2172.57	-1086.28	7486.28	5712.25
3	9132.01	-4800.00	4332.01	-2166.01	6966.01	4642.96
4	9702.65	-3200.00	6502.65	-3251.32	6451.32	3756.04
5	10280.12	-1600.00	8680.12	-4340.06	5940.06	3020.94
6	10483.47	0.00	10483.47	-5241.73	5241.73	2328.61
7	11056.77	0.00	11056.77	-5528.39	5528.39	2145.31
8	11616.13	0.00	11616.13	-5808.06	5808.06	1968.77
9	12151.21	0.00	12151.21	-6075.61	6075.61	1798.97
10	16967.49	0.00	16967.49	-8483.74	8483.74	2194.28

NET PRESENT VALUE   14821.32

YOU HAVE A WINNER. THE NET PRESENT VALUE OF THE INVESTMENT IS
POSITIVE. WOULD YOU LIKE TO EVALUATE ANOTHER NVESTMENT (YES=1,
NO=2) ?2

FIGURE 10-2 (continued)

THE EFFECT OF DEPLETION ALLOWANCES

Income from investment in certain natural resources is subject to depletion allowances before income taxes are computed. Under certain conditions, notably where the taxpayer is in a relatively high tax bracket, depletion provisions in the tax law can provide considerable savings.

As an example, consider the case of a single individual who has a present net taxable income of $200,000. He spends $800,000 to drill and develop a new oil well that has an estimated reserve of 200,000 barrels. Oil is produced and sold in accordance with the schedule shown in column 2 of Table 10-12 to produce the gross income shown in column 3. Column 4 shows the net cash flow after production costs have been deducted.

The depletion allowance that can be deducted in a given year may be based on a fixed percentage of the gross income (22% for oil), provided that the deduction does not exceed 50% of the net income before such deduction. Depletion, computed in this manner, is shown in column 7 of Table 10-12. Another method is to base depletion on the estimated investment cost of the product. In this case the estimated 200,000 barrels of oil in the well cost $800,000. Depletion may therefore, if desired, be charged at the rate of $4.00 per barrel, as shown in column 6 of the same table.

The net taxable income resulting from the most favorable application of these depletion allowances is shown in column 8. The advantage of depletion allowances stems from the fact that the total depletion that can be claimed often exceeds the depreciable capital investment. In this case, the summation of the deductions claimed (denoted a in column 7 and b in column 5) is $870,500, even though the total investment was only $800,000. Inasmuch as the taxpayer has a net taxable income of $200,000 even before returns from the oil well are considered, use of Table 10-2 shows that the incremental tax rate is 50% in 1984, thus giving the income tax shown in column 9 of Table 10-12. Column 10 shows the net after-tax cash flow provided to the investor. By equating the sum of the present values of the cash flows shown to the $800,000 of investment, we find that the I.R.R. after taxes is roughly 42%. It is of interest to note that the before-tax I.R.R.—by use of the net cash flow information in column 4— is roughly 76%.

TAX-EXEMPT INCOME

Interest received on bonds or other obligations of states, U.S. territories or the District of Columbia, or political subdivisions thereof (such as cities and counties) is exempt from federal income tax. Obviously, where one alternative in an economy study involves income upon which no tax is paid and the income from another alternative is taxable, these facts must be given proper consideration. Ordinarily, the interest rate of such tax-exempt securities is from 7 to 10%. An individual who derives his entire income from such securities would thus pay no income tax. In some instances the net income so derived, despite the rela-

TABLE 10-12 Capital Recovery Provided by an Oil Well, with Cost and Percentage Depletion Allowances Used in Computation of Income Taxes

(1) Year	(2) Barrels of Oil Sold	(3) Gross Income (Cash Flow)	(4) Net Income (Cash Flow) Before Income Taxes	(5) 50% of Net Income	(6) Cost Depletion at $4.00 per Barrel	(7) Depletion Allowance at 22% of Gross Income	(8) [= (4) − either (6) or (7)] Net Taxable Income	(9) [= −(8) × 50%] Income Tax	(10) [= (4) + (9)] Net Cash Flow After Taxes
1	70,000	$1,400,000	+$800,000	$400,000	$280,000	$308,000[a]	$492,000	−$246,000	$554,000
2	60,000	1,200,000	+700,000	350,000	240,000	264,000[a]	436,000	−218,000	482,000
3	45,000	900,000	+480,000	340,000	180,000	198,000[a]	282,000	−141,000	339,000
4	20,000	400,000	+240,000	120,000	80,000	88,000[a]	152,000	−76,000	164,000
5	5,000	85,000	+25,000	12,500[b]	20,000	18,700	12,500	−6,250	18,750

[a] Percentage depletion shown deducted.
[b] Only amount shown can be deducted because of percentage depletion exceeding 50% of net income.

tively low interest rates received, may be greater than the income that would be obtained from investments yielding larger interest rates but subject to income tax. For an individual with a moderately high income, tax-exempt investments may be quite attractive.

EXAMPLE 10-15

Consider the case of a married individual who already has a taxable income of $44,000 and is planning to invest $5,000 of capital. Opportunity A offers a prospective before-tax return of 14% per year on a nondepreciable investment. Opportunity B is to buy tax-exempt bonds that pay interest at the rate of 10%. From Table 10-2 it is seen that this individual in 1984 will have to pay income taxes at the rate of 33% on the incremental taxable income. Her net income, after taxes, from opportunity A would be as follows:

$$
\begin{aligned}
\text{Gross income} &= \$5,000 \times 0.14 & \$700 \\
\text{Additional income tax} &= \$700 \times 0.33 & -231 \\
\text{NET INCOME AFTER TAXES} & & \$469
\end{aligned}
$$

This net after-tax income is a return of only $469/\$5,000 = 9.38\%$, so the after-tax 10% offered by opportunity B is higher. ∎

In most business investment problems, however, the alternative of investing capital in tax-exempt securities is not of sufficient attractiveness to merit much consideration. Nevertheless, investment in tax-exempt securities is a possibility that should not be overlooked, particularly if low risk of loss is desired.

PROBLEMS

Unless instructed to the contrary, use 1986 A.C.R.S. percentages from Table 9-3 (where appropriate) and other provisions set forth by ERTA of 1981.

10-1 If the incremental federal income tax rate is 46% and the incremental state income tax rate is 10%, what is the effective combined income tax rate (t) when state income taxes are deductible from federal income taxes?

10-2 A firm's total operating revenues in 1984 were $11,240,000. The sum of all expenditures (excluding capital expenditures) was $5,890,000. Depreciation for the entire year came to a total of $3,415,000.
(a) Compute the federal income taxes payable in 1984.
(b) What was this firm's average income tax rate in 1984?
(c) What was this firm's net income after taxes?
(d) Determine the after-tax cash flow for the firm.

10-3 Consider a firm that in 1984 had a taxable income of $90,000 and total operating revenues of $220,000. By referring to tax tables in Chapter 10, answer these questions.
(a) What amount of federal income taxes was paid in 1984?
(b) What was the net income (after federal income taxes) in 1984?
(c) What was the total amount of deductible expenses (e.g., materials, labor, fuel, interest) and depreciation claimed in 1984?

10-4 A married woman, having a taxable income of $30,000 per year, is considering reinvesting $15,000 of her capital, which now is in a savings and loan association drawing 7% effective annual interest. One possibility is to invest it in tax-exempt bonds that would pay $4\frac{1}{2}\%$ interest. Another possibility is to invest it in some property that probably could be sold at the end of 1 year for $16,000. Capital gains tax, at the minimum rate, would have to be paid on the gain. Which use for the capital will be most profitable for her?

10-5 A corporation has estimated that its taxable income will be $57,000 in 1985. It has the opportunity to invest in a business venture that is expected to add $8,000 to next year's taxable income. How much in federal taxes will be owed *with* and *without* the proposed venture?

10-6 Bob Brown had a taxable income for 1984 of $30,000, derived from his salary and interest from investments. For the same year, Joe Jones had a taxable income of $27,000 composed of $12,000 from salary and a long-term capital gain of $15,000. Both men are married.
(a) Which man had the greater after-tax income?
(b) What effective tax rate did Joe pay on the $15,000 capital gain?

10-7 Greta Goodtone, who is married, had such a large income that she was in the 50% income tax bracket. Her business manager convinced her to purchase a farm for $75,000, with the idea of improving it over a 4-year period and then selling it for a capital gain. Greta bought the farm, and she immediately spent an additional $15,000 for structural improvements. During the next 4 years the annual operating costs of the farm exceeded the revenues by $15,000 per year. However, Greta was able to deduct these annual losses on her annual income tax returns and also to claim $5,000 each year as depreciation expense. At the end of the fourth year she sold the farm for $150,000.
(a) What was the before-tax internal rate of return on the investment in the farm?
(b) What was the after-tax internal rate of return if the whole gain was taxed at a rate of 20%?
(c) What was the after-tax internal rate of return if she was charged 50% for "recovery of previously charged depreciation" and the remainder of the gain was taxed at 20%? (See Appendix 10-A.)

10-8 (a) Amanda Plumrose bought some common stock 10 years ago for $5,000. She decides to sell the stock now for a long-term capital gain amounting to $3,000. Furthermore, the taxable income from her job is $8,000. At the end of the year, how much in taxes will she owe the federal government?
(b) A certain taxpayer has a taxable income that is subject to an average tax rate of 38% and an incremental rate of 45%. He has the chance to either invest in tax-free school bonds yielding 8% or to lend money at interest to a company. If he lends to the firm, what interest rate is necessary to yield him the same after-tax return as the municipal bonds?

10-9 An individual who was, and is, in the 50% income tax bracket, purchased some non-depreciable property for $20,000. Two weeks later he had an opportunity to sell it for $23,000. He held the property for 4 years and then sold it for $35,200. In the meantime he paid annual property taxes of $600. His objective in keeping the property was to achieve a 10% return on his capital, after taxes. Did he achieve his objective?

10-10 A piece of office equipment is being purchased for $20,000 and has a 5-year recovery period for A.C.R.S. depreciation purposes. The double declining balance method of depreciation can also be used with a useful life of 5 years. Compare the present worth of depreciation write-offs for both methods when the after-tax M.A.R.R. is 10%. Which method is better?

10-11 A corporation built a warehouse at a cost of $100,000, estimating that it would have a depreciable life of 10 years and a salvage value of $50,000 at that time. It charged depreciation on a straight line basis. At the end of 5 years it had no further need for the warehouse and was able to sell it for $120,000. Assuming that the corporation was in the 50% ordinary income tax bracket, what was the net after-tax result from the sale? Long-term capital gains are taxed at 28%. Refer to Appendix 10-A concerning "recovery of previously charged depreciation."

10-12 A small, local TV repair shop is planning to invest in some new circuit testing equipment. The details of the proposed investment are:

first cost = $5,000 (does not qualify for an investment tax credit)

salvage value = 0

extra revenue = $2,000/year

extra expenses = $800/year

expected life = 5 years (also equal to A.C.R.S. recovery period)

effective income tax rate = 30%

(a) If A.C.R.S. depreciation is used, calculate the present worth of after-tax cash flows when the M.A.R.R. (after taxes) is 12%. Should the equipment be purchased?

(b) If double declining balance depreciation is used, calculate after-tax cash flows and recommend whether the new equipment should be purchased if the after-tax M.A.R.R. is 12%. Use the present worth method.

10-13 A firm must decide between two systems, A and B. Their effective income tax rate is 50% and sum-of-the-years'-digits depreciation is used. Ignore the possibility of investment tax credits and capital gains/losses. If the after-tax desired return on investment is 10%, which system should be chosen?

	System A	System B
1. Initial cost	$10,000	$20,000
2. *Tax* life	4 years	4 years
3. *Useful* life	5 years	6 years
4. Salvage value for tax purposes, realized at end of *useful* life	+$10,000	+$10,000
5. Annual revenues less disbursements	$2,000	$4,000

10-14 You have a piece of equipment with a present book value of $192,000. Next year's depreciation will be $96,000. You can sell the equipment now for $80,000 or you can sell it one year from now for the same amount. If you do not sell it now, you will definitely sell it next year. How much before-tax cash flow must the equipment produce for you over the next year (assume that it all comes at the end of the year) to justify keeping the equipment for one more year? Assume an after-tax M.A.R.R. of 25%. Depreciation recapture, if any, is taxed at an effective income tax rate of 50%.

10-15 A corporation purchases a $60,000 asset with a CLADR guideline life of 3 years and a salvage value of $10,000. No investment tax credit can be claimed.

(a) If after 3 years the asset is sold for $20,000, what is the recaptured depreciation when the sum-of-the-years'-digits depreciation method is used? Assume that all accounting adjustments between book value and terminal market value occur at the time the asset is sold.

(b) If $t = 50\%$, calculate the after-tax present worth of this investment when the after-tax M.A.R.R. is 15%.

10-16 A company currently has an *average* income tax rate of 28%. Determine the taxable income that corresponds to this value, assuming that the company has only *ordinary* taxable income. Use Table 10-1 and the 1983 rates shown there to answer this question.

10-17 A certain rental property has a first cost of $50,000. If this property is purchased, it is believed that the property will be held for 10 years and then sold for an estimated $30,000. The estimated receipts from rentals are $10,000 a year throughout the 10 years. Estimated annual upkeep costs are $3,000 the first year and will increase by $300 each year to a figure of $5,700 in the tenth year. In addition, it is estimated that there will be a single nonrecurrent outlay of $2000 for maintenance overhaul at the end of the fifth year. Assume A.C.R.S. depreciation with a 10-year recovery period and an effective income tax rate of 50%. Assume that the $2,000 overhaul cost can be treated for tax purposes as a current expense in the fifth year.

If an investor has a 10% after-tax minimum attractive rate of return, should he purchase the rental property? Show all calculations.

10-18   Currently, a firm has annual operating revenues of $190,000, cost of sales is $50,000, and accounting depreciation charges are running at $40,000 annually. A new project is proposed that will raise revenues by $30,000 and cost of sales will be increased by $10,000. If this new project necessitates a total capital cost of $50,000 which can be depreciated to zero salvage value at the end of its 5-year life, what is the external rate of return on this proposed project after federal income taxes are paid? Assume that A.C.R.S. depreciation is used and a 10% investment tax credit can be claimed. Recovered funds from the project will be invested outside the firm at an after-tax rate of 10%.

10-19   A company must purchase a particular asset for $10,000. The A.C.R.S. recovery period of the asset is 5 years, and it will have a salvage value of zero in computing its depreciation charges. An investment tax credit of 10% can be applied to this asset. However, management of this company believes the asset will have an actual market value of $2,000 at the end of its 5-year life. The annual operating and maintenance costs are $2,000 the first year and increase by $200 per year thereafter. With M.A.R.R. = 12% after taxes and an effective income tax rate of 0.50, calculate the after-tax equivalent annual cost of this asset. Assume that the company is profitable in its other activities, and that recaptured depreciation is taxed at 50%.

10-20   You can purchase your own machine for $12,000 to replace a rented machine. The rented machine costs $4,000 per year. The machine that you are considering would have a life of 8 years and a $5,000 salvage value at the end of the life. By how much could annual operating expenses of the purchased machine increase and still provide a return of 10% after taxes? Assume that you are in a 50% tax bracket and the revenues produced with either machine are identical. Assume that straight line depreciation will be utilized to recover the investment in the machine and the CLADR class life is 8 years.

10-21   Your boss asked you to determine whether the company can justify the purchase of a special piece of earth-moving equipment. The data for the problem are as follows:

First cost of equipment:	$20,000
Depreciation:	Straight line with $2,000 salvage value, 10-year life
Property taxes and insurance:	$500/year
Maintenance cost:	$400 the first year, increasing $40 per year thereafter
Operating savings (due to increased productivity):	$7,000 the first year, decreasing $300 per year thereafter

Even though government (IRS) guidelines suggest that the salvage value should be no less than 10% of the equipment's first cost (i.e., $2,000) the boss expects there to be *no* salvage value in reality. Assume that the boss is correct, and make an after-tax study for the equipment when the effective income tax rate is 0.50, after-tax M.A.R.R. = 10%, and capital losses can be used to offset capital gains, which are taxed at 28%. Utilize a 10% investment tax credit in your analysis.

10-22   (a) Refer to Example 10-12 and determine the best press using the E.R.R. method when $e = 10\%$.
   (b) Refer to Example 10-13. Rework it on an after-tax basis with an inflation rate of zero and a real M.A.R.R. (i_r) of 10%. Use the present worth method.

10-23   An injection molding machine can be purchased and installed for $90,000. It has an A.C.R.S. recovery period of 5 years (use 1986 percentages) but will be kept in service for 8 years. The salvage value for calculating depreciation is 0, but it is believed that $10,000 can be obtained when the machine is disposed of at the end of year 8. An investment tax credit of 10% can be taken at the time of purchase. The net annual "value added" (i.e., receipts less disbursements) that can be attributed to this machine is constant over 8 years and amounts to $15,000 in actual dollars. An effective income tax rate of 50% is used by the company and the after-tax M.A.R.R. (which is an i_c) equals 15%. The general inflation rate (f) is expected to average 6% over the next 10 years.

(a) What is the approximate value of the company's *before-tax* M.A.R.R.?
(b) What is the amount of the investment tax credit that can be claimed?
(c) Determine the A.C.R.S. depreciation amounts in years 1 to 5.
(d) What is the taxable income at the end of year 8 that is capital-investment related?
(e) What is the firm's real (inflation free) after-tax M.A.R.R.?
(f) Set up a table and calculate the after-tax cash flows for this machine. Should a recommendation be made to purchase the machine?

10-24 Individual industries will use energy as efficiently as it is economical to do so, and there are several incentives to improve the efficiency of energy consumption. One incentive for the purchase of more energy-efficient equipment is to reduce the time allowed to write off the initial cost. Another "incentive" might be to raise the price of energy in the form of an energy tax. To illustrate these two incentives, consider the selection of a new centrifugal pump for a refinery. The pump is to operate 8,000 hours per year. Pump A costs $1,600, consumes 10 hp, and has an overall efficiency of 65% (it delivers 6.5 hp). The other available alternative, pump B, costs $1,000, consumes 13 hp, and has an overall efficiency of 50% (delivers 6.5 hp). Compute the after-tax rate of return on extra investment in pump A, assuming an effective income tax rate of 50%, a tax life of 10 years [part (a) only], zero salvage values, and straight line depreciation for each of these situations:

(a) The cost of electricity is $0.04/kWh.
(b) A 6-year depreciation write-off period is allowed, the expected life of both pumps is still 10 years, and the cost of electricity is $0.04/kWh.

10-25 (a) Suppose that you have just completed the mechanical design of a high-speed auto-mated palletizer that has an investment cost of $3,000,000. The existing palletizer is quite old and has no salvage value. The new palletizer can be depreciated over 5 years with the A.C.R.S. method, and the equipment's expected *market value* at that time is $300,000. A 10% investment tax credit can be claimed at the time the equipment is purchased. The effective income tax rate is 50%. One million pallets will be handled by the palletizer each year during the expected 5-year project life.

What net annual savings per pallet (i.e., total savings less disbursements) will have to be generated by the palletizer to justify this purchase in view of an after-tax M.A.R.R. of 15%? Ignore the effects of inflation in this problem.

(b) Referring to the situation described in part (a), suppose you have estimated that these incremental savings and costs will be realized after the automated palletizer is installed:

(1) Twenty-five operators will no longer be needed. Each operator earns an average of $15,000 per year in direct wages, and company fringe benefits are 30% of direct wages.
(2) Property taxes and insurance amounting to 5% of the palletizer's installed cost will have to be paid over and above those paid for the present system.
(3) Maintenance costs relative to the present system will increase by $15,000 per year.
(4) Orders will be filled more efficiently with the automated palletizer (less waste, quicker response). These savings are estimated at $100,000 per year.

The effective income tax rate is 50%, and the equipment's life, depreciation method, after-tax M.A.R.R., and so on, are the same in part (a). If inflation is ignored and the existing system has no salvage value, what is the maximum amount of money that can be invested in the automated palletizer (in view of the savings/costs itemized above) so that a 15% after-tax return on this investment is realized?

10-26   Your company has to obtain some new production equipment for the next 6 years and leasing is being considered. You have been directed to accomplish an actual dollar after-tax study of the leasing approach. The pertinent information for the study is as follows:

> *Lease costs*: First year, $80,000; second year, $60,000: third through sixth years, $50,000 per year. Assume that a 6-year contract has been offered by the lessor that fixes these costs over the 6-year period.
>
> *Other costs* (not covered in contract): $4,000 in year 0 dollars, and estimated to escalate 10% each year.
>
> *Effective income tax rate*: 50%.

(a) Develop the actual dollar after-tax-cash-flow (A.T.C.F.) for the leasing alternative.
(b) If the real M.A.R.R. (i_r) after taxes is 5% and the annual inflation rate (f) is 9.524%, what is the actual dollar after-tax A.W.-C. for the leasing alternative?

10-27   Your company expects the market to be good over the next five-years for a new product. Thus, additional investment in new production equipment is being considered. The estimated actual dollar before-tax cash flow is shown below for the 5-year period, and the market value of the equipment at the end of 5 years is estimated to be 20 percent of initial cost:

Year k	Actual $ B.T.C.F.
0	− 200,000
1	60,000
2	′
3	′
4	′
5	60,000
5	40,000

The effective corporate income tax rate is 50 percent; book value of the equipment at end of the 5-year A.C.R.S. recovery period is $0 and 1986 recovery percentages are to be used.

(a) What is the after-tax internal rate of return (I.R.R.)?
(b) What is the after-tax external rate of return (E.R.R.)? Assume the external reinvestment rate is 12 percent.
(c) Assume the general inflation rate (f) is 10 percent. Develop the real dollar after-tax-cash flows.

10-28   A company has two different machines it can purchase to perform a specified task. Both machines will perform the same job. Machine A costs $150,000 initially while machine B (the deluxe model) costs $200,000. It has been estimated that costs will be $1,000 for machine A and $500 for machine B in the first year. Management expects these costs to increase with inflation, which is expected to average 10% per year. The company uses a 10-year study period and their effective income tax rate is 50%. Both machines qualify as 5-year A.C.R.S. property. Which machine should the company choose?

10-29   Because of tighter safety regulations, an improved air filtration system must be installed at a plant that produces a highly corrosive chemical compound. The investment cost of the system is $260,000 in today's (late 1985) dollars. The system has a useful life of 10 years and a tax life of 5 years. A.C.R.S. depreciation is used with a zero salvage value for tax purposes. However, it is expected that the market value of the system at the end of its 10-year life will be $50,000 in today's dollars. Costs of operating and maintaining (O & M) the system, estimated in today's dollars, are expected to be $6,000 per year.

Annual property tax is 4% of the investment cost and *does not inflate*. Assume that the plant has a remaining life of 20 years, and that all O & M and replacement costs inflate at 6% per year from the time of initial installation. If the effective income tax rate is 50%, set up a table to determine the after-tax cash flow over a 20-year period. The after-tax return desired on investment capital is 12%. What is the net present worth of cost of this system after income taxes have been taken into account? Assume that a 10% investment tax credit can be claimed.

CHAPTER 10 APPENDIX: TAX COMPUTATIONS INVOLVING CAPITAL GAINS AND LOSSES OF CORPORATIONS

Table 10-2 summarized tax treatment of capital gains and losses. The following discussion provides further explanation and numerical examples.

Corporations, as well as individuals, treat a net short-term capital gain the same as ordinary income for tax purposes. Further, any net long-term capital gain should be taxed at the full rate for ordinary income or, alternatively, at a flat rate of 28%, whichever is smaller.

A corporation may deduct capital losses only to the extent of its capital gains in any given year. However, any capital loss in the current year that exceeds its capital gains for that year may be carried over to subsequent years until it is completely absorbed. When carried over, the loss will retain its original character as long term or short term; thus a long-term capital loss carried over from a previous year will offset long-term gains of the current year before it offsets short-term gains of the current year.

EXAMPLE 10-A-1

A corporation had a taxable income of $120,000 in 1984, which included a $30,000 net capital gain computed as follows.

Long-term capital gain		$65,000
Long-term capital loss		(18,000)
Net long-term capital gain,		$47,000
Short-term capital gain	$10,000	
Short-term capital loss	(27,000)	
Net short-term capital loss		($17,000)
Excess of net long-term capital gain over net short-term capital loss		$30,000

What is the total income tax to be paid?

Solution
Regular tax computation:

First $25,000 at 15%	$ 3,750
Next $25,000 at 18% (taxable income in excess of $25,000)	4,500
Next $25,000 at 30% (taxable income in excess of $50,000)	7,500
Next $25,000 at 40% (taxable income in excess of $76,000)	10,000
Remaining ($120,000 − $100,000) at 46%	9,200
TOTAL TAX	$34,950

APPLICATIONS OF ENGINEERING ECONOMY

Alternative tax computation:

Taxable income	$120,000
Less: excess of net long-term capital gain over net short-term capital loss (see above)	30,000
Ordinary income	$90,000
Tax on ordinary income:	
First $25,000 at 15%	3,750
Next $25,000 at 18% (taxable income in excess of $25,000)	4,500
Next $25,000 at 30% (taxable income in excess of $50,000)	7,500
Remaining ($90,000 − $75,000) at 40%	6,000
TOTAL	$21,750
Tax on net capital gain:	
28% of $30,000	8,400
TOTAL TAX	$30,150

Since the alternative tax of $30,150 is less than the regular tax of $34,950, the corporation would choose to pay the alternative tax. ∎

When depreciable property such as buildings and equipment used in a business is sold, it can result in gains that are taxed partly at capital gain rates and partly at ordinary income rates, according to how much depreciation has been charged. The determination of how such a gain or loss affects taxes can be complicated and can be explained best by reference to Figure 10-A-1 and by assuming that straight line depreciation has been used for accounting and tax purposes.† If we

†Some complications exist for assets purchased prior to 1964 and for cases in which accelerated depreciation methods are used. When such conditions are encountered, the specific IRS rules should be consulted.

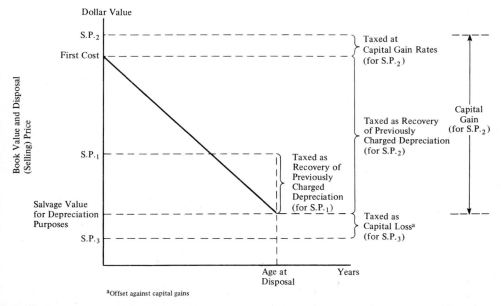

ᵃOffset against capital gains

FIGURE 10-A-1   Tax Categories for Various Possible Gains and Losses on Disposal of Depreciable Property Used in a Business.

sell the asset for a price less than the original cost but greater than the book value (S.P.$_1$ in Figure 10-A-1), the excess of selling price over book value represents a recovery of previously charged, and deducted, depreciation cost. Consequently, this excess is considered to be ordinary income and is taxed as such, since this amount had been deducted from ordinary income in prior years. If the selling price is greater than the original cost (S.P.$_2$ in Figure 10-A-1), the difference between the original cost and the book value again is recovery of previously charged depreciation, and it is taxed as ordinary income, but the excess of the selling price above the original cost is taxed in a manner similar to a capital gain. On the other hand, if we sell the asset for a price that is less than the book value (S.P.$_3$ in Figure 10-A-1), a loss results which is treated as a capital loss.

EXAMPLE 10-A-2

A corporation has equipment that cost $200,000 5 years ago. At that time a straight line depreciation schedule was set up based on an estimated 15-year life and $20,000 salvage value. The incremental tax rate on ordinary income is 46% and on capital gains or losses is 28%. Determine the income taxes to be paid (or saved) on the gain (or loss) if the property is sold now for (a) $160,000; (b) $240,000; (c) $90,000.

Solution

Total depreciation charged to now $= (\frac{5}{15})(\$200,000 - \$20,000) = \$60,000$
Book value now $\qquad\qquad = \$200,000 - \$60,000 = \$140,000$

(a) $160,000 - $140,000 = $20,000 gain due to recovery of previously charged depreciation. $20,000 $\times$ 46% = $9,200 taxes to be paid.

(b) $240,000 - $140,000 = $100,000 gain. $200,000 - $140,000 = $60,000 is recovery of previously charged depreciation. $240,000 - $200,000 = $40,000 is gain subject to capital gain rate. $60,000 $\times$ 0.46 + $40,000 $\times$ 0.28 = $38,800 taxes to be paid.

(c) $140,000 - $90,000 = $50,000 capital loss. $50,000 $\times$ 0.28 = $14,000 taxes saved. ∎

CHAPTER 11

Replacement Studies

Business firms and individuals must constantly decide whether existing assets should be continued in service or whether available new assets will better and more economically meet current and future needs. These decisions must be made with increasing frequency as the dynamic pace of business quickens and technology produces more rapid changes. Thus the *replacement problem*, as it commonly is called, requires careful engineering economy studies if we are to arrive at sound decisions.

Unfortunately, replacement problems sometimes are accompanied by unpleasant financial facts. Often it turns out that earlier decisions, particularly concerning the anticipated useful life of existing assets, were not as good as might be desired, especially when hindsight can be applied. Frequently, new and apparently superior assets unexpectedly appear on the scene. Sometimes replacement situations arise because there are completely unanticipated, and even unreasonable, changes in market conditions. Consequently, there is a tendency to regard the entire area of replacement as an emotional "bugbear," whereas in fact it often represents economic opportunity.

Economic studies of replacement alternatives are performed in the same manner as economic studies of any two alternatives—the only difference is that one alternative is to keep an *existing* (*old*) *asset*, which can be descriptively called the *defender*; and there are one or more alternative *replacement* (*new*) *assets*, which can be called *challengers*.

Replacement decisions are critically important to a firm. A hasty decision to get rid of old equipment or faddishly to buy the most modern or elaborate available equipment can be a serious drain on capital available. On the other hand, a firm may go to the other extreme by delaying replacements until it becomes noncompetitive or there is no other way to continue production. A key question to be answered in this chapter is whether the defender should be kept one or more years or immediately replaced with the best available challenger.

REASONS FOR REPLACEMENT

It should be recognized that not all existing assets replaced are physically scrapped or even disposed of by the owner. Frequently, new assets are acquired to perform the services of existing assets, with the existing assets not retired but merely transferred to some other use—often an ''inferior'' use, such as standby service. Of course, many assets are sold by their present owners and are used by one or more other owners before reaching the scrap heap.

The four major reasons for replacement are as follows:

1. *Physical impairment:* The existing asset is worn out, owing to normal use or accident, and no longer will render its intended function unless extensive repairs are made.
2. *Inadequacy:* The existing asset does not have sufficient capacity to fill the current and expected demands. Here, clearly, the requirements have changed from those that were anticipated at the time the asset was acquired. This condition does not necessarily imply physical impairment, but to meet the new demands the asset either must be supplemented or replaced.
3. *Obsolescence:* This may be of two types: (a) functional, or (b) economic. Both types result in loss of profits. In the case of functional obsolescence, there has been a decrease in the demand for the output of the asset; thus a loss in revenue follows. For example, the market may wish a higher quality product than it previously demanded. Economic obsolescence is the result of there being a new asset that will produce at lower cost than can be obtained with the old asset.
4. *Rental or lease possibilities:* This is a variation of obsolescence, except that the replacement asset does not necessarily have to be different, in any respect, from the existing asset. The possible economic advantage is due to advantageous financial factors that sometimes may accrue from leasing. These usually involve income tax considerations.

For purposes of replacement studies, the following is a distinction between economic life and other types of lives for typical assets.

Economic life (also known as service life) is the period of time extending from date of installation to date of retirement (by demotion or disposal) from the primary intended service. The need for retirement is signaled in an engineering economy study when the equivalent cost of a new asset (challenger) is less than the equivalent cost of keeping the present asset (defender) for an additional period of time.

Ownership life is the period between date of acquisition and date of disposal by a specific owner. A given asset may have several service lives for a given owner. For example, a car may serve as the primary family car for several years and then serve only for local job commuting for several more years.

Physical life is the period between original acquisition and final disposal of an asset over its succession of owners. For example, the car above may have several owners over its existence.

Depreciable life is the period over which depreciation charges are scheduled to be made.

INCOME TAXES IN REPLACEMENT STUDIES

The replacement of assets often results in capital gains or losses, or gains or losses from the sale of land or *depreciable property*, as was discussed in Chapter 10. Consequently, to obtain an accurate economic analysis in such cases the studies must be made on an *after-tax basis*. In specific cases the situation may be complicated by whether the organization has had other such gains or losses within the tax year, inasmuch as such losses must be offset against any gains before the full tax-credit benefits apply. However, because it often is difficult to determine the existence of such gains or losses throughout an organization without undue loss of time, many companies follow the practice of considering each case separately and *applying the full tax benefits or penalties* to the gains or losses associated with each study, unless the amounts involved are very large.

For simplicity, all problems discussed in this chapter utilize an ordinary income tax rate of 50% (unless otherwise noted) and a capital gains tax rate of 28%.

It is evident that the existence of a taxable gain or loss, in connection with replacement, can have a considerable effect on the results. A prospective gain from the disposal of assets will be reduced by 50% in most cases. Correspondingly the net after-tax effect of a loss from the disposal of an asset is substantially diminished when 28% of the loss is recovered through a tax credit. Consequently, one's normal propensity for disposal or retention of existing assets can be influenced considerably by these tax effects.

The correct consideration of the income tax aspects of retirement and replacement situations may well result in conclusions exactly opposite from the conclusions one may reach by intuition or even by conducting a before-tax study. For example, intuition often favors disposal at an apparent profit. However, a taxable gain reduces the net after-tax cash realized from such a disposal. Further, although intuition often runs counter to disposal of an asset at an apparent loss, a loss that is deductible from taxable income increases the net after-tax cash realized from such a disposal.

FACTORS THAT MUST BE CONSIDERED IN REPLACEMENT STUDIES

Following is a list of frequently encountered factors that must be considered in replacement studies.

1. Recognition and acceptance of past "error."
2. The possible existence of a sunk cost.
3. Remaining life of the old asset (defender).
4. Economic life for the proposed replacement asset (challenger).
5. Method of handling unamortized values.
6. Possible capital gains or losses.

Once a proper viewpoint has been established with respect to these items, little difficulty is experienced in making replacement studies and in arriving at sound decisions.

RECOGNITION OF PAST "ERRORS" AND SUNK COSTS

When an asset's book value is greater than its current market value, the difference frequently has been designated as a past "error." Such "errors" also arise when capacity is inadequate, maintenance costs are higher than anticipated, and so forth. This designation is unfortunate because in most cases these differences are not the result of errors but of honest inability to foresee future conditions. The distinction is important in establishing a proper perspective in replacement analyses:

Any unamortized values (i.e., undistributed depreciation of an asset's initial cost) arising from an existing asset under consideration for replacement are strictly the result of *past* decisions—the initial decision to invest in that asset and decisions as to the method and number of years to be used for depreciation purposes. Such losses are *sunk costs* and thus have no relevance to the replacement decisions that must be made (*except to the extent that they result in a tax saving*).

Acceptance of these facts may be made easier by posing a hypothetical question: "What will be the costs of my competitor whose similar property is completely depreciated and who therefore has no past 'errors' to consider?" In other words, we must decide whether we wish to live in the *past*, with its errors and discrepancies, or to be in a sound competitive position in the *future*. A common reaction is: "I can't afford to take the loss in value of the existing asset that will result if the replacement is made." The fact is that the loss already has occurred, whether or not it could be afforded, and it exists whether or not the replacement is made.

It must be remembered, of course, that the existence of a sunk cost may mean that there will be a resulting income tax saving that should be considered. This is illustrated in the after-tax solution to Example 11-2 to be presented shortly.

INVESTMENT VALUE OF EXISTING ASSETS FOR REPLACEMENT STUDIES

Recognition of the nonrelevance of book values and sunk costs leads to the proper value to be placed on existing assets for replacement study purposes. Clearly, the value of an existing asset at any time is determined by either (1) what amount of money will be tied up in it if it is kept rather than disposed of, or (2) the present worth of any income that can be obtained in the future through its profitable use. For replacement study purposes the second measure of value is not relevant; we do not yet know whether the asset should be retained—a requirement for it to earn future profits. Instead, we wish to determine whether it should be retained or replaced. Consequently, it is certain that the *minimum* present value upon which capital recovery costs (depreciation and profit) can be based is the amount of capital that will be tied up in the asset if it is retained, but that could be recovered immediately if it were replaced. Thus the *present realizable* market value (modified by any income tax effects) is the correct investment amount to be assigned to an existing asset in replacement studies. A good way to reason that this is true is to make use of the *opportunity cost* or *opportunity forgone* principle. That is, if it should be decided to keep the existing asset, one is giving up the opportunity to obtain the net realizable market value at that time. Thus this represents the *opportunity cost* of continuing to keep the defender.

There is one addendum to the rationale stated above: If any new investment expenditure (such as for overhaul) is needed to upgrade the existing asset so that it will be competitive in level of service with the challenger, the extra amount should be added to the present realizable market value to determine the total investment in the existing asset for replacement study purposes.

Application of the foregoing principle does not rule out the possibility that the true value of an existing asset, *as an operating property*, may turn out to be greater than the investment value assigned to it for economy study purposes. If this happens, then one can say that, on the basis of the investment value used for replacement study purposes, the asset turned out to earn a greater rate of profit than anticipated.

EXAMPLE 11-1

The investment cost of a machine purchased 5 years ago was $20,000. It has been depreciated by the straight line method at $2,000 per year and its current book value is $10,000. The market value of the machine, if sold now, is $5,000 and it would cost $2000 to overhaul the machine to make it serviceable for another 5 years. What is the investment cost and unamortized value of the existing machine?

Solution

The investment in an existing asset is its present realizable market value plus any required expenditures to make it serviceable (and comparable) relative to new machines that may be available. Therefore, the investment cost of keeping the present machine is $5,000 + $2,000 = $7,000. If this machine were sold for $5,000, the unamortized value would be $10,000 − $5,000 = $5,000. ∎

METHODS OF HANDLING LOSSES DUE TO UNAMORTIZED VALUES

Although unamortized values are sunk costs and have no place in before-tax replacement studies, they do cause considerable concern on the part of business owners and managers. Therefore, an understanding of how they are handled may provide some peace of mind to those making engineering economy studies. Several methods are used for dealing with unamortized values. Some are sound; others are incorrect.

One incorrect method is, in effect, a pretense that no unamortized loss has occurred. This is accomplished by using book value as the investment cost for the old equipment in the replacement study. People who do this usually justify their action by saying that if the old equipment is kept, they can continue to charge depreciation on it until the entire amount of invested capital is recovered. Actually, such wishful thinking could only be realized if one had no competition. If a competitive situation exists, one's competitor may obtain the more economical new equipment and be able to sell at lower prices.

A second incorrect method of handling unamortized values involves adding them to the cost of the challenger and then computing future depreciation costs with this total if the replacement is made. Users of this procedure believe that they can thus require the challenger to repay the unamortized value "that the replacement has caused." This line of reasoning is fallacious on two counts. First, the unamortized value is a fact, regardless of whether the replacement is made or not. Second, if one has competition, we probably will not be able to make the challenger pay the sunk cost. (One's selling price usually is dictated by competition and not by one's desire to recover a sunk cost.)

The correct handling of unamortized values requires that they be recognized for what they are—losses of capital. One of the most common procedures is to charge them directly to the profit and loss account. This considers them a loss in the current operating period and thus deducts them from current earnings. This is a satisfactory procedure except for the fact that unusually large losses will have a serious effect upon the stated accounting profits of the current period.

A more satisfactory method is to provide a surplus account against which such losses may be charged. Thus a certain sum is set aside each accounting period to build up and maintain this surplus. When any losses occur from unamortized values, they are charged against the surplus account and current profits are not affected seriously. The periodic contribution to the surplus account is in the nature of an insurance premium to prevent profits from being affected by losses from unamortized values.

A TYPICAL REPLACEMENT SITUATION

The following is a typical replacement situation used to illustrate a number of factors that must be considered in replacement studies. It is first solved on a before-tax basis. Then an after-tax analysis is performed to include A.C.R.S. depreciation for the new (challenger) asset and a pre-1981 ERTA depreciation method for the defender. Such a mixture of depreciation methods will be rather common throughout the 1980s.

EXAMPLE 11-2

In early 1984 an operating manager of a large chemical plant became concerned about the performance of an important pump in a cascading process. He presents his chief engineer with the following information and requests that a replacement analysis be performed.

Five years ago the plant purchased pump A, including a driving motor, for $17,000. At the time of purchase it was estimated that it would have a CLADR life of 10 years from Table 9-1 and a salvage value at the end of that time of 10% of the initial cost. Depreciation has been charged in the accounting records on this basis by use of the *straight line* procedure. Considerable difficulty has been experienced with the pump; annual replacement of the impeller and bearings has been required at a cost of $1,750. Normal annual operating and maintenance costs have been $3,250. Annual taxes and insurance are 2% of the first cost. It appears that the pump will continue to operate for another 10 years, the foreseeable demand period, if the present maintenance and repair practice is continued. It is estimated that if it is continued in service its ultimate market, or salvage, value will be about $200.

An alternative to keeping the existing pump in service is to immediately sell it and to purchase a new and different type of pump (pump B) for $16,000. A cash market value of $750 could be obtained for the existing pump. A 10-year useful life would be assigned to the new pump and an estimated market value would be 10% of the initial cost at the end of year 10. The new pump has an A.C.R.S. recovery period of 5 years and qualifies for A.C.R.S. depreciation. Operating and maintenance costs for the new pump are estimated at $3,000 per year. Annual taxes and insurance would total 2% of first cost.

This company is in a 46% effective income tax bracket on ordinary income and pays 28% on long-term capital gains. It has a M.A.R.R. of 12% on its capital before taxes and 6% after taxes. No state or local income taxes are applicable. The data for Example 11-2 are summarized in Table 11-1 on the following page.

In a before-tax analysis of the defender and challenger, care must be taken to identify correctly the investment amount in the existing pump. From the problem statement, this would be the current market value of $750. Notice that the investment amount of pump A ignores the original purchase price of $17,000 and the present book value of $9,350.

By using the principles discussed thus far in this chapter, a before-tax analysis of *equivalent annual costs* (A.C.) of pump A and pump B can be made.

TABLE 11-1 Summary of Information for Example 11-2

Existing pump A (defender)

Investment when purchased 5 years ago		$17,000
Estimated useful life when originally purchased		10 years
Estimated salvage value at the end of 10 years		$1,700
Annual depreciation $(17,000 - 1,700)/10$		$1,530
Annual disbursements:		
Replacement of impeller and bearings	$1,750	
Operating and maintenance	3,250	
Taxes and insurance: $17,000 \times 2\%$	340	
		$5,340
Present market value		$750
Market value at the end of 10 additional years		$200
Current book value $[\$17,000 - 5(\$1,530)]$		$9,350
Book value at the end of 10 additional years		$1,700

Replacement pump B (challenger)

Investment		$16,000
Estimated useful life		10 years
A.C.R.S. recovery period		5 years
Investment tax credit (10%)		$1,600
Estimated market value at the end of 10 years		$1,600
Annual disbursements:		
Operating and maintenance	$3,000	
Taxes and insurance: $16,000 \times 2\%$	320	
		$3,320

Effective income tax rate = 46%
M.A.R.R. (before taxes) = 12%
  and M.A.R.R. (after taxes) = 6%

Solution of Example 11-2 by A.C. Method (Before Taxes)

	Keep Old Pump A	Replacement Pump B
Annual costs:		
Disbursements	$5,340	$3,320
Capital recovery		
$(\$750 - 200)(A/P, 12\%, 10) + \$200(0.12)$	121.40	
$(\$16,000 - 1,600)(A/P, 12\%, 10) + \$1,600(0.12)$		2,740.80
TOTAL	$5,461.40	$6,060.80

Since $5,461.40 < \$6,060.80$, the replacement pump apparently is not justified and the defender should be kept at least one more year. We could also make the analysis using other methods (e.g., I.R.R.) and the indicated choice would be the same. ∎

It now must be acknowledged that a before-tax analysis often is not a valid basis for a replacement decision because of the effect of depreciation and any significant capital gain or capital loss (unamortized value) on income taxes. An after-tax analysis is a valid basis for such decisions and is shown below.

Solution of Example 11-2 by A.C. Method (After Taxes)

Table 11-2 and 11-3 show computations of the after-tax cash flows for the existing and replacement pumps of Example 11-2. Note in Tables 11-2 and 11-3 that the amounts for the alternative "Keep old pump A" on the line for year 0 have reversed signs in parentheses. This is because, as a computational convenience, the amounts first were determined on the basis of selling or replacing the existing pump A. *The reversed signs in parentheses indicate the opposite effect, which would result from keeping pump A.* Also shown in Tables 11-2 and 11-3 are the after-tax investment in the replacement pump B and the after-tax market values for both alternatives.

After the one time and annual effects have been determined in Tables 11-2 and 11-3, the third step in an after-tax replacement study involves the equivalence calculations, which basically are no different from any other study of alternatives, except, of course, that an after-tax minimum attractive rate of return is used. The following is the after-tax A.C. analysis for Example 11-2.

$$\text{A.C. of defender (pump A)} = \$3,158(A/P, 6\%, 10) + \$2,180$$
$$+ \$704(F/A, 6\%, 5)(A/F, 6\%, 10)$$
$$- \$620(A/F, 6\%, 10)$$
$$= \$2,863$$

$$\text{A.C. of challenger (pump B)} = \$14,400(A/P, 6\%, 10)$$
$$+ \$689(P/F, 6\%, 1)(A/P, 6\%, 10)$$
$$+ \cdots - \$864(A/F, 6\%, 10)$$
$$= \$2,848$$

Because $2,848 is less than $2,863, the extra investment in the replacement pump is shown to be barely justified. However, other considerations, such as the improved reliability of the new pump, may cause pump B to be chosen

TABLE 11-2 After-Tax Cash Flow Computations for the Defender (Existing Pump A) in Example 11-2

Year, k	Before-Tax Cash Flow	Straight Line Depreciation	Taxable Income	Cash Flow from Income Tax (46%)	After-Tax Cash Flow
0	$(-)\$\ 750$		$(+) - \$8,600^a$	$(-) + \$2,408^a$	$(-)\$3,158$
1–5	$-\ 5,340$	$-1,530$	$-\ 6,870$	$+\ 3,160$	$-\ 2,180$
6–10	$-\ 5,340$	0	$-\ 5,340$	$+\ 2,456$	$-\ 2,884$
10	$\$\ 200$		$-\$1,500^b$	$+\$\ 420^b$	$+\ \$\ 620$

^aIf sold for $750 there would be an $8,600 long-term capital loss on disposal. If $8,600 long-term capital gains elsewhere were offset by this loss, there would be no tax on the capital gain, saving 28% × $8,600 = $2,408. If pump A is not sold, this loss is not realized and the resulting income taxes on long-term capital gains would be $2,408 higher than if the pump had been sold.

^bIf sold for $200 there would be a $1,500 ($200 − $1,700) long-term capital loss on disposal. This would offset 28% × $1,500 = $420 in capital gains taxes.

TABLE 11-3 After-Tax Cash Flow Computations for the Challenger (Replacement Pump B) in Example 11-2

Year, k	Before-Tax Cash Flow	Investment	×	Depreciation 1984 A.C.R.S. Percentages	=	Amount	=	Taxable Income	Cash Flow from Income Tax (46%)	After-Tax Cash Flow
0	−$16,000								+$1,600[a]	−$14,400
1	− 3,320	$16,000	×	15%	=	−$2,400		−$5,720	+ 2,631	− 689
2	− 3,320	16,000	×	22%	=	− 3,520		− 6,840	+ 3,146	− 174
3	− 3,320	16,000	×	21%	=	− 3,360		− 6,680	+ 3,071	− 249
4	− 3,320	16,000	×	21%	=	− 3,360		− 6,680	+ 3,071	− 249
5	− 3,320	16,000	×	21%	=	− 3,360		− 6,680	+ 3,071	− 249
6–10	− 3,320					0		− 3,320	+ 1,527	− 1,793
10	+ 1,600							+ 1,600[b]	− 736[b]	+ 864

[a]Investment tax credit of 10% for a 5-year A.C.R.S. asset.
[b]If sold for $1,600, there is a $1,600 recovery of depreciation taxable at 46%.

regardless of its economic advantage. The after-tax annual costs of both alternatives are considerably less than the before-tax annual costs.

Also notice that the after-tax analysis *reverses* the results of the before-tax analysis for the same problem. As suspected, identical results would not necessarily always be obtained, particularly for analyses in which one or more alternatives involve large unamortized values that are taxed at rates markedly different from the rates on ordinary income. ∎

We will now describe how to determine economic lives of depreciable assets and how economic analyses should be performed in the frequent situation when the economic life of the defender differs from that for the challenger. Several of the examples illustrating these principles do not explicitly include income taxes. However, the principles used in determining economic lives apply on an after-tax basis whenever income taxes are applicable.

DETERMINING THE ECONOMIC LIFE OF A NEW ASSET OR CHALLENGER

In all engineering economy studies up to this point, we have made the analyses with assumed "useful lives" for all alternatives under consideration, including challengers in replacement studies. For any new asset, an *economic life* can be computed if various operating and maintenance costs are known (or can be estimated) in addition to its year-by-year salvage values. The economic life minimizes the equivalent cost of owning and operating an asset, and it is often shorter than ownership and/or physical life. *Determination of an economic life for the challenger(s) and defender is important in a replacement analysis because of the fundamental principle that new and existing assets should be compared over their economic lives.* One reason this is often not done in practice lies in the difficulty of obtaining accurate future cost and salvage value estimates for an alternative.

Determination of economic lives through formal methods can be accomplished with before-tax or after-tax cash flow estimates. Example 11-3 illustrates one such formal method for determining the life, N^*, of a new asset at which before-tax equivalent annual costs of ownership and operation are minimized. This life is called the economic life.

EXAMPLE 11-3
A new forklift truck will require an investment of $20,000 and is expected to have year-end salvage values and disbursements as shown in columns 2 and 5, respectively, of Table 11-4. If the before-tax M.A.R.R. is 10%, how long should the asset be retained in service?

Solution
The solution to this problem is obtained by completing columns 3, 4, 6, and 7 of Table 11-4. In the solution the customary year-end occurrence of all cash flows is assumed. The depreciation for any year is the difference between the

TABLE 11-4 Determination of the Economic Life (N*) of a New Asset (Example 11-3)

(1) Year	Cost of Service for *k*th Year					Equivalent Annual Cost (A.C.) if Kept *N** Years
	(2) Salvage Value at Year End	(3) Depreciation During Year	(4) Cost of Capital at 10%	(5) Annual Disbursements	(6) [= (3) + (4) + (5)] Total Cost for Year, $T.C._k$	(7) $A.C._{\cdot N^*} = \left[\sum\limits_{k=1}^{N^*} T.C._k (P/F,\ 10\%,\ k) \right](A/P,\ 10\%,\ N^*)$
0	$20,000	—	—	—	—	—
1	15,000	$5,000	$2,000	$ 2,000	$ 9,000	$9,000
2	11,250	3,750	1,500	3,000	8,250	8,643
3	8,500	2,750	1,130	4,620	8,500	8,600 ← minimum
4	6,500	2,000	850	8,000	10,850	9,082
5	4,750	1,750	650	12,000	14,400	9,965

beginning and year-end salvage values. In this example, depreciation is not computed according to any formal method but rather results from expected forces in the marketplace. The opportunity cost of capital in year k is 10% of the capital unrecovered after cumulative depreciation at the beginning of each year has been subtracted from the original investment. The values in column 7 are the equivalent annual costs that would be incurred each year if the asset were retained in service for the number of years shown and replaced at the end of the year. The minimum annual cost occurs at end of year N^*.

From the values shown in column 7 it is apparent that the asset will have minimum annual cost if it is kept in service only 3 years (i.e., $N^* = 3$). ■

The computational approach in the example above, as shown in Table 11-4, was to determine the total annual cost for each year (sometimes called *marginal* or *year-by-year cost*) and then to convert these into an equivalent annual cost (A.C.). The A.C. for any life can also be calculated using the more familiar capital recovery formulas. For example, for a life of 2 years, the A.C. can be calculated as

$$(\$20,000 - \$11,250)(A/P, 10\%, 2) + \$11,250(10\%)$$
$$+ [\$2,000(P/F, 10\%, 1)$$
$$+ \$3,000(P/F, 10\%, 2)](A/P, 10\%, 2) = \$8,643$$

which checks with the corresponding row in column 7 of Table 11-4.

THE ECONOMIC LIFE OF A DEFENDER

In replacement analyses, we must also determine the economic (or service) life that is most favorable to the defender. When a major outlay for defender alteration or overhaul is needed, the life that will yield the least annual equivalent cost is likely to be the period that will elapse before the next major alteration or overhaul will be needed. Alternatively, when there is no defender market value now or later (and no outlay for alteration or overhaul) and when defender operating disbursements are expected to increase annually, the remaining life that will yield the least annual cost will be 1 year (or possibly less).

When market values are greater than zero and expected to decline from year to year, it may be necessary to calculate the "apparent" remaining economic life. This is done in the same manner as in Example 11-3 for a new asset. The only difference is that the present realizable market value of the defender is considered to be its investment value.

Regardless of how the "apparent" economic remaining life for the defender is determined, whenever a decision is made to keep the defender this does not mean that it should be kept only for this period of time. Indeed, the defender should be kept longer than the "apparent" economic life as long as its *marginal cost* (total cost for an additional year of service) is less than the minimum A.C. for the best challenger.

EXAMPLE 11-4

Suppose that it is desired to determine how much longer an old forklift truck should remain in service before it should be replaced by the new truck for which data were given in Example 11-3 and Table 11-4. The old truck (defender) is 2 years old, originally cost $13,000, and has a present realizable market value of $5,000. If kept, its salvage value and operating disbursements are expected to be as follows:

Year from Now	Salvage Value	Operating Disbursements
1	$4,000	$5,500
2	3,000	6,600
3	2,000	7,800
4	1,000	8,800

Determine the most economical period to keep the old truck before replacing it (if at all) with the present challenger of Example 11-3.

Solution

Table 11-5 shows the calculation of total cost for each year (marginal cost) and A.C. for each year for the defender using the same format as that used in Table 11-4. Note that the minimum A.C. of $7,000 corresponds to keeping the old truck for 1 more year. However, the marginal cost for keeping the truck for the second year is $8,000, which is still less than the minimum A.C. for the challenger (i.e., $8,600, from Example 11-3). The marginal cost for keeping the defender the third year and beyond is greater than the $8,600 minimum A.C. for the challenger, so it would be most economical to keep the defender for 2 more years and then to replace it with the challenger. This is shown graphically in Figure 11-1. ∎

Example 11-4 assumes that there is only one challenger alternative available. In this situation if the old asset (defender) is retained beyond the point where its marginal (year-by-year) costs exceed the minimum A.C. for the challenger, the difference in costs continues to grow and replacement becomes more urgent, as illustrated to the right of the intersection in Figure 11-1.

TABLE 11-5 Determination of the Economic Life of an Old Asset (Example 11-4)

(1) Year from Now	(2) Depreciation for year	(3) Cost of Capital at 10%	(4) Operating Disbursements	(5) Total Cost for Year (Marginal Cost)	(6) Equivalent Uniform Annual Cost
1	$1,000	$500	$5,500	$ 7,000	$7,000
2	1,000	400	6,600	8,000	7,475
3	1,000	300	7,800	9,100	7,966
4	1,000	200	8,800	10,000	8,406

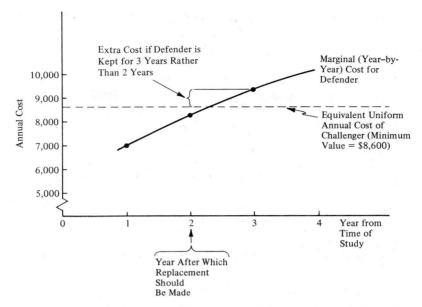

FIGURE 11-1 Defender Versus Challenger Fork Trucks (Based on Examples 11-3 and 11-4).

Figure 11-2 illustrates the effect of improved new challengers in the future. If an improved challenger X becomes available before replacement with the new asset of Figure 11-1, then a new replacement study probably should take place to consider the improved challenger. If there is a possibility of a further-improved challenger Y as of, say, 4 years later, it may be better still to postpone replacement until that challenger becomes available. Thus retention of the old

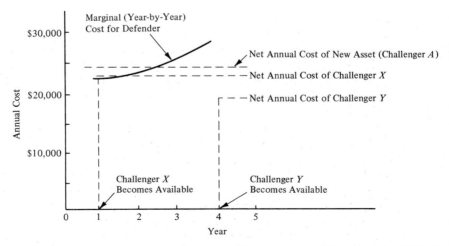

FIGURE 11-2 Old Versus New Asset Costs with Improved Challengers Becoming Available in the Future.

asset beyond its breakeven point with the best available challenger has a cost that may well grow with time. However, this cost of waiting can, in some instances, be worthwhile if it permits purchase of an improved asset having economies that offset the cost of waiting. Of course, a decision to postpone a replacement may "buy time and information" also. Because technological change tends to be sudden and dramatic rather than uniform and gradual, new challengers with significantly improved features can arise sporadically and can change replacement plans substantially.

When replacement is not signaled by the economy study, more information may become available before the next "challenge" to the defender, and hence the next comparison should include that additional information. *Postponement* generally should mean a postponement of the decision "when to replace," not the decision to postpone replacement until a specified future date.

ECONOMIC COMPARISONS IN WHICH DEFENDER ECONOMIC LIFE DIFFERS FROM THAT OF THE CHALLENGER

When the lives of the replacement alternatives are equal, as in Example 11-2, economic comparisons of alternatives can be made by any of the theoretically correct methods without complication. However, when the expected economic lives of the defender and challenger(s) differ, this should be taken into account and the analysis may be complicated.

Chapter 6 described the two assumptions that are commonly used for economic comparisons of alternatives having different lives: (1) *repeatability*, and (2) *cotermination*. The repeatability assumption involves two main subassumptions as follows:

1. The period of needed service for which the alternatives are being compared is either indefinitely long or a length of time equal to a common multiple of the lives of the alternatives.
2. What is estimated to happen in the first life span will happen in all succeeding life spans, if any, for each alternative.

For replacement analyses, the first subassumption may be acceptable, but normally the second subassumption is not reasonable for the defender. The defender is typically an older piece of equipment with a modest current realizable market value (selling price). An identical replacement, even if it could be found, probably would have an installed cost far in excess of the current market value of the defender.

Failure to meet the second subassumption can be circumvented if the period of needed service is assumed to be indefinitely long and if we recognize that the analysis is really to determine if *now* is the time to replace the defender. When the defender is replaced, it will be by the challenger—the best available replacement.

Example 11-4 involving the defender versus challenger fork trucks made use of the *repeatability* assumption. That is, it was assumed that the challenger would have a minimum A.C. regardless of when it replaces the defender. Figure 11-3 shows time diagrams of the cost consequences of keeping the defender for 2 more years versus adopting the challenger now, with the challenger costs to be repeated into the indefinite future. *It can be seen in Figure 11-3 that the only difference between the alternatives is in years 1 and 2.*

The *repeatability* assumption, applied to replacement problems, simplifies their economic comparison by permitting direct comparison of the alternatives by any correct analysis method. Whenever the *repeatability* assumption is not applicable, the *coterminated* assumption may be used, which involves using a finite study period for all alternatives.

Use of the *coterminated* assumption involves detailing what, and when, cash flows are expected to occur for each alternative and then determining which is most economical, using any of the correct economic analysis methods. When the effects of inflation are to be considered in replacement studies, it is recommended that the coterminated assumption be used.

To simplify the calculations associated with replacement analysis, the Machinery and Allied Products Institute (MAPI) has published a set of worksheets and graphical aids. A brief discussion of the MAPI method of replacement analysis is provided in Appendix D.

EXAMPLE 11-5

Suppose that we are faced with the same replacement problem as in Example 11-3 except that the period of needed service (or the firm's planning horizon) is (a) 3 years, and (b) 4 years.

Keep Defender for Two Years:

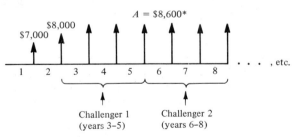

Replace with Challenger Now:

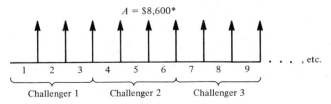

FIGURE 11-3 Effect of *Repeatability* Assumption Applied to Alternatives in Example 11-4.

Solution

(a) For a planning horizon of 3 years, one might intuitively think that either the defender should be kept for 3 years or it should be replaced immediately by the challenger to serve the next 3 years. From Table 11-5 the A.C. for the defender for 3 years is $7,966, and from Table 11-4 the A.C. for the challenger for 3 years is $8,600. Thus, following this reasoning, the defender would be kept for 3 years. However, this is not quite right. Focusing on the "Total Cost for Year" column, one can see that the defender has the lowest cost in the first 2 years, but in the third year its cost is $9,100, whereas the cost of the first year of service for the challenger is only $9,000. Hence it would be economic to replace the defender after the second year. This conclusion can be confirmed by enumerating all replacement possibilities and their respective costs and then computing the A.C. for each, as will be done for the 4-year planning horizon in part (b).

(b) For a planning horizon of 4 years, the alternatives and their respective costs for each year and the A.C. of each are given in Table 11-6. Thus the most economic alternative is to keep the defender for 2 years and the challenger for 2 years. The decision to keep the defender for 2 years happens to be the same as when the repeatability assumption was used, but of course this would not be necessarily true in general. ∎

REPLACEMENT VERSUS AUGMENTATION

Conditions frequently arise wherein an existing asset is not of sufficient capacity to meet increased demands that have developed. In such cases there usually are two alternatives open: (1) to augment the existing asset, or (2) to replace it with a new asset of sufficient capacity to meet the new demands.

Example 11-6 provides a situation involving augmentation versus replacement and includes the consideration of income taxes. The period of required service is coterminated at 5 years, and inflation during this time must be considered in the analysis. Furthermore, both alternatives qualify for A.C.R.S. depreciation.

TABLE 11-6 Determination of Economic Life of Defender for a Planning Horizon of 4 Years

Keep Defender for:	Keep Challenger for:	Total (Marginal) Costs for Each Year					A.C. at 10% for 4 Years
		1	2	3	4		
0 years	4 years	$9,000	$8,250	$8,500	$10,850		$9,082
1	3	7,000	9,000	8,250	8,500		8,301
2	2	7,000	8,000	9,000	8,250	(min)→	8,005
3	1	7,000	8,000	9,100	9,000		8,190
4	0	7,000	8,000	9,100	10,000		8,406

EXAMPLE 11-6

In 1982 (5 years ago) a construction company purchased an air compressor for $2,000. Its useful life was estimated to be 10 years, but its A.C.R.S. recovery period was 5 years. Annual out-of-pocket disbursements for operations, maintenance, taxes, and insurance are running at approximately $1,050.

The demands for compressed air have increased to the point where either an additional small compressor, costing $1,000, will have to be added, or the old compressor will have to be replaced with a larger compressor, which will cost $4,000. It is estimated that the new, small compressor would have annual disbursements of $300, and that the annual disbursements for the larger, more modern compressor would be $500. The new small (augmenting) unit would have a 5-year A.C.R.S. recovery period and a $550 estimated market value at the end of year 5. Depreciation charges for the large unit and the small unit will be based on a 5-year A.C.R.S. recovery period, and it is expected that the actual market value of the large unit at the end of the 5-year study period (or period of needed service) would be $2,000. If the existing compressor is replaced, it will bring a present market value of $200. If the old compressor is used for another 5 years, it is believed that its market value at that time will be negligible.

The company earns about 20% on its capital before income taxes and about 10% after taxes, being in a 50% income tax bracket. Inflation will affect all operating costs and market values at an average rate of 6% per year. Should the existing compressor be augmented or replaced based on an after-tax annual cost analysis and a 5-year study period?

Solution

Table 11-7 shows the after-tax cash flow computations for Example 11-6. This example includes numerous realistic conditions that are present in today's economic analyses: augmentation of an existing asset versus its complete replacement; inflation of cash flows; A.C.R.S. depreciation; an investment tax credit; and an opportunity cost of keeping an existing asset. A brief description of some of the main points in Table 11-7 follows.

If the existing compressor is sold for $200, this would be a gain of capital (depreciation recovery) taxed at the ordinary income rate of 50% since the book value after 5 years with A.C.R.S. depreciation would be zero. If the old compressor is *kept*, the opportunity cost is $200 and the income taxes avoided would be $200(0.50) = $100. A 10% investment tax credit on the purchase of the small (augmenting) compressor is also shown under "Income Taxes" in year 0. Before-tax cash flows are inflated by 6% per year as shown, and this includes the estimated market value of the small unit. Because a zero salvage value is built into the A.C.R.S. method of depreciation, the entire amount of this inflated market value is subject to ordinary income taxation at 50%. The after-tax equivalent annual cost at 6% of this alternative is $897.

If the new, larger compressor is purchased, a 10% investment tax credit could again be taken. The A.C.R.S. depreciation method would result in a book value of zero at the end of year 5, producing an actual dollar depreciation recovery amounting to $2,676. The after-tax cash flows that result from this alternative have an equivalent annual cost at 6% of $611.

TABLE 11-7 After-Tax Cash Flow Computations for Replacement Versus Augmentation Alternatives in Example 11-6

Alternative	Year, k	Before-Tax Cash Flow	A.C.R.S. Depreciation (1987)	Taxable Income	Income Taxes	After-Tax Cash Flow
Augment old compressor	0	$-\$200^a - \$1,000$		$+\$\ 200^b$	$+\$100^c + \$100^d$	$-\$1,000$
	1	$-(1,050 + 300)(1.06)^1$	$-\$200^e$	$-1,631$	$+816$	$-\ 615$
	2	$-(1,050 + 300)(1.06)^2$	-320	$-1,837$	$+919$	$-\ 598$
	3	$-(1,050 + 300)(1.06)^3$	-240	$-1,848$	$+924$	$-\ 684$
	4	$-(1,050 + 300)(1.06)^4$	-160	$-1,864$	$+932$	$-\ 772$
	5	$-(1,050 + 300)(1.06)^5$	-80	$-1,887$	$+944$	$-\ 863$
	5	$+\ 550(1.06)^5$		$+\ 736^f$	-368	$+\ 368$
New, larger compressor	0	$-\$4,000$			$+400^g$	$-\$3,600$
	1	$-500(1.06)^1$	$-\$\ 800^h$	$-\$1,330$	$+665$	$+\ 135$
	2	$-500(1.06)^2$	$-1,280$	$-1,842$	$+921$	$+\ 359$
	3	$-500(1.06)^3$	-960	$-1,556$	$+778$	$+\ 182$
	4	$-500(1.06)^4$	-640	$-1,271$	$+636$	$+\ \ 5$
	5	$-500(1.06)^5$	-320	-989	$+495$	$-\ 174$
	5	$+2,000(1.06)^5$		$+2,676^i$	$-7,338$	$+1,338$

[a] Market value of existing unit.
[b] Recovered depreciation for the existing unit.
[c] The 10% investment tax credit.
[d] Income tax avoided on recovered depreciation.
[e] The existing unit has been completely depreciated.
[f] Inflated value of depreciation recovery.
[g] The 10% investment tax credit.
[h] A.C.R.S. depreciation.
[i] Inflated value of depreciation recovery.

Since $611 < $897, the new larger compressor would be the recommended choice. ∎

RETIREMENT WITHOUT REPLACEMENT

In some cases it may be decided that although an existing asset is to be retired from its current use, it will not be replaced or removed from all service. Although it may not be able to compete economically in its existing usage, it may be desirable and even economical to retain it as a standby unit or for some different use. The cost to retain it under such conditions may be quite low, owing to its relatively low present realizable resale value and perhaps low costs for operation and maintenance. There often are income tax considerations that also bear on the true cost of retaining the asset. This type of situation is illustrated in the following example.

EXAMPLE 11-7

What would be the cost of retaining, as a standby unit for 5 years, the existing compressor discussed in Example 11-6? Assume that the before-tax annual cost for operation and maintenance as a standby would be $155.

Solution

First, it is evident that if the compressor is retained, the company will forego an after-tax recovery of $100, which would have resulted from its disposal. Thus the annual capital recovery cost is $100(A/P, 10\%, 5) = $26.38. The $155 for operation and maintenance (the depreciation charge for the existing asset is $0 per year) results in a total annual reduction in taxable income of $155. At the 50% rate, taxes would be reduced by $77.50 and the net effect is to have an annual after-tax cost of $77.50.

Thus the total equivalent annual cost of keeping the old compressor on standby service for 5 years is $26.38 + $77.50 = $103.88. ∎

DETERMINATION OF VALUE BY REPLACEMENT THEORY

As was pointed out in Chapter 9, it frequently is necessary to determine (estimate or assess) the value of old assets, often a considerable number of years after their acquisition. During the years since the assets were acquired, price levels will have changed and technological progress have occurred. As a result, the valuation process is difficult.

If proper cost data are available, a reasonably accurate value of old assets can be obtained by applying replacement theory. As has been pointed out in this chapter, an asset has economic value only if it can be operated profitably in competition with the most economically efficient asset available. The economic value of an existing asset is the maximum amount on which depreciation and interest can be charged while permitting it to compete on an even basis with the most efficient new asset. Such a comparison may be illustrated as follows.

EXAMPLE 11-8

An old asset could last for 5 more years and experience annual disbursements of $17,000. A potential new asset would require an investment of $50,000 and have annual disbursements of $14,000 over a 5-year life. If either alternative is expected to have zero salvage value at the end of 5 years and capital should earn 20% before taxes, what is the present value of the old asset at which it is equally as economical as the new asset?

Solution

Letting V = value of the old asset, we obtain:

	Old Asset	New Asset
Capital recovery amount		
$V(A/P, 20\%, 5)$	$0.3344	
$50,000(A/P, 20\%, 5)$		$16,719
Annual disbursements	17,000	14,000
TOTAL ANNUAL COSTS	$17,000 + $0.3344V	$30,719

Equating the total annual costs and solving, we find that V = $41,026. This is a good measure of the present economic value of the old asset if the new asset is the best alternative available.

Because the alternatives have the same economic lives, it would also be easy to obtain the same answer by equating present worths of costs over the 5-year period. Thus

$$V + \$17,000(P/A, 20\%, 5) = \$50,000 + \$14,000(P/A, 20\%, 5)$$

$$V = \$41,026 \quad \blacksquare$$

A REPLACEMENT CASE STUDY

The Operations Engineering Group of the Power Engineering Corporation has been studying the possibility of replacing a large boiler in one of their petroleum refining processes. A new boiler with roughly the same size and reliability can be purchased from a well-known U.S. manufacturer. It has been determined to be the best "challenger" currently available. The existing boiler ("defender") was purchased in 1979 (5 years ago) and has experienced what the firm's engineers feel to be excessive operation and maintenance (O&M) expenses. Cost projections by the Operations Engineering Group (in actual, or inflation-adjusted, dollars) are summarized below in addition to other relevant information and assumptions.

Defender: Original cost was $200,000; CLADR guideline life (see Table 9-1) for straight line depreciation was 10 years; original estimate of salvage value for computing depreciation was $22,000; O&M expenses for the next year (sixth year of life for the defender) are $80,000 and will increase by approximately $10,000 per year thereafter; present estimated net market value is $20,000; modification cost now to upgrade the defender to make it comparable in performance to the challenger is $100,000; estimated market value of the modified defender in 5 years is negligible; the modification costs are fully depreciable over 5 years by using the straight line method.

Challenger: Initial cost (including purchase price, freight, installation, and related expenses) is $400,000; A.C.R.S. guideline period is 5 years; salvage value for tax purposes is $0; estimated net market value at the end of 5 years is $165,000; annual O&M expenses are $48,000.

Assumptions: Five-year service requirement (planning horizon); depreciable amount of the modified defender equals its *book value* at the end of year 5 less original salvage value, *plus modification costs*; an investment credit of 10% can be claimed on new assets; effective tax rate on ordinary income is 50%; capital gains (losses) create a tax liability (savings) which is taxed at 28%; the firm's after-tax M.A.R.R. is 18%; sufficient capital is available to implement the challenger (if it is selected).

Assignment: You have been requested to compare the two alternatives on an after-tax basis in view of the information given above. Use a tabular setup and show all calculations.

Solution

The determination of after-tax cash flow for the *defender* is summarized in Table 11-8.

The year 0 row has two separate entries—the first showing the opportunity cost of keeping the defender. The current book value of the defender is $200,000 - 5($17,800) = $111,000, and the associated *capital loss* would be $91,000 *if* the defender is sold. Because we are considering the possibility of retaining the defender, this entry would be reversed in sign so that we experience

TABLE 11-8 After-Tax Analysis of Existing Boiler (Defender)

End of Year	Before-Tax Cash Flow	Depreciation[a]	Taxable Income	Cash Flow for Taxes	After-Tax Cash Flow
0	(−)$ 20,000		(+)$ 91,000	(−)$25,480	(−)$ 45,480
0	− 100,000				− 100,000
1	− 80,000	− $37,800	− 117,800	+ 58,900	− 21,100
2	− 90,000	− 37,800	− 127,800	+ 63,900	− 26,100
3	− 100,000	− 37,800	− 137,800	+ 68,900	− 31,100
4	− 110,000	− 37,800	− 147,800	+ 73,900	− 36,100
5	− 120,000	− 37,800	− 157,000	+ 78,900	− 41,100
5	0		− 22,000	+ 6,160	+ 6,160

[a]Depreciation $= \dfrac{\$200,000 - \$22,000}{10} + \dfrac{\$100,000}{5} = \$37,800/\text{yr.}$

an increase in taxable income relative to the firm's position if the defender had been sold. Similarly, the income tax saving of 28% × $91,000 that results *if* the defender had been sold gives rise to an "opportunity tax liability" of $25,480 when the defender is kept. (In a large firm the absense of a tax savings of $25,480 due to keeping the defender will create an additional tax liability for the year.) A $22,000 loss on disposal at the end of year 5 would be claimed because the market value of the modified defender is $0 while its book value is $22,000. A tax credit of 28% × $22,000 = $6,160 will result. From Table 11-8, the present-worth of after-tax cash flow for the defender is computed at a M.A.R.R. of 18% and found to be − $234,927.

The computation of after-tax cash flow for the *challenger* is provided in Table 11-9. An investment tax credit of 10% × $400,000 = $40,000 is taken in year 0. Depreciation is determined with 1984 A.C.R.S. percentages. Because the book value with the A.C.R.S. method is zero at the end of year 5, the estimated market value of $165,000 at that time is fully taxable as "recovery of previously charged depreciation" (see Appendix 10-A). The income tax owed is 50% × $165,000 = $82,500. At a M.A.R.R. of 18%, the present worth of the challenger is − $276,383.

From the foregoing analysis, you would probably recommend keeping the defender and modifying the existing boilers. Because the comparison is close, nonmonetary considerations could easily swing the decision in favor of the challenger. A list of these considerations might include operating safety, reliability of the system, and improved control over the boiler effluents.

SUMMARY

There are six factors that must be considered to varying degrees in replacement decisions. The *first* is the possibility that if replacement is deferred, assets available in the future may be improvements over those presently available. When

TABLE 11-9   After-Tax Analysis of a New Boiler (Challenger)

End of Year	Before-Tax Cash Flow	Depreciation[a]	Taxable Income	Cash Flow for Taxes	After-Tax Cash Flow
0	− $400,000			+ $40,000	− $360,000
1	− 48,000	− $60,000	− $108,000	+ 54,000	+ 6,000
2	− 48,000	− 88,000	− 136,000	+ 68,000	+ 20,000
3	− 48,000	− 84,000	− 132,000	+ 66,000	+ 18,000
4	− 48,000	− 84,000	− 132,000	+ 66,000	+ 18,000
5	− 48,000	− 84,000	− 132,000	+ 66,000	+ 18,000
5	+ 165,000		+ 165,000	− 82,500	+ 82,500

[a]Use 1984 recovery percentages: 15% in year 1, 22% in year 2, 21% in year 3, 21% in year 4, and 21% in year 5.

technology is changing rapidly, we hesitate to acquire an asset that very soon may be obsolete economically. Consequently, in making replacement decisions, a knowledge of the state and trend of technological development in the area involved is important.

The *second* factor relates to probable variations in the future market value of an asset. If study shows that the most likely disposal date is a considerable number of years distant, moderate changes in the assumed disposal value are not likely to make any substantial change in the analysis. On the other hand, if only a few years are involved, prospective changes in the market value can have a considerable effect on the year-to-year economy.

Where the trade-in or disposal value is changing rapidly, we must be certain that a sufficient number of years is considered in determining the economy of the old asset. For example, on the basis of keeping the old asset 1 more year, it might appear to be more economical to make the replacement now. If, however, a 3- or 4-year retention period is used, in order to gain the smaller amounts of depreciation during the additional years, the old asset might be more economical than a new one. Thus, in general, slow rates of decrease in market value, with increasing operation and/or maintenance costs, tend to favor early replacement, whereas rapid rates of decrease in market value favor later replacement.

A *third* factor is the probability that the enterprise will grow in the immediate future. If it is evident that such growth will require increased output unobtainable from the existing asset, it is obvious that replacement or augmentation will have to be made in the near future. The resulting problem is basically to determine whether replacement or augmentation is cheaper, and at what time either should occur. The only complicating factor is the possibility that the desired equipment may not be available when needed if replacement is deferred, or that the increased demand may develop earlier than anticipated. To avoid these possibilities, companies sometimes prefer to make the replacement somewhat earlier than necessary and then attempt to develop the market.

A *fourth* factor is somewhat counter to the previous one. Replacement frequently results in excess capacity, since new machines usually are more efficient than those they replace. If there is no actual use for such excess capacity, it has no value and thus should not be given any consideration. On the other hand, if the excess capacity can be used for some function not rendered by the old asset, this should be considered in the replacement study. This can easily be done by deducting the annual cash flows *after income taxes* derived from the added service from the annual cost flow after income taxes for the new machine.

The possibility that the purchase price of the replacement asset will change in the future is a *fifth* factor that must be considered. Although it is difficult to predict what price changes will occur in the future, the long history of inflation that has occurred causes many to believe that the trend is apt to continue. Such a condition tends to favor earlier replacement. However, this practice should not be followed blindly, particularly where the need for the service is expected to continue for many years. Two additional facts must also be considered. First, the earlier the replacement, the sooner the new asset will have to be replaced— at higher first cost. Second, it is likely that technological improvements will occur, and by deferring replacement it is possible that a more efficient asset may be obtained. Therefore, before too much emphasis is given to probable price increases, other possible and probable future changes also should be considered.

Budgetary and personnel considerations are a *sixth* factor that frequently affects replacement decisions. Most companies do not have unlimited funds, and many projects are usually competing for the funds that they do have. In such cases replacement studies supply information on the basis of which decisions can be made regarding the timing of replacements as funds become available.

PROBLEMS

11-1 A small plant has four space heaters that were purchased 3 years ago for $1,300 each. They will last 10 more years, and will have no salvage value at that time. An expansion to the plant can be heated by two additional space heaters that will last 10 years, have no salvage value, and cost $1,800 each. An alternative method of heating is to install a central heating system that will cost $6,400, last 10 years, and have a salvage value of $4,000. If central heating is installed, the present four space heaters can be sold for $200 each. Compare the equivalent uniform annual cost of heating with space heaters and with central heating. Assume that fuel costs will be the same for either system and that the minimum attractive rate of return (before taxes) for the company is 15%. What would you recommend?

11-2 One year ago a machine was purchased at a cost of $2,000 to be useful for 6 years. However, the machine has failed to perform properly and has cost $500 per year for repairs, adjustments, and shutdowns. A new machine is available to accomplish the functions desired and has an initial cost of $3,500. Its maintenance costs are expected to be $50 per year during its service life of 5 years. The approximate market value of the presently owned machine has been estimated to be roughly $1,200. If the operating costs (other than maintenance) for both machines are equal, show whether it is economical to purchase the new machine. Perform a before-tax study using an interest rate of 12%, and assume that terminal salvage values will be negligible.

11-3 Suppose that you have an old car, which is a real gas guzzler. It is 10 years old and could be sold to a local dealer for $400 cash. The annual maintenance costs will average $800 per year into the foreseeable future, and the car only averages 10 miles per gallon. Gasoline costs $1.50 per gallon and your driving averages 15,000 miles per year. You now have an opportunity to replace the old car with a brand new one that costs $8,000. If you buy it you will pay cash. Because of a 2-year warranty, the maintenance costs are expected to be negligible. This car averages 30 miles per gallon. Use the incremental internal rate of return method and specify which alternative you should select. Utilize a 2-year comparison period and assume the new car can be sold for $5,000 at end of year 2. Ignore the affects of income taxes and let your M.A.R.R. be 15%. State any other assumptions you make.

11-4 A pipeline contractor is considering the purchase of certain pieces of earthmoving equipment to reduce his costs. His alternatives are as follows.

Plan A: Retain a backhoe already in use. The contractor still owes $10,000 on this machine but can sell it on the current open market for $15,000. This machine will last another 6 years, with maintenance, insurance, and labor costs totaling $28,000 per year.

Plan B: Purchase a trenching machine and a highlift to do the trenching and backfill work and sell the current backhoe on the open market. The new machines will cost $55,000, will last 6 years with a salvage value of $15,000, and will reduce annual maintenance, insurance, and labor costs to $20,000.

If money is worth 8% before taxes to the contractor, which plan should he follow? Use the present worth procedure to evaluate both plans on a before-tax basis.

11-5 A 3-year-old asset that was originally purchased for $4,500 is being considered for replacement. The new asset under consideration would cost $6,000. The engineering department has made the following estimates of the operating and maintenance costs of the two alternatives.

Year	Old Asset	New Asset
1	$2,000	$ 500
2	2,400	1,500
3	—	2,500
4	—	3,500
5	—	4,500

The dealer has agreed to place a $2,000 trade-in value on the old asset if the new one is purchased now. It is estimated that the salvage value for either of the assets will be zero at any time in the future. If the before-tax rate of return is 12%, make an annual cost analysis of this situation and recommend the best course of action that can be taken now. Use the repeatability assumption and give a short written explanation of your answer.

11-6 You have a machine that cost $30,000 2 years ago—it has a present market value of $5,000. Operating costs total $2,000 per year so long as the machine is in use. In 2 more years it will no longer be useful and you can sell it for $500 scrap value. You are considering replacing this machine with a new model incorporating the latest technology—this new model will cost you $20,000 now. It has an annual operating cost of $1,000 with a useful life of 8 years and negligible salvage value. If you delay the purchase of the new model, the cost will be $24,000 because of the installation difficulties that do not exist now. When your present machine is retired, you have no hopes of getting another one like it, since it was a very limited model. If money is worth 10% before taxes, what should you do?

11-7 A diesel engine was installed 10 years ago at a cost of $50,000. It has a present realizable market value of $14,000. If kept, it can be expected to last 5 more years, have operating disbursements of $14,000 per year, and to have a salvage value of $8,000 at the end of the 5 years. This engine can be replaced with an improved version costing $65,000 and having an expected life of 20 years. This improved version will have estimated annual operating disbursements of $9,000 and an ultimate salvage value of $13,000. It is thought that an engine will be needed indefinitely and that the results of the economy study would not be affected by the consideration of income taxes. Using a before-tax M.A.R.R. of 15%, make an annual-cost analysis to determine whether to keep or replace the old engine.

11-8 (a) The replacement of a boring machine is being considered by the Reardorn Furniture Company. The new improved machine will cost $30,000 installed, will have an estimated useful life of 12 years, and a $2,000 salvage value at that time. It is estimated that annual operating and maintenance costs will average $16,000 per year. The present machine has a book value of $6,000 and a present market value of $4,000. Data for the present machine for the next 3 years are as follows:

Year	Salvage Value at End of Year	Book Value at End of Year	Operating and Maintenance Costs During the Year
1	$3,000	$4,500	$20,000
2	2,500	3,000	25,000
3	2,000	1,500	30,000

Using a before-tax interest rate of 15%, make an *annual-cost comparison* to determine whether it is economical to make the replacement now.

(b) If the operating and maintenance costs for the present machine had been estimated to be $15,000, $18,000, and $23,000 in years 1, 2, and 3, respectively, what replacement strategy should be recommended?

11-9 A diesel engine was installed 5 years ago at a cost of $80,000. Its present market value is $45,000. If the present engine is kept, it can be expected to last 5 more years, have operating disbursements of $12,000 per year, and have a market value of $20,000 at that time. This engine is depreciated by the straight line method using a 10-year life and a terminal book value of $10,000 for tax purposes. The old engine can be replaced at the present time with a rebuilt diesel engine that will cost $60,000, have operating disbursements of $9,000, per year and have a life of 5 years. The salvage value for tax purposes is $15,000 and sum-of-the-years'-digits depreciation will be used. Its market value at the end of 5 years is expected to be zero. The effective income tax rate is 50%. Depreciation recapture is taxed at 50% and capital losses produce tax savings at 28% of the loss. If the desired after-tax M.A.R.R. is 12%, which alternative would you recommend?

11-10 Suppose that it is desired to make an after-tax analysis for the situation posed in Problem 11-7. The existing engine is being depreciated by the straight line method over 15 years; an estimated salvage value of $8,000 is used for depreciation purposes. Assume that if the replacement is made, the improved version will be depreciated by the A.C.R.S. method, using a 5-year recovery period and a zero salvage value. Also, assume that annual disbursements will affect taxes at 50%, and that any capital gain or capital loss will affect taxes at 28%. A 10% investment credit can be taken if the new engine is purchased. Use the annual equivalent cost method to determine if the replacement is justified by making an after-tax M.A.R.R. of 10% or more.

11-11 Use the present worth method to select the better alternative. Let the minimum attractive rate of return equal 10% after taxes and the effective income tax rate be 50%. State any assumptions you make. Capital gains (losses) are taxed at 28%. The firm is known to be profitable in its overall operation.

Alternative A: Retain an already owned machine in service for 8 more years.
Alternative B: Sell old machine and rent a new one for 8 years.

Data	Alternative A	Alternative B
Labor per year	$300,000	$250,000
Material cost/year	$250,000	$100,000
Insurance and property taxes/year	4% of first cost	None
Maintenance/year	$8,000	None
Rental cost/year	None	$100,000

Alternative A:

Cost of old machine 5 years ago = $500,000

Book value now = $350,000

Depreciation with straight line method, 15-year life = $30,000/year

Estimated salvage value at end of useful life = $50,000

Present market value = $150,000

11-12   You have a machine that was purchased 4 years ago and was set up on a 5-year sum-of-the-year's-digits depreciation schedule with no salvage value. Original cost was $150,000. The machine can last for 10 years or more. A new machine is now available at a cost of only $100,000—it can be depreciated over 5 years with the A.C.R.S. method—no salvage value. Annual operating cost of the new machine is only $5,000 while operating cost of the present one is $20,000/year. The new machine has a useful life greater than 10 years. You find that $40,000 is the best price you can get if you sell the present machine now. Your best projection for the future is that you will need the service provided by either of the two machines for the next 5 years. The market value of the old machine is estimated at $2,000 in 5 years, but the market value of the new machine is estimated at $5,000 in 5 years. If the after-tax rate M.A.R.R. is 10%, should you sell the old machine and purchase the new one? You do not need both. Assume that the company is in a 50% tax bracket, and that long-term capital gains (losses) are taxed at 28%. A 10% investment tax credit can be taken.

11-13   Robert Roe has just purchased a 4-year-old used car, paying $3,000 for it. A friend has suggested that he should determine in advance how long he should keep the car so as to assure the greatest overall economy. Robert has decided that, because of style changes, he would not want to keep the car longer than 4 years, and he has estimated the out-of-pocket costs and trade-in values for years 1 through 4 as follows:

	Year 1	Year 2	Year 3	Year 4
Operating costs	$ 950	$1,050	$1,100	$1,550
Trade-in value at end of year	2,250	1,800	1,450	1,160

If Robert's capital is worth 12%, at the end of which year should he replace the car?

11-14   Determine the economic service life of a machine that has a first cost of $5,000 and estimated operating costs and year-end salvage values as follows. Assume that the interest rate is (a) 0%; (b) 10%.

Year, k	Operating Cost for Year, c_k	Salvage Value, S_N
1	$ 800	$4,000
2	900	3,500
3	1,100	3,000
4	1,100	2,000
5	1,300	1,500
6	1,400	1,000

11-15   Find the economic service life for a tractor with an expected disbursement pattern of the form shown on page 422. Assume that the salvage value is $3,500 - 250t$ at the end of any given year and suppose that the interest rate is 10%.

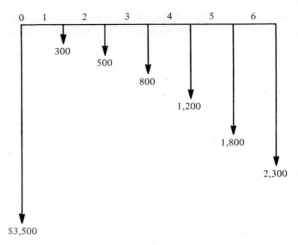

PROBLEM 11-15.

11-16 Consider a piece of equipment that initially cost $8,000 and has these estimated annual operating and maintenance costs and salvage values:

End of Year	Operating and Maintenance Costs for Year	Salvage Value
1	$3,000	$4,700
2	3,000	3,200
3	3,500	2,200
4	4,000	1,450
5	4,500	950
6	5,250	600
7	6,250	300
8	7,750	0

If the cost of money is 7%, determine the most economical time to replace this equipment.

11-17 A motorcycle can be purchased for $2,500 when new. There follows a schedule of annual operating expenses for each year and trade-in values at the end of each year. Assume that these amounts would be repeated for future replacements, and that the motorcycle will not be kept more than 3 years. If capital is worth 10% before taxes, determine at which year's end the cycle should be replaced so that costs will be minimized.

	Year 1	Year 2	Year 3
Operating expenses for year	$ 850	$ 950	$1,025
Trade-in value at end of year	1,700	1,400	1,000

11-18 The company for which you are working must improve certain of its facilities to meet increasing sales. The existing facilities, which have been fully depreciated, can be used for another 5 years provided that some new materials-handling equipment, costing $25,000, is added. If this procedure is followed, neither the old nor the new equipment would have any salvage value at the end of 5 years, when it all would have to be replaced. An alternative is to dismantle the existing facilities and build entirely new ones at a cost of $600,000. The new installation would have a depreciable life of 20 years, and it would result in a reduction of at least $109,000 per year in out-of-pocket disbursements, as compared with the first alternative. The company has a total income tax rate of 50% and has a 20% M.A.R.R. before taxes and a 10% M.A.R.R. after taxes. It uses straight line depreciation (instead of A.C.R.S.) for accounting and tax purposes and finds that its book values agree quite well with actual salvage values. A 10% investment tax credit can also be taken. (a) Make a before-tax analysis and recommendation, using the I.R.R. procedure. (b) Make an after-tax analysis using the I.R.R. procedure. Would your recommendation be the same in part (a)?

11-19 A steel highway bridge must either be reinforced or replaced. Reinforcement would cost $22,000 and would make the bridge adequate for an additional 5 years of service. If the bridge is torn down, the scrap value of the steel would exceed the demolition cost by $14,000. If it is reinforced, it is estimated that its net salvage value would be increased by $16,000 at the time it is retired from service. A new prestressed concrete bridge would cost $140,000 and would meet the foreseeable requirements of the next 40 years. Such a bridge would have no scrap or salvage value. It is estimated that the annual maintenance costs of the reinforced bridge would exceed those of the concrete bridge by $3,200. Assume that money costs the state 10% and that the state pays no federal income taxes. What would you recommend?

11-20 Four years ago the Attaboy Lawn Mower Company purchased a piece of equipment for their assembly line. Because of increasing maintenance costs for this equipment, a new piece of machinery is being considered. The cost characteristics of the defender (present equipment) and the challenger are shown below:

Defender	Challenger
Original cost = $9,000	Purchase cost = $13,000
Maintenance = $300 in year 1, increasing by 10% per year thereafter.	Maintenance = $100 in year 1, increasing by 10% per year thereafter.
	Market value = $3,000 at the end of year 5.
Original estimated salvage value for tax purposes = 0	A.C.R.S. life = 5 years
Original estimated life = 9 yrs.	A 10% ITC can be taken

Suppose that a $3,200 market value is available now for the defender. Perform an after-tax analysis, using an after-tax M.A.R.R. of 10% to determine which alternative to select. The effective income tax rate is 50% on ordinary income, capital gains (losses) are taxed at 28% and straight line depreciation is applied to the defender, while the challenger qualifies for A.C.R.S. depreciation.

11-21 A company is considering replacing a turret lathe with a single-spindle screw machine. The turret lathe was purchased 6 years previously at a cost of $80,000 and depreciation has been figured on a 10-year, straight line basis, using zero salvage value. It can now be sold for $15,000, and if retained would operate satisfactorily for 4 more years and have zero salvage value. The screw machine is estimated to have a useful life of 10 years. A.C.R.S. depreciation would be used with a recovery period of 5 years. It would require only 50% attendance of an operator who receives $7.50 per hour. The machines would have equal capacities and would be operated 8 hours per day, 250 days per year. Maintenance on the turret lathe has been $3,000 per year; for the screw machine it is estimated to be $1,500 per year. Taxes and insurance on each machine would be 2% of the first cost annually. If capital is worth 10% to the company, after taxes, and the company has a 50% tax rate, what is the maximum price it can afford to pay for the screw machine? Capital gains (losses) are taxed at 28%, and an investment tax credit of 10% can be taken if the screw machine is purchased. Use the repeatability assumption.

11-22 Ten years ago a corporation built a warehouse at a cost of $400,000 in an area that since has developed into a major retail location. At the time the warehouse was constructed it was estimated to have a depreciable life of 20 years, with no salvage value, and straight line depreciation has been used. The corporation now finds it would be more convenient to have its warehouse in a less congested location, and can sell the old warehouse for $250,000. A new warehouse, in the desired new location, would cost $500,000, have an A.C.R.S. recovery period of 10 years with no salvage value, and there would be an annual saving of $4,000 per year in the operation and maintenance costs. Taxes and insurance on the old warehouse have been 5% of the first cost per year, while for the new warehouse they are estimated to be only 3% per year. The corporation has a 50% income tax rate, and a 28% long-term capital gains tax rate. Capital is worth not less than 12% after taxes. What would you recommend on the basis of an after-tax I.R.R. analysis?

11-23 Five years ago an airline installed a baggage conveyor system in a terminal, knowing that within a few years it would move to a new section of the terminal and that this equipment would then have to be moved. The original cost of the installation was $120,000, and through accelerated depreciation methods the company has been able to write off the entire cost. It now finds that it will cost $40,000 to move and reinstall the conveyor. This cost could be depreciated over the next 5 years, which the airline believes is a good estimate of the remaining useful life of the system if moved. It finds that it can purchase a somewhat more efficient conveyor system for an installed cost of $120,000, and this system would result in an estimated reduction in annual operating and maintenance costs of $6,000 during its estimated 5-year depreciable life. Annual operating and maintenance costs are expected to inflate by 6% per year. In both alternatives, straight line depreciation is used with a salvage value of $0. A small airline, which will occupy the present space, has offered to buy the old conveyor for $90,000.

Annual property taxes and insurance on the present equipment have been $1,500, but it is estimated that they would increase to $1,800 if the equipment is moved and reinstalled. For the new system it is estimated that these would be about $2,750 per year. All other costs would be about equal for the two alternatives. The company is in the 46% income tax bracket and pays 28% on any long-term capital gains. It wishes to obtain at least 10%, after taxes, on any invested capital. What would you recommend?

11-24 It has been decided to replace an existing machine process with a newer, more productive process which costs $80,000 and has an estimated market value of $20,000 at the end of its service life of 10 years. Installation charges for the new process will amount to $3000—this is not added to first cost but will be an expensed item during the first year of operation. A.C.R.S. depreciation will be utilized, and the recovery period is 5 years. A 10% investment credit will also be taken. The new process will reduce direct costs (labor, insurance, maintenance, rework, etc.) by $10,000 in the first year and this amount is expected to inflate by 5% each year thereafter during the 10-year life. It is also known that the book value of the *old* machine process is $10,000 but that its fair market value is $14,000. Capital gains are taxed as ordinary income, and the effective income tax rate is 50%. Determine the prospective *after-tax* rate of return for the incremental cash flows associated with the *new* machine process if it is believed that the existing process will perform satisfactorily for 10 more years.

11-25 A large firm desires to automate one of its high-volume production lines. Thus, it proposes to substitute capital for labor. Formulate this general replacement situation as a multiattributed decision problem (refer to Chapter 7) and discuss how the firm might go about justifying the investment.

CHAPTER 12

Economy Studies for Public Projects

Public projects are those authorized, financed, and operated by governmental agencies—federal, State, or local. Such public works are numerous, may be of any size, and frequently are much larger than private ventures. Because they require the expenditure of capital, such projects have economy aspects with respect to their design, acquisition and operation. However, because they are public projects, a number of important and special factors exist which ordinarily are not found in privately financed and operated businesses. To some extent these problems are based on personal philosophies and desires that are subject to change with time. Thus, to make and interpret economy studies of public projects, it is essential to have some understanding of these unique problems.

Some basic differences between privately owned and publicly owned projects may be noted by considering the following factors:

	Private	Public
1. Purpose	Provide goods or services at a profit Provide jobs Promote technology Improve living standards	Protect health Protect lives and property Provide services (at no profit)
2. Sources of capital	Private investors and lenders	Taxation Private lenders

	Private	Public
3. Method of financing	Individual ownership Partnerships Corporations	Direct payment from taxes Loans without interest Loans at low interest Self-liquidating bonds Indirect subsidies Guarantee of private loans
4. Multiple purposes	Rarely	Common, such as electrical power, flood control irrigation, and recreation
5. Life of individual projects	Usually relatively short (5 to 20 years)	Usually relatively long (20 to 60 years)
6. Relationship of suppliers of capital to project	Direct	Indirect, or none
7. Conflict of purposes	Usually none	Quite common (dam for flood control and power)
8. Conflict of interests	Usually none	Very common (between agencies)
9. Effect of politics	Little to moderate	Frequent factors; short-term tenure of decision makers Pressure groups Financial and residential restrictions, etc.
10. Measurement of efficiency	Rate of return on capital	Very difficult; no direct comparison with private projects

As a consequence of these differences, it often is not possible to make economy studies and investment decisions for public works projects in exactly the same manner as for privately owned projects. Different decision criteria must be used, and this creates problems for the public, who pays the bill, for those who must make the decisions, and for those who must manage public works projects.

In this chapter attention is devoted to projects financed by the federal government, primarily because the basic principles are more readily set forth. Later these principles are related to state and local projects, where the problems usually are somewhat simpler.

SELF-LIQUIDATING PROJECTS

The term *self-liquidating project* is often applied to a governmental project that is expected to earn direct revenue sufficient to repay the cost in a specified period of time. Such projects usually provide utility services, such as water, electric power, sewage disposal, or irrigation water. In addition, toll bridges, tunnels, and highways are built and operated in this manner.

Self-liquidating projects are not expected to earn profits or pay income taxes. Neither do they pay property taxes, although in some cases they do make *in-lieu* payments in place of the property or franchise taxes that would have been paid to the cities, counties, or states involved had the project been built and operated by private ownership. For example, the U.S. government agreed to pay the States of Arizona and Nevada $300,000 each annually for 50 years in lieu of taxes that might have accrued if Hoover Dam had been privately constructed and operated. In most cases the in-lieu payments are considerably less than the actual property or franchise taxes would be. Furthermore, once the in-lieu payments are agreed upon, usually at the time the project is originated, they virtually never are changed thereafter. Such is not the situation in the case of property taxes.

Whether self-liquidating projects should or should not pay the same taxes as privately owned projects is not within the scope of this text. Such payments undoubtedly would make it much easier to compare the economy of similar publicly and privately operated activities. However, that public projects do not have to earn profits or pay certain taxes is a fact that must be given proper consideration in making economy studies of such projects and in making decisions regarding them.

MULTIPLE-PURPOSE PROJECTS

Many governmental and private projects have more than one purpose or function. A governmental project, for example, may be intended for flood control, irrigation, and generation of electric power. A private project may utilize waste gases in a petroleum refinery to generate steam and electricity for refinery use and to cogenerate electricity for public sale. Such projects commonly are called *multiple-purpose projects*. By designing and building them to serve more than one purpose, greater overall economy can be achieved. This is very important in projects involving very large sums of capital and utilization of natural resources, such as rivers. It is not uncommon for a public project to have four or five purposes. This usually is desirable, but, at the same time, it creates economy and managerial problems because of overlapping utilization of facilities and, sometimes, conflict of interest between the several purposes and agencies involved. An understanding of the problems involved in such situations is essential to anyone who wishes to make economy studies of such projects or to understand the cost data and political issues arising from them.

A number of basic problems may arise in connection with multipurpose public works. These may be illustrated by the simple example of a dam, shown in Figure 12-1, which is to be built in a semiarid portion of the United States, primarily to provide control against serious floods. It is at once apparent that, if the flow of the water impounded behind the dam could be regulated and diverted onto the adjoining land to provide irrigation water, the value of the land would be increased tremendously. This would result in an increase in the nation's resources, and it thus appears desirable to expand the project into one having two purposes: flood control and irrigation.

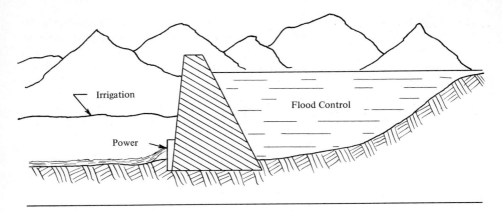

FIGURE 12-1   Schematic Representation of a Multiple-purpose Project Involving Flood Control, Irrigation, and Power.

The existence of a dam with a high water level on one side and a much lower level on the other side, at least during a good portion of the year, immediately suggests that some of the nation's resources will be wasted unless some of the water is permitted to run through turbines to generate electric power that can be sold to customers in the adjacent territory. Thus the project is expanded to have a third purpose—the generating of power.

In the semiarid surroundings the creation of a large lake, such as would exist behind the dam, would provide valuable recreation facilities for hunting, fishing, boating, camping, and so on. This gives a fourth purpose to the project. All of these purposes have desirable economic and social worth, and what started as a single-purpose project has ended by having four purposes. Not to develop the project to fill all four functions probably would mean that valuable national resources would be wasted. On the other hand, however, the potential loss of land (e.g., for mining minerals and retaining ecological balance, etc.) must be factored into the analysis as disbenefits.

If the project is built for four purposes, the important fact that one dam will serve all of them will result in at least three basic problems. The *first* is the allocation of the cost of the dam to the four purposes that it will serve. Assume, for example, that the dam will cost $10,000,000. What portion of this amount should be assigned to flood control? What amounts to irrigation, power generation, and recreation? If the identical dam will serve all four purposes, deciding how much of its cost should be assigned to any one of the purposes obviously might present considerable difficulties. One extreme would be to decide that, since the dam was required for flood control, its entire cost should be assigned to this purpose. If this were done, the cost of the electric power and the irrigation water would obviously be far less than if a considerable part or all of the $10,000,000 had been assigned to these functions.

Another extreme would be to assign each purpose with an amount equal to the total cost of the dam. Such a procedure would have two effects. First, it would undoubtedly make some of the purposes so costly—for example, recreation and irrigation—that they could not be undertaken. Second, by using this procedure, the cost of the dam might be returned to the government several times. Although many taxpayers would appreciate such a profitable occurrence, it is hardly feasible.

The *second* basic problem in multiple-purpose projects is the matter of conflict of interest among the several purposes. This may be illustrated by the matter of the water level maintained behind the dam. For flood control it would be best to have the reservoir empty most of the year to provide maximum storage capacity during the season when floods might occur. Such a condition would be most unsatisfactory for power generation. For this purpose it would be desirable to maintain as high a level as possible behind the dam during most of the year. For recreational purposes a fairly constant level also would be desirable. The requirements for irrigation probably would be somewhere between those for flood control and for power generation. Furthermore, while the lake created might be ardently approved by boating and recreation enthusiasts, it might be opposed with equal vigor by ecology-minded groups. Thus some very definite conflicts of interest usually arise in connection with multiple-purpose works. As a result, compromise decisions must be made.

A *third* problem connected with multiple-purpose public works is the matter of politics. Since the various purposes are likely to be desired or opposed by various segments of the public and by various interest groups that will be affected, it is inevitable that such projects frequently become political issues. This often has an effect upon cost allocations and thus on the overall economy of these projects.

The net result of these three factors is that the cost allocations made in multiple-purpose public projects tend to be arbitrary. As a consequence, production and selling costs of the services provided also are arbitrary. Because of this, they cannot be used as valid ''yardsticks'' with which similar private projects can be compared to determine the relative efficiencies of public and private ownership.

WHY ECONOMY STUDIES OF PUBLIC WORKS?

In view of the various problems that have been cited, engineers and other individuals sometimes raise the question as to whether economy studies of governmental projects should be attempted. It must be admitted that such economy studies frequently cannot be made in as complete and satisfactory a manner as in the case of studies of privately financed projects. However, decisions regarding the investment and use of capital in public projects must be made—by the public, by elected or appointed officials and, very importantly, by managers, who usually cannot use the same measures of operating efficiency that are available to managers of private ventures. The alternative to basing such decisions on the best possible economy studies is to base them on hunch and expediency. It is therefore essential that economy studies be made in the best possible manner, but with a sound understanding of the nature of such activities and of all the background, conditions, and limitations connected with them.

It must be recognized that economy studies of public works can be made from several viewpoints, each applicable for certain conditions. Many studies are made from the point of view of the governmental body involved. In comparing alternative structures or methods for accomplishing the same objective, such a point of view is satisfactory. The point of view of the citizens in a restricted area is used in other studies, such as local self-liquidating or service projects where those who must pay the costs are the direct beneficiaries of the project. In other cases the entire nation may be affected, so the economy study should reflect a very broad point of view. Such studies are apt to involve many ''irreducible'' costs and benefits that cannot be assigned to specific persons or groups. In each case as many factors as possible should be evaluated in monetary terms, because this usually will be most helpful in making decisions regarding the expenditure of public funds. However, it must also be recognized that non-monetary factors often exist and are frequently of great importance in final decisions.

THE DIFFICULTIES INHERENT IN ECONOMY STUDIES OF PUBLIC WORKS

There are a number of difficulties inherent in public works which must be considered in making economy studies and economic decisions regarding them. Some of these are as follows:

1. There is no profit standard to be used as a measure of financial effectiveness. Most public works are intended to be nonprofit.
2. There is no easy-to-quantify monetary measure of many of the benefits provided by a public project.
3. There frequently is little or no direct connection between the project and the public, who are the owners.
4. Whenever public funds are used, there is apt to be political influence. This can have very serious effects on the economy of projects, from conception through operation, and knowledge of its actual or possible existence is an important factor to be considered.
5. The usual profit motive as a stimulus to effective operation is absent. This may have a marked effect on the effectiveness of a project, from conception through operation. This does not mean that all public works are inefficient or that many managers and employees of them are not trying to do, and are actually doing, an effective job. But the fact remains that the direct stimuli present in privately owned companies are lacking and that this may have a considerable effect on project economics.
6. Public works usually are much more circumscribed by legal restrictions than are private companies. Their ability to obtain capital usually is restricted. Frequently, their area of operations is restricted as, for example, where a municipally owned power company cannot sell power outside the city limits.
7. In many cases decisions concerning public works, particularly their conception and authorization, are made by elected officials whose tenure of office

is very uncertain. *As a result, immediate costs and benefits may be stressed, to the detriment of long-range economy.*

WHAT INTEREST RATE SHOULD BE USED FOR PUBLIC PROJECTS?

The interest rate should play the same formal role in the evaluation of public sector investment projects that it does in the private sector. The rationale for its use is somewhat different. In the private sector, it is used because it leads directly to the private sector goal of profit maximization or cost minimization. In the public sector, its basic function is similar, in that it should lead to a maximization of social benefits, provided that these have been appropriately measured. Choice of an interest rate will lead to determination of how available funds may be allocated best among competing projects.

Three main considerations may bear on what interest rate to use in governmental economy studies:

1. The interest rate on borrowed capital.
2. The opportunity cost of capital to the governmental agency.
3. The opportunity cost of capital to the taxpayers.

In general, it is appropriate to use the borrowing rate only for cases in which money is borrowed specifically for the project(s) under investigation and use of that money will not cause other worthy projects to be forgone.

The opportunity cost (interest rate) encompasses the annual rate of profit (or other social or personal benefit) to the constituency served by the government agency or the composite of taxpayers who will eventually pay for the government project. If projects are chosen so that the return rate on all accepted projects is higher than that on any of the rejected projects, the interest rate used in the economic analysis is that associated with the best opportunity forgone. If this is done for all projects and investment capital available within a government agency, the result is a *governmental opportunity cost.* If, on the other hand, one considers the best opportunities available to the taxpayers if the money were not obtained through taxes for use by the governmental agency, the result is a *taxpayers' opportunity cost.*

Theory suggests that in usual governmental economic analyses the interest rate should be the largest of the three listed above. Generally, the taxpayers opportunity cost is the highest of the three. As an indicative example, a federal government directive† in 1972 specified that an interest rate of 10% should be used in economy studies for a wide-range of federal projects. This 10%, it can be argued, is at least a rough approximation of the average return taxpayers could be obtaining from the use of that money. In any case, it is greater than typical rates of interest that the federal government has been paying for the use of borrowed money.

†Office of Management and Budget, *Circular No. A-94* (rev.), March 27, 1972. The 10% was an inflation-free interest rate, which means that the *combined* interest rate would be higher.

THE BENEFIT–COST (B/C) RATIO

Basically, an economy study of a public project is no different from any other economy study. Any of the methods of Chapter 5 could be used for such a study. From a practical viewpoint, however, because profit almost never is involved and most projects have multiple benefits, some of which cannot be measured precisely in terms of dollars, many economy studies of public projects are made by comparing equivalent costs or by determining the ratio of the equivalent benefits to the equivalent costs.

The *benefit–cost (B/C) ratio* can be defined as the ratio of the equivalent worth of benefits to the equivalent worth of costs. The equivalent worths are usually A.W.'s or P.W.'s, but they can also be F.W.'s. In the case of governmental agencies, the benefits normally accrue to the user public and the net costs of supplying the benefits accrue to the government agency. The B/C ratio is also referred to as the *savings-investment ratio* (SIR) by some governmental agencies.

Two commonly used formulations of the B/C ratio (here expressed in terms of equivalent *annual worths*) are as follows:

1. Conventional B/C ratio:

$$B/C = \frac{\text{A.W. (net benefits to user)}}{\text{A.W. (total net costs to supplier)}} = \frac{B}{C.R. + (O + M)} \quad (12\text{-}1)$$

where A.W.$(\cdot)$ = annual worth of $(\cdot)$

B = annual worth of net benefits (gross benefits minus costs) to the user

C.R. = capital recovery cost or the equivalent annual cost of the initial investment, including any salvage value

$O + M$ = uniform annual net operating and maintenance costs to the supplier

2. Modified B/C ratio:

$$B/C = \frac{B - (O + M)}{C.R.} \quad (12\text{-}2)$$

The numerator of the modified B/C ratio expresses the equivalent worth of the net benefits minus operating and maintenance costs; the denominator includes only the investment costs. A project is acceptable when the ratio of benefits to costs is greater than or equal to 1.0,

The conventional B/C ratio method appears to be used more often than the modified B/C ratio method. Although both B/C methods give consistent answers regarding whether the ratio is greater than or equal to 1.0 and both yield the same recommended choice when comparing mutually exclusive investment alternatives, they can yield different rankings (e.g., which is best, second best, etc.) for independent investment opportunities. In this regard when comparing mutually exclusive investment *alternatives*, an incremental approach is required.

Independent *proposals* are deemed worthwhile investments if the B/C ratios are ≥ 1.0. Both of these situations are illustrated in the following sections.

COMPUTATION OF B/C RATIOS FOR A SINGLE PROJECT

EXAMPLE 12-1

Determine the B/C ratio for the following project.

First cost	$20,000
Project life	5 years
Salvage value	$4,000
Annual benefits	$10,000
Annual O + M disbursements	$4,400
Interest rate	8%

Solution

The conventional B/C ratio and modified B/C ratio, based on equivalent annual worths, are computed as follows:

$$\text{C.R.} = (\$20,000 - \$4,000)(A/P, 8\%, 5) + \$4,000(0.08)$$

$$= \$4,327$$

$$\text{conventional B/C ratio} = \frac{B}{\text{C.R.} + (O + M)} = \frac{\$10,000}{\$4,327 + \$4,400}$$

$$= 1.146 > 1.0$$

$$\text{modified B/C ratio} = \frac{B - (O + M)}{\text{C.R.}} = \frac{\$10,000 - \$4,400}{\$4,327}$$

$$= 1.294 > 1.0$$

Since B/C ≥ 1.0 (by either ratio), the individual investment opportunity is a worthwhile investment. ■

COMPARISON OF INDEPENDENT OPPORTUNITIES BY B/C RATIOS

EXAMPLE 12-2

It is desired to compare the two investment projects in Example 6-6 by the B/C ratio method, with annual revenue to be considered as annual benefits. Assume that the two are *independent* opportunities rather than mutually exclusive alternatives. The minimum attractive rate of return is 10%, and a restatement of the problem is as follows:

	Alternative	
	A	B
Investment	$3,500	$5,000
Annual net benefits to user	1,900	2,500
Annual net O + M expenses to supplier	645	1,383
Estimated life	4 years	8 years
Net salvage value	0	0

Solution

	Alternative	
	A	B
Annual net benefits minus net O + M expenses:		
$1,900 − $645	$1,255	
$2,500 − $1,383		$1,117
C.R. costs:		
$3,500(A/P, 10%, 4)	1,104	
$5,000(A/P, 10%, 8)		937
Modified B/C ratio:		
$1,255/$1,104	1.14	
$1,117/$937		1.19

Thus investment opportunity B is the better of the two, but both are satisfactory because the B/C ratios are > 1.0. ∎

COMPARISON OF MUTUALLY EXCLUSIVE ALTERNATIVES BY B/C RATIOS

Mutually exclusive projects and alternative levels of investment in a given project also can be evaluated by using B/C ratios.

When different levels of investment and cost may be employed in carrying out a specific objective, usually with somewhat different levels of results, confusion and error sometimes exist in interpreting the corresponding benefit-cost ratios and in making the proper decision as to which alternative to adopt. The proper selection between the alternatives using the B/C ratio involves the same principles of incremental return discussed in Chapter 6. Specifically:

1. Each increment of cost should justify itself by a sufficient B/C ratio (generally ≥ 1.0) on that increment.
2. Compare higher cost alternative against a lower cost alternative only if that lower cost alternative is justified.

3. Choose the alternative requiring the highest cost for which funds are available and for which each increment of cost is justified (by a B/C ratio $\geq$ 1.0).

A typical problem and the correct method of analysis follow.

EXAMPLE 12-3

In dealing with a certain problem, seven alternatives are available, the first being to take no corrective action and thus to accept an annual loss of $500,000 per year. The annual benefits and costs and the B/C ratios are as shown in Table 12-1. Because all the B/C ratios exceed 1, it is not at once apparent which alternative should be selected. We might be tempted to select alternative C because it has the highest B/C ratio and its annual costs are much less than some of the other alternatives. On the other hand, if a project is justified when its B/C ratio exceeds 1, then we might be tempted to select alternative G because it produces the largest annual benefits. Actually, neither of these decisions would be correct. Which is the best alternative?

Solution

If the data are computed and arranged according to increasing annual cost as shown in Table 12-2, the proper decision becomes clear. In accordance with the principles set forth in Chapter 6, alternative A is the basic alternative because it involves the minimum expenditure of capital. The incremental benefit and incremental costs, *compared with the acceptable alternative having the next lower annual costs*, are computed. Then the *incremental* B/C ratios are computed, as shown in the next-to-last column of Table 12-2. It will be noted that, as in the case of computing incremental rates of return, whenever an incremental B/C ratio is less than 1.0, that alternative is eliminated from further consideration, and each successive alternative is compared with the last, previous alternative having an acceptable B/C ratio. Thus, in Table 12-2, alternatives D and G are not acceptable. Based on the incremental B/C ratios, and assuming capital is not limited, it is clear that alternative F should be selected because it provides maximum benefits while having a B/C ratio greater than 1.0 on each increment of cost. ■

TABLE 12-1 Annual Benefits, Costs, and Benefit–Cost Ratios for Seven Alternatives

Alternative	Annual Benefits ($000s)	Annual Costs ($000s)	B/C
A	—	500	—
E	900	600	1.50
C	1,600	800	2.00
B	1,660	850	1.95
D	1,200	900	1.33
F	1,825	1,000	1.83
G	1,875	1,100	1.70

TABLE 12-2 Incremental Benefit–Cost Ratios for Seven Alternatives

Increment Considered	Incremental Benefit ($000s)	Incremental Cost ($000s)	Incremental B/C Ratio	Justified?
A	—	—	—	—
A → E	900	100	9.0	Yes
E → C	700	200	3.5	Yes
C → B	60	50	1.2	Yes
B → D	−460	50	Negative	No
B → F	165	150	1.1	Yes
F → G	50	100	0.5	No

EXAMPLE 12-4

Suppose that a certain government agency is considering the same set of three *independent* projects described in Example 6-12. The budget for sponsoring these public projects is $200 million and a 10% interest rate is required in such undertakings. By using benefit–cost analysis, which project(s) should be initiated? A restatement of relevant data is as follows:

Project	Investment, P	Net Annual Benefits, A	Useful Life, N
A	$93 million	$13.0 million	15 years
B	55 million	9.5 million	10 years
C	71 million	10.4 million	30 years

Solution

From three independent projects, a total of $2^3 = 8$ mutually exclusive alternatives (MEAs) can be developed. By using the procedure described in Chapter 6 (see Example 6-13), the following table of MEAs results for the three projects under consideration.

Mutually Exclusive Alternative (MEA)	Project A	Project B	Project C	Total P.W. (Investment Costs)	Total P.W. of Benefits	B/C Ratio
1	1	0	0	93.0	98.9	1.0634
2	0	1	0	55.0	58.4	1.0618
3	0	0	1	71.0	98.0	1.3803
4	1	1	0	148.0	157.3	1.0628
5	1	0	1	164.0	196.9	1.2000
6	0	1	1	126.0	156.4	1.2413
7 (Eliminate)	1	1	1	219.0	255.3	1.1658
8	0	0	0	0	0	—

Alternative 7 is eliminated because its budget exceeds the $200 million available.

An *incremental analysis* must be performed to make the correct choice. The comparisons are as follows in order of increasing investment cost:

Comparison of MEAs	$\Delta B/\Delta C$	Decision
8 (do nothing) $\rightarrow$ 2	1.0634	Select alternative 2
2 $\rightarrow$ 3	2.4750	Select alternative 3
3 $\rightarrow$ 1	0.0409	Keep alternative 3
3 $\rightarrow$ 6	1.0618	Select alternative 6
6 $\rightarrow$ 4	0.0409	Keep alternative 6
6 $\rightarrow$ 5	1.0658	Select alternative 5

As a result of an incremental analysis, the decision would be to select alternative 5 (projects A and C). Notice that this choice agrees with that of Example 6-12. It is also important to observe that the correct alternative *was not chosen* by the total B/C ratio computed for each. Based on maximizing the B/C ratio, alternative 3 would have been incorrectly chosen. The B/C ratio suffers the same drawbacks that characterize rate of return methods when used to select among mutually exclusive alternatives. Therefore, an incremental analysis is required when utilizing the B/C ratio method (and rate of return methods) to evaluate mutually exclusive investment alternatives. ■

Before a public project is undertaken, some governing body—the Congress in the case of federal projects—must approve that it has social usefulness. Its economic justification then is based on the B/C ratio. Typically, a project may be designed and carried out at different levels. If only a portion of the project is undertaken, the resulting B/C ratio may be less than 1.0. If developed more extensively, the ratio may rise rapidly and be substantially greater than 1.0. Further expansion of the project may result in a decrease in the B/C ratio, with the project ultimately becoming uneconomic if overexpanded. Diminishing returns are just as applicable to public projects as to private projects.

LIFE-CYCLE COSTS

Comparing mutually exclusive projects which satisfy a common objective on the basis of their *life-cycle costs* has become popular recently in various government agencies.† The life-cycle costing procedure sums the *equivalent* worths of design, procurement, installation, operating, and dismantling costs for each alternative over its useful life and permits the alternative having the *minimum* life-cycle cost to be identified. Typically, an interest rate of 10% is utilized in the calculation of equivalent life-cycle costs of each alternative. As seen in

†For example, see R.T. Ruegg et al., *Life-Cycle Costing*, National Bureau of Standards Science Series 113, September 1978.

Example 12-5, this procedure is identical to the cost-only evaluation methods (A.W.-C., P.W.-C., and F.W.-C.) discussed in Chapter 5 and 6.

EXAMPLE 12-5

A government agency is considering the replacement of an existing coal-fired furnace in one of its buildings by a large heat pump that will cost $15,000 to purchase and install. An overhaul cost of $3,000 would be required to make the furnace competitive with a heat pump. Data representing most likely estimates for both alternatives are shown below. Compare the alternatives by using the life-cycle costing procedure.

	Coal-Fired Furnace[a]	Electric Heat Pump[a]
First cost	$3,000	$15,000
Maintenance cost/year	$1,200	$1,000
Energy cost/year	$3,800 (coal)	$2,500 (electricity)
Salvage value (at end of useful life)	0	$3,000
Useful life	15 years	15 years
Interest rate	10%	10%

[a]All cost estimates are in real dollars at the present time, and 10% is an inflation-free real interest rate. Coal escalates at a differential rate of 2% per year and electricity escalates at 5% per year.

Solution

Because this is a government-sponsored project, income taxes do not need to be considered. The most straightforward approach to determining life cycle costs of each alternative is to work with real dollars and $i_r = 10\%$. The present worth and annual worth of costs for each plan are computed as follows.

Coal-Fired Furnace:

$$\text{P.W.-C.} = \$3,000 + \$1,200(P/A, 10\%, 15)$$

$$+ \$3,800\left(P/A, \frac{10\% - 2\%}{1.02}, 15\right)$$

$$= \$44,967$$

$$\text{A.W.-C.} = \$44,967(A/P, 10\%, 15) = \$5,912/\text{year}$$

Electric Heat Pump:

$$\text{P.W.-C.} = \$15,000 + \$1,000(P/A, 10\%, 15)$$

$$+ \$2,500(P/A, \frac{10\% - 5\%}{1.05}, 15) - \$3,000(P/F, 10\%, 15)$$

$$= \$48,260$$

$$\text{A.W.-C.} = \$48,260(A/P, 10\%, 15) = \$6,345/\text{year}$$

Calculating a present worth of the geometric gradient for escalating annual energy costs has been discussed in Chapters 4 and 8. From the analysis above, it is clear that the coal-fired furnace minimizes the life-cycle costs to the government agency. ■

SUMMARY

From the discussion and examples of public projects presented in this chapter, it is apparent that because of the methods of financing, the absence of the tax and profit requirements, and political and social factors, the same criteria frequently *cannot* be applied to such works as are used in evaluating privately financed projects. Neither should public projects be used as "yardsticks" with which to compare private projects. Nevertheless, whenever possible, public works should be justified on an economic basis to assure that the public obtains the maximum return from the tax money that is spent. Whether an engineer is working on such projects, is called upon as a consultant to them, or only fills the role of a taxpayer, he or she is bound by professional ethics to do the utmost to see that these projects are carried out in the best possible manner within the limitations of the legislation enacted for their authorization.

PROBLEMS

12-1   (a) What is a self-liquidating project?
(b) What is a multiple-purpose project?
(c) Why is it difficult to make a logical assignment of costs in a multiple-purpose project?
(d) Why is it difficult to assess the efficiency of operation in most multiple-purpose projects?

12-2   Discuss some of the considerations that go into determining an interest rate to be used in evaluating alternatives in the public sector.

12-3   Consider the following mutually exclusive alternatives:

Alternative	Equivalent Annual Cost of Project	Expected Annual Flood Damage	Annual Benefits
I. No flood control	0	$100,000	0
II. Construct levees	$ 30,000	80,000	$112,000
III. Build small dam	100,000	5,000	110,000

Which alternative would be chosen according to these decision criteria:
(a) Maximum benefit?
(b) Minimum cost?
(c) Maximum benefits minus costs?
(d) Largest investment having an incremental B/C ratio larger than 1.0?
(e) Largest B/C ratio?
Which project *should* be chosen?

12-4   A municipality is considering installing a sewage treatment plant which will cost $500,000, and require annual out-of-pocket costs of $25,000 for an expected life of 25 years. If the plant is installed, it is thought that it will result in annual savings of $15,000 in the required filtration and chlorination of the water supply. The project is to be financed by 8% bonds and the $500,000 for repayment after 25 years is to be accumulated in an 8% sinking fund. Determine the following:

(a) The additional annual tax money required to just pay the cost of this project.

(b) The minimum annual value which is implicitly put on the intangible benefits such as lessened stream pollution and increased natural beauty caused by the installation of the treatment plant.

12-5   An area on the Colorado River is subject to periodic flood damage, which occurs, on the average, every 2 years and results in $2,000,000 loss. It has been proposed that the river channel should be straightened and deepened, at a cost of $2,500,000, to reduce the probable damage to not over $1,600,000 for each occurrence during a period of 20 years before it would have to again be deepened. This procedure also would involve annual expenditures of $80,000 for minimal maintenance. One legislator in the area has proposed that a better solution would be to construct a flood-control dam at a cost of $8,500,000, which would last indefinitely with annual maintenance costs of not over $50,000. He estimates that this project would reduce the probable annual flood damage to not over $450,000. In addition, this solution would provide a substantial amount of irrigation water that would produce an annual revenue of $175,000 and recreational facilities, which he estimates would be worth at least $45,000 per year to the adjacent populace. A second legislator believes that the dam should be built and that the river channel also should be straightened and deepened, noting that the total cost of $11,000,000 would reduce the probable annual flood loss to not over $350,000, while providing the same irrigation and recreational benefits. If the state's capital is worth 10%, determine the B/C ratios and the incremental B/C ratio. Recommend which alternative should be adopted.

12-6   Ten years ago the port of Secoma built a new pier containing a large amount of steel work, at a cost of $300,000, estimating that it would have a life of 50 years. The annual maintenance cost, much of it for painting and repair caused by the environment, has turned out to be unexpectedly high, averaging $27,000. The port manager has proposed to the port commission that this pier be replaced immediately with a reinforced concrete pier at an initial cost of $600,000. He assures them that this pier will have a life of at least 50 years, with annual maintenance costs of not over $2,000. He presents the following figures as justification for the replacement, having determined that the net salvage value of the existing pier is $40,000:

Annual Cost of Present Pier		Annual Cost of Proposed Pier	
Depreciation: $300,000/50	$ 6,000	Depreciation: $600,000/50	$12,000
Maintenance cost	27,000	Maintenance cost	2,000
TOTAL	$33,000	TOTAL	$14,000

He has stated that since the port earns a net profit of over $3,000,000 per year, the project could be financed out of annual earnings and there would thus be no interest cost, and an annual saving of $19,000 would be obtained by making the replacement.

(a) Comment on the port manager's analysis.

(b) Make your own analysis and recommendation regarding the proposal.

12-7   A state Resources Development Department has proposed building a dam and hydro-electric project that will remedy a flood situation on a mountain river, generate power, provide water for irrigation and domestic use, and provide certain recreational facilities for boating and fishing. The construction costs would be:

Dam	$40,000,000
Access roads	2,000,000
Power plant	4,000,000
Transmission lines	1,500,000
Fish ladders and elevators	800,000
Irrigation and water canals	3,000,000

It is proposed to finance the project by issuing 8%, tax-exempt, 40-year bonds.

It is estimated that the annual operation and maintenance costs will be $1,250,000 for the power generating and distributing facilities and $750,000 for all other portions of the project. In addition, the state will pay $400,000 annually to the county where the project is located in lieu of property taxes. Estimates of the annual benefits and revenues are as follows:

Flood control	$ 900,000
Sale of power	2,700,000
Sale of water	1,600,000
Recreation benefits	1,800,000
Income from sports concessions	200,000

(a) Determine the B/C ratio for the project, using the estimated values of the benefits as stated.

(b) Would the elimination of the benefits that you consider to be intangible leave the project justified economically?

12-8 It is being proposed that a new toll bridge be built over an inlet of a bay, to be financed by the state highway department by means of a self-liquidating, 20-year 10% bond issue. The bridge would reduce the travel distance into a city by 8 miles. A careful traffic survey indicates that the $40,000,000 bridge would be used by an average of 40,000 private cars and 3,000 trucks and commercial vehicles each day. A policy has been established that trucks and commercial vehicles should pay five times the charge for private vehicles on such state toll bridges. If the estimated annual out-of-pocket costs for operation and maintenance for the bridge are $1,000,000, what toll fees would have to be paid by private vehicles and by commercial vehicles in order for the bridge to be self-liquidating?

12-9 A state-sponsored Forest Management Bureau is evaluating alternative routes for a new road into a formerly inaccessbile region. Three mutually exclusive plans for routing the road provide different benefits, as indicated in the following table:

Route	Construction Cost	Annual Maintenance Cost	Annual Savings in Fire Damage	Annual Recreational Benefit	Annual Timber Access Benefit
A	$185,000	$2,000	$ 5,000	$3,000	$ 500
B	220,000	3,000	7,000	6,500	1,500
C	290,000	4,000	12,000	6,000	2,800

The roads are assumed to have an economic life of 50 years, and the interest rate is 8%. Which route should be selected according to the B/C ratio method?

12-10 In developing a publicly owned, commercial, waterfront area, five possible independent plans are being considered. Their costs and estimated benefits are as follows:

	Present Worth ($000s)	
Plan	Costs	Benefits
A	123,000	139,000
B	135,000	150,000
C	99,000	114,000

Which plan should be adopted, if any, if the controlling board wishes to invest any amount required provided that the B/C ratio on the required investment is at least 1.0? Solve by use of incremental analysis.

12-11 Five mutually exclusive alternatives are available for developing a certain public project. The following tabulation shows the annual benefits and costs for each:

Alternative	Annual Benefits	Annual Costs
A	$1,800,000	$2,000,000
B	5,600,000	4,200,000
C	8,400,000	6,800,000
D	2,600,000	2,800,000
E	6,600,000	5,400,000

(a) Assume that the projects are of the type for which the benefits can be determined with considerable certainty and that the agency is willing to invest money as long as the B/C ratio is at least 1. Which alternative should be selected?

(b) If the projects involved intangible benefits which required considerable judgment in assigning their values, would this affect your recommendation?

12-12 A small, municipal airport is overcrowded and must increase its capacity to handle takeoffs and landings of planes. One plan is to purchase adjacent land and build a new landing strip and off-ramp. This would cost $200,000 and would add $16,000 per year in maintenance costs. A second plan is to widen and lengthen the existing runways at a cost of only $80,000, but this plan would seriously interrupt the operation of the airport, causing much public and user inconvenience, and would result in an estimated loss of revenue of $5,000 per month for the next 12 months. Furthermore, the maintenance costs with this plan are estimated to be $2,000 per year higher than for the other plan during the first year and would increase by $500 per year for each of the following 9 years. Use a study period of 10 years and an interest rate of 10% per year in determining the life cycle costs of both plans. Which one should be chosen?

12-13 With reference to Example 12-5, what is the amount of change in each of these factors that would cause the life-cycle costs of the electric heat pump to be lower than those of the furnace:

(a) Salvage value of the heat pump?

(b) Escalation rate of electricity?

(c) Energy cost per year for coal?

(d) First cost of the heat pump?

12-14   Rework Example 12-4 by using the life-cycle costing procedure. Should your result agree with that determined with the B/C ratio method?

12-15   Select some public works project which currently is being proposed or carried out in your area. Obtain the economic analysis for the project. Determine the following:

(a) Are there any intangibles included in the benefits?

(b) If there are intangibles, would the project be economically justified if the intangible benefits were omitted?

(c) Would most people in the affected area agree on the values assigned to these benefits?

CHAPTER 13

Economy Studies
for Public Utilities

Privately owned, regulated public utilities are an important part of the United States economy and constitute a unique form of business enterprise. Economy studies in such firms, because of their unique characteristics, are often performed in a manner slightly different than for the nonregulated sector of the economy. To understand these differences, and the reasons for them, it is helpful to have an understanding of the nature of privately owned, regulated utilities. It is the purpose of this chapter to point out the characteristic features of such utility firms, and then to show how these characteristics affect their engineering economy studies.

Some utility firms are owned and operated by governmental bodies, usually by cities. However, some are state-owned or even federally owned, such as the Tennessee Valley Authority. They are basically the same as any other governmentally owned and operated activity, and their engineering economy studies were previously discussed in Chapter 12.

THE NATURE OF PUBLIC UTILITIES

Public utilities provide services, such as gas, electric power, water, telephone and radio communication; and transportation, including air, rail, bus, and pipeline. To provide these services, public property, such as streets, highways, or air space, must be utilized, or the utility must be given the right of *eminent domain* to acquire property where and when needed.

Because a utility must have unusually large amounts of capital invested in fixed plant and equipment,† economy of operation for the company and lower rates to the public can be brought about only by high use factors for such assets. This means that unnecessary duplication of such facilities would not be economical or in the public interest. For example, two competing electric power companies in the same area would not provide the most economical service. If each had customers on the same street, power-line poles, transformers, distribution lines, and many clerical functions would have to be duplicated. One of the companies could serve all the customers with very little increase in its investment in fixed assets.

In some types of utility services it is virtually impossible for satisfactory service to be obtained by the public except through a single utility, or at least a group of closely coordinated and noncompeting companies. For example, if two competing telephone companies served the same area and one company would not accept calls from the customers of the other, a completely unsatisfactory situation would result.

Recognizing the advantages to the public—the utility's customers—of avoiding wasteful duplication and competition, the public—again the customers—usually grants a utility an exclusive franchise for its particular service in a given geographical area. Thus most utilities have, in effect, monopolies with respect to the particular service that they render. However, it should be noted that many regulated utilities are not devoid of competition, in that there may be different, but competing, services. For example, a gas company may compete with an electric company in respect to domestic water heating and cooking.

Because the public—acting as cities, states, or the federal government—grants the monopolistic position to the utility, *it retains the right to regulate the utility.* The public controls the rates the utility may charge, so that no undue profit will be made as a result of the privilege it has granted. It is obvious that without competition the price to be charged for the services cannot be established through the usual process of competition. Therefore, a public regulatory body is established by the people to exercise the desired control through various types of measures. As a consequence, there is the unique situation of customers determining, by direct regulation, what the supplier of a service can charge.

HOW UTILITIES ARE REGULATED

Intrastate public utilities usually are regulated by a state agency, commonly known as a public utilities commission. The members of the commission may be appointed or elected, depending upon the laws of the individual state. Utilities

---

†For example, the Edison Electric Institute estimates that electric utilities need to spend over $300 billion during the 1980s to convert from high-cost petroleum fuels and replace aging, inefficient plants.

that operate interstate, such as railroads, bus lines, air lines, telephone companies, and pipeline companies, are regulated by Federal agencies.

Governmental regulation was established originally to prevent public utility companies from discriminating between customers as to service provided and prices charged. The functions of regulatory agencies have been expanded so that at present they are concerned with the setting of rates and the establishment and maintenance of standards of service.

Basic to the setting of rates for public utility services is the concept that the utility companies must be able to earn profits and to pay dividends sufficient to assure the obtaining of capital necessary to provide assets and working capital that are required for rendering the service. If an adequate rate of profit is not obtained, the necessary capital will not be forthcoming from investors, and, as a result, the public will not be able to have the desired utility service. On the other hand, the utility operates at the permission of the public, through necessary franchises or other grants of permission, and thus is granted a monopolistic position by its customers. The public, therefore, has a right to expect that it should pay no more for the services than will permit an efficiently operated utility to earn the minimum required rate of profit that will assure the continuity of service at the level of quality desired. Thus the regulatory commissions have a delicate task of sensing the desires of the public as to the level of service desired and its willingness to pay for it, the price that must be paid for capital in the current money market, and the efficiency of the plans and operations of the utility.

Utility commission control is exercised primarily through the setting or approving of rates that may be charged for utility services. These are set so as to permit the utility to earn the required rate of profit on the capital utilized in rendering the service. This amount of capital is sometimes called the *rate-base value*.

Although the matter of public utility regulation is complex and sometimes seems to be quite ponderous, it has worked rather well. Some of the best-managed companies in this country can be found among public utilities in states where there are strong, but fair, regulatory bodies. Services of excellent quality are consistently provided by these companies at low cost. The greatest difficulties are experienced where regulatory bodies are weak, are not farsighted or aware of current developments, or tend to impose managerial restrictions rather than merely regulation.

CHARACTERISTICS OF PUBLIC UTILITIES

Because of the inherent nature of the services they render, their monopolistic position, and the regulation to which they are subject, public utilities have a number of economic characteristics differentiating them from other businesses which must be taken into account in making economy studies. The major ones are discussed in the following paragraphs.

1. The investment per worker and the ratio of fixed costs to variable costs are very high. Correspondingly, the investment per dollar of gross revenue for

typical electric utilities is about three times that of steel companies and about five times that of general manufacturing companies. This results in high fixed costs. It is not uncommon for the fixed costs to be 70% of the total unit cost. This means that careful attention must be given to investment problems and to assuring an adequate flow of capital for expansion.

2. Utilities *must* render whatever service is demanded by customers, within established rate schedules. Subject to regulatory safeguards, a utility must expand to meet the growth of the community it serves.

3. A utility is *required* to keep abreast of technical developments in its field which would permit reduction in the cost to the customer for its services or would improve the quality of the service if such improvement is demanded by the customers. A utility should be prepared to improve the quality of the service, even though not demanded immediately by the customers, in order to maintain public goodwill and to protect its monopolistic position.

4. The rates charged for a utility's services are based on total costs, including a fair return, after income taxes, on the rate-base value of its property. The so-called rate base is roughly equal to the book value of a utility's plant and equipment.

5. The earnings of a public utility are virtually limited by the rate base. As a result, profit on sales is of very little significance. If sales income is increased as the result of increased operating costs—for example, by increased and more effective advertising—it may not produce any long-term profit increase. Profits for the current year might be realized, but if the increase were to result in a return that was judged by the regulatory commission to be greater than necessary, a rate reduction would be ordered. Thus the benefits of the improved operation in terms of financial gain to the company would be eliminated. The same situation exists with respect to increased earnings due to improved efficiency of operation.

This situation might appear to, and possibly does, reduce the incentive of a public utility to improve the efficiency of its operations. However, it is a condition that a utility accepts in return for the preferred position granted to it by the public.

6. Utilities have much greater stability of income than other companies. The upper limit of earnings, after income taxes, usually is not permitted to exceed about 12 to 16% on equity capital. It should be noted that, although there is a maximum limit put on earnings, there is no guarantee of any such profits, and there is no assurance against loss. However, if the utility can show that it is operating efficiently, it usually can obtain permission to increase its rates when needed to produce a fair profit, and thus be able to attract needed capital. For the well-managed utility company there is a stability of earnings that does not exist for the nonutility company, and this has considerable effect in lowering the rate the utility must pay for both borrowed and equity capital.

7. Because of the stable nature of their business and earnings, utilities commonly employ a higher percentage of borrowed capital than do nonutility companies. Whereas most nonutilities seldom use more than about 30% of debt capital, many utilities use 50 to 60% of borrowed capital.

8. The assets of a utility, on the average, involve longer write-off periods than those of nonutilities. This is caused by the physical nature of the assets

and by the fact that the monopolistic situation results in less functional depreciation.

9. Utilities must rely on a larger proportion of new capital for expansion than do other companies. This follows from the fact that earnings are so regulated that they are sufficient to pay only for the cost of capital. Therefore, if the earnings are just high enough to meet the payment of dividends demanded by investors, it follows that very few profits can be retained as surplus to provide for expansion. For example, one utility having equity capital of over $1,375,000,000 has only $104,000,000 in earned surplus, accumulated over more than 40 years.

10. Public utilities are much less limited as to the availability of capital than are nonutility companies. This is because of their greater stability of revenues and earnings and the fact that the regulatory agencies recognize that they must be permitted to earn a return that will assure an adequate flow of capital.

SOME GENERAL CONCEPTS REGARDING PUBLIC UTILITY ECONOMY STUDIES

Because of the characteristics of regulated public utilities which have just been discussed, there are some general features that usually are contained in engineering economy studies of such companies. These are as follows:

1. Whereas economy studies of nonutility companies usually reflect the viewpoint of the owners, those of regulated public utilities more nearly reflect the interests of the customer.

2. Public utility economy studies usually involve alternative ways of, or alternative programs for, *doing* something. Because a utility is obligated to provide the service demanded by its customers, studies are seldom made of the economy of doing versus not doing. Instead, it is more often a matter of how to do it most economically.

3. Administrative and general supervision costs frequently are not included. Because these expenses will be about the same for each alternative, they may be omitted.

4. Revenue data seldom are included. Here, again, the revenue ordinarily will be the same, regardless of the method of providing the service.

5. The cost of money, depreciation, income taxes, and property taxes are usually expressed in terms of the capital invested.

A popular procedure among investor-owned utilities for conducting engineering economy studies that utilizes the above concepts is the *revenue requirements method*. It is closely related to the equivalent worth methods described in Chapter 5.

The revenue requirements method is a basis for comparing mutually exclusive alternatives. The method calculates revenues that a given project must provide to just meet all the costs associated with it, including a minimum acceptable return to investors. This is accomplished by equating the present worth of all project costs to the stream of required revenues. Figure 13-1 shows the relationship between a project's revenue requirements and its costs.

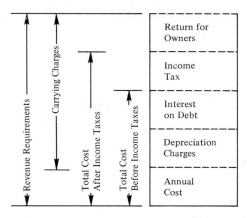

FIGURE 13-1 Relationship of Revenue Requirements and Costs for a Public Utility.

The revenue requirements method can be used to rank alternative investment opportunities, and it is also employed in establishing approximate rates (prices) for goods and/or services provided by a proposed project. In the case of mutually exclusive alternatives designed to accomplish a particular function, the alternative having the lowest revenue requirements should be chosen.

The following assumptions are made in using this method.

1. Total investment in an asset during any year is equal to its beginning-of-year book value.
2. Amount of debt capital invested in an asset during any year will be a constant fraction of its book value during that year, and this fraction will remain constant throughout the asset's life.
3. Equity and debt capital involve constant rates of return throughout the life of the project.
4. Book depreciation charges are used to retire capital stock and bonds each year in proportion to debt–equity mix of financing employed.
5. The effective income tax rate is constant over the life of the project.

DEVELOPMENT OF THE REVENUE REQUIREMENTS METHOD†

The interest paid on borrowed capital (debt) is tax deductible. Therefore, the after-tax cost of debt, i'_a, is

$$i'_a = i'_b - ti'_b$$

$$= (1 - t)[(1 + i_b)(1 + f) - 1]$$

(13-1)

†The notation in Chapter 13 differs from that used in the rest of the book because various costs of debt and equity capital and resultant costs of total capital are required here. Many of these same concepts are utilized in other chapters but are not developed and employed as rigorously as in Chapter 13.

where i_b = real before-tax cost of borrowed capital

i'_b = inflation adjusted cost of borrowed capital

$= [(1 + i_b)(1 + f) - 1]$

t = effective incremental income tax rate

f = annual inflation rate

The firm's cost of capital depends on the cost of both debt and equity capital.‡ The *after-tax cost of capital*, K'_a, including an adjustment for inflation, is

$$K'_a = \lambda i'_a + (1 - \lambda)e'_a \qquad (13\text{-}2)$$

$$= \lambda(1 - t)i'_b + (1 - \lambda)e'_a$$

where λ = fraction borrowed money in the utility's total capitalization

$(1 - \lambda)$ = fraction equity capital in total capitalization

e'_a = inflation adjusted equity rate $= [(1 + e_a)(1 + f) - 1]$

The real after-tax cost of capital, K_a, is

$$K_a = [(1 + K'_a)/(1 + f)] - 1 \qquad (13\text{-}3)$$

$$= \lambda(1 - t)i_b + (1 - \lambda)e_a - \lambda t f/(1 + f)$$

where e_a is the real equity rate. The *composite cost of capital*, K'_{ba}, includes the before-tax cost of borrowed capital. Its inflation-adjusted value is determined with Equation 13-4:

$$K'_{ba} = \lambda i'_b + (1 - \lambda)e'_a \qquad (13\text{-}4)$$

The real composite cost of capital, K_{ba}, is

$$K_{ba} = \lambda i_b + (1 - \lambda)e_a \qquad (13\text{-}5)$$

The minimum revenue requirement consists of carrying charges (CC) resulting from capital investments that must be amortized plus all associated expenses that occur periodically as can be seen in Figure 13-1. Carrying charges include the following;

1. Book depreciation
2. Return on investment.
3. Income taxes.

The following equation is used to find the annual carrying charges in year k, CC_k:

$$CC_k = D_{B_k} + [(1 - \lambda)e_a + \lambda i_b]UI_k + T_k \qquad (13\text{-}6)$$

where D_{B_k} = book depreciation at the beginning of year k, $1 \le k \le N$

UI_k = unrecovered investment in year k

T_k = income taxes in year k

---

‡There is some controversy over whether an investor-owned utility should use a tax-adjusted cost of capital in its determination of present worth of revenue requirements. An after-tax cost of capital is recommended in this book. (See Oso, 1978 and Ward and Sullivan, 1981 in Appendix B.)

Since tax depreciation and interest paid on debt are tax-deductible, the income tax in any given year is determined by the following equation when $f = 0$:

$$T_k = t(CC_k - \lambda i_b UI_k - D_{T_k})$$ (13-7)

where D_{T_k} is the tax depreciation in year k.

Note that carrying charges (CC_k) are a function of income taxes (T_k) in Equation 13-6 and income taxes (T_k) are a function of carrying charges (CC_k) in Equation 13-7. This can be seen clearly in Figure 13-1. The revenue requirement can be determined if the income taxes are known and, similarly, the income taxes can be computed if the revenue requirement is known. There are two equations and two unknowns (i.e., CC_k and T_k). Solving for T_k, we find

$$T_k = [t/(1 - t)][(1 - \lambda)e_a UI_k + D_{B_k} - D_{T_k}]$$ (13-8)

The revenue requirement in year k, (RR_k), is

$$RR_k = CC_k + C_k$$ (13-9)

where C_k represents all recurring annual expenses in year k.

ILLUSTRATION OF TABULAR PROCEDURE FOR CALCULATING REVENUE REQUIREMENTS

The following example demonstrates how to find the annual revenue requirements of a project in tabular form. Column headings of the table do not necessarily have to correspond to the format used in this example. Columns can be added or deleted according to the components of revenue requirement as given in Figure 13-1. For instance, if there had been a property tax, an additional column for this component would be used in Example 13-1.

EXAMPLE 13-1

The details of a particular project are as follows:

Project life, N	4 years
Book life	4 years
Tax life	4 years
Initial investment	$7,500
Salvage value	$1,500
Annual cost (O&M)	$500
Real cost of borrowed money, i_b	0.05
Real return to equity, e_a	0.1607
Debt ratio, λ	0.3
Effective income tax rate, t	0.5
Book depreciation	Straight line
Tax depreciation	Straight line
Average inflation rate, f	0

Based on unrecovered investment, the revenue requirement from Table 13-1 in year 1 is calculated as follows.

Column 1   Unrecovered investment at the beginning of year 1, UI_1, is the initial investment

Column 2   Book depreciation in year 1 is

$$D_{B_1} = (\$7{,}500 - \$1{,}500)/4 = \$1{,}500$$

Column 3   Tax depreciation in year 1 is

$$D_{T_1} = (\$7{,}500 - \$1{,}500)/4 = \$1{,}500$$

Column 4   Annual cost is given to be $500

Column 5   Return on debt is

$$\lambda i_b UI_1 = (0.3)(0.05)(\$7{,}500)$$

$$= \$112.50$$

Column 6   Return on equity is

$$(1 - \lambda)e_a UI_1 = (1 - 0.3)(0.1607)(\$7{,}500)$$

$$= \$843.68$$

Column 7   Income tax is found using Equation 13-8:

$$T_1 = [t/(1 - t)][(1 - \lambda)e_a UI_1 + D_{B_1} - D_{T_1}]$$

$$= [0.5/(1 - 0.5)][(1 - 0.3)(0.1607)(\$7{,}500)$$

$$+ 1{,}500 - \$1{,}500]$$

$$= \$843.68$$

Column 8   Revenue requirement in year 1 is the sum of columns 2, 4, 5, 6, and 7.

$$RR_1 = \$1{,}500 + \$500 + \$112.5 + \$843.68 + \$843.68$$

$$= \$3{,}799.86$$

TABLE 13-1   Annual Revenue Requirements (RR) with Straight Line Depreciation

Year, k	(1) Unre-covered Invest. UI_k	(2) Book Depr., D_{B_k}	(3) Tax Depr., D_{T_k}	(4) Annual Cost, C_k	(5) Debt Return, $\lambda i_b UI_k$	(6) Equity Return $(1 - \lambda)e_a UI_k$	(7) Income Tax, T_k	(8) RR_k = cols. 2 + 4 + 5 + 6 + 7
1	$7,500	$1,500	$1,500	$500	$112.50	$843.68	$843.68	$3,799.86
2	6,000	1,500	1,500	500	90	674.94	674.94	3,439.88
3	4,500	1,500	1,500	500	67.50	506.21	506.21	3,079.92
4	3,000	1,500	1,500	500	45	337.47	337.47	2,719.94

The revenue requirement in year 2 is found as shown above:

Column 1   Book depreciation during the first year in the amount of $1,500 is recovered. Therefore, the unrecovered investment at the beginning of year 2 is

$$UI_2 = UI_1 - D_{B_1} = \$7,500 - \$1,500 = \$6,000$$

Column 2 $D_{B_2} = \$1,500$

Column 3 $D_{T_2} = \$1,500$

Column 4   Annual cost is given to be $500.
Column 5   Debt return $= (0.3)(0.05)(\$6,000) = \$90$.
Column 6   Equity return $= (0.7)(0.1607)(\$6,000) = \$674.94$.

Column 7 $T_2 = [0.5/(1 - 0.5)][(0.7)(0.1607)(\$6,000)]$

$$= \$674.94$$

Column 8 $RR_2 = \$1,500 + \$500 + \$90 + \$674.94 + \$674.94$

$$= \$3,439.88$$

The remaining revenue requirements in Table 13-1 are computed similarly. At the end of year 4, book depreciation of $1,500 is recovered as well as a salvage value of $1,500. Hence there is no unrecovered investment at the end of year 4.

The tax-adjusted cost of capital, K_a, is determined with Equation 13-3 when $f = 0$:

$$K_a = \lambda(1 - t)i_b + (1 - \lambda)e_a - \lambda tf/(1 + f)$$

$$= 0.3(1 - 0.5)(0.05) + (1 - 0.3)(0.1607) - 0$$

$$= 0.12 \qquad \blacksquare$$

DETERMINATION OF EQUIVALENT ANNUAL REVENUE REQUIREMENT

The equivalent uniform annual revenue requirement in Example 13-1, also referred to as the "levelized" revenue requirement, is computed at the tax-adjusted cost of capital to be

$$\overline{RR}(12\%) = [\$3,799.86(P/F, 12\%, 1) + \$3,439.88(P/F, 12\%, 2)$$

$$+ \$3,079.92(P/F, 12\%, 3)$$

$$+ \$2,719.94(P/F, 12\%, 4)](A/P, 12\%, 4)$$

$$= \$3,310.70$$

or

$$\overline{RR}(12\%) = \$3{,}799.86 - \$359.98(A/G, 12\%, 4)$$

$$= \$3{,}310.70$$

The stream of annual revenue requirements, RR_k, or the "levelized" revenue requirement, $\overline{RR}$, over the life of the project is just sufficient to recover all costs including recovery of capital and a return on the unrecovered investment. If the actual revenues generated by a project exceed the minimum revenue requirement found in Example 13-1, the project is "profitable" after all cost and return obligations have been met. If the actual uniform equivalent revenue requirement is the same as the "minimum" revenue requirement determined as in Table 13-1, the project is acceptable because it still provides investors with a minimum acceptable return on their capital. If the actual uniform equivalent revenue requirement is less than the minimum revenue requirement, the project is unacceptable and should not be selected on economic grounds.

MODIFICATIONS TO EXAMPLE 13-1

Example 13-1 is now used to show that the tabular procedure for finding revenue requirements (i.e., Table 13-1) has many advantages over the so-called "fixed-charge" method of computing an equivalent uniform annual revenue requirement of a project.

The fixed-charge rate (FCR) is the annual carrying charge of a proposed alternative expressed as a fraction of the initial investment. As federal tax laws change and other complications (e.g., inflation) enter into the analysis, the FCR must be revised to provide valid comparisons among alternatives. Many utilities now favor the computation of annual revenue requirements, as illustrated in Table 13-1, over the fixed charge method.

In Example 13-2 these four cases are considered:

(a) The fixed-charge method.
(b) Table 13-1 with sum-of-the-years'-digits (SYD) depreciation for tax purposes.
(c) Table 13-1 with SYD depreciation and a 3-year tax life.
(d) Table 13-1 with A.C.R.S. depreciation, zero salvage value, and a 3-year tax life.

EXAMPLE 13-2
(a) Solution of Example 13-1 by the Fixed-Charge Method
This example utilizes a discount rate equal to the composite cost of capital (Equation 13-5), following the practice of several investor-owned utilities. The annual cost of income taxes as a proportion of the initial investment for the conditions given in Example 13-1 can be determined as follows.†

---

†Refer to P. H. Jeynes, *Profitability and Economic Choice* (Ames, Iowa: Iowa State University Press, 1968), pp. 181–223, for derivations of these equations.

$$T = [t/(1 - t)][1 - (\lambda i_b/K_{ba})][K_{ba} + (A/F, K_{ba}, N)(1 - c) - (1 - c)/N]$$

$$= \varphi[K_{ba} + (A/F, K_{ba}, N)(1 - c) - (1 - c)/N] \qquad (13\text{-}10)$$

$$= \varphi(\text{minimum required profit} + \text{sinking fund}$$
$$\qquad \text{depreciation} - \text{straight line depreciation})$$

where

$$\varphi = [t/(1 - t)][1 - (\lambda i_b/K_{ba})] \qquad (13\text{-}11)$$

$$(\varphi \text{ is termed the adjusted income-tax factor})$$

$$c = \text{salvage value proportion}$$

$$(= \text{salvage value/investment})$$

$$N = \text{life of project}$$

For Example 13-2(a), K_{ba} equals 0.1275 and the value of φ is

$$\varphi = [0.5/(1 - 0.5)][1 - (0.3)(0.05)/0.1275]$$

$$= 0.88235$$

In Table 13-2 it is seen that the fixed-charge rate is multiplied by the capital investment to determine an annual equivalent carrying charge. The equivalent uniform annual revenue requirement is then the sum of the annual carrying charges and the annual O&M costs and equals $3,313.79, i.e. 0.3752 ($7,500) + $500 = $3,313.79.

The remainder of examples in this chapter are solved by using the after-tax cost of capital, K_a, as the discount rate.

TABLE 13-2 Fixed-Charge Method of Calculating a Levelized Revenue Requirement

	Amount	Fixed Charge Rate (FCR)
Depreciation[a]		
($7,500 − $1,500)(A/F, 12.75%, 4)	$1,241.70	$1,241.70 ÷ $7,500 = 0.1656
Minimum required profit[a]		
$7,500(12.75%)	956.25	$956.25 ÷ $7,500 = 0.1275
Income tax[b]		
φ (minimum required profit + sinking fund depreciation − straight line depreciation) =		$615.84 ÷ $7,500
0.88235 ($956.25 + $1,241.70 − $1,500)	615.84	= 0.0821
Annual carrying changes	$2,813.79	0.3752
Annual O&M costs	500.00	
Equivalent annual revenue requirement	$3,313.79	

[a]The sum of depreciation, as computed, and minimum required profit is the capital recovery amount.
[b]Computed from equation 13-10 ■

(b) Solution of Example 13-1 with SYD Tax Depreciation

From Table 13-3, the levelized revenue requirement using the tax-adjusted cost of capital (K_a) is

$$\overline{RR}(12\%) = \$2,899.85 + \$240.03(A/G, 12\%, 4)$$

$$= \$3,226.02 \qquad\blacksquare$$

(c) Solution of Example 13-1 with a Tax-Life of 3 Years and SYD Tax Depreciation

The levelized revenue requirement from Table 13-4 at K_a is

$$\overline{RR}(12\%) = \$2,299.85 + \$640.03(A/G, 12\%, 4)$$

$$= \$3,169.56 \qquad\blacksquare$$

To avoid time-consuming redevelopments of the fixed-charge equations (i.e., equations 13-10 and 13-11) for numerous complications that arise in practice, it is recommended that the tabular procedure above be used either manually or with a digital computer.

(d) Solution of Example 13-1 with A.C.R.S. Depreciation, a Recovery Period of 3 Years, and Zero Salvage Value

From Table 13-5, the levelized revenue requirement using the tax-adjusted cost of capital is

$$\overline{RR}(12\%) = [\$3,574.86(P/F, 12\%, 1) + \$2,224.90(P/F, 12\%, 2)$$

$$+ \$3,499.93(P/F, 12\%, 3) + \$4,699.98(P/F, 12\%, 4)]$$

$$\times (A/P, 12\%, 4)$$

$$= \$10,444(0.3292) = \$3,438$$

TABLE 13-3   **Annual Revenue Requirements with SYD Depreciation**[a]

Year, k	(1) Unrecovered Invest.	(2) Book Depr.	(3) Tax Depr.	(4) Annual Cost	(5) Debt Return	(6) Equity Return	(7) Income Tax	(8) RR = cols. 2 + 4 + 5 + 6 + 7
1	$7,500	$1,500	$2,400	$500	$112.50	$843.68	$-\$$ 56.33	$2,899.85
2	6,000	1,500	1,800	500	90	674.94	374.94	3,139.88
3	4,500	1,500	1,200	500	67.50	506.21	806.21	3,379.91
4	3,000	1,500	600	500	45	337.47	1,237.47	3,619.94

[a]Columns 1, 2, 4, 5, and 6 are found as in Example 13-1.

Column 3:   Tax depreciation in year 1 is,

$$D_{T_1} = 4/[4(4 + 1)/2]\,(\$7,500 - \$1,500) = \$2,400$$

Column 7: $T_1 = [t/(1 - t)][(1 - \lambda)e_a\,UI_1 + D_{B_1} - D_{T_1}]$
$$= -\$56.33$$

Column 8: $RR_1 = \$1,500 + \$2,400 + \$500 + \$112.5 + \$843.68 + (-\$56.33)$
$$= \$2,899.85$$

TABLE 13-4 Annual Revenue Requirements with a Shortened Tax Life

Year, k	(1) Unre-covered Invest.	(2) Book Depr.	(3) Tax Depr.	(4) Annual Cost	(5) Debt Return	(6) Equity Return	(7) Income Tax	(8) RR = Cols. 2 + 4 + 5 + 6 + 7
1	$7,500	$1,500	$3,000	$500	$112.50	$843.68	−$ 656.33	$2,299.85
2	6,000	1,500	2,000	500	90	674.94	174.94	2,939.88
3	4,500	1,500	1,000	500	67.50	506.21	1,006.21	3,579.91
4	3,000	1,500	0	500	45	337.47	1,237.47	4,219.94

The value of $\overline{RR}$ increases in Example 13-2(d) compared to 13-2(c), for example, because the carrying charges increase when the salvage value is assumed to be zero instead of $1,500. ∎

Subsequent examples demonstrate how the tabular format of Example 13-1 can be extended to longer-lived studies that encompass several different types of project comparisons.

ALTERNATIVE NEW INSTALLATIONS

EXAMPLE 13-3

A public utility must extend electric power service to a new shopping center. A decision must be made as to whether a pole-line or an underground system should be used. The pole-line system would cost only $15,800 to install, but because of numerous changes that are anticipated in the development and use of the shopping center, it is estimated that annual maintenance costs would be $2,900. An underground system would cost $31,500 to install, but the annual maintenance costs would not exceed $550. Annual property taxes are 1.5% of first cost. The company operates with 33% borrowed capital, on which it pays an interest rate of 8%. Capital should earn about 11% after taxes. For this

TABLE 13-5 Annual Revenue Requirements with A.C.R.S. Depreciation

Year, k	(1) Unre-covered Invest.	(2) Book Depr.	(3) Tax Depr.[a]	(4) Annual Cost	(5) Debt Return	(6) Equity Return	(7) Income Tax	(8) RR = Cols. 2 + 4 + 5 + 6 + 7
1	$7,500	$1,875	$2,475	$500	$112.50	$843.68	+$ 243.68	$3,574.86
2	5,625	1,875	3,375	500	84.38	632.76	− 867.25	2,224.89
3	3,750	1,875	1,650	500	56.25	421.84	+ 646.84	3,499.93
4	1,875	1,875	0	500	28.13	210.92	+2,085.92	4,699.98

[a]Depreciable amount = $7,500. Year 1, 7,500 × 0.33 = 2,475; year 2, 7,500 × 0.45 = 3,375; year 3, 7,500 × 0.22 = 1,650. These are 1986 A.C.R.S. percentages.

problem, the after-tax return of 11% is interpreted as the value of K_a. The utility is in the 46% income tax bracket. A 20-year study period is to be used, and the effects of inflation on cash flows are to be ignored.

Solution

It is assumed that both book and tax depreciation methods are straight line and that the book life of 20 years equals the tax life. From Equation 13-2, the required return to equity can be determined when $f = 0$:

$$0.11 = 0.33(1 - 0.46)(0.08) + 0.67(e_a)$$

$$e_a = 0.1429$$

From Table 13-6, the levelized revenue requirement for the pole-line system using the tax-adjusted cost of capital is

$$\overline{RR}(11\%) = \$7,145.49 - \$160.93(A/G, 11\%, 20)$$

$$= \$6,138.23$$

From Table 13-7 the levelized revenue requirement for the underground system is

$$\overline{RR}(11\%) = \$9,014.10 - \$320.83(A/G, 11\%, 20)$$

$$= \$7,006.03$$

The equivalent uniform annual revenue requirement for the pole-line system is less than that of the underground system. Therefore, there would be a substantial saving in favor of the pole-line system. Of course, this is a case in which intangibles such as appearance and resistance to storm damage might swing the decision to the underground system. ∎

TABLE 13-6   Annual Revenue Requirements for the Pole-Line System

Year, k	(1) Unre- covered Invest.	(2) Book Depr.	(3) Tax Depr.	(4) Annual Cost[a]	(5) Debt Return	(6) Equity Return	(7) Income Tax	(8) RR = Cols. 2 + 4 + 5 + 6 + 7
1	$15,800	$790	$790	$3,137	$417.12	$1,512.74	$1,288.63	$7,145.49
2	15,010	790	790	3,137	396.26	1,437.10	1,224.20	6,984.56
3	14,220	790	790	3,137	375.41	1,361.47	1,159.77	6,823.65
.								
.								
.								
20		790	790	3,137	20.86	75.64	64.43	4,087.93

[a]Annual cost is sum of annual maintenance cost and annual property tax.

TABLE 13-7 Annual Revenue Requirements for the Underground System

Year, k	(1) Unre-covered Invest.	(2) Book Depr.	(3) Tax Depr.	(4) Annual Cost[a]	(5) Debt Return	(6) Equity Return	(7) Income Tax	(8) RR = Cols. 2 + 4 + 5 + 6 + 7
1	$31,500	$1,575	$1,575	$1,022.50	$831.60	$3,015.90	$2,569.10	$9,014.10
2	29,925	1,575	1,575	1,022.50	790.02	2,865.11	2,440.65	8,693.28
3	28,350	1,575	1,575	1,022.50	748.44	2,714.31	2,312.19	8,372.44
.								
.								
.								
20	1,575	1,575	1,575	1,022.50	41.58	150.79	128.46	2,918.33

[a]Annual cost is sum of annual maintenance cost and annual property tax.

A UTILITY REPLACEMENT PROBLEM

EXAMPLE 13-4

Because of earth slippage, a flume operated by a privately owned company requires annual maintenance costing $10,000. It is estimated that this flume will not be needed beyond 5 years because of changes that are contemplated in the system. A question has arisen as to whether it would be advisable to replace the flume immediately with a new steel pipe costing $40,000, which would require no maintenance expense during the next 5 years, or to continue the high maintenance costs of the old flume. The old flume has been fully depreciated and has no scrap value but is assessed for property tax purposes at $10,000. The new pipeline would have to be written off in 5 years and probably would have no salvage or scrap value because of the high cost of removal. Property taxes are 1% of new or assessed value. The company uses 25% borrowed capital for which it pays an interest rate of 8% before income taxes. After-tax cost of total capital is 12%. This company experiences a 50% effective income tax rate. It is desired to compare the levelized revenue requirements of both alternatives.

Solution

It is assumed that both book and tax depreciation methods are straight line and that book life equals tax life. Solve for e_a, given $K_a = 0.12$:

$$e_a = 0.146666 \ldots$$

From Table 13-8, the levelized revenue requirements using the tax-adjusted cost of capital is

$$\overline{RR}(12\%) = \$10,100$$

From Table 13-9, the levelized revenue requirements of the replacement is

$$\overline{RR}(12\%) = \$18,000 - \$1,920(A/G, 12\%, 5)$$

$$= \$14,593$$

TABLE 13-8 Annual Revenue Requirements for Retaining the Old Flume

Year, k	(1) Unre- covered Invest.	(2) Book Depr.	(3) Tax Depr.	(4) Annual Cost[a]	(5) Debt Return	(6) Equity Return	(7) Income Tax	(8) RR = Cols. 2 + 4 + 5 + 6 + 7
1	0	0	0	$10,100	0	0	0	$10,100
2	0	0	0	10,100	0	0	0	10,100
3	0	0	0	10,100	0	0	0	10,100
4	0	0	0	10,100	0	0	0	10,100
5	0	0	0	10,100	0	0	0	10,100

[a]Annual cost is sum of annual maintenance cost and annual property tax.

Year, k	(1) Unre- covered Invest.	(2) Book Depr.	(3) Tax Depr.	(4) Annual Cost[a]	(5) Debt Return	(6) Equity Return	(7) Income Tax	(8) RR = Cols. 2 + 4 + 5 + 6 + 7
1	$40,000	$8,000	$8,000	$400	$800	$4,400	$4,400	$18,000
2	32,000	8,000	8,000	400	640	3,520	3,520	16,080
3	24,000	8,000	8,000	400	480	2,640	2,640	14,160
4	16,000	8,000	8,000	400	320	1,760	1,760	12,240
5	8,000	8,000	8,000	400	160	880	880	10,320

[a]Annual cost is sum of annual maintenance cost and annual property tax = 0.01($40,000).

Comparison of uniform annual revenue requirements of both alternatives shows an annual cost advantage ($4,493) in keeping the old flume. ■

In Table 13-8 there is no income tax shown in column 7 because no additional capital is required for the "retain old flume" alternative. This is readily apparent by examining equation 13-8. In this alternative the annual revenue requirement of $10,100 is the before-tax amount needed to exactly meet operating expenses. When capital must be invested as in Table 13-9, several additional obligations that affect annual revenue requirements are encountered.

IMMEDIATE VERSUS DEFERRED INVESTMENT

Many engineering economy studies in utility companies involve immediate versus deferred investment to meet future demands, since utilities must always be prepared to meet the demands for service placed on them. The following is an example of this type.

EXAMPLE 13-5

A water company must decide whether to install a new pumping plant now and abandon a gravity-feed system, which has been fully depreciated, or wait 5 years, at which time it would have to be installed because of deterioration of the piping in the gravity-feed system. Annual costs for operation, maintenance, and taxes for the gravity system are $45,000. The pumping plant will cost $375,000 to install, and it is estimated that it would have a salvage value of 5% of its initial cost at the time of removal from service 20 years hence, when a new and larger system will be installed. Annual operation, maintenance costs, and property taxes for the proposed plant would be $30,000. The gravity-feed system has no salvage value now or later.

If the pumping plant is installed now, it would have an economic life of 20 years. If installed 5 years hence, its economic life would be only 15 years, but its salvage value would still be 5% of its initial cost. Using the revenue requirements method, determine which alternative is better. Straight line depreciation is assumed for both book and tax purposes. The company operates with 50% borrowed capital, on which it pays an interest rate of 7%. The equity rate is expected to be about 14% and the company pays an effective income tax rate of 50%.

Solution

From Table 13-10, the levelized revenue requirement of the new pumping plant using the tax-adjusted cost of capital is

$$\overline{RR}(8.75\%) = \$113,437.5 - \$3,117.18(A/G, 8.75\%, 20)$$

$$= \$92,135$$

From Table 13-11, the levelized revenue requirement of the deferred installation is

$$\overline{RR}(8.75\%) = \{\$45,000(P/A, 8.75\%, 5)$$

$$+ [\$119,375(P/A, 8.75\%, 15) - \$4,156.25(P/G, 8.75\%, 15)]$$

$$\times (P/F, 8.75\%, 5)\}(A/P, 8.75\%, 20)$$

$$= \$74,876$$

Comparison of equivalent uniform annual revenue requirements for both alternatives shows that it is more economical to defer building the pumping plant for 5 years. ■

REVENUE REQUIREMENTS ANALYSIS UNDER CONDITIONS OF INFLATION

Difficulty arises when we consider inflation in engineering economy studies because of depreciation and other fixed actual dollar annuities which are not sensitive to inflation. This difficulty also arises in connection with the revenue requirements method. Consider Example 13-1 when annual costs inflate at 10% per year and when the cost of borrowed money and returns to equity also increase due to inflation. It is assumed further that the salvage value is not responsive to inflation. The after-tax cost of capital from Equation 13-2 is now

$$K_a' = \lambda(1 - t)i_b' + (1 - \lambda)e_a'$$

$$= 0.3(1 - 0.5)(0.155) + (1 - 0.3)(0.27677)$$

$$= 0.216989$$

$$\cong 21.7\%$$

where $i_b' = (1.05)(1.10) - 1 = 0.155$ and $e_a' = (1.1607)(1.10) - 1 = 0.27677$.

EXAMPLE 13-6

We now reconsider Example 13-1 when an inflation rate of 10% affects annual expenses and the cost of capital.

TABLE 13-10 Install New Pumping Plant Now

Year, k	(1) Unrecovered Invest.	(2) Book Depr.	(3) Tax Depr.	(4) Annual Cost	(5) Debt Return	(6) Equity Return	(7) Income Tax	(8) RR = Cols. 2 + 4 + 5 + 6 + 7
1	$375,000	$17,812.50	$17,812.50	$30,000	$13,125	$26,250	$26,250	$113,437.50
2	357,187.50	17,812.50	17,812.50	30,000	12,501.56	25,003.13	25,003.13	110,320.32
3	339,375	17,812.50	17,812.50	30,000	11,878.13	23,756.25	23,756.25	107,203.13
.								
.								
.								
20	36,562.50	17,812.50	17,812.50	30,000	1,279.69	2,559.38	2,559.38	54,210.95

TABLE 13-11 Defer Installation of New Pumping Plant for 5 Years

Year, k	(1) Unrecovered Invest.	(2) Book Depr.	(3) Tax Depr.	(4) Annual Cost	(5) Debt Return	(6) Equity Return	(7) Income Tax	(8) RR = Cols. 2 + 4 + 5 + 6 + 7
1	0	0	0	$45,000	0	0	0	$ 45,000
2	0	0	0	45,000	0	0	0	45,000
.								
.								
5	0	0	0	45,000	0	0	0	45,000
6	$375,000	$23,750	$23,750	30,000	$13,125	$26,250	$26,250	119,375
7	351,250	23,750	23,750	30,000	12,293.75	24,587.50	24,587.50	115,218.75
8	327,500	23,750	23,750	30,000	11,462.50	22,925	22,925	111,062.50
.								
.								
.								
20	42,500	23,750	23,750	30,000	1,487.50	2,975	2,975	61,187.50

TABLE 13-12 Solution of Example 13-1 with f = 10% and SYD Tax Depreciation[a]

Year, k	(1) Unrecovered Invest.	(2) Book Depr.	(3) Tax Depr.	(4) Annual Cost	(5) Debt Return	(6) Equity Return	(7) Income Tax	(8) RR = cols. 2 + 4 + 5 + 6 + 7
1	$7,500	$1,500	$2,400	$550	$348.75	$1,453.04	$ 553.04	$4,404.83
2	6,000	1,500	1,800	605	279	1,162.43	862.43	4,408.86
3	4,500	1,500	1,200	665.50	209.25	871.83	1,171.83	4,418.41
4	3,000	1,500	600	732.50	139.50	581.22	1,481.22	4,433.99

[a]Columns 1, 2, and 3 are found as in Example 13-2(b).

Column 4: Annual cost in year $k = \$500(1 + f)^k$

Column 5: Debt return in year $k = \lambda i_b' \, UI_k$

Debt return in year $1 = (0.3)(0.155)(\$7,500)$
$= \$348.75$

Column 6: Equity return in year $k = (1 - \lambda)e_a' \, UI_k$

Equity return in year $1 = (1 - 0.3)(0.27677)(\$7,500)$
$= \$1,453.04$

Column 7: Income tax $T_k = [t/(1 - t)][(1 - \lambda)e_a' \, UI_k + D_{Bk} - D_{Tk}]$

$T_1 = (0.5)/(1 - 0.5)[col.6 + col.2 - col.3]$
$= \$553.04$

Solution

From Table 13-12, the levelized revenue requirement using the tax-adjusted cost of capital is

$$\overline{RR}(21.7\%) = [\$4,404.83(P/F, 21.7\%, 1) + \$4,408.86(P/F, 21.7\%, 2)$$

$$+ \$4,418.41(P/F, 21.7\%, 3) + \$4,433.99(P/F, 21.7\%, 4)]$$

$$\times (A/P, 21.7\%, 4)$$

$$= \$4,414.22$$

Without inflation, the uniform annual revenue requirement was $3,169.56. Thus a 10% rate of average inflation increases the revenue requirement by almost 40%! Notice that most of this increase in cost is attributable to the inflation-adjusted interest rates. ∎

PROBLEMS

13-1 (a) Describe the types of regulation to which public utilities, but not private industries, are subject. Why is regulation necessary?
(b) How would economy studies differ in a government-owned utility as opposed to a privately owned utility?

13-2 (a) What advantages to the public result from utility companies?
(b) What disadvantages might there be to public utilities?
(c) How is regulation of public utilities that operate solely within an individual state achieved as opposed to those that provide services to many states (e.g., telephone companies and gas pipelines)?

13-3 Briefly summarize the basic characteristics that distinguish public utilities from non-regulated industries such as steel, automobile, and chemical manufacturing.

13-4 Why are most public utilities heavily financed with borrowed capital? What characteristics of this industry make it possible to attract large amounts of borrowed capital, and what advantages (disadvantages) are associated with the use of borrowed money?

13-5 Explain why it may be in the best interest of the consuming public for a regulatory agency to permit a public utility to charge sufficiently high rates to allow it to earn an adequate return on its capital.

13-6 (a) In a certain state a member of the Public Utilities Commission said, "I will oppose all rate increases. I am interested only in the rates the customers have to pay today." Comment on the results that could follow if all members of the Commission rigidly followed this concept.
(b) Comment on this statement: "No company that provides an exclusive and required service, such as electric power, should be permitted to make a profit."

13-7 Is there justification for a privately owned, regulated public utility being permitted to include in its rates the cost of advertising that encourages the public to increase utilization of its service?

Note: Solve the remaining problems by using the after-tax cost of capital, K_a.

13-8 A telephone company can provide certain facilities having a 10-year life and zero salvage value by either of two alternatives. Alternative A requires a first cost of $70,000 and $3,000 per year for maintenance. Alternative B will have a first cost of $48,000 and will require $6,000 annually for maintenance. Property taxes and insurance would be 4% of first cost per year for either alternative. The after-tax cost of capital is 10%, with 30% being borrowed at a 6% interest rate. The effective income tax rate is 50%. Which alternative will provide the lower annual equivalent revenue requirement? An A.C.R.S. tax life of 5 years is used, and book depreciation is computed with the straight line method over 10 years.

13-9 A gas company must decide to build a new meter-repair and testing facility now or wait 3 years before doing so. It estimates that until the new facility is built its annual costs for these functions will be $90,000 greater than when the new facility is completed. The new facility would cost $900,000 and would not be needed beyond 20 years from the present time. The ultimate salvage value would be $200,000 at that time. The company uses 40% borrowed capital, paying 8% interest (before taxes) for it, and the regulatory body permits it to earn 13.8% on its equity capital. Assuming that the company has a 46% income tax rate, determine the equivalent annual revenue requirement for both options and recommend which is better. Assume that book (and tax) depreciation over 20 years is computed with the straight line method.

13-10 A utility company can construct a modern power plant which can generate power at 24 mils ($0.024) per kilowatthour at a 70% load factor. The 24 mils covers all costs including profit on capital and also income taxes. A large industrially owned power plant will soon make wholesale power available. To take advantage of the wholesale power, it will cost $180 per kilowatt of capacity to build the necessary transmission line, which will experience a 70% load factor. Annual maintenance costs for this line will be $0.90 per kilowatt of capacity, and it is to be fully depreciated for book purposes over a 30-year period. A 15-year A.C.R.S. recovery period will be utilized for calculating depreciation for income tax purposes. The cost of money for the company is 12%, with 40% borrowed capital at an interest rate of 7%. The income tax rate is 50%. At what price must the company be able to purchase the power in order to be as economical as generation with the new modern plant?

13-11 An electric utility company has an opportunity to build a small hydroelectric generating plant, of 20,000-kilowatt capacity, on a mountain stream where the flow is quite seasonal. As a consequence, the annual output of energy would be only 40,000,000 kilowatthours. The initial cost would be $2,000,000, and it is estimated that the annual operation and maintenance costs would be $32,000 during its estimated 30-year economic life. It is believed that the property would have a salvage value of $200,000 at the end of the 30-year period. An alternative is to build a geothermal generating plant, which would have the same annual capacity, at a cost of $1,600,000. Because it would have to pay the owners of the property for the geothermal steam, the estimated annual cost for the steam and operation and maintenance is $120,000. A 30-year contract can be obtained on the steam supply, and it is believed that this period is realistic for the economic life of the plant but that the salvage value at that time would be little more than zero. Property taxes and insurance on either plant would be 2% of first cost per year. The company employs 40% borrowed capital, for which it pays 8.5% interest. It earns 13% after taxes on total capital, and it has a 50% effective income tax rate. Which development should be undertaken? State all assumptions that you make.

13-12 (a) Using an A.C.R.S. tax life of 3 years and 1986 cost recovery percentages, rework Example 13-4. The public service commission has decided to allow a straight line recovery of investment over 5 years for rate-setting purposes. Ignore inflation.
   (b) Rework part (a) when a general inflation rate of 6% per year affects both the recurring annual costs of operation and the cost of capital.

13-13 Use the revenue requirements methodology to compare alternatives A and B in Problem 13-8 when the average inflation rate on maintenance is 6% per year. Assume that property taxes do not respond to inflation, and adjust the cost of capital to account for inflation.

13-14 A natural gas pipeline company is considering two plans to provide service required by present demand and the forecasted growth of demand for the coming 18 years. Alternative A requires an immediate investment of $700,000 in property that has an estimated life of 18 years with 10% terminal salvage value. Annual disbursements for operation and maintenance will be $25,000. Annual property taxes will be 2% of first cost. Alternative B requires an immediate investment of $400,000 in property that has an estimated life of 18 years with 20% terminal salvage value. Annual disbursements for its operation and maintenance during the first 8 years will be $42,000. After 8 years, an additional investment of $450,000 will be required in property having an estimated life of 10 years with 50% terminal salvage value. After this additional property is installed, annual disbursements (for years 9 to 18) for operation and maintenance of the combined properties will be $72,000. Annual property taxes will be 2% of the first cost of property in service at any time. The regulatory commission is allowing a 10% "fair return" on depreciated book value to cover the cost of money (K_a) to the utility. Assume that this rate of return will continue throughout the 18 years. The utility company's effective tax rate is 50%. Straight line depreciation is to be used for book purposes in setting rates and SYD depreciation is to be used for income tax purposes. Half of the utility's financing is by debt with interest at 8%. Determine which plan minimizes equivalent annual revenue requirements after property and income taxes have been taken into account.

13-15 In making its forecast of requirements in a certain area for the next 30 years, a telephone company has determined that a 600-pair cable is required immediately and a total of 1,000 pairs will be required by the end of 15 years. An underground conduit of sufficient size to handle the cable needs is being installed now at a cost of $10,000. If a 1,000-pair cable is installed now, it will cost $30,000. As an alternative it can install a 600-pair cable immediately at a cost of $20,000 and install an additional 400-pair cable at the end of 15 years at an estimated cost of $16,000. Because of technical obsolescence, it is company policy to consider the economic life of either installation to be 30 years from the present time. Annual property taxes on either alternative would be 2% of the installed cost, and the salvage value of all cable and conduit at the end of the 30-year period is estimated to be 10% of the first cost. The company uses 40% borrowed capital, for which it pays 8%. It earns 12% after taxes on total capital and has a 50% effective income tax rate. Which alternative would you recommend? Assume that depreciation for tax and book purposes is computed with the straight line method over 15 years for both alternatives.

CHAPTER 14

Capital Budgeting

Proper financing, budgeting, and management of capital represents the basic top management function of a firm and is crucial to its welfare. Capital budgeting may be defined as the series of decisions by individuals and firms concerning how much and where resources will be obtained and expended to meet future objectives. The scope of capital budgeting includes:

1. How the capital is acquired and from what sources (i.e., *financing*).
2. How individual capital projects (and combinations of projects) are identified and evaluated.
3. How standards for project acceptability are set.
4. How final project selections are made.
5. How postinvestment reviews are conducted.

CAPITAL FINANCING

The method by which capital is to be obtained to initiate engineering and business ventures, and whether it is equity or borrowed capital, may be of importance in an engineering economy study. Many well-engineered projects have failed because of improper or too costly financing.

It is possible to make economy studies from two distinctly different viewpoints relative to the capital used. One considers the *total* capital used without regard for its source; this method, in effect, evaluates the *project* rather than the interests of any group of capital suppliers. The other looks at a proposed venture from the viewpoint of the suppliers of the *equity* capital (stockholders) and thus is concerned with interests of the present owners of a business. Therefore, the engineer who is to make economy studies and provide economic recommendations should have an understanding of the various methods by which equity and debt capital are obtained and the consequences of the financing method used.

The illustrations and problems in this book normally evaluate the *project* because in most engineering economy analyses the choice between alternatives can be made independently of sources of funds to be used. Hence, up to this point, the firm's overall "pool of capital" has been regarded as the source of investment funds and its "cost" in Chapter 10, for example, was the after-tax M.A.R.R. However, the sources of capital may be important in investment analyses in which different sources can be used for different alternatives (e.g., leasing versus borrowing versus equity), as well as when it is desired to know the profitability or cost of various projects to the owners of a firm who have equity capital at stake.

BASIC DIFFERENCE BETWEEN EQUITY AND BORROWED CAPITAL

In previous chapters various differences in the uses of equity and borrowed capital have been mentioned. These can be summarized as follows:

1. Equity capital is supplied and used by its owners in the expectation that a *profit* will be earned. There is no assurance that a profit will, in fact, be gained or that the invested capital will be recovered. Similarly, there are no limitations placed on the use of the funds except those imposed by the owners themselves. There is no *explicit* cost for the use of such capital, in the ordinary sense of a tax-deductible cost.

2. When borrowed funds are used, a fixed rate of interest must be paid to the suppliers of the capital, and the debt must be repaid at a specified time. The suppliers of debt capital do not share in the profits resulting from the use of the capital; the interest which they receive, of course, comes out of the firm's revenues. In many instances the borrower pledges some type of security to assure that the money will be repaid. Quite frequently, the terms of the loan may place some restrictions on the uses to which the funds may be put, and in some cases restriction may be put on further borrowing. Interest paid for the use of borrowed funds is a tax-deductible expense for the firm.

THE CORPORATION

The corporation is a form of organization that was originated to avoid many of the disadvantages of the individual and partnership forms of ownership of business enterprises. A corporation is a fictitious being, recognized by law, that can engage in almost any type of business transaction in which a real person could occupy himself or herself. It operates under a charter that is granted by a state and is endowed by this charter with certain rights and privileges, such as perpetual life without regard to any change in the person of its owners, the stockholders. In payment for these privileges and the enjoyment of legal entity, the corporation is subject to certain restrictions. It is limited in its field of action by the provisions of its charter. In order to enter new fields of enterprise, it must apply for a revision of its charter or obtain a new one. Special taxes are also assessed against it.

The capital of a corporation is acquired through the sale of stock. The purchasers of the stock are part owners, usually called stockholders, of the corporation and its assets. In this manner the ownership may be spread throughout the entire world, and as a result enormous sums of capital can feasibly be accumulated. With few exceptions, the stockholders of a corporation, although they are the owners and are entitled to share in the profits, are not liable for the debts of the corporation. They are thus never compelled to suffer any loss beyond the value of their stock. Because the life of a corporation is continuous or indefinite, long-term investments can be made and the future faced with some degree of certainty. This makes debt capital (particularly long-term) easier to obtain and generally at a lower interest cost for corporations than for individual and partnership types of business organizations.

The widespread ownership that is possible in corporations usually results in the responsibility for operation being delegated by the owners to a group of hired managers. Frequently, the management may own very little or no stock in the corporation. At the same time, individual stockholders may exercise little or no significant influence in the running of the corporation and may be interested only in the annual dividends that they receive. As a result, the management sometimes tends to make decisions on the basis of what is best for themselves rather than what is best for the stockholders.

One factor that does not favor the corporation form of business organization is that, except in limited circumstances, the profits of a corporation are subject to double taxation of income.† That is, after the corporation income tax is paid, any remaining profits that are distributed as dividends to stockholders are again taxed as income to those stockholders. Thus, if the corporation's income tax rate is 50%, and all remaining profits are distributed to stockholders paying an average of 30% income taxes, $1.00 earned by the corporation becomes $1.00(1 - 0.50) = $0.50 distributed to the stockholder, which becomes $0.50(1 - 0.30) = $0.35 net after taxes. Hence the actual total tax rate is ($1.00 - $0.35)/$1.00 = 0.65 = 65%.

---

†Profits of certain small corporations with no more than 35 stockholders can be taxed as only individual income to those stockholders, thereby avoiding the double taxation. These corporations are commonly known as "Subchapter S Corporations."

Stock certificates are issued as evidence of stock ownership. The value of stock commonly is measured in three ways. Market value is the price it will bring if sold on the market. Book value is determined by dividing the net equity (net worth) of the corporation by the number of shares outstanding, assuming that all the stock is of one type. A third measure is the price–earnings ratio, which is the price divided by the annual after-tax net income (earnings).

There are a number of types of stock, but two are of primary importance. These are *common stock*, which represents ordinary ownership without special guarantees of return on investment; and *preferred stock*, which has certain privileges and restrictions not available for common stock. For instance, dividends on common stock are *not* paid until the fixed percentage return on preferred stock has been paid.

FINANCING WITH DEBT CAPITAL

There are many situations where the use of borrowed capital is preferable to the use of equity capital. Expansion through the use of equity capital requires either the existing owners to supply more capital or the sale of additional stock to others, which results in decreasing the percentage ownership of the existing stockholders. If the additional capital is needed for a fairly definite period of time and there is considerable assurance that the existing or future cash flow can readily pay the costs and provide for the repayment of borrowed capital, it may be to the advantage of the existing owners to obtain the needed capital by borrowing. It is common for 5 to 30% of the total capital of private, competitive corporations to be from debt sources.

If additional debt capital is needed only for a short period of time, usually less than 5 years and more frequently less than 2 years, it may be borrowed from a bank or other lending agency by the signing of a short-term note. Such a note is merely a promise to repay the amount borrowed, with interest, at a fixed future date or dates. The lending agency may require something of tangible value as security for the loan, or at least it will make certain the financial position of the borrowing organization is such that there is minimal risk involved.

If capital is obtained through short-term borrowing, the corporation is faced with the necessity either of not needing the capital for long or of refinancing the loan every 2 years or so. Obviously, this prevents long-rang planning and investment in projects that have long lives and that may ultimately be quite profitable but may not produce much cash flow during the first few years. Under such conditions there is considerable uncertainty as to whether the money for repayment of a short-term loan will be available when required, or whether refinancing will be available at reasonable cost if this is needed. Because long commitments of capital are required in most projects, corporations usually resort to bond issues for obtaining long-term debt capital.

FINANCING WITH BONDS

A *bond* is essentially a long-term note given to the lender by the borrower, stipulating the terms of repayment and other conditions. In return for the money loaned, the corporation promises to repay the loan and interest upon it at a specified rate. In addition, the corporation may give a deed to certain of its assets that becomes effective if it defaults in the payment of interest or principal as promised. Through these provisions the bondholder has a more stable and secure investment than does the holder of common or preferred stock. Because the bond merely represents corporate indebtedness, the bondholder has no voice in the affairs of the business, at least for as long as his interest is paid, and of course he is not entitled to any share of the profits.

Bonds usually are issued in units of from $100 to $1,000 each, which is known as the *face value* or *par value* of the bond. This is to be repaid the lender at the end of a specified period of time. When the face value has been repaid, the bond is said to have been *retired* or *redeemed*. The interest rate quoted on the bond is called the *bond rate*, and the periodic interest payment due is computed as the face value times the bond interest rate per period.

A description of what happens during the normal life cycle of a bond can be illustrated by the diagram and three-step explanation of Figure 14-1.

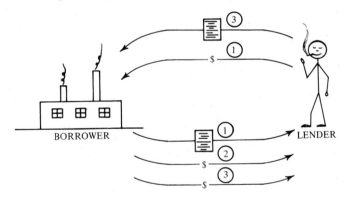

Description of Step	Supplementary Comments
① Bond sold by borrower to lender. Lender gets bond certificate.	Bonds are issued in even denominations (face values) like $250, $1,000, $10,000, etc., but amount paid by lender is determined by market supply and demand. The transaction is usually done through a broker.
② Periodic bond interest payments are made to lender.	The amount of each bond interest payment is computed as face value times bond interest rate.
③ Borrower "redeems" bond by paying principal and getting bond certificate back.	Usually done at end of the stated bond life and the amount paid back is usually the face value.

FIGURE 14-1 Life Cycle of Financing a Bond.

Interest may be paid in either of two ways. The name of the owner of *registered bonds* is recorded on the record books of the corporation, and interest payments are sent to the owner as they become due without any action on his part. *Coupon bonds* have a coupon attached to the bond for each interest payment that will come due during the life of the bond. When an interest payment is due, the holder clips the corresponding coupon from the bond and can convert it into cash, usually at any bank. A registered bond thus requires no action on the part of the holder, but it is not as easily transferred to new ownership as is the coupon bond.

BOND RETIREMENT

Because stock represents ownership, it is unnecessary for a corporation to make absolute provision for payments (dividends) to the stockholders. If profits remain after the operating expenses are paid, part or all of these can be divided among the stockholders. Bonds, on the other hand, represent debt, and the interest upon them is a cost of doing business. In addition to this periodic cost, the corporation must look forward to the day when the bonds become due and the principal must be repaid to the bondholders. Provision may be made for repaying the principal by two different methods.

If the business has prospered and general market conditions are good when the bonds come due, the corporation may be able to sell a new issue of bonds and use the proceeds to pay off the holders of the old issue. If conditions are right, the new issue may bear a lower rate of interest than the original bonds. If this is the case, the corporation maintains the desired capital at a decreased cost. On the other hand, if business conditions are bad when the time for refinancing arrives, the bond market may not be favorable, and it may be impossible to sell a new bond issue, or possible only at an increased interest rate. This could be a serous handicap to the corporation. In addition, the bondholders wish to have assurance that provision is being made so that there will be no doubt concerning the availability of funds with which their bonds will be retired.

When it is desired to repay long-term loans and thus reduce corporate indebtedness, a systematic program frequently is adopted for repayment of a bond issue when it becomes due. Some such provision, planned in advance, gives assurance to the bond holders and makes the bonds more attractive to the investing public; it may also allow the bonds to be issued at a lower rate of interest.

In many cases the corporation periodically sets aside definite sums which, with the interest they earn, will accumulate to the amount needed to retire the bonds at the time they are due. Because it is convenient to have these periodic deposits equal in amount, the retirement procedure becomes a sinking fund. This is one of the most common uses of a sinking fund. By its use the bondholders know that adequate provision is being made to safeguard their investment. The corporation knows in advance what the annual cost for bond retirement will be.

If a bond issue of $100,000 in 10-year bonds, in $1,000 units, paying 10% nominal interest in semiannual payments, must be retired by the use of a sinking fund that earns 8% compounded semiannually, the semiannual cost for retirement will be as follows:

$$A = F(A/F, i, N)$$

$$F = \$100,000$$

$$i = \tfrac{8}{2} = 4\% \text{ per period}$$

$$N = 2 \times 10 = 20 \text{ periods}$$

Thus

$$A = \$100,000(0.0336) = \$3,358$$

In addition, the semiannual interest on the bonds must be paid. This would be calculated as follows:

$$I = \$100,000 \times \frac{0.10}{2} = \$5,000$$

$$\text{total semiannual cost} = \$3,358 + \$5,000 = \$8,358$$

$$\text{annual cost} = \$8,358 \times 2 = \$16,716$$

The total cost for interest and retirement of the entire bond issue over 20 periods (10 years) will be

$$\$8,358 \times 20 = \$167,160$$

BOND VALUE

A bond provides an excellent example of commercial value being the present worth of the future net cash flows that are expected to be received through ownership of a property, as discussed in Chapter 5. Thus the value of a bond, at any time, is the present worth of the future payments that will be received. For a bond, let

Z = face, or par, value

C = redemption or disposal price (usually equal to Z)

r = bond rate (nominal interest) per interest period

N = number of periods before redemption

i = bond yield rate per period

V_N = value of the bond N periods prior to redemption

The owner of a bond will receive two types of payments. The first consists of the series of periodic interest payments he or she will receive until the bond is retired. There will be N such payments, each amounting to rZ. These constitute an annuity of N payments. In addition, when the bond is retired or sold, he will receive a single payment in amount equal to C. The present value of the bond is the present value of these two types of payments at the bond's yield rate:

$$V_N = C(P/F, i\%, N) + rZ(P/A, i\%, N) \qquad (14\text{-}1)$$

EXAMPLE 14-1

Find the cost (present value) of a 10-year 6% bond, interest payable semianually, that is redeemable at par, if bought to yield 10% per year. The face value of the bond is $1,000.

$$N = 10 \times 2 = 20 \text{ periods}$$

$$r = \tfrac{6}{2} = 3\% \text{ per period}$$

$$i = \tfrac{10}{2} = 5\% \text{ per period}$$

$$C = Z = \$1,000$$

Solution

Using Equation 14-1, we obtain

$$V_N = \$1,000(P/F, 5\%, 20) + \$1,000(0.03)(P/A, 5\%, 20)$$

$$= \$376.89 + \$373.87 = \$750.76 \qquad \blacksquare$$

FINANCING WITH EQUITY CAPITAL

The primary sources of equity capital to a corporation are new issues of preferred and/or common stock, depreciation reserves and retained earnings. New issues of common stock and the expected return to equity investors are briefly discussed in this section.

The opportunity cost of equity to a corporation may be approximated by the annual return that stockholders expect on their investment. A general approach to valuation of the opportunity cost of equity capital is based on assumptions regarding future dividends paid to stockholders and rate of increase in the price of the security. Because corporations are assumed to have perpetual lives, the following expression for the current selling price (P) of one share of common stock can be related to the expected annual dividend (D) and the cost of equity (e'_a) with *no* growth in the future price for the stock:

$$e'_a = \frac{D}{P} \qquad (14\text{-}2)$$

When the future price of the security is assumed to grow at a rate of g each year, the cost of equity can be approximated by Equation 14-3.

$$e'_a = \frac{D}{P} + g \qquad (14\text{-}3)$$

Suppose that a share of common stock is priced at $100 and a dividend of $8 is currently paid annually. The expected annual growth in price is 4% per year. If an investor is willing to purchase this security based on the assumption that dividends remain constant and the price grows at 4% annually, the expected return to him or her is about $8/$100 + 0.04 = 0.12, or 12% per year. A second less risky security being considered may sell for $100 and pay a dividend of $10 annually with $g = 0$. If the investor is indifferent between the two securities, an additional expected return of 2% is required to compensate for the extra risk associated with the first investment. In both cases the opportunity cost of equity is 12% and 10%, respectively.

The determination of the cost of all types of equity capital is difficult in practice. Even from a theoretical standpoint, there is considerable disagreement regarding how to assess the cost of equity.† For purposes of this book, the opportunity cost principle and Equations 14-2 and 14-3 provide a basic, though oversimplified, point of departure for approximating this quantity.

DEPRECIATION FUNDS AS A SOURCE OF CAPITAL

As was explained in Chapter 9, the funds that are set aside out of revenue as an allowance for depreciation are a part of the net cash flow and usually are retained and used in a business. These funds are available for reinvestment and must be utilized to the best advantage. Thus they are an important internal source of capital for financing new projects as was seen in Figure 1-2.

Because one of the purposes of depreciation accounting is to provide for replacement of a property when required, we might conclude that depreciation funds provide only for such replacement and never for new equipment of a different type. This, however, is only partially true. In a great many instances when a particular property or piece of equipment has become of no further economic value, the function for which it was originally purchased no longer exists. Under such conditions, we do not wish to replace it with a similar piece of equipment. Instead, other needs have developed, and different equipment or property is needed. The accumulated depreciation funds may thus be used to meet the new needs.

In other instances a piece of equipment may continue to be used after its original value has been recovered through normal depreciation procedures. Here again the accumulated funds are available for other use until the original equipment must be replaced. If depreciation procedures are used that provide for the recovery of a large portion of the initial cost during the first few years of life, there usually will be excess funds available before the equipment must be replaced.

In effect, the depreciation funds provide a revolving investment fund that may be used to the best possible advantage. The funds are thus an important source of capital for financing new ventures within an existing enterprise. They are particularly important because, not being subject to income taxes, they are available in their entirety. Obviously, the depreciation funds must be managed so

†See, for example, Franco Modigliani and Menton H. Miller, ''The cost of capital, corporation finance, and the theory of investment,'' *American Economic Review*, June 1958, pp. 261–297 and D. Durand, ''The cost of capital in an imperfect market: a reply to Modigliani and Miller,'' *American Economic Review*, September 1959, pp. 639–655.

that required capital is available for replacing essential equipment when the time for replacement arrives.

FINANCING THROUGH RETAINED PROFITS

Another important source of internal capital for expansion of existing enterprises is retained profits that are reinvested in the business instead of being paid to the owners. Although this method of financing is used by most companies, there are three factors that tend to limit its use.

Probably the greatest deterrent is the fact that the owners (the stockholders in the case of a corporation) usually expect and demand that they receive some profits from their investment. Therefore, it *usually* is necessary for a large portion (maybe 70% or more) of the profits to be paid to the owners in the form of dividends. This is essential to assure the continued availability of equity (ownership) capital when it is needed. However, it usually is possible to obtain part of the needed capital for expansion by retaining a portion of the profits. Although such a retention of profits reduces the immediate amount of the dividends per share of stock, it increases the book value of the stock and should also result in greater future dividends and/or market resale value for the stock.

Many investors prefer to have some of the profits retained and reinvested so as to help in increasing the value of their stock. The preferential treatment given to *capital gains†*—profits derived through the sale of stock for more than its cost—by the income tax laws is no small factor in this preference.

The second, also serious, limitation on the use of retained profits is the fact that income taxes must be paid upon them and deducted from them before they may be used. With federal taxes taking as much as 50% of the individual's income and 46% of the taxable income of most corporations, this severely limits the amount available after dividends and taxes have been paid.

A third, and less important, deterrent is the fact that as profits are retained and used, the total annual profits and the profits per share of stock should increase. This gives the impression that a company is able to pay higher wage rates. Such an implication frequently is used by unions in wage negotiations without acknowledgment that the larger profits are due to the investment of increased amounts of capital by each shareholder through retained profits. This disadvantage may be quite easily avoided by issuing stock dividends in lieu of the retained profits, thereby maintaining a fairly constant rate of profit per share.

It should be noted here that regulated public utilities usually cannot retain a large percentage of their profits, since they are permitted to charge rates only sufficiently high to provide profits that will assure a flow of required capital into the business.

---

†Income tax provisions for capital gains are discussed in Chapter 10.

THE AVERAGE COST OF CAPITAL
AND THE EFFECT OF INCOME TAXES

Most businesses do not operate entirely on either equity capital or borrowed capital. Instead, in most instances good practice calls for most of the capital to be obtained from equity sources, a smaller portion being borrowed. To obtain the desired amount of equity capital, a sufficient rate of dividends must be maintained to make the investment attractive to investors in view of the risks involved. Because of the absence of security, the rate that must be paid for equity capital is virtually always greater than must be paid to obtain borrowed capital.

One factor that is usually important to decisions regarding debt versus equity capital is the probable effect of income taxes on the cost of that capital to the firm. Any interest paid for the use of money borrowed by a firm (or individual) is deductible from income or profits reported for federal and state income tax purposes. Hence taxes are saved when borrowed funds are used to finance a project.

For example, suppose that $10,000 is borrowed by a certain firm at an annual interest rate of 8%, and the effective (combining federal and state) income tax rate is 60%. The annual interest is

$$I = P \times i = \$10,000 \times 8\% = \$800$$

This $800 reduces the net profits or income on which taxes must be paid and hence results in *saving* income taxes in the amount of $800 × 60% = $480. Therefore, the net real after-tax cost of the interest is $800 − $480 = $320, and the after-tax interest cost is $320/$10,000 = 3.2%. In other words, if there are other profits so that $480 of the interest cost can be charged against positive taxable income, the $10,000 of debt capital need produce only $320 of added income rather than $800 to satisfy the demands of the suppliers of the capital. Without other taxable profits, the maximum the $10,000 would have to earn, in order to satisfy the interest demands, would be $800.

On the other hand, dividends paid by a firm for the use of equity capital are not deductible from the income or profits of the firm, on which income taxes must be paid. Hence, if the firm plans to pay only 7% dividends in order to satisfy the stockholders who supply $10,000 of equity capital, the investment would have to earn $1,750[= $10,000 × 7% ÷ (1.0 − 0.6)] before taxes. At a 60% income tax rate, $1,750 × 0.60 = $1,050 would be paid in income taxes, leaving $700 to satisfy the stockholders. Consequently, in this case the before-tax cost of equity *to the firm* is 17.5%, and the after-tax rate is 7%. Recall that the before-tax cost of borrowed capital was 8%.

It is apparent that if the effects of income taxes are ignored, borrowed capital at 8% costs more than equity capital at 7%. However, the much more meaningful comparison, *with income taxes being considered*, shows that the borrowed capital at 3.2% (under most conditions) is much less costly for financing a project than equity capital at 7% from the viewpoint of this firm.

When both types of capital are used for the general financing of an enterprise, the "bare" cost of the capital is not that paid for equity capital or for borrowed capital. Instead, it is some intermediate rate, depending upon the proportions of each type of capital used. For example, suppose that for the case above involving $10,000 capital that might be borrowed for 8% (with an income tax rate of 60%) or obtained from stockholders for 7%, one is considering various mixes of debt and equity capital. Table 14-1 shows the average cost of several combinations of debt and equity capital both before and after considering the effect of income taxes.

In many economy studies it is clear that only borrowed capital or only equity capital will be employed. In such cases the corresponding cost of capital is sometimes used in interest and equivalence calculations. In most economy studies, however, all money is assumed to come from a common investment pool, and considerations of financing are taken up separately from the project economy study. The following section discusses several approaches that firms utilize to determine their minimum attractive rate of return, which is then used as the interest rate in equivalence calculations.

TABLE 14-1 Example of Average Cost of Capital for Various Percentages of Equity and Borrowed Capital, Both Before and After Income Taxes

| | Borrowed Capital | | | | | |
	0%	20%	40%	60%	80%	100%
(1) Total capital	$10,000	$10,000	$10,000	$10,000	$10,000	$10,000
(2) Borrowed capital [(1) × % borrowed]	0	2,000	4,000	6,000	8,000	10,000
(3) Before-tax interest cost [(2) × 8%]	0	160	320	480	640	800
(4) After-tax interest cost [(3) × (1 − 0.60)]	0	64	128	192	256	320
(5) Equity capital [(1) − (2)]	10,000	8,000	6,000	4,000	2,000	0
(6) Dividends cost [(5) × 7%]	700	560	420	280	140	0
(7) Before-tax earnings required to pay dividends [(6)/(1 − 0.60)]	1,750	1,400	1,050	700	350	0
(8) Total cost before taxes [(3) + (7)]	1,750	1,560	1,370	1,180	990	800
(9) Total cost after taxes [(4) + (6)]	700	624	548	472	396	320
(10) Average rate before taxes [(8)/(1)]	17.5%	15.6%	13.7%	11.8%	9.9%	8.0%
(11) Average rate after taxes [(9)/(1)]	7.00%	6.24%	5.48%	4.72%	3.96%	3.20%

DETERMINING THE MINIMUM ATTRACTIVE RATE OF RETURN

The minimum attractive rate of return (M.A.R.R.) is usually a policy issue resolved by the top management of an organization in view of numerous considerations. Among these considerations are:

1. The amount of money available for investment and the source and cost of these funds (i.e., equity funds or borrowed funds).
2. The number of good alternatives available for investment and their purpose (i.e., whether they sustain present operations and are *essential* or expand on present operations and are *elective*).
3. The amount of perceived risk that is associated with investment opportunities available to the firm, and the projected cost of administering projects over short planning horizons versus long planning horizons.
4. The type of organization involved (i.e., government, public utility, or competitive industry).

In theory the M.A.R.R. should be chosen to maximize the economic well-being of an organization, subject to the types of considerations listed above. How an individual firm accomplishes this in practice is far from clear cut and is frequently open to criticism. One popular approach to establishing a M.A.R.R. involves the *opportunity cost* viewpoint described in Chapter 4, and it results from the phenomenon of *capital rationing*†

Rationing of capital is necessary when the amount of available capital is insufficient to sponsor all worthy investment opportunities. A simple example of capital rationing is given in Figure 14-2, where the cumulative investment requirements of seven independent projects are plotted against the prospective internal rate of return of each. Here there is shown a limit of $6 million on available capital. In view of this limitation the last funded project would be "E," with a prospective I.R.R., or "hurdle rate," of 19%. In this case the minimum attractive rate of return by the opportunity cost principle would be 19%. By *not* investing $1 million in project E, the firm would presumably be forgoing the chance to realize a 19% return. As the amount of investment capital and opportunities available change over time, the firm's M.A.R.R. would also change.

Superimposed on Figure 14-2 is the approximate cost of obtaining the $6 million to illustrate that project E is acceptable only as long as its I.R.R. exceeds the cost of raising the last $1 million. As seen in Figure 14-2, the cost of capital will tend to increase gradually as larger sums of money are acquired through increased borrowing and/or new issuances of common stock (equity).‡ One last observation in connection with Figure 14-2 is that the perceived risk

---

†Donaldson, G., "Strategic hurdle rates for capital investment," *Harvard Business Review*, March–April 1972.

‡This is because the investor in bonds might require a higher return as compensation for greater risk caused by a smaller proportion of equity capital available to absorb capital and operating losses. Similarly, the equity investor might expect a higher return because this capital is more speculative in nature (there is a greater likelihood of reduced earnings in economic downturns).

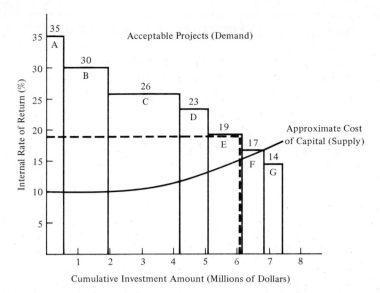

FIGURE 14-2 Determination of the M.A.R.R. Based on the Opportunity Cost Viewpoint.

associated with financing and undertaking the seven projects has been determined by top management to be acceptable.

Representative M.A.R.R.'s for government agencies, public utilities and private industries in 1983 are indicated in Figure 14-3. Even though it is not possible to provide consistent trends in M.A.R.R.'s throughout selected industries with a high degree of confidence, typical M.A.R.R.'s utilized and several underlying causes for their differences among organizations are depicted on the four scales shown in Figure 14-3.

For municipalities and government agencies described in Chapter 12, the M.A.R.R.'s are usually based on the before-tax cost of borrowed money, which ranges from about 8 to 10%. There is little risk to the investor and almost all capital is obtained through borrowing.

In the case of regulated public utilities (electricity, telephone, gas, water, etc.), the after-tax cost of capital is frequently used to establish a M.A.R.R. as discussed in Chapter 13. From Figure 14-3 it can be seen that these M.A.R.R.'s lie in the neighborhood of 11 to 14% and that borrowed funds constitute 40 to 60% of a utility's capitalization. Because of the stable nature of these companies, a large fraction of funds acquired is customarily borrowed and the associated risk borne by the equity investor in the utility industry is low.

Private, competitive industries frequently employ the opportunity cost viewpoint towards choosing an after-tax M.A.R.R. (Engineering economy studies in competitive industry as well as in public utilities are nearly always made on an after-tax basis.) These M.A.R.R.'s typically cover a rather broad range of from 15 to 20% or higher, principally due to the small amount of borrowed money in the firm's capitalization that typically ranges from around 35% to none at all! As more equity capital is utilized by a firm, it is customary to experience higher after-tax return and/or capital gain expectations by owners of the equity capital.

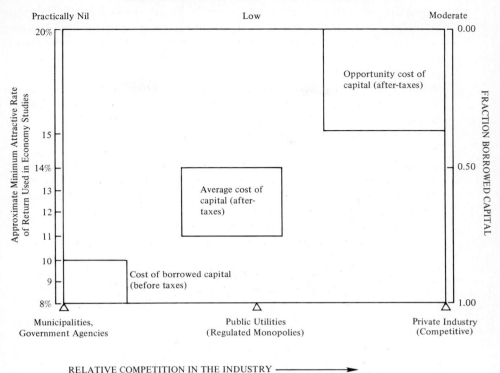

FIGURE 14-3 **Representative M.A.R.R.'s Utilized by Various Types of Industries.**

LEASING VERSUS PURCHASING OF ASSETS

The decision to lease versus purchase an asset represents a situation where the source of capital may affect which alternative is eventually chosen. Leasing is a source of capital generally regarded as a long-term liability similar to a mortgage, whereas the purchase of an asset typically uses funds from the firm's overall pool of capital (much of which is equity). Before considering examples of lease–buy problems, background information about leases is first provided.

Rent is the amount that one has to pay for the use of property that he or she does not own. A lease is a type of contract by which an individual or a corporation acquires the use of an asset for a specified period of time at a specified *rent.* A large number of leasing methods has been utilized in real estate financing for hundreds of years, but only within the last 35 years has leasing of industrial equipment become popular. Most frequently, automobiles, trucks, buildings, or various types of equipment are involved. Such alternatives may arise either in connection with a consideration of the purchase of completely new assets or in connection with the possible replacement of existing assets. In most cases it is essential to consider the income tax implications so that we can compare accurately the alternatives of purchasing versus leasing or retention versus leasing.

For corporations, the rent paid on leased assets used in their trade or business is generally deductible as a business expense. To determine if lease payments are deductible as rent, the contract must be for a true leasing arrangement instead of a conditional sales agreement. In a true lease the corporation using the property (lessee) does not acquire ownership in or title to the asset, whereas a conditional sales contract will transfer to the lessee an equity interest or title to the asset being leased. Hence the test of whether or not lease payments qualify as business expenses lies in distinguishing between a "true" lease and a conditional sale.†

In this section we shall assume that a true lease exists and that an asset may be acquired through *leasing* or *purchasing*. Before the decision is made regarding leasing or purchasing, however, the services of the asset in question must have been shown to be economically justifiable. That is, from an investment standpoint the equipment will "pay its own way," and the decision regarding how best to finance it constitutes the "lease versus purchase decision."

The reasons companies decide to lease rather than purchase are not as simple and clear as might be expected. We quite naturally would think that a decision in favor of leasing would be based on the fact that it was more economical, yet this very often is not the basis for the decision. One study involving companies that had leased fleets of automobiles and trucks revealed that in the majority of cases, no detailed comparison of the costs of leasing versus purchasing had been made. The major reasons given by these companies were, in order of frequency, as follows:

1. Leasing freed needed working capital, or enabled needed equipment to be acquired without the company going into debt.
2. Leasing was thought to have some income tax advantages.
3. Leasing reduced maintenance and administrative problems; for example, it eliminated the necessity for a company to be a fleet operator, leaving this to the specialist.
4. Leasing was cheaper.

Inasmuch as leasing is only one of several ways of obtaining working capital, a decision to lease should consider the cost of obtaining capital by all possible methods. In the study cited, it was found that very few of the companies had made such a comparison.

A number of studies have shown that there is no real income tax advantage in leasing. This is particularly true since accelerated methods (e.g., A.C.R.S.) have been permitted for depreciation. Assuming a given purchase price, the firm offering a lease contract (lessor) can charge no more for depreciation than can the owner of assets. If assets are leased, the annual lease payments are deducted in computing income taxes; if the assets are purchased, annual depreciation is deducted. Most companies have now come to realize that leasing does not offer tax advantages, but the myth still persists in some quarters.

There may or may not be savings in maintenance costs through leasing. This will depend on the actual circumstances, which should be carefully evaluated in each case. There is no doubt that leasing usually does simplify maintenance

---

†For further information, see *Tax Guide for Small Business*, U.S. Internal Revenue Service Publication 334, published annually.

problems, and this may be an important factor. Also, many indirect costs, which frequently are difficult to determine, will usually be associated with ownership.

In many instances leasing can be shown to be cheaper than purchasing, but the actual comparative costs and all other factors should be considered before a decision to lease is made.

The following examples illustrate correct methods of handling a lease versus purchase study on an after-tax basis.

EXAMPLE 14-2

A company is considering building a small office building at a cost of $100,000 on land that would cost $20,000. It is estimated that the land could be sold at its cost at the end of 50 years, and that the building, although fully depreciated on a straight line basis over the 50 years, probably would have a salvage value of $20,000. Annual disbursements for maintenance, property taxes, insurance, and so on, are estimated at $5,250. An alternative is to lease a building for $16,500 per year for a 50-year period. The company ordinarily obtains about 20% on its capital before taxes and, being in a 50% tax bracket, about 10% after taxes. What should it do?

Solution (Using I.R.R. Method)

The following table for computing after-tax cash flows will facilitate analysis of the two alternatives.

Year	(A) Before-Tax Cash Flow	(B) Depreci- ation	(C) = (A) + (B) Taxable Income	D = −(C) × 50% Cash Flow for Income Taxes	(E) = (A) + (D) After-Tax Cash Flow
			PURCHASE BUILDING		
0	− $120,000				− $120,000
1–50	− 5,250	− $2,000	− $ 7,250	+ $ 3,625	− 1,625
50	+ 40,000		+ 20,000^a	− 10,000	+ 30,000
			LEASE BUILDING		
1–50	− $ 16,500		− $16,500	+ $ 8,250	− $ 8,250

^{a}Capital gain on building.

The I.R.R. on investment required to purchase rather than to lease the building is the $i'\%$ at which

$$-\$120,000 + (-\$1,625 + \$8,250)(P/A, i'\%, 50)$$
$$+ \$30,000(P/F, i'\%, 50) = 0$$

The $i'\%$ can be found to be approximately 5.0%. Since $5.0\% < 10\%$, this means that it would be better to lease the building. Of course, any of the other theoretically correct methods would result in the same recommendation. ∎

EXAMPLE 14-3

An industrial forklift truck can be purchased for $30,000 or leased for a fixed amount of $9,200 per year payable at the *beginning* of each year. The lease contract provides that maintenance expenses are borne by the lessor. Regardless of whether the truck is purchased or leased, the study period is 6 years. If purchased, annual maintenance expenses are expected to be $1,000 in year 0 purchasing power and they will inflate at 5% per year over the study period. The market value of the truck is expected to be negligible after 6 years of normal use. Depreciation is determined with the A.C.R.S. method over a 5-year recovery period, and a 10% investment credit can be taken. The effective income tax rate is 50% and the after-tax M.A.R.R., which includes an allowance for inflation, is 15%.

Use the annual worth method and determine whether the forklift truck should be purchased or leased. This firm is profitable in its overall business activity.

Solution

The effects of inflation and income taxes on the after tax cash flows of both alternatives are shown in the following table:

Year	(A) Before-Tax Cash Flow	(B) Deprecia- tion[a]	(C) = (A) + (B) Taxable Income	D = −(C) × 50% Cash Flow for Income Taxes	(E) = (A) + (D) After-Tax Cash Flow
	PURCHASE TRUCK (A$ STUDY)[b]				
0	− $30,000			+ $3,000	− $27,000
1	− 1,050	− $6,000	− $ 7,050	+ 3,525	+ 2,475
2	− 1,102	− 9,600	− 10,702	+ 5,351	+ 4,249
3	− 1,158	− 7,200	− 8,358	+ 4,179	+ 3,021
4	− 1,216	− 4,800	− 6,016	+ 3,008	+ 1,792
5	− 1,276	− 2,400	− 3,676	+ 1,838	+ 562
6	− 1,340	0	− 1,340	+ 670	− 670
	LEASE TRUCK (A$ STUDY)[c]				
0	− $ 9,200		− $ 9,200	+ $4,600	− $ 4,600
1–5	− 9,200	0	− 9,200	+ 4,600	− 4,600
6	0	0	0	0	0

[a]A.C.R.S. (1986 cost recovery percentages).
[b]The net A.W. at M.A.R.R. = 15% is − $4,924.
[c]The net A.W. at M.A.R.R. = 15% is − $5,290.

From the analysis above, the purchase alternative is less costly than the lease alternative and would probably be selected. If capital were not readily available, the firm may elect to lease the forklift truck because the difference in annual equivalent worth is small. Furthermore, if estimates of maintenance expenses and inflation are believed to be inaccurate, the firm would tend to favor leasing as a hedge against an uncertain future. ■

AFTER-TAX EVALUATIONS
BASED ON EQUITY INVESTMENT

Most problems encountered in previous chapters have not distinguished between the types of funds (i.e., equity versus debt capital) that comprise the pool of capital available for investment in engineering and business projects. It has been assumed that the finance department has obtained funds (capital financing) from internal or external sources as shown in Figure 1-2. Next, as a result of the capital rationing process in Figure 14-2, a relatively fixed amount of capital is available for investment in new equipment, an enlarged building, an upgraded process, and so on. Engineers and other technical personnel generally do not concern themselves with how these funds have been sourced because their mission is to ascertain the best (most profitable) uses of the capital, given that it is available. Consequently, this book has dealt with problems from the standpoint of a firm's *overall* pool of investment capital rather than a subset of it. Hence, the interest rate (i.e., the M.A.R.R.) used for time-value-of-money calculations typically reflects both the cost of borrowed capital and equity capital.

In this section we shall demonstrate how to analyze a capital investment problem from the viewpoint of the *equity investor* (i.e., the stockholder). The reason why this may be desired is that a particular undertaking may be quite risky, and equity investors would like an explicit assessment of the project's profitability in terms of *their* investment. Another motivation for looking at an investment from an equity viewpoint stems from the basic aim of private enterprise, which is to *maximize the future wealth* of the owners of the firm. To meet this objective, a project's after-tax cash flows must be evaluated from the viewpoint of a firm's stockholders using an interest rate equal to their opportunity cost of capital.

To perform after-tax economic analyses on equity capital rather than total capital required for a project, one merely needs to convert all cash flows into *after-tax equity cash flows*. This can be done by adding a column for "Loan and Interest" in the standard after-tax cash flow table presented in Chapter 10, as is illustrated in the following example.

EXAMPLE 14-4

This example is the same as Example 10-8 except that information is given on borrowed capital and it is desired to determine the after-tax N.P.W. of equity capital. It is restated with additional information as follows. Certain new machinery is estimated to cost $180,000 installed. It is expected to reduce net annual operating disbursements by $36,000 per year (not including interest) for 10 years and to have $30,000 salvage value at the end of the tenth year. Fifty thousand dollars of the investment is to be borrowed at 10% annual interest, with all the principal to be repaid at the end of the tenth year. Show a tabulation to determine the after-tax equity cash flow and calculate the after-tax N.P.W. of equity capital. The opportunity cost of equity capital has been set at 18% per year.

Solution

Table 14-2 shows the recommended tabular format with the column headings indicating how the calculations are made. Notice that interest on borrowed capital is deductible as a business expense and hence reduces before-tax savings to $31,000. From the results of the right-hand column, the after-tax N.P.W. of equity investment can be calculated:

TABLE 14-2 Computations to Determine the After-Tax Equity Cash Flow for Example 14-4

Year	(A) Before-Tax Cash Flow	(B) Depreciation	(C) Loan and Interest	(D) = (A) + (B) + Interest Portion of (C) Taxable Income	E = -(D) × 50% Cash Flow for Income Taxes	(F) = (A) + (C) + (E) After-Tax Equity Cash Flow
0	-$180,000		+ $50,000			-$130,000
1-10	+ 36,000	-$15,000	- 5,000	+$16,000	-$8,000	+ 23,000
10	+ 30,000		- 50,000			- 20,000

$$\text{N.P.W.} = -\$130{,}000 + \$23{,}000(P/A,\ 18\%,\ 10)$$
$$- \$20{,}000(P/F,\ 18\%,\ 10)$$
$$= -\$30{,}457$$

Because the N.P.W. is negative, this investment appears to be a poor one based on after-tax equity cash flows. ∎

EXAMPLE 14-5

A firm is considering purchasing an asset for $10,000 with half of this amount coming from retained earnings (equity) and the other half being borrowed for 3 years at an effective interest rate of 12% per year. Uniform annual payments, consisting of interest and loan principal, will be utilized to repay the $5,000 loan. Straight line depreciation over a 3-year period can be claimed, and the asset's salvage value at this time is expected to be $4,000. Before-tax income from the asset is estimated to be $5,000 per year, which does not include the cost of borrowed money. Finally, the firm's effective income tax rate is 50%.

If the opportunity cost of capital to equity investors is 20% per year after taxes, is this a profitable investment from the stockholder's viewpoint?

Solution

The amount of the uniform loan repayment is $5,000(A/P, 12\%, 3) = \$2,081.75$. As described in Chapter 5, this capital recovery amount consists of repayment of borrowed money (principal) and interest on the amount of unpaid principal at the beginning of each year. Because interest is a deductible business expense but repayment of loan principal is not, a schedule of annual loan principal and interest must be developed as shown below (to the nearest dollar):

End of Year	Interest		Principal
1	$5,000(0.12)$	$= \$600$	$2,082 - \$600 = \$1,482$
2	$(5,000 - 1,482)(0.12)$	$= 422$	$2,082 - 422 = 1,660$
3	$(5,000 - 1,482 - 1,660)(0.12)$	$= 223$	$2,082 - 223 = 1,859$

The after-tax equity cash flow for the asset can now be determined as indicated in the heading of Table 14-2:

	(A)	(B)	(C) Loan Cash Flow:		(D)	(E)	(F)
Year	Before-Tax Cash flow	Deprecia-tion	Interest[a]	Principal[b]	Taxable Income	Cash Flow for Income Taxes	After-Tax Equity Cash Flow
0	$-\$10,000$			$+\$5,000$			$-\$5,000$
1	5,000	$-\$2,000$	$-\$600$	$-\$1,482$	$\$2,400$	$-\$1,200$	$1,718^{c}$
2	5,000	$-2,000$	-422	$-1,660$	$2,578$	$-1,289$	1,629
3	5,000	$-2,000$	-223	$-1,859$	$2,777$	$-1,389$	1,529
3	4,000						4,000

[a]Interest is deductible from column A.
[b]Principal is an after-tax cost deducted from column F.
[c]$\$5{,}000 - \$600 - \$1{,}200 - \$1{,}482 = \$1{,}718$.

The present worth of column F at 20% is $763. Therefore, because the N.P.W. is greater than zero, this investment is attractive from the viewpoint of the equity investor. ∎

After-tax evaluations based on equity investment frequently utilize an annual schedule of loan interest and principal such as the one illustrated in Example 14-5. To formalize the development of such a schedule, the following discussion and equations are offered.

Consider a loan repayment scheme (e.g., a home mortgage) involving equal uniform *monthly* payments of $A = P_0(A/P, i_b, N)$, where typically i_b is a nominal loan interest rate per month, N is the number of months over which the loan is repaid, and P_0 is the initial lump-sum amount of the loan at time 0. In after-tax problems that include different debt–equity mixes in the financing of a project, the *interest* on a loan must be broken out of A because it is a deductible business expense whereas the remainder of A is repayment of principal (i.e., equity) and is not deductible in calculating a firm's taxable income. Graphically, the relationship between loan equity and loan interest has this general appearance:

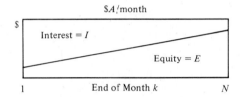

A/month

Interest $= I$

Equity $= E$

1   End of Month k N

As a function of month k ($1 \leq k \leq N$), it is desired to be able to determine quickly the interest in month k (I_k) and the equity in month k (E_k). It will always be true that $I_k + E_k = A$. The interest repaid during month 1 is $i_b P_0$, and $E_1 = A - i_b P_0$. In general, the interest repaid in any month is the amount of loan principal, or equity, owed at the beginning of that month multiplied by i_b. The loan principal owed at the beginning of month, k, or the end of month $k - 1$, is P_{k-1} and can be easily determined:

$$P_{k-1} = A(P/A, i_b, N-k+1) \tag{14-4}$$

Thus

$$I_k = P_{k-1}(i_b) \tag{14-5}$$

and

$$E_k = A - I_k \tag{14-6}$$

For example, suppose that an asset is purchased for $5,000 and a 20% down payment is made. The remaining $4,000 is financed at $1\frac{1}{2}\%$ per month (on the unpaid balance) over 24 months. How much interest and equity are repaid during the eighteenth month?

$A = \$4,000(A/P, 1\frac{1}{2}\%, 24) = \$199.70$

$P_{17} = \$199.70(P/A, 1\frac{1}{2}\%, 24 - 18 + 1) = \$1,317.66$ (from Equation 14-4)

$I_{18} = \$1,317.66(0.015) = \$19.76$ (from Equation 14-5)

$E_{18} = \$199.70 - 19.76 = \$179.94$ (from Equation 14-6)

In many problems the present worth of before-tax and/or after-tax interest costs of a debt-financed project must be calculated over the entire loan period. The appropriate discount rate is usually a firm's M.A.R.R. expressed over the same interval as i_b. Each I_k could be computed as shown above and the present worth then determined as indicated below.

Before-tax cost of interest (i_{BT}): $\text{P.W.}(i_{BT}) = \sum_{k=1}^{N} I_k(P/F, i_{BT}, k)$ (14-7)

After-tax cost of interest (i_{AT}): $\text{P.W.}(i_{AT}) = \left[\sum_{k=1}^{N} I_k(P/F, i_{AT}, k) \right](1 - t)$

(14-8)

where t is the effective income tax rate. To save time with computations, Equation 14-8 can be reduced as follows:

$$\text{P.W.}(i_{AT}) = (1 - t)A\left[(P/A, i_{AT}, N) - \frac{(P/F, i_b, N) - (P/F, i_{AT}, N)}{i_{AT} - i_b} \right]$$

(14-9)

Referring to the loan example above, what is the present worth of after-tax interest cost if $t = 0.46$ and $i_{AT} = 2\%$ per month? By using Equation 14-9, this result is obtained:

$\text{P.W.}(i_{AT}) = (1 - 0.46)(\$199.70)$

$\qquad \times \left[(P/A, 2\%, 24) - \frac{(P/F, 1.5\%, 24) - (P/F, 2\%, 24)}{0.02 - 0.015} \right]$

$\qquad = (1 - 0.46)(\$199.70)\left[18.9139 - \frac{0.6995 - 0.6217}{0.005} \right]$

$\qquad = \$361.68$

CAPITAL BUDGETING—SELECTION AMONG INDEPENDENT CAPITAL PROJECTS

Most companies constantly are presented with a number of opportunities in which they can invest capital. In most cases the amount of capital available is limited, or additional amounts can be obtained only at increasing incremental cost as shown in Figure 14-2. They thus have a problem of budgeting, or allocating, the available capital to the various possible uses.

One popular approach to capital budgeting utilizes the N.P.W. criterion and was previously demonstrated in Chapter 6. If project risks are about equal, the procedure was to compute the N.P.W. for each investment opportunity and then to determine the combination of projects that maximized present worth subject to various constraints on resource availability. The following example is offered as a brief review of this procedure.

EXAMPLE 14-6

Consider these five independent projects and determine the best allocation of capital among them if no more than $300,000 is available to invest.

Independent Project	Initial Capital Outlay	N.P.W.
A	− $100,000	+ $25,000
B	− 125,000	+ 30,000
C	− 150,000	+ 35,000
D	− 75,000	+ 40,000

Solution

All possible combinations of these projects taken two, three, and four at a time are shown together with the total N.P.W. and initial capital outlay of each. After eliminating those combinations that violate the $300,000 funds constraint, the proper selection of projects would be ABD and the maximum present worth is $95,000. The process of enumerating combinations of projects having nearly identical risks is best accomplished with a computer when large numbers of projects are being evaluated. To facilitate this search for the best capital allocation, linear programming models are especially helpful, as described in Chapter 17.

Combinations	Total N.P.W.	Total Capital Outlay
AB	55	225
AC	60	250
AD	65	175
BC	65	275
BD	70	200
CD	75	225
ABC	90	375
ACD	100	325
BCD	105	350
ABD	95	300* Best
ABCD	130	450

Methods for determining to which possible projects available funds should be allocated seem to require the exercise of judgment in most realistic capital budgeting problems. Such a problem, and possible methods of solution, can be illustrated with a simple example.

TABLE 14-3   Prospective Projects for a Firm^a

TABLE 14-3   Prospective Projects for a Firm[a]

Project	Investment	Life (years)	Rate of Return (%)
A	$40,000	5	7
B	15,000	5	10
C	20,000	10	8
D	25,000	15	6
E	10,000	4	5

[a]Here we assume that the indicated rates of return for these projects can be repeated indefinitely by subsequent "replacements." If this assumption is not acceptable, the present worth method should be utilized for "nonrepeating" projects as discussed in Chapter 6.

EXAMPLE 14-7

Assume that a firm has five investment opportunities available, which require the indicated amounts of capital and which have economic lives and prospective after-tax internal rates of return as shown in Table 14-3. Further, assume that the five ventures are independent of each other—investment in one does not prevent investment in any other, and none is dependent upon another being undertaken.

Now let it be assumed that the company has unlimited funds available, or at least sufficient funds to finance all these projects, and that capital funds cost the company 6% after taxes. For these conditions the company probably would decide to undertake all projects that offered a return of at least 6%, and thus projects A, B, C, and D would be financed. However, such a conclusion would assume that the risks associated with each project are reasonable in the light of the prospective internal rate of return or are no greater than those encountered in the normal projects of the company.

Unfortunately, in most cases the amount of capital is limited, either by absolute amount or increasing cost. If the total of capital funds available is $60,000, the decision becomes more difficult. Here it would be helpful to list the projects in order of decreasing profitability in Table 14-4 (omitting the undesirable project E). Here it is clear that a complication exists. We quite naturally would wish to undertake those venutres that have the greatest profit potential. However, if projects B and C are undertaken, there will not be sufficient capital for financing project A, which offers the next greatest rate of return. Projects B, C, and D could be undertaken and would provide an annual return of $4,600

TABLE 14-4   Prospective Projects of Table 14-3 Ordered According to Internal Rate of Return

Project	Investment	Life (years)	Rate of Return (%)
B	$15,000	5	10
C	20,000	10	8
A	40,000	5	7
D	25,000	15	6

TABLE 14-5 Prospective Projects of Table 14-4 Ordered According to Overall Desirability

Project	Investment	Life (years)	Rate of Return (%)	Risk Rating
C	$20,000	10	8	Lower
A	40,000	5	7	Average
B	15,000	5	10	Higher
D	25,000	15	6	Average

(= \$15,000 $\times$ 10% + \$20,000 $\times$ 8% + \$25,000 $\times$ 6%). If project A were undertaken, together with either B or C, the total annual return would not exceed \$4,600.† A further complicating factor is the fact that project D involves a longer life than the others. It thus is apparent that we might not always decide to adopt the alternative that offers the greatest profit potential.

The problem of allocating limited capital becomes even more complex when the risks associated with the various available projects are not the same. As mentioned in Chapters 5 and 6, many companies calculate the *payback period* as a preliminary index of a given project's risk. More reliable measures of risk are obtained, when warranted, with techniques described in Chapter 7. Assume that the risks associated with project B are determined to be higher than the average risk associated with projects undertaken by the firm, and that those associated with project C are lower than average. The company thus might rank the projects according to their desirability as in Table 14-5. Under these conditions the company might decide to finance projects C and A, thus avoiding one project with a higher-than-average risk and another having the lowest prospective return and longest life of the group. ∎

SELECTION AMONG INDEPENDENT CAPITAL PROJECTS USING RISK CATEGORIES

Another means of budgeting capital while explicitly considering risk is to place projects into two or more risk categories and to determine in advance what approximate proportion of the available capital will be invested in each category. Then one can provisionally pick which projects earn the highest returns within the capital available for each risk category. After this is done, if one judges that a certain provisionally rejected project with a certain risk and rate of return is more desirable than some other project with a relatively higher risk and higher rate of return (or relatively lower risk and lower rate of return), then a trade-off can be made. This is possible as long as the total capital required and the resulting amount of capital invested in each risk category is reasonably in balance relative to the objectives of the firm and the risks and returns involved.

---

†This assumes that the leftover capital could earn no more than 6% per year.

TABLE 14-6 Prospective Projects Within Three Risk Categories

High Risk			Medium Risk			Low Risk		
Project	Invest-ment	Rate of Return (%)	Project	Invest-ment	Rate of Return (%)	Project	Invest-ment	Rate of Return (%)
F	$2,000	30	K	$3,500	25	N	$1,500	21
G	2,000	28	L	1,500	24	O	5,000	19
H	1,000	23	R	1,000	22	P	3,500	17
J	1,500	16				Q	6,500	14
M	2,500	15						

EXAMPLE 14-8

Table 14-6 illustrates projects F through R categorized according to high, medium, or low risk. (The consideration of these projects is completely separate from the previous consideration of projects A through E in Example 14-7.) Their prospective after-tax internal rates of return are also indicated.

Suppose that the firm has $15,000 capital and management desires to invest approximately $\frac{1}{3}$, or $5,000, in each risk category as long as the return seems commensurate with the risk. Thus the firm would conditionally accept projects F, G, and H in the high-risk category and projects K and L in the medium-risk category. In the low-risk category the firm could conditionally accept either projects N and P or project O. Suppose it is judged that low-risk project N having an internal rate of return of 21% on an investment of $1,500 is preferred to the provisionally accepted medium-risk project L earning 24% on the same investment amount. Hence project O is the apparent remaining provisional choice in the low-risk category. Other trade-offs between risk categories can now be considered for any combination of projects, keeping in mind the $15,000 total capital constraint. Suppose that after consideration of many combinations, the only remaining trade-off that is judged wise is to accept medium-risk project R, earning 22%, rather than high-risk project H, earning 23%.

Summarizing the results of these judgmental decisions, we find that the final acceptances are projects F and G, requiring $4,000 in the high-risk category; projects K and R, requiring $4,500 in the medium-risk category; and projects N and O, requiring $6,500 in the low-risk category. Thus the final allocation of capital among the risk categories is a bit more conservative than the initially planned one-third split. ∎

AN OVERVIEW OF ECONOMIC ANALYSIS

The remainder of this chapter is devoted to the following considerations that are vital to successful capital budgeting and investment analysis.

1. Preliminary planning and screening of investment alternatives.
2. Use of standard and acceptable economy study methods.

3. Estimating and reliability of data.
4. Postaudits of actual versus estimated performance.
5. Objectivity of the analyst.

Preliminary Planning and Screening

The most effective guarantee of the quality and usefulness of an economy study is to make sure that adequate preliminary planning and screening of alternatives are performed.

Preliminary planning includes study of the objectives to be satisfied, review of the facts pertinent to the problem, and identification of all logical plans or alternatives that will meet the objectives. Often this will result in a very large number of alternatives and subalternatives. These can be screened by preparing preliminary estimates and making rough economic comparisons, and dropping obviously inferior plans at once from further evaluation. The analyst then can make detailed economy studies of those alternatives that appear most promising.

Economy studies of alternatives, no matter how well performed, cannot result in the selection of the best plan if that potentially best plan is not identified and considered in the first place. There is a reasonable limit to the number of alternative plans that can be subjected to detailed economy studies. Hence it is vitally important that adequate preliminary planning which includes all viable alternatives be accomplished, and it makes good sense that these alternatives be screened before detailed economy studies are made.

The Use of Standard and Acceptable Economy Study Methods

This book has concentrated on the use of theoretically correct and acceptable methods for performing engineering economy studies. Many firms utilize standard study procedures and forms to facilitate uniformity of analysis and ease of review by decision makers. As an example, the following is an illustrative segment of the excellently done *Instruction Manual-DCF Plan* by Giddings & Lewis, Inc.†

The full manual contains a complete set of instructions on what to consider when comparing alternatives and how to complete the analysis forms line by line as needed. The use of the manual facilitates the determination of an after-tax internal rate of return (I.R.R.) for a given project. The method of adjusting for the effect of depreciation on after-tax cash flows is a variation of that previously illustrated in this book, but the results are the same. (The reader should be able to verify that Equation 10-8 has been utilized to calculate an after-tax depreciation "credit" in Figure 14-6.) The illustrated procedure should not be interpreted as necessarily the best available, and it certainly should not be emulated by any other firm without careful consideration of whether modifications are needed to meet the needs of the firm.

---

†Excerpts from *Instruction Manual-DCF Plan* are reproduced by permission of Giddings & Lewis, Inc., Fond du Lac, Wis.

EXAMPLE 14-9 (Using Forms in Figures 14-4 to 14-6)

The illustrative example, for which all details will not be given, is an economic analysis of a machine requiring a $270,500 investment to be depreciated over 15 years, an additional investment of $15,000 to be declared as an expense in the year of expenditure, and an additional $2,000 for working capital, making a total investment of $287,500. The primary purpose here is to demonstrate the basic procedure and forms used by a particular company to make its after-tax engineering economy evaluations. A before-tax analysis of a similar problem was presented as a case study in Chapter 5.

Once the capital expenditures have been estimated, the next step is to determine the year-by-year estimates of before-tax cash flows as shown for the first year in Figure 14-4 and called "Pre-Tax Income from Operations." The net result on line 31 shows a net income or cash flow of $77,900. This $77,900 is broken down into $19,475 for 0 to 3 months and $58,425 for 3 to 12 months (details not shown) and the latter amounts entered in the second column of Figure 14-5. Similar analyses are done using separate forms such as in Figure 14-4 for each of the years 2, 3, 4, and 5, and the results entered in the second column of Figure 14-5. Because identical cash flows are assumed at end of years 6 through 10, the totals for that period can be lumped together as the $440,000 entry near the bottom of the second column of Figure 14-5.

The third column of Figure 14-5 converts pretax incomes (cash flows) into after-tax incomes using 47%, which is the residual after an assumed combined federal and state tax rate of 53%.

The present worths can then be calculated using the 10%, 15%, 25%, and 40% present worth factors provided on the form in Figure 14-5. The 10-year totals are shown at the bottom for each trial percentage.

The present worths of income from operations over a 10-year period at 0% (which is the actual income or cash flow), 10%, 15%, 25%, and 40% are then entered on line 5 of the form in Figure 14-6. Other cash inflows are then entered in the "Trial No. 1—0% Interest Rate" column of Figure 14-6. A $7,950 tax credit is entered in line 7 for the expensed investment items. The $15,650 on line 8 is the after-tax proceeds from the sale of former assets (sold 8 months after time 0). Line 10 shows a depreciation tax credit of $143,365. Expected after-tax proceeds from the hypothetical liquidation of the proposed assets at the close of the comparison are shown as $79,975 on line 13. The total of all cash inflows at 0% interest is $654,181.

The investment requirements stated at the beginning of this example are entered in lines 1, 2, 3, and 4 of Figure 14-6.

The forms in Figures 14-5 and 14-6 utilize single-sum present worth factors which happen to be based on *continuous* compounding. Line 10 of Figure 14-6 contains specially derived present worth factors which adjust for the difference between the economic comparison period of 10 years and the years over which depreciation is to be charged, which is 15 years. Since all investment outlays occur within 3 months of time 0, the present worth factors for these costs (on lines 1, 2, and 3) are 1.0 at all interest rates.

The next step in Figure 14-6 is to multiply all cash flows by the present worth factors for all interest rates and to sum the present worths of cash outflows (line 4, called "A") and of cash inflows (line 14, called "B") for each interest rate.

GIDDINGS & LEWIS Discounted Cash Flow Capital Investment Evaluation Procedure | Page No.

LONG FORM METHOD

5—L —

SUMMARY SHEET : Pre-Tax INCOME from OPERATIONS | Period _____

INCREASE	DECREASE	Item No.	ANALYST _____ DATE _____
			Effect of Project on REVENUE
_____	_____	1.	PRODUCT QUALITY (Potential loss or gain of profit through change is product quality.)
12,250		2.	PRODUCTION CAPACITY (Potential loss or gain as result of change in capacity.)
_____	_____	3.	FLOOR DEMONSTRATION Hrs./Mo. (Potential profit from having G&L machine available for Customer Demonstration.)
12,250x	Y	4.	TOTAL
			Effect of Project on DIRECT COSTS
_____	30,000	5.	Direct Labor Hrs. x (D.L. Rate + Fringes)
		6.	Direct Material
_____	7,500	7.	Sub-Contracting
		8.	Other
Y	37,500x	9.	TOTAL
			Effect of Project on INDIRECT COSTS
_____	6,500	10.	Indirect Labor (Incl. Fringe) (HANDLING)
_____	1,500	11.	Overtime (Premium, etc.)
		12.	Indirect Materials & Supplies
_____	14,000	13.	Tooling (FIXTURES)
		14.	Tool Repairs
_____	2,000	15.	Scrap & Rework
_____	4,250	16.	Inspection
		17.	Maintenance
_____	2,400	18.	Down Time
		19.	Power & Utilities
_____	2,200	20.	Floor Space & Occupancy
4,700		21.	Property Taxes & Insurance
		22.	Inventory
_____	_____	23.	Safety
_____	_____	24.	Flexability
_____	_____	25.	Working Conditions
_____	_____	26.	Other
4,700 Y	32,850x	27.	TOTAL
			COMBINED EFFECT
	12,250	28.	Net Increase in REVENUE (4x-4y)
	37,500	29.	Net Decrease in DIRECT Costs (9x-9y)
	28,150	30.	Net Decrease in INDIRECT Costs (27x-27y)
	77,900	31.	NET INCOME FROM OPERATIONS (28+29+30)

LJM Nov. 1968

FIGURE 14-4   Example of a Form for Determining Revenues and Expenses (Pretax Income) for a Given Year.

As explained in the lower part of Figure 14-6, the DCF rate of return (I.R.R.) is the rate at which the present worth of inflows (A) equals the present worth of outflows (B). A convenient interpolation chart is furnished for determining the final answer as the rate at which the ratio A/B = 1.0, which is 17% for the case illustrated.   ∎

GIDDINGS & LEWIS Discounted Cash Flow Capital Investment Evaluation Procedure

LONG FORM METHOD: <u>Present Worth</u> of INCOME from OPERATIONS during Comparison Period.

COMPARISON PERIOD ___10___ Yrs   ZERO POINT (Date) ___6-1-80___

PERIOD AFTER ZERO POINT	INCOME FROM OPERATIONS — PRE-TAX (Pages 5L)	INCOME FROM OPERATIONS — AFTER TAX (X 47%)	0% FACTOR	10% FACTOR	10% PRESENT WORTH	15% FACTOR	15% PRESENT WORTH	25% FACTOR	25% PRESENT WORTH	40% FACTOR	40% PRESENT WORTH
0 - 3 Mo.	19,475	9,153	1.000	1.000	9,153	1.000	9,153	1.000	9,153	1.000	9,153
3 - 12 Mo.	58,425	27,460		.952	26,142	.929	25,510	.885	24,302	.824	22,627
2nd Yr.	81,500	38,305		.861	32,481	.799	30,606	.689	26,392	.553	21,182
3rd Yr.	86,420	40,617		.779	31,641	.688	27,944	.537	21,811	.370	15,028
4th Yr.	88,050	41,384		.705	22,126	.592	18,579	.418	13,119	.248	7,783
5th Yr.	92,600	43,522		.638	27.767	.510	22,196	.326	14,188	.166	7,225
6th Yr.				.577		.439		.254		.112	
7th Yr.				.522		.378		.197		.075	
8th Yr.				.473		.325		.154		.050	
9th Yr.				.428		.280		.119		.034	
10th Yr.				.387		.241		.093		.023	
11th Yr.				.350		.207		.073		.015	
12th Yr.				.317		.178		.057		.010	
13th Yr.				.287		.154		.044		.007	
14th Yr.				.259		.132		.034		.005	
15th Yr.				.235		.114		.027		.003	
16th Yr.				.212		.098		.021		.002	
17th Yr.				.192		.084		.016		.001	
18th Yr.				.174		.073		.013		.001	
19th Yr.				.157		.062		.010		.001	
20th Yr.				.142		.054		.008		.001	
Sub-Totals											
5-Year Periods											
0 - 5 Yrs.				.787	•	.704		.571		.432	
5 - 10 Yrs.	440,000	206,800		.477	98,644	.332	68,658	.164	33,915	.0585	12,098
10 - 15 Yrs.				.290		.157		.047		.0079	
15 - 20 Yrs.				.176		.074		.013		.0011	•
20 - 25 Yrs.				.107		.0350		.0038		.0001	
25 - 30 Yrs.				.0646		.0165		.001		--	
30 - 35 Yrs.				.0392		.0078		.0003		--	
35 - 40 Yrs.				.0238		.0037		.0001		--	
TOTALS		$407,241			$248,454		$202,646		$142,280		$95,096

*By Definition Here, "Income from Operations" excludes all tax credits, credit for salvage value of old asset, and terminal credits.

LJM Nov. 1968

ANALYST_____   DATE_____

FIGURE 14-5 Example of a Form for Calculating the Present Worth of Income from Operations.

Based on Figure 14-6, note that the project's net present worth can be easily obtained, for a given interest rate, as B − A. A benefit–cost ratio is simply computed as B ÷ A. For instance at i = 15%, the N.P.W. is $27,071 and the B/C ratio is 1.094. The versatility of the procedure described in Example 14-9 should be apparent.

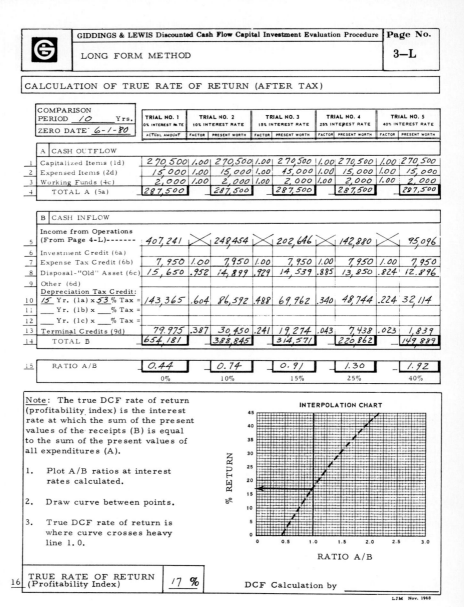

GIDDINGS & LEWIS Discounted Cash Flow Capital Investment Evaluation Procedure		Page No.
Ⓖ	LONG FORM METHOD	3—L

CALCULATION OF TRUE RATE OF RETURN (AFTER TAX)

COMPARISON PERIOD _10_ Yrs. ZERO DATE 6-1-80		TRIAL NO. 1 0% INTEREST RATE ACTUAL AMOUNT	TRIAL NO. 2 10% INTEREST RATE FACTOR / PRESENT WORTH	TRIAL NO. 3 15% INTEREST RATE FACTOR / PRESENT WORTH	TRIAL NO. 4 25% INTEREST RATE FACTOR / PRESENT WORTH	TRIAL NO. 5 40% INTEREST RATE FACTOR / PRESENT WORTH
A	CASH OUTFLOW					
1	Capitalized Items (1d)	270,500	1.00 270,500	1.00 270,500	1.00 270,500	1.00 270,500
2	Expensed Items (2d)	15,000	1.00 15,000	1.00 45,000	1.00 15,000	1.00 15,000
3	Working Funds (4c)	2,000	1.00 2,000	1.00 2,000	1.00 2,000	1.00 2,000
4	TOTAL A (5a)	287,500	287,500	287,500	287,500	287,500
B	CASH INFLOW					
5	Income from Operations (From Page 4-L)	407,241	248,454	202,646	142,880	95,096
6	Investment Credit (6a)					
7	Expense Tax Credit (6b)	7,950	1.00 7,950	1.00 7,950	1.00 7,950	1.00 7,950
8	Disposal-"Old" Asset (6c)	15,650	.952 14,899	.929 14,539	.885 13,850	.824 12,896
9	Other (6d)					
10	Depreciation Tax Credit: _15_ Yr. (1a) x _53_ % Tax =	143,365	.604 86,592	.488 69,962	.340 48,744	.224 32,114
11	___ Yr. (1b) x ___ % Tax =					
12	___ Yr. (1c) x ___ % Tax =					
13	Terminal Credits (9d)	79,975	.387 30,450	.241 19,274	.043 7,438	.023 1,839
14	TOTAL B	654,181	388,845	314,571	220,862	149,889
15	RATIO A/B	0.44	0.74	0.91	1.30	1.92
		0%	10%	15%	25%	40%

Note: The true DCF rate of return (profitability index) is the interest rate at which the sum of the present values of the receipts (B) is equal to the sum of the present values of all expenditures (A).

1. Plot A/B ratios at interest rates calculated.

2. Draw curve between points.

3. True DCF rate of return is where curve crosses heavy line 1.0.

INTERPOLATION CHART

% RETURN vs RATIO A/B

16	TRUE RATE OF RETURN (Profitability Index)	17 %	DCF Calculation by _____

LJM Nov. 1968

FIGURE 14-6 Example of a Form for Summarizing the Present Worth of Cash Outflows and Cash Inflows to Calculate the After-Tax Rate of Return.

Estimating and the Reliability of Data

The most critical and usually most difficult aspect of economy studies is the estimating of the variables or factors—such as revenue, operating costs, and life—which are vital for those studies. The term *estimate*, when applied to economic analyses, can have a multiplicity of meanings. At one extreme it can be used to indicate a carefully considered computation of some quantity for which the exact magnitude cannot be determined. At the other extreme, it can be used to denote what are actually just offhand approximations that are little better than outright guesses.

The first part of Chapter 8 discussed sources of estimates and ways estimates are accomplished. Although engineers often need to obtain estimates and data from others who are more knowledgeable or are in the best position to make particular estimates, they should consider themselves responsible for the validity of the data they use. They should evaluate the basic data they obtain for reasonableness, investigate any indication of unreliability, and confer with those most knowledgeable for further clarification if needed.

In making estimates for engineering economy studies, one must be wary of averages, standard data, and other regular published sources. One is generally evaluating specific plans, not average plans, and hence needs specific estimates rather than average estimates. For example, average building maintenance cost may be $1.00 per square foot per year, but for a particular case it may be as low as $0.40 or as high as $2.50, depending on specific conditions. Average cost data can, however, be quite useful in *verifying* the reasonableness of specific cost data used for an economy study.

In evaluating the adequacy of an estimate or the extent to which effort ought to be devoted to generate the estimate, one should judge the importance of the estimate to the study in which it is used and the importance of the study results to the firm. Chapter 7 discussed the use of sensitivity concepts for judging the relative importance of one or more estimated variables on study results. The importance to the firm of a particular study depends on many ramifications of the study—the effect on objectives; the effect on the organization, facilities, and finances; and the risks assumed.

Obviously, the more sensitive an engineering economy study result is to a given estimate and the greater the importance of the study to the firm, the greater is the effort or attention justified for that estimate. This general principle must be moderated by the realities of just how much valuable information is potentially obtainable from additional estimating effort or attention, what resources are available for making estimates, and the degree of urgency for completion of the estimate studies.

Postinvestment Reviews of Actual Versus Estimated Performance

After projects have been undertaken or completed, the firm should conduct one or more postcompletion reviews or audits. These audits, sometimes called investment performance reports, can serve four purposes:

1. To verify any resulting savings or profit.
2. To reveal reasons for any project failure.
3. To check on the soundness of various managers' and engineers' proposals.
4. To aid in economy studies of future capital expenditure proposals.

The first two reasons are vitally important and are often sufficient justification for the postcompletion audit. The latter two reasons bear on the quality of economy studies, and hence should be examined more closely.

Managers and engineers who know they will be held to account for the results of their proposals tend to make every effort to ensure their reasonableness and accuracy. However, this accountability should not be overemphasized, for it will cause the managers and engineers to become overcautious and to avoid proposing projects that are really needed rather than risk exposure to censure.

The last-stated purpose of postaudits is to provide information for avoiding past pitfalls and for making better estimates of similar future projects. Such audits reveal the tendency of various managers or engineers to be overly cautious or pessimistic in their estimates and thus provide a basis for adjustments to bring estimates for future projects closer into line with reality.

Objectivity of the Economy Study Analyst

The objectivity of the economy study analyst (and any person furnishing him or her estimates) is a matter that is hardly challengeable in principle but sometimes compromised in practice.

It is too easy for one to consciously or unconsciously develop a bias favoring a particular project or alternative. One may then be convinced that it is in the firm's best interest to demonstrate the superiority of that particular alternative, thereby causing the person to slant his or her estimates in favor of that alternative. Such actions violate professional standards and destroy the integrity of engineering economy studies.

The only condition under which economy studies can fulfill their intended function is for the analyst and others furnishing estimates to be wholly detached and let the results of the study be viewed objectively. The purpose of the study is accomplished only if it provides the best chance for selection of the plan or alternative that actually will turn out to be the most advantageous—not if it results in the "proving in" of some other favorite or preconceived best alternative.

PROBLEMS

14-1 What are five possible sources of capital to a corporation for financing new projects?

14-2 (a) What is equity capital, and explain how it is different from debt capital.

(b) Why do bondholders, on the average in the long-term, receive a lower return than do holders of common stock in the same corporation?

(c) Why would a fast-growing company possibly *not* desire to use large amounts of debt capital in its operation?

14-3 (a) List at least four characteristics of a corporation.

(b) Give an example of how the profits of a corporation are "double taxed" and discuss how this may retard investment in a firm's stock.

14-4 A corporation sold an issue of 20-year bonds, having a face value of $5,000,000, for $4,750,000. The bonds bear interest at 10%, payable semiannually. The company wishes to establish a sinking fund for retiring the bond issue and will make semiannual deposits that will earn 8%, compounded semiannually. Compute the annual cost for interest and redemption of these bonds.

14-5 **(a)** A company has issued 10-year bonds, with a face value of $1,000,000, in $1,000 units. Interest at 8% is paid quarterly. If an investor desires to earn 10% nominal interest on $10,000 worth of these bonds, what would the selling price have to be?
(b) If the company plans to redeem these bonds in total at the end of 10 years and establishes a sinking fund that earns 8%, compounded semiannually, for this purpose, what is the *annual* cost of interest and redemption?

14-6 A 20-year bond with a face value of $5,000 is offered for sale at $3,800. The rate of interest on the bond is 7%, paid semiannually. This bond is now 10 years old (i.e., the owner has received 20 semiannual interest payments). If the bond were purchased for $3,800, what effective rate of interest would be realized on this investment opportunity?

14-7 You bought a $1,000 bond that paid interest at the rate of 7%, payable semiannually, and held it for 10 years. You then sold it at a price that resulted in a yield of 10% nominal on your capital. What was the selling price?

14-8 The Yog Company, a privately owned business, has an opportunity to buy the Small Company, from which it purchases a large amount of its raw materials. The purchase price is to be $500,000. It is estimated that the Yog Company can realize a saving of $75,000 annually from operating the Small Company instead of having to purchase its raw materials from it. The anticipated profit of $75,000 is exclusive of any financial expense that might be involved in the transaction but includes provision for writing off the investment over a 20-year period. The Yog Company does not have $500,000 available that it can use to buy the Small Company. However, it can sell a $500,000 bond issue and use the proceeds for this purpose. The bonds will be issued as 20-year 10% bonds. Should the Yog Company issue the bonds and buy the Small Company?

14-9 The Yog Manufacturing Company's common stock is presently selling for $32 per share and annual dividends have been constant at $2.40 per share. If an investor believes that the price of a share of common stock will grow at 5% per year into the foreseeable future, what is the approximate cost of equity to Yog? What assumptions did you make?

14-10 Reconsider the situation of problem 14-8 when $500,000 in *new* common stock is issued rather than a bond issue. The cost of equity is that obtained in Problem 14-9. Should Yog purchase the Small Company with equity funds? What difficulties could this pose to present stockholders?

14-11 During the first 5 years of its life, a small corporation, which has a capitalization of $2,000,000 represented by 2,000 shares of common stock, has paid no dividends, in order to finance its expansion out of retained profits. It now needs $1,000,000 in additional capital to finance and stock two new warehouses. Discuss briefly the advantages and disadvantages of obtaining the required capital from (a) selling additional common stock, (b) borrowing on a 5-year bank loan at 10% interest, and (c) issuing 8%, 10-year bonds which would contain a provision that the corporation could not incur further indebtedness until the bond issue was retired.

14-12 Consider a small manufacturing firm in which there are 1,000 shares of common stock divided among three owners as follows:

Owner 1: 250 shares

Owner 2: 375 shares

Owner 3: 375 shares

Each share has been "valued" at $500, based on what the owners know they can get for their stock. The firm needs to raise an additional $125,000 to purchase some new equipment, repay short-term bank notes, and increase substantially raw materials inventories. Their current after-tax return on equity is 12% and the effective income tax rate is 32%.

Plan I: Owners 2 and 3 agree to each make 125 shares of their stock available at $500/share.

Plan II: Instead of selling stock (i.e., ownership in the firm), the three owners believe they can sell 8-year mortgage bonds, bearing 10% interest each year, to raise the needed

$125,000. The bonds would be sold in $1000 units and redeemed in full at the end of 8 years. A 6% sinking fund will be used to guarantee repayment of the loan.

By spending the $125,000 as indicated above, the owners are certain they can increase revenues by $40,000 per year (on the average), exclusive of financing costs. Assume that depreciation associated with the purchase of new equipment is negligible in this situation. Which plan would you recommend for acquiring the needed $125,000? Show all calculations, and state any assumptions that you feel are required.

14-13 Determine the before-tax and after-tax average cost of capital for a firm that has this capital structure:

Amount	Source of Capital	Rate of Return (%)	Amount per Year
$3 million	Short-term bank loans	10	$0.30 million
$7 million	Mortgage bonds	7	$0.49 million
$4 million	Preferred stock	8	$0.32 million
$11 million	Common stock and retained earnings	13	$1.43 million

Assume that the firm's effective tax rate is 46% and that a 13% rate of return to purchasers of common stock represents a satisfactory opportunity cost of equity capital.

14-14 The after-tax cost of capital (K_a') to a large utility company is defined as follows:

$$K_a' = (1 - t)(\lambda)i_b' + (1 - \lambda)e_a'$$

where t = effective income tax rate
λ = fraction of total capitalization in borrowed funds
i_b' = marketplace before-tax cost of borrowed funds
e_a' = after-tax opportunity cost of equity funds

If the *real* (inflation-free) annual return on borrowed funds and equity capital are roughly constant at 2% and 5%, respectively, what is the value of K_a' when inflation averages 7% per year? Let $\lambda = 0.40$ and $t = 0.50$.

14-15 Refer to Example 14-3. If annual maintenance expenses can range from $800 to $1,300 per year and inflation can vary from 3% to 8% per year, determine whether the forklift truck should be purchased or leased for each combination of extreme values.

Annual Maintenance	Annual Inflation Rate (%)	Recommendation
$ 800	3	?
800	8	?
1,300	3	?
1,300	8	?

14-16 An existing piece of equipment has been performing poorly and needs replacing. More modern equipment can be *purchased* for cash using retained earnings (equity funds) or it can be *leased* from a reputable firm. If purchased, the equipment will cost $20,000 and have a depreciable life of 5 years with no salvage value. Straight line depreciation is used by the firm. Because of improved operating characteristics of the equipment, raw materials savings of $5,000 per year are expected to result relative to continued use of the present equipment. However, labor costs for the new equipment will most likely increase by $2,000 per year and maintenance will go up by $1,000 per year. To lease the new equipment requires a refundable deposit of $2,000 and the yearly leasing fee is $6,000. Annual materials savings and extra labor costs will be the same when purchasing or leasing the equipment, but the company will provide maintenance for their equipment as part of the leasing fee. The desired after-tax rate of return (I.R.R.) is 15% and the effective income tax rate is 50%. If purchased, it is believed that the equipment can be sold at the end of 5 years for $1,500 even through $0 was used in calculating depreciation. An investment credit of 10% can be used to offset income taxes at the time of the purchase (year 0). Determine whether the company should buy or lease the new equipment, assuming that it has been decided to replace the present equipment.

14-17 Determine the more economical means of acquiring a business machine if you may either (1) purchase the machine for $5,000 with a probable resale value of $1,000 at the end of 5 years, or (2) rent the machine at an annual rate of $900/year for 5 years with an initial deposit of $500 refundable upon returning the machine in good condition. If you own the machine, you will depreciate it for tax purposes at an annual rate of $800. All leasing rental charges are deductible for income tax purposes. As owner or lessee you will pay all expenses associated with the operation of the machine.
   (a) Compare these alternatives by use of the annual cost method. The after-tax minimum attractive rate of return is 10% and the effective income tax rate is 50%.
   (b) How high could the annual rental be such that the leasing option is still the more desirable alternative?

14-18 Suppose that a machine costing $11,000 can be financed entirely by borrowed funds or by equity capital. With borrowed funds, the loan is to be repaid at the rate of $2,000 at the end of each year for the first 4 years and $3,000 at the end of the fifth year. Interest charges are computed at 10% of the unpaid, beginning-of-year balance of the loan. Depreciation is calculated on a straight line basis, the depreciable life is 5 years, and the estimated salvage value is $1,000. The expected before-tax cash flow attributable to the machine *before* deducting interest charges and operating costs is $10,000, and the effective income tax rate is 50%. Operating costs will amount to $3,000 per year.
   (a) Determine the after-tax cash flows of both financing plans, and compute the present worth of each at a minimum attractive rate of return of 15%.
   (b) Compute the I.R.R. on after-tax equity cash flow for both financing plans.
   (c) When borrowed funds are used to finance a project, what problems can arise when the I.R.R. method is utilized to compare alternative financing plans?

14-19 A firm is considering the introduction of a new product in 1986. The marketing department has estimated that the product can be sold over a period of 5 years at a price of $7.00 per unit. Sales are estimated to be 10,000 units the first year and will increase by 2,000 units each year. For the life of the product, equipment necessary to produce the item will cost $200,000. It is estimated that this equipment can be sold for $50,000 at the end of year 5. A.C.R.S. depreciation with a 5-year recovery period will be used. The equipment will be financed by borrowing $160,000 at 10% compounded on the unpaid balance each year. This debt is to be repaid in five equal end-of-year payments of $32,000 each. The $40,000 balance will be financed from equity funds. Operating and maintenance costs (not including taxes) will be $50,000 the first year and *decrease* by $3,000 each year thereafter. The firm's effective tax rate is 40%, and its minimum attractive rate of return is 15%. Calculate the *present worth* of the after-tax equity cash flow if the corporation is profitable in its other activities.

14-20 Suppose that you borrow $10,000 at $i_b = 15\%$ per year and agree to repay this amount in 3 equal end-of-year payments. What is your after-tax cost of interest in each of the 3 years, and what is the present worth of the after-tax interest expense? Let $t = 0.32$ and $i_{AT} = 18\%$ per year.

14-21 Rework Example 14-4 when the $50,000 worth of borrowed funds, at 10% interest, is to be repaid in equal end-of-year amounts such that no principal remains after the tenth payment is made. Be careful to note that interest and repaid principal will vary in amount from year to year in this situation.

14-22   A company has $290,000 for investment in new projects during the coming year. Projects currently being considered are as follows:

Project	Capital Required	Life (years)	Estimated Annual Rate of Profit (%)	Risk
A	$ 60,000	5	10	Low
B	100,000	3	20	Average
C	150,000	5	8	Average
D	110,000	8	15	High

The company follows a general policy of not committing capital for a longer period than 8 years, and it prefers 5 years or less. Uncommitted capital would remain temporarily in a bank, where it would earn at least 8%. Which projects do you recommend?

14-23   A company has $20,000 to invest in elective projects and has taken the position that at least half of these funds should be spent on medium- and low-risk projects. Based on the investments shown below, categorized by risk, what subset of these independent capital investment opportunities should be recommended?

Project	Investment	Prospective Annual Return After Taxes (%)
HIGH RISK		
M	$ 4,000	32
N	2,000	28
O	6,000	22
P	5,500	17
MEDIUM RISK		
S	8,000	22
T	7,500	20
U	10,000	16
V	5,000	12
W	6,200	10
LOW RISK		
C	12,000	16
D	8,000	14
E	10,000	13
F	5,000	8

14-24   **(a)** List some of the factors that a high-technology electronics firm might consider in establishing a minimum attractive rate of return for its investments.
(b) What other objectives, in addition to maximizing the future worth (or present worth) of the firm, would a high-technology company consider in its capital investment decision making?
(c) What additional types of data might a high-technology electronics firm include in its version of the Giddings & Lewis investment analysis manual?

PART III

OTHER USEFUL METHODS FOR MINIMIZING RESOURCE REQUIREMENTS

CHAPTER 15

Minimum-Cost Formulas

This and the following three chapters deal with various types of engineering economy problems in which study periods tend to be rather short (1 year or less, usually). As a result, the compounding of interest is not one of the more important concerns that underlies these problems. Earlier, in Chapter 3, such problems were treated as present-economy studies. The subject of this chapter is the description and illustration of certain mathematical models that have been used in practice to determine minimum costs of alternatives having cost functions expressed in one (or more) variable. The topic of value analysis is described briefly in Chapter 16. Linear programming formulations of special types of resource allocation problems are presented in Chapter 17. In Chapter 18 present economy studies associated with critical path methods are illustrated.

MINIMUM-COST ECONOMY

In the day-to-day operations of business organizations, many economy studies are made for the purpose of enabling costs to be minimized, where the total cost for an item or operation is the sum of a fixed cost, a variable cost that increases *directly* with respect to some design variable such as the number of units produced, and a second variable cost that is *inversely* proportional to the same design variable. Thus the total cost C is

$$C = ax + \frac{b}{x} + k \qquad (15\text{-}1)$$

where a, b, and k are constants and x is the design variable that can be controlled. A number of real situations commonly experienced in engineering practice can be dealt with by relating them to the theoretical relationship expressed in Equation 15-1. It is the purpose of this chapter to examine several typical problems of this type, including determination of economic order quantities, the most economical size of electrical conductors, and the most economical replacement interval for assets subject to sudden failures.

The necessary condition for minimum cost can readily be determined by differentiating Equation 15-1 (the cost function) with respect to the design variable x, equating to zero, and solving for the value of x'. Thus

$$\frac{dC}{dx} = a - \frac{b}{x^2} = 0 \qquad (15\text{-}2)$$

$$x' = \sqrt{\frac{b}{a}} \qquad (15\text{-}3)$$

The second derivative, d^2C/dx^2, equals $2b/x^3$ and is positive since b and x are both positive. This is the sufficient condition for a *minimum* optimum value of x'.

The optimum value of the design variable, x', is the point at which directly varying costs are equal to the inversely varying costs. This may be shown by noting that the increasing variable cost at x' is

$$ax' = a\sqrt{\frac{b}{a}} = \sqrt{ab}$$

and the decreasing variable cost is

$$\frac{b}{x'} = \frac{b}{\sqrt{\dfrac{b}{a}}} = \sqrt{ab}$$

Figure 15-1 depicts the cost relationships for the case just considered, where the costs are directly and inversely proportional to the decision variable.

ECONOMIC PURCHASE ORDER SIZE

A common situation involving the minimum-cost concept is the purchasing of batches of goods for use or sale. Ordinarily, the goods or supplies are purchased periodically throughout the year in lots, and each lot is received at one time and put into storage. Thereafter, the supply is used as needed, sometimes at a fairly uniform rate, until the supply is exhausted or diminished to a predetermined

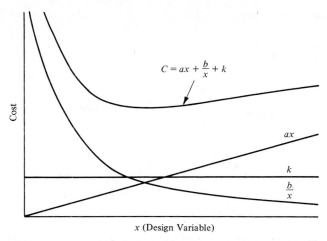

$$C = ax + \frac{b}{x} + k$$

ax

k

$\frac{b}{x}$

x (Design Variable)

FIGURE 15-1 General Relationship of the Fixed Cost (k), Directly Variable Cost (ax), Inversely Variable Cost (b/x), and Total Cost (C) in Equation 15-1.

minimum inventory quantity, at which time a new order is placed. This situation is depicted in Figure 15-2. In such a situation the cost of originating, placing, and paying for an order is essentially constant, regardless of the quantity ordered. Consequently, the order cost *per unit* varies inversely with the size of the order. On the other hand, a number of costs increase as the size of the order increases: for example, the amount of storage space and its costs, the interest on the average inventory in stock, insurance, and taxes all increase directly with the order quantity. These directly varying costs are commonly known as *carrying* or *holding costs*. The cost per piece of the material purchased is, in many cases and for a considerable range of quantities, independent of order size. (Variable price schedules are discussed later.) Clearly, under these conditions there will

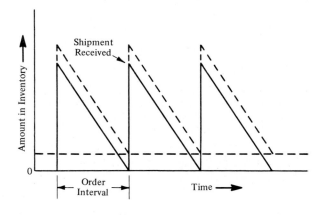

Key: ——— Without safety stock
        --- With safety stock

FIGURE 15-2 Variation in Inventory, With and Without a Safety Stock, When Lot Quantities Can Be Obtained Instantaneously and Use is at a Constant Rate.

be some order size that will be the most economical, and this *economic purchase order size* can be determined by the application of Equation 15-3. Thus, if

Q = most economic order quantity (lot size)

U = number of units used per year (at a constant rate)

S = cost of placing an order (setup cost)

C = commodity cost per unit

I = inventory carrying cost per year as a proportion of inventory value

the total cost for one year's supply will be the sum of the carrying cost, the order cost, and the commodity cost. If no minimum inventory is maintained, the average inventory will be $Q/2$, and the carrying cost will be $QIC/2$. This, of course, assumes that exclusive storage space does not have to be reserved for each inventory item; the space for one item can be used for another as the stock diminishes. Similarly, the number of orders per year will be U/Q, and the annual order cost will be SU/Q. The commodity cost will, of course, be CU and thus will be unaffected by order size. The total cost (T.C.) of acquiring and storing one year's worth of the commodity thus can be written as

$$\text{T.C.} = \left(\frac{IC}{2}\right) Q + \frac{SU}{Q} + CU$$

It will be noted that this expression has the same form as the right-hand side of Equation 15-1, with $IC/2$ corresponding to a, SU corresponding to b, and CU corresponding to k. Consequently, the most economical order quantity may be obtained from Equation 15-3 as

$$Q = \sqrt{\frac{\text{inversely varying cost coefficient}}{\text{directly varying cost coefficient}}}$$

or

$$Q = \sqrt{\frac{2SU}{IC}} \qquad (15\text{-}4)$$

EXAMPLE 15-1

Determine the most economic purchase order size for the conditions applicable to Equation 15-4 when annual usage is 8,000 units, the unit commodity cost is $2, the cost of placing an order is $5, and the annual inventory carrying cost is 30% of the average inventory.

Solution

$$Q = \sqrt{\frac{2 \times \$5 \times 8,000}{0.3 \times 2}} = 365 \text{ units}$$

To provide the required 8,000 units in 1 year, almost 22 lots of 365 each would have to be ordered. If exactly 8,000 units were required, this could be obtained in 21 lots of 365 each and 1 lot of 335 units. The total cost per year is

$$\frac{0.3(\$2)(365)}{2} + \frac{\$5(8,000)}{365} + \$2(8,000)$$

$$\text{T.C.} = \$16,219/\text{year} \quad \blacksquare$$

ECONOMIC ORDER SIZE
FOR VARIABLE PRICE SCHEDULES

Price schedules that vary with lot size also may have a pronounced effect upon the economic size for lots that are purchased. For example, assume that the item mentioned in Example 15-1 can be purchased under the following price schedule:

Lot Size	Unit Price
1–200	$2.10
201–500	2.00
501 and over	1.85

When lot sizes of 1 to 200 are considered, 40 or more purchase orders per year are required. The minimum T.C. in this case is

$$\frac{0.3(\$2.10)(200)}{2} + \frac{\$5(8,000)}{200} + \$2.10(8,000) = \$17,063/\text{year}$$

For lot sizes of 201 to 500, orders placed per year range from 16 to 39, and T.C., which is almost constant in this range, is in the neighborhood of $16,250 per year. Finally, when orders for more than 500 units are placed, 16 or less full lots are received each year and illustrative T.C.'s for the $1.85 unit price schedule appear below:

Number of Lots	Lot Size	Total Cost
14	571	$15,029
10	800	15,072
5	1,600	15,269

Figure 15-3 shows the cost of obtaining the year's requirement of 8,000 units under the indicated price schedule. As can be seen from this graph, minimum cost would be obtained by purchasing in lots of 501 pieces, requiring 16 lots.

Because there is never any assurance that the theoretical economical order quantity in any price range will fall within the lot size range for the particular price, or that this lot size will give the minimum total cost, problems of this type must usually be solved by a multiple-step procedure. (Of course, if curves such as are shown in Figure 15-3 are drawn, the correct solution is at once apparent.) The economical order quantity for one or more of the prices in the

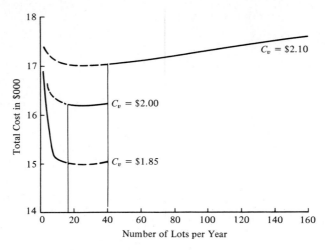

Key: −−− Price schedule does not apply

FIGURE 15-3 Cost of Obtaining a Year's Requirement of 8,000 Units Under a Lot Size Pricing Schedule.

price–quantity schedule is computed until a result is obtained that falls within the corresponding range of the schedule. The annual cost resulting from the use of the corresponding number of lots is then computed. Next the annual cost that would result from using the smallest lot size at the next lower price is determined. If this cost is less, the theoretical lot size for this price range should be computed and the annual cost at this theoretical lot size ascertained, in order to determine whether the total cost curve is increasing or decreasing. If the cost increases as the lot size increases, the entire procedure—determining the cost for the most economical lot size and comparing it with the cost for the minimum lot size of the next lower price—must be repeated until the lowest annual cost is obtained.

ECONOMIC PRODUCTION LOT SIZE

In many plants the parts that are used are produced internally. Typically, they are produced at a rate that is higher than the use rate, yet an entire production lot cannot be produced instantaneously. The resulting inventory situation is as depicted in Figure 15-4.

This condition is similar to that for determining economic purchase order quantities, with two exceptions. First, each time a lot is produced there are certain setup costs incurred. These correspond to the order costs S, where goods are purchased. Second, inasmuch as the entire lot is not obtained instantaneously and some items are being used before the entire lot is produced, a correction must be made to account for this condition. If P = the production rate per year,

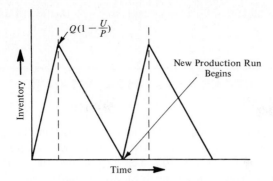

FIGURE 15-4 Inventory Quantity When Production is in Lots at a Uniform Rate Considerably Greater Than the Uniform Use Rate and No Minimum (Safety) Stock is Maintained.

with production at a constant rate, the average inventory will be

$$\frac{Q}{2}\left(1 - \frac{U}{P}\right) \qquad (15\text{-}5)$$

Again using the concept expressed in Equation 15-3, and assuming nonexclusive storage space, we find that the most economic lot size to produce is

$$Q = \sqrt{\frac{2SU}{I(1 - U/P)C}} \qquad (15\text{-}6)$$

$$Q = \sqrt{\frac{2S}{I[(1/U) - (1/P)]C}} \qquad (15\text{-}7)$$

THE EFFECT OF RISK AND UNCERTAINTY ON LOT SIZE

When a lot of goods is purchased or manufactured there may be uncertainty as to whether the entire quantity will in fact be used. Such uncertainty is due to several factors. Future demand for the item may be terminated. Design changes in a product may make parts for the old design obsolete. Consequently, it frequently is desirable to take such uncertainties into account in economic lot size formulas.

A little thought leads us to realize that the net effect of the risks just mentioned will be to make us want to have fewer parts on hand if there is a danger of their becoming obsolete. Numerous formulas have been derived that take such risks into account. All of them have the usual lot size form, but with some increase occurring in the denominator so as to decrease the economic lot size. A typical formula of this type is

$$Q = \sqrt{\frac{2SU}{(I + \alpha)C}} \qquad (15\text{-}8)$$

where α is an annual obsolescence cost factor expressed as a fraction of the unit commodity cost.

THE IMPORTANCE OF THE CARRYING COST RATE IN LOT SIZE FORMULAS

A great amount of discussion has centered around the interest rate that should be used in economic lot size determinations. There are two aspects of this matter that should be considered. First, we should make certain that the holding or carrying cost does, in fact, cover the costs involved. Many studies have indicated that the charges used frequently are too low. The second aspect relates to the effect that interest rate can have on the economic lot size. Example 15-2 provides an illustration of the effect that a change in interest rate can have.

EXAMPLE 15-2

Consider again Example 15-1 where a carrying cost rate of 30% resulted in an economic lot size of 365 units. What would be the economic lot size if the carrying cost rate were reduced to 20%?

Solution

$$Q = \sqrt{\frac{2 \times \$5 \times 8,000}{0.2 \times \$2}} = 448 \text{ units}$$

Thus a decrease in the interest charge will increase the economic lot size, as might be expected, since the inventory holding costs are reduced. In common terms used for sensitivity studies, a one-third decrease in interest rate increases the economic lot size by $(448 - 365)/365 = 23\%$. ∎

MINIMUM-COST SITUATIONS WITHOUT LINEAR COST RELATIONSHIPS

In problems thus far considered in this chapter, the cost relationships have been assumed to be either directly or inversely proportional to the number of items purchased or manufactured. Obviously, there are many situations where costs increase with some design variable and other costs decrease, but not in direct or inverse proportion. Such a condition exists in many "conductor" problems, where electric current or some fluid is forced through a conductor. Energy must be used to overcome the friction between the fluid, or current, and the conductor. If the size of the conductor is increased, with accompanying investment cost, the frictional loss, and the corresponding cost of overcoming it, will be reduced. Thus one cost increases, as conductor size is increased, and another decreases, and the total cost is the sum of the two. Since the two costs usually are not directly or inversely proportional to conductor size, Equation 15-3 cannot be applied. However, the costs can readily be computed and tabulated and the total cost and most economical conductor size determined.

TABLE 15-1 Annual Cost of Lost Power and Investment Charges for Various Sizes of Copper Wire (per 1,000 feet)

	A Wire size	000	00	0	1	2
B	Resistance (ohms at 20°C)	0.0618	0.0779	0.0983	0.124	0.156
C	Weight (lb)	508	403	320	253	201
D	Kilowatthours at 50 amperes and 4,500 hours per year	695	876	1,107	1,397	1,757
E	Cost of lost energy at $0.015 per kilowatthour	$10.43	$13.14	$16.60	$20.95	$26.36
F	Investment at $0.35 per pound	$177.80	$141.05	$112.00	$88.55	$70.35
G	Investment costs (depreciation, taxes, interest, and insurance) at 16% of line F	$28.45	$22.57	$17.92	$14.17	$11.26
H	Total annual cost (line E + line G)	$38.88	$35.71	$34.52	$35.12	$37.61

Table 15-1 shows the cost relationships for a particular copper conductor that is to transmit 50 amperes for 4,500 hours per year.† It is shown that size 0 conductor would be most economical for this case. Figure 15-5 shows the cost curves for this electrical conductor problem, and it will be noted that the minimum-cost point does not occur at the point where the two component costs are equal.

†This problem illustrates the well-known *Kelvin's law*, which was derived for a somewhat idealized transmission-line situation and then generalized.

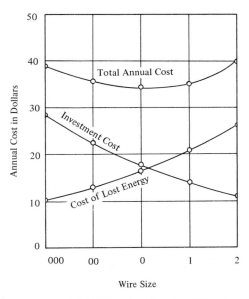

FIGURE 15-5 Breakeven Chart for Determining the Most Economical Size of Copper Conductor. (Data from Table 15-1.)

PRODUCTION TO MEET A VARIABLE DEMAND

Quite often a business must manufacture to meet a variable demand, which may be very low during certain parts of the year and very high at other times. Such a situation requires a decision to be made as to how the demand will be met. A single production unit may be able to meet the annual demand by operating continuously throughout the year, with the output being stored during the off-peak periods so as to meet the demands of the peak periods. Such a practice reduces the investment cost in production equipment, but the inventory and storage costs are increased. An alternative procedure would be to utilize more production units during a shorter period of time, thus increasing investment and possibly operation costs, but reducing inventory and storage costs. Clearly, this common type of situation presents a minimum-cost possibility by balancing increasing investment costs against decreasing inventory costs.

EXAMPLE 15-3

A company produces containers for the fruit-packing industry. The demand occurs during 3 months of the year, as shown in the first two columns of Table 15-2. It is customary to shut down the plant during October for vacations and overhaul of equipment, so that production occurs during only 11 months of the year. A machine can produce 250,000 units per month. The fixed costs, including depreciation, taxes, insurance, profit on capital, building charges, and so on, are $3,000 per year per machine, regardless of the amount of use. In addition, there are variable costs for each machine amounting to $700 per month. The containers have a value of $30 per thousand, and the annual inventory and storage costs amount to 20% of the value of the product.

The month-end inventories, resulting from using one, two, and three machines, are shown in Table 15-2, with the inventory flow depicted in Figure

TABLE 15-2 Monthly Demand and Month-End Inventory of Containers (in 000s) When Produced by One, Two, or Three Machines (Example 15-3)

Month	Demand	One Machine Production	One Machine Inventory	Two Machines Production	Two Machines Inventory	Three Machines Production	Three Machines Inventory
Jan.	0	250	1,000	0	0	0	0
Feb.	0	250	1,250	0	0	0	0
Mar.	0	250	1,500	250[a]	250	0	0
Apr.	0	250	1,750	500	750	0	0
May	0	250	2,000	500	1,250	500[b]	500
June	700	250	1,550	500	1,050	750	550
July	1,100	250	700	500	450	750	200
Aug.	950	250	0	500	0	750	0
Sept.	0	0	0	0	0	0	0
Oct.	0	250	250	0	0	0	0
Nov.	0	250	500	0	0	0	0
Dec.	0	250	750	0	0	0	0
	AVERAGE FOR YEAR		937.5		312.5		104.2

[a]One machine.
[b]Two machines.

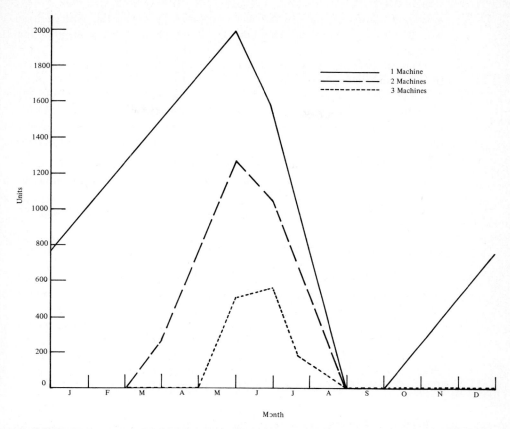

FIGURE 15-6 Monthly Inventories Resulting from Using 1, 2, or 3 Production Units to Meet a Seasonal Demand. (Data from Table 15-2.)

15-6. It is desired to determine the total annual cost for each number of machines.

Solution

For one machine:

Fixed costs	$ 3,000
Variable costs: $700 × 11	7,700
Inventory costs: $30 × 937.5 × 0.20	5,625
TOTAL	$16,325

In the same manner, the annual cost using two machines would be $15,575, and using three machines would be $17,325. Thus it is clear that the use of two machines would give minimum annual cost. ∎

REPLACEMENT OF ASSETS THAT FAIL SUDDENLY

Certain types of assets fail suddenly while in service and do not exhibit significant deterioration in their useful capabilities until they fail. Examples of such assets are transistors, light bulbs, and jet engines. However, this type of asset is usually subject to an increasing rate of failure as its cumulative usage or age increases.

There are two general classes of costs involved in deciding whether to replace (or maintain) individual items upon failure or to replace (maintain) an entire group of these items at the same time: (1) costs that vary directly with time: for example, annual costs of maintenance for individually serviced items; and (2) costs that vary indirectly with time, such as annual costs of periodic group maintenance on all items. The aim in this type of replacement analysis is to determine answers for the following questions.

1. Should a group of assets subject to sudden failure be replaced in entirety, or should they be individually replaced upon failure?
2. If group replacement is the best policy, what is the most economical group replacement interval?

The general solution procedure for obtaining answers to these questions involves the comparison of cost of group replacement (maintenance) at different replacement intervals with the cost of individually replacing items as they fail in service. Often group replacement at a stated interval will prove to be the more economical of these two policies.

To illustrate how one might solve this type of replacement problem, an example problem is worked that demonstrates the general solution procedure. The time value of money is often omitted as an important consideration in these problems because optimum replacement intervals are frequently less than a year. They are, therefore, problems of the present-economy type.

EXAMPLE 15-4

A small air cargo service has 20 airplanes of the same make and each airplane has two engines on it. The research department of the manufacturer of these planes has collected past data on engine breakdowns based on 250 hours of flying time each month:

Months After Maintenance	1	2	3	4	4
Probability of Engine Breakdown	0.2	0.1	0.1	0.2	0.4

The cost of remedial maintenance after a failure is $1,000 per engine, and if both engines of all planes were maintained as a group, it would cost $250 per engine. Assume that the maintenance schedule has no adverse affect on meeting air shipment schedules or revenues generated.

Solution

Let us assume that breakdowns occurring during a month are tallied at the beginning of the following month. Thus maintenance that occurs at the beginning of, say, the third month will be 1 month old at the beginning of the fourth month. During the first t time intervals, all failures are replaced as they occur. At the end of the tth month, all units are replaced regardless of their ages. The problem is to find that value of t which will minimize total cost per month. Because it is assumed that the entire replacement interval in question is of short duration such that the timing of money can be neglected, the total cost from time of group installation through the end of t months can be given by

$$K(t) = NC_1 + C_2 \sum_{x=1}^{t} f(x) \qquad (15\text{-}9)$$

where $K(t)$ = total cost for t months

C_1 = unit cost of replacement in a group

C_2 = unit cost of individual replacement after failure

$f(x)$ = number of failures in the xth month

N = number of units in the group

Hence the objective is to find the value of t that minimizes $K(t)/t$. The relationship of the various costs is shown in Figure 15-7.

Table 15-3 illustrates how the total cost of maintaining aircraft engines during each month is calculated. Here $P_1, P_2, \ldots, P_5$ represents the probability of engine failure within $1, 2, \ldots, 5$ months after the previous maintenance was performed. For the data given above, and by using Table 15-3, an average cost of maintenance per month can be calculated as shown in Table 15-4. Here it is apparent that the minimum value of $K(t)/t$ occurs in month 3. Thus the most economic policy is to perform group maintenance on aircraft engines every 3 months and individually repair engines as they fail within this interval of time. ∎

LIMITATIONS IN THE USE OF FORMULAS

Caution should be observed in the use of minimum-cost formulas as a method of making economy studies. Most economy study formulas are complex because of the necessity of making provision for all the possible cost factors that may be encountered. Some are so complex that electronic computers must be used to obtain a solution in a reasonable time.

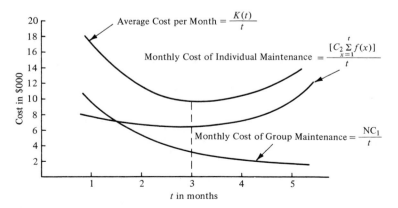

FIGURE 15-7   Replacement Costs for Example 15-4.

TABLE 15-3 Computational Setup for Determining Total Replacement Costs for Aircraft Engines in Example 15-4

Month, t	$f(x)$	$K(t)$
1	$f(1) = NP_1$	$K(1) = NC_1 + C_2 f(1)$ or $K(1) = NC_1 + C_2 NP_1 = N(C_1 + C_2 P_1)$
2	$f(2) = NP_2 + f(1)P(1)$	$K(2) = NC_1 + C_2 f(1) + C_2 f(2)$ $= NC_1 + C_2[f(1) + f(2)]$ or $K(2) = NC_1 + C_2 NP_1 + C_2 NP_2 + C_2 f(1)P_1$ $= NC_1 + C_2[N(P_1 + P_2) + f(1)P_1]$
3	$f(3) = NP_3 + f(2)P_1 + f(1)P_2$	$K(3) = NC_1 + C_2[f(1) + f(2) + f(3)]$ or $K(3) = NC_1 + C_2[N(P_1 + P_2 + P_3)$ $+ f(1)(P_1 + P_2) + f(2)P_1]$
4	$f(4) = NP_4 + f(3)P_1 + f(2)P_2 + f(1)P_3$	$K(4) = NC_1 + C_2[f(1) + f(2) + f(3) + f(4)]$ or $K(4) = NC_1 + C_2[N(P_1 + P_2 + P_3 + P_4)$ $+ f(1)(P_1 + P_2 + P_3)$ $+ f(2)(P_1 + P_2)$ $+ f(3)P_1]$
5	$f(5) = NP_5 + f(4)P_1 + f(3)P_2 + f(2)P_3$ $+ f(1)P_4$	$K(5) = NC_1 + C_2[f(1) + f(2) + f(3) + f(4) + f(5)]$ or $K(5) = NC_1 + C_2[N(P_1 + P_2 + P_3 + P_4 + P_5)$ $+ f(1)(P_1 + P_2 + P_3 + P_4)$ $+ f(2)(P_1 + P_2 + P_3)$ $+ f(3)(P_1 + P_2)$ $+ f(4)P_1]$

TABLE 15-4 Calculations of Replacement Costs for Aircraft Engines in Example 15-4

Month	Average Number of Engines to Be Maintained, $f(x)$	Cost of Individual Maintenance, $C_2 f(x)$	Cumulative Cost of Individual Maintenance, $C_2 \sum f(x)$	Averge Cost of Individual Maintenance per Month, $[C_2 \sum f(x)]/t$	Average Cost of Group Maintenance per Month, $NC_1 t$	Average Total Cost of Maintenance per Month, $K(t)/t$
1	$40(0.2) = 8$	$ 8,000	$ 8,000	$ 8,000	$10,000	$18,000
2	$40(0.1) + 8(0.2)$ $= 5.6$	5,600	13,600	6,800	5,000	11,800
3	$40(0.1) + 5.6(0.2)$ $+ 8(0.1) = 5.92$	5,920	19,520	6,506	3,333	9,839 ← minimum
4	$40(0.2) + 5.92(0.2)$ $+ 5.6(0.1) + 8(0.1)$ $= 10.544$	10,544	30,064	7,516	2,500	10,016
5	$40(0.4) + 10.544(0.2)$ $+ 5.92(0.1)$ $+ 5.6(0.1) + 8(0.2)$ $= 20.861$	20,861	50,925	10,185	2,000	12,185

Proponents of formulas usually justify them on the basis of the following:

1. They save time.
2. They can be used by persons who do not understand the principles on which they are based and who otherwise could not solve economy study problems.
3. They provide some assurance that pertinent factors will not be omitted from consideration.

Although all these claims *can* be true, they are not always so, and the following precautions should be kept in mind.

1. Formulas do not always save time.
2. The use of formulas by those who do not completely understand them can often be dangerous. (For example, how can they know whether a factor should be included or omitted?) Analyses or decisions seldom should be made by those who do not fully understand their implications.
3. There is no assurance that a given formula will make provision for all the factors that exist in a particular situation.
4. Formulas sometimes reveal cost relationships only at the breakeven point. There are many cases where the effect of not operating at the theoretically correct point is of prime importance in making a decision. Graphical solutions frequently are of great help in such problems.
5. Formulas give no consideration to intangible factors. Where there are important nonmonetary factors, formulas should be used with great caution.

Thus formulas are of value, but they should be used by those who thoroughly understand their implications and underlying assumptions.

PROBLEMS

15-1 Use the economic lot size formula to determine the most economical size of a purchase order under these conditions: order preparation cost is $140, the cost of each item is $0.20, the annual demand for the item is 100,000 pieces, and the annual carrying charges are 25% of average inventory. What is the total cost of acquiring and storing 1 year's worth of this commodity?

15-2 When the economic lot size formula is used, what relationship exists between the annual setup (or order) cost and the annual carrying (inventory) cost at the minimum-cost point?

15-3 The annual requirements for a certain type of plastic wrapping material are 80,000 pounds on a continuing basis. It is estimated that the average price of the material during the next few years will be about $0.42 per pound. At present it costs $80.00 to initiate and receive an order, and the percentage factor to cover interest, taxes, insurance, and storage is 0.30. An improved system and equipment for initiating and receiving orders would cost $1,700, and it would reduce the "order" cost by $30.00 per order and reduce the percentage carrying charge by $\frac{1}{4}$. This investment would have to be recovered with interest in four years. If capital is worth 12% before taxes, and neglecting income tax effects, what do you recommend? No minimum inventory is maintained.

15-4 For a given set of conditions, how does each of the following affect the economic lot size?
(a) Reducing the setup, or order, cost by one-fourth.
(b) Tripling the cost of carrying inventory.
(c) The possibility of pilferage and/or obsolescence.

15-5 The Attaboy Lawn Mower Company can purchase component parts for one of its high-volume mowers according to the following price schedule:

Quantity per Order	Price per Unit
1–2,000	$5.00
2,001–5,000	4.00
Over 5,000	3.80

This company expects to use 20,000 of these parts per year on a continuing basis. The incremental cost of processing an order is $50 and the annual carrying cost per unit is 15% of the purchase price. If the company maintains no appreciable minimum inventory reserve, what is the most economical order quantity?

15-6 A company manufactures the deep-drawn metal cases for one of its products and uses them at a nearly constant rate of 8,000 per year. It can produce them at a rate of 40,000 per year. Because of the necessity for changing, aligning, and "proving" the dies each time a setup is made, the cost for making a setup and tearing it down at the conclusion of a run is $200. The variable costs, including material, die wear, and labor, are $4 per unit, and the annual "holding" cost is 25% of the unit variable cost. Because of the high production rate, the company maintains no appreciable inventory. Determine the most economical lot size for the production of the cases.

15-7 If the company in Problem 15-6 will invest $1,200 to modify the drawing dies used in making the cases, it can reduce the setup cost by $100. Assuming that (a) the dies have a 5-year life, (b) annual taxes and insurance amount to 3% of the first cost on all equipment and tooling, (c) the dies are used on three different presses, and (d) the company requires a before-tax return of at least 15% on such investments, would the modification be justified? Neglect possible income tax effects.

15-8 A small manufacturing firm can produce 400 ceramic fixtures each day, and the firm operates 250 days per year. These fixtures are required in the assembly of a product that the company sells at a uniform rate of 40,000 per year. If the setup cost for initiating production of these fixtures is $500, variable cost per piece is $40, and carrying charges are 30% of the average inventory, what is the most economical production quantity?

15-9 In Problem 15-8, suppose that obsolescence is expected to be high because of possible style changes. To account for this factor, it has been estimated that the annual obsolescence cost is roughly 35% of variable cost per fixture. How does this affect the most economic lot size?

15-10 In a local textile plant, 50-yard bolts of a commonly used knitted fabric can be purchased according to this price schedule:

Quantity per Order	Price per Unit (Bolt)
1–100	$17.00
101–500	14.75
Over 500	14.25

This plant expects to use 3,000 bolts of this particular fabric on a continuing basis during the coming year. The incremental cost of processing a purchase order is $150 and the annual carrying cost per unit (interest on tied-up dollars and storage cost in a shared area) is 40% of the purchase price. The company maintains no minimum inventory for this fabric. What is the most economical order quantity? State all assumptions you make in your analysis.

15-11 The Sureshot Rifle Works sells rifle barrels to other companies that assemble the final product. They sell rifle barrels at a roughly constant rate of 100 per day and manufacture them at a rate of 300 per day. Setup costs for a production run are $2,000 and storage costs in a shared area are $0.024 per barrel per day. The cost of each barrel is broken out as follows: *labor* = $6.20 per barrel, *materials* = $4.23 per barrel, and *overhead* = $3.10 per barrel. The cost of capital (interest charges) is 30% per year.

Assume a 300-day work year and use average inventory for determining storage and interest charges.

(a) Find the minimum-cost batch size of rifle barrels to manufacture.

(b) Determine the total cost per year of the inventory function, exclusive of commodity costs.

15-12 You are required to make a manufacture versus purchase decision for your company involving a part which until now has always been manufactured. Ten thousand of the parts are used each year, and they have been produced in most economical lot sizes based on a setup of $80, all other increment costs of $5 per unit, and a percentage charge of 20% to cover the cost of carrying and storing inventory. No minimum inventory has been maintained, since the production and use rates are the same. A supplier has offered to make the parts under a subcontract at a price of $6 per unit. It is estimated that the cost for placing and receiving an order would be $20, and it is planned to order them in economical lot size quantities. What should be done?

15-13 Assume that the prices and resistances of various sizes of electrical cable are the following:

Size	Resistance (ohms per 1,000 feet)	Price (per 1,000 feet)
1	0.1240	$ 750
0	0.0983	972
00	0.0779	1,086
000	0.0618	1,200

The cost of electrical energy to an industrial user of this cable is 2.5 cents per kilowatt-hour. Annual capital recovery cost is based on a 25-year life with no salvage value, and the minimum attractive rate of return is 10%. What is the most economical size of cable to transmit 200 amperes for 1,800 hours each year?

15-14 Suppose that for a group of 10,000 electronic parts subject to sudden failure, the net cost of group replacement is $0.40 per unit, while the unit cost of individual replacement is $2.00. Further, the expected number of failures each period is shown in the following table. All failures that occur during each period are replaced only at the end of that time period.

Total Failures (Replacements) in Each Period t for 10,000 Electronic Parts

Period, t	Replacements Current $f(t)$	Cumulative, $\Sigma f(t)$
1	100	100
2	400	500
3	1,100	1,600
4	1,200	2,800
5	2,500	5,300
6	2,300	7,600
7	2,600	10,400
8	2,500	12,900

Determine whether group replacement is economical and, if so, what the optimum replacement interval is.

CHAPTER 16

Value Engineering

Since the early days of mechanized industry, there has been a tendency for devices and products to be designed and made more complex and elaborate than was necessary. Inventors and designers ordinarily are intent primarily on achieving some functional objective, with insufficient attention being given to keeping the device simple, economical, or easy to use and service. The engineer's job has traditionally been involved with balancing *cost* and *quality* of a product— getting the most for the money. As Arthur Wellington observed in 1887, "The engineer is one who can do with a dollar what any bungler can do with two."

From the beginning of this century industrial engineers have given much successful attention to improving and simplifying methods, processes, and systems, but not much attention was given to rational simplification of designs— frequently because the product already was designed and in production. In recent years a procedure has developed for analyzing products and devices, from a design viewpoint, to determine and improve their economic value. This procedure has come to be known as *value analysis* or *value engineering*. It is widely used by many companies and agencies, some of which have reported savings of over $1,000,000 per year resulting from its application.

Briefly, value engineering is a method for examining the value of a product or service in relation to its cost with the aim of providing the required function(s) at the lowest overall cost. Ideally, value engineering seeks to provide the necessary function(s) during the design phase of a product's creation at the lowest cost, without lowering quality. Hence engineering economy studies, usually at the present-economy level, are an inherent and essential part of value engineering. Consequently, the purpose of this chapter is to introduce the reader to the basic concepts of value engineering and to illustrate the relationship between value engineering and engineering economy.

The origin of value engineering goes back to the late 1940s when the General Electric Company initiated a large-scale program to identify how material substitutions in many of their products would affect the functions performed by the products in addition to their costs and associated market value. It was discovered that many of the substitutions resulted in an improved product at a lower cost. Furthermore, lower costs were also achieved by closely examining rigidly enforced design specifications and standards to determine whether they were overly conservative. The person credited with the success of General Electric's program was L.D. Miles, who has since become known as the "father of value analysis." Later in 1954 the Bureau of Ships patterned its value engineering program after that developed at General Electric. Substantial savings in the cost of building ships for the U.S. Navy resulted and prompted the Department of Defense to require its prime contractors to initiate value engineering procedures. Subsequently, many nondefense industries were able to effectively utilize the principles of value engineering to improve the quality of their products while holding costs constant or even reducing them.

TYPES OF VALUE

In Chapter 2 it was pointed out that goods have value because of their utility, and that consumers pay the purchase price for goods because they satisfy a need. However, consumers often do not analyze a product to determine exactly what needs it satisfies or what portion of the purchase price is paid for each need that it satisfies. For example, in early 1971 a large department store sold fur-trimmed "hot pants" for $24. What utility did they possess? One could list several, functions that they clearly fulfilled—warmth (limited), provide a certain degree of modesty, attract attention to the wearer. But what portion of the $24 purchase price would the buyer attribute to each function; would each purchaser make the same allocation; and could the same functions be achieved at a lesser cost?

In value engineering work, two types of value are recognized. *Use value* pertains to the properties and qualities that accomplish a use, work, or service. This type of value exists because without these properties and qualities useful work or service functions could not be achieved. *Esteem value*, on the other hand, pertains to the properties or qualities that make people want to possess the product or service. In somewhat oversimplified terms, we might say that use values cause a product to perform, and esteem values cause it to sell.

As examples of these types of values, consider a man's suit. The primary function of such a suit is to cover the body and to provide warmth and protection against the elements. A secondary function is to provide pockets in which objects may be carried. Fulfilling these functional requirements causes a suit to have use value, and a very plain, ill-fitting garment, made from cheap, coarse, colorless cloth could adequately provide these functions. However, it is doubtful that such a garment would sell very well in the United States. Most men prefer that their suits be made of better cloth, fit reasonably well, have good tailoring details and style, hold a press, and be attractive in color. These are items that make the garment desired; they are esteem values. Thus esteem values can be valuable attributes of goods or services, but only in fact if the customer will pay for them.

The objective of value engineering is to determine the most economical way of providing required use values, consistent with proper functional, safety, reliability, and quality standards and, at minimum cost, that degree of esteem value which the customer demands and for which he will pay.

METHODOLOGY OF VALUE ENGINEERING

Value engineering is sometimes referred to as "just plain old-fashioned, everyday cost reduction with a new name." This sentiment indicates a lack of understanding of both value engineering and the full meaning of cost reduction. The best product at minimum cost to the manufacturer can be delivered through the *simultaneous* use of conventional cost reduction and value engineering.

Before discussing the essential features of value engineering, consider some of the elements of a successful cost reduction program:

1. Systems and procedures for planning and scheduling work.
2. Organizational planning and analysis.
3. Methods improvement and work simplification.
4. Establishment of labor and materials standards.
5. Optimization of manufacturing processes.
6. Control of raw materials, in-process inventories, and finished-goods inventories.
7. Preventive maintenance programs.

These activities are primarily concerned with reduction of costs per se.

In contrast to conventional cost reduction programs, value engineering focuses attention on the inherent worth of the end product with a view toward better satisfying the user's functional requirements at the lowest overall cost to the firm. Thus value engineering is heavily oriented toward assuring the customer that his essential technical requirements are achieved at the minimum cost. Because of its close attention to function, quality, and worth of the product delivered, value engineering generally is not regarded as synonymous with cost reduction.

The value engineering methodology is based on these six fundamental questions:

1. What is it (e.g., the product or design being evaluated)?
2. What does it do (i.e., what functions are provided)?
3. How much does it cost?
4. Are the functions necessary?

5. How else could the functions be accomplished?
6. What would these alternatives cost and are any of them less expensive than the present design?

A certain amount of creativity and "free thinking" are required to explore each of these questions fully. Brainstorming sessions have successfully been used to challenge acceptance of the status quo and to overcome roadblock excuses for not delving into unfamiliar and untried design possibilities.

At this point some "idea stimulators" such as the following are often helpful in initiating the generation of information during a value engineering brainstorming session.

1. Why does it (the present or proposed product) have this shape?
2. How much of this design is the result of custom? Opinion? Tradition?
3. What else would do the job (what *is* the job)?
4. Suppose this were left out?
5. Describe what the product is *not*.
6. Can it be made safer and/or easier to use?
7. What can be done to give extra value to the customer?
8. Is there a less costly part that will perform the same function?
9. Are there newly developed materials that could be used?
10. Are all machined surfaces necessary?
11. Are tolerances closer than they need to be?
12. Is this the best manufacturing process?
13. Why are we making (buying) it?

These questions may seem simplistic, but when seriously considered they provide essential information for the three-step procedure that is often in value engineering work. These three steps are:

1. *Identify the functions required.*
2. *Determine value of the functions by comparison.*
3. *Develop feasible value alternatives.*

Value engineering starts with an assumption that a device or service is wanted. With this assumption accepted, the *first step* is to identify the functions, primary and secondary, that the device or service is to render. The primary function nearly always can be expressed in two words, such as *open valve, grind garbage, support flywheel, cover mechanism, exclude water,* or *provide light.* If a device does not satisfy the primary functional need, no matter how good it may be otherwise, it will not be produced. However, there frequently are secondary functions. The primary function of a household electric refrigerator is to preserve food; secondary functions are to make ice cubes, condition butter, provide storage space, and so on. The primary function of a room air conditioner is to provide cooling; secondary functions are to filter air and to dehumidify it. Sometimes secondary functions may be extremely important. For example, an electrical-appliance cord has the primary function of conducting electricity, but two secondary functions, which are of absolute importance, are that it be flexible and that it provide adequate electrical insulation. However, it is clear, in these examples, that ability to completely satisfy the secondary functions would not justify production of the devices if the primary need were not fulfilled.

When we identify and describe the primary and secondary functions of complex devices, it frequently is helpful to divide the components into functional groups, such as mechanical parts, electrical parts, enclosures, and so on. This procedure often will make clearer the primary and secondary functions of the parts. It also will help to expose useless components.

In the *second step*, determining value by comparison, we postulate that the real, or basic, value cannot exceed the cost of accomplishing the function by the least expensive method that also will meet the safety and reliability requirements. This least expensive method may be either one that exists or that can be developed. Obviously, these value comparisons require present economy studies, involving selection between alternative materials, designs, or processes. For example, portions of a shipping crate were held together by a bolt, a washer, and a nut. These cost $0.018 per set. It was determined that the parts could be held together satisfactorily by an ordinary nail that was clinched on the end. Such a nail cost $0.002. Obviously, the labor cost for using the bolt and nut differed from that for using the nail, and a hole also had to be drilled. The total-cost comparison was as follows:

	Bolt and Nut	Nail
Material	$0.018	$0.002
Labor	0.030	0.020
TOTAL	$0.048	$0.022

Because the nail appeared to be the lowest-cost method available, the maximum value that could be assigned to the fastening function was $0.022. This amount would be used as a target in evaluating all alternative methods that might be developed and considered for fastening the crate members. Any cost above the $0.022 target figure would have to be justified on some other basis than functional value.

Often a device may have more than one primary function and, consequently, more than one target alternative may be required for satisfying these functions. The spacer stud shown in Figure 16-1 is such an example. This stud had two

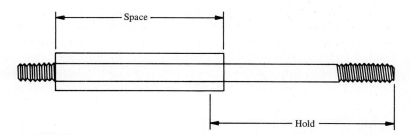

FIGURE 16-1   Original Design and Functions of a Spacer Stud.

primary functions: to hold and to space. It was made on a screw machine from hexagonal stock. The hexagonal portion of the part served the secondary function of providing a convenient means of holding it with a wrench while tightening the nuts that went on the ends. The cost of this spacer stud was $0.08. By value comparison it was found that the holding function could be achieved by a steel screw, costing $0.005. Therefore, the holding function of the spacer stud could not be worth more than $0.005. Similarly, it was found that the spacing function could be provided either by a cut-off length of tubing or by a rolled spacer, each of which could be obtained for $0.0025. Thus the total functional value provided by the spacer stud at a cost of $0.08 could be obtained by other means for $0.0075. Hence the target value was $0.0075.

The spacer stud example also illustrates the *third step* in the value-analysis process—causing value alternatives to be developed and selecting the most feasible alternative. While it is desirable to develop a number of alternatives, in most cases some will not be practicable, and most of those that are realistic possibilities will not have equal costs. Further, it is at this stage that esteem values become relevant. An alternative that might be completely satisfactory from a functional viewpoint might be lacking in sales appeal. One must decide what changes and additions are necessary to provide the required esteem value. But one also must be certain that the required esteem value is obtained at a minimum cost, and that no more is added than the customer is willing to pay for. In both decisions present economy studies are necessary.

In the case of the spacer stud, it was quite evident that if a satisfactory new cost, which would be only about one-tenth that of the old cost, was to be achieved, a screw-machine product could not be used and an entirely different type of design would have to be developed. Also, the use of a long screw and a piece of tubing was not an ideal solution, since two pieces were involved and the absence of a hexagonal section and threads on both ends would necessitate undesirable changes in the methods used in assembling the device in which the spacer stud was used. Thus a new and feasible design had to be developed, which would have a cost as close as possible to the target value.

In this instance, attention first was directed to obtaining, at minimum cost, the main body of the stud. An eight-penny nail was almost the same size in diameter and length, and it could be obtained for $0.001. It also had an upset head on one end. If the head could be moved from the end a sufficient distance to leave room for a thread, and if it could be made hexagonal in shape, and if threads then were rolled on each end, the primary function of holding and the secondary function of providing a wrench grip would be satisfied. The spacing function could be satisfied by upsetting a round head at the proper place along the shaft. Investigation proved that such a device, shown in Figure 16-2, could readily be produced, at a cost of $0.008. Thus a feasible alternative, close in cost to the target value, was achieved by the procedure of clearly defining the

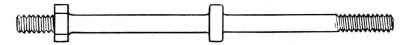

FIGURE 16-2 Modified Design for Spacer Stud Resulting from Value Analysis.

functions, evaluating the comparative value of these functions, and then developing a feasible alternative, based on means that would provide the required functions at or near target-value costs.

In this case, as frequently occurs in producers' goods, esteem value was of no consequence. Yet frequently, designs of such components needlessly incorporate details which can only be classified as adding esteem value, but for which the customer will not pay, if he has a choice. It is in this area that value analysis has made possible very large savings.

Lawrence D. Miles has applied colorful terms, "blast, create, and refine," to describe this three-step approach. The *blast* step involves determination of the comparative cost of something that will perform the major part of the functions and that will provide the base cost of the materials in simple form. This frequently requires radical departure from the existing design. The *create* step is the addition of items or features to the base so that the total functional requirements will be satisfied. The third step, *refine*, requires getting practical and realistic and making modifications that will take into account esteem values, reliabilities, ease of use, and so on. In effect, what is required is good, sound, economical engineering design, which should *always* be put into practice, Unfortunately, there are numerous examples to testify that such frequently is not the case.

In the development of feasible alternatives, it is extremely important that the person making the analysis be thoroughly acquainted with the application for which the device is intended. For example, in some applications a rivet might be a completely satisfactory and more economical substitute for a bolt and nut. On the other hand, if ease of disassembly were a requirement, such a substitution would not be feasible. For certain applications a low-cost Plexiglas cover might be a suitable substitute for a more costly molded-glass cover; in applications where high temperatures are involved, such a substitution would not be proper. Thus a thorough knowledge of the application not only aids in properly defining functions, but also helps to assure that the various alternatives will be satisfactory. However, we must make certain that familiarity with the application does not result in unwillingness to consider alternatives that may differ radically from the existing one. This attitude is one of the greatest deterrents to effective value analysis.

IMPORTANT COST CONCEPTS IN VALUE ENGINEERING

There are certain cost concepts that are of great importance in value engineering work. These basically are closely related to fixed and increment costs, which were considered in detail in Chapter 8.

Accurate and relevant cost records are a necessity in value engineering work. The analyst also needs to know the real reasons for design features and decisions. These are not always easily determined. A design decision that appears to have originated in the engineering department may actually have had its origin in the sales or purchasing department. Without a knowledge of the real reasons behind design and process decisions, it is difficult to determine the true costs or savings that may be associated with value alternatives.

Ordinary accounting costs seldom are of much help in such decisions. What is needed are *decision costs*. In *make versus buy* decisions we normally must consider labor costs, material costs, the variable overhead, and possibly some portion of the fixed overhead. Whether or not any fixed overhead should be included in the cost of production may vary in different cases. Frequently, this decision depends on the degree of plant utilization. If the facilities presently are not being used to capacity, some idleness or "stretch-out" probably exists. As a consequence, additional work can be absorbed with little or no actual increase in fixed overhead. Similarly, a decision to eliminate a product or process might result in "stretch-out" with no actual decrease in fixed overhead. On the other hand, if activity is at or near capacity any increase in output will add overhead.

In determining the cost of making a product one way versus the cost of making it another way, such as in value engineering analyses, there usually is little effect on fixed overhead.

One frequent and misleading cost situation that may occur is that where partially idle skilled workers may be used for operations that do not require their degree of skill—so as not to have high-cost workers idle. Such a situation results in excessively high, and unrealistic, costs, that may go undetected until a proper value analysis is made. A similar cost situation can result where a product is run in odd-sized lots when certain machines or workers are idle. The resulting frequent setup costs may be much too high. Thus in value engineering work and in engineering economy studies, it is most important that the true decision costs be obtained.

VALUE ANALYSIS IN PURCHASING

Although in this chapter most of the discussion has been in connection with its obvious application to the design of products and components, value engineering has an important role relative to purchasing. Obviously, it is much easier and cheaper to make design changes in products while they are on the drafting board or when writing purchase specifications than after they are in existence. Consequently, those who write purchase specifications, or who approve designs before orders are placed, are in a position to analyze designs from a value engineering viewpoint and to insist on changes so as to eliminate costly features which do not contribute to basic use value and required esteem value. By demonstrating that such features can be eliminated, the customer is in a position to force quoted prices to be reduced. The need to seek new suppliers and designs then becomes apparent.

Value engineering practitioners have been outstandingly successful in getting both designers and purchasing personnel to accept and apply value analysis and to make the necessary present-economy studies. This undoubtedly is because they have done a good job of selling and have shown the dollars-and-cents results that can be achieved. Value engineering concepts have even permeated certain governmental agencies to the point where they not only are practicing them, but they also are requiring their contractors to make value analyses of their products and operations. Certainly, every economy study analyst should applaud, aid, and abet these practices, recognizing the key role that engineering economy plays in these procedures.

PROBLEMS

16-1 Explain the relationship between value engineering and engineering economy.

16-2 Why may the criterion of minimum production cost not be a proper basis for evaluating a design?

16-3 Why have companies and government agencies been eager to have value analysis applied at the design stage?

16-4 What are the basic differences between use value and esteem value?

16-5 (a) Explain why esteem values may knowingly be incorporated into products.
(b) What basic economic principle should be applied to esteem values in a given product?

16-6 Analyze a front bumper on a typical American automobile in terms of use and esteem values. Do the same for a stereo phonograph enclosed in a cabinet.

16-7 Name the primary use function and possible secondary use functions of the following: (a) shoes, (b) a sidewalk, (c) a desk clock, and (d) a man's necktie.

16-8 Select some simple article that is on your desk or in your room.
(a) Identify its primary use and other functions.
(b) Evalaute the worth of these functions.
(c) Determine the least expensive alternative for providing the primary use function
(d) Would this least-cost alternative be satisfactory to you? If not, why?

16-9 (a) What percentage of the selling price of a butane-fuel cigarette lighter would you assign to use and esteem values?
(b) What absolute amount would you assign as the use value of such a lighter?

CHAPTER 17

Economy Studies Based on Linear Programming

Many industrial problems involve the allocation of limited resources for the purpose of obtaining the best possible results from their use. "Best possible results" usually means that the aim is to maximize profits or to minimize costs. It also implies that several alternatives exist from which to choose to accomplish a specific goal. Thus the problem is to allocate fixed and known amounts of resources in satisfying a given goal such that we maximize profits (or minimize costs) for feasible alternatives under consideration. This is really a general statement of what previous chapters have been dealing with.

For certain types of resource allocation problems, a technique known as *linear programming* (LP) can be used to great advantage. The purpose of this chapter, therefore, is to introduce the reader to linear programming formulations of selected economy problems and to indicate how these problems can be solved with graphical methods and/or enumerative methods. References are given for solution methods applicable to more complicated problems.

As we shall soon see, linear programming adequately represents a wide variety of real-world problems and can be quickly encoded for solution with digital computers. The *simplex method* of linear programming, developed in 1947 by George Dantzig, made the solution of large problems computationally tractable. Today linear programming is routinely used by many industries, including agriculture, steel, chemicals, airlines, petroleum, and utilities.

Several conditions must be met before linear programming can become a reliable tool. First, we are concerned with specifying nonnegative values of a set of variables that optimize a linear function expressed in terms of these variables. Second, the optimization of this function must also satisfy one or more linear constraints that mathematically take into account the availability of resources. Linearity implies, for example, that profit (or cost) per unit of output remains constant regardless of production level. Similarly, total resources consumed are assumed to be a linear function of the production level.

A SIMPLE, ILLUSTRATIVE PRODUCTION PROBLEM

To illustrate how a simple resource allocation problem can be formulated as a linear programming situation and solved graphically, consider the manufacture of item a and item b by a small machine shop. Each unit produced requires a certain amount of machining time (i.e., standard time per operation) in each of three departments, as follows:

Item	Time (min) in Department:		
	1	2	3
a	40	24	20
b	30	32	24

Each department works a standard day consisting of 480 minutes, so it is clear that with no overtime there is a limit to the availability of machining time in each department. Because we cannot produce negative amounts of item a and item b, nor can we utilize a negative amount of time in their manufacture, none of the factors present in this problem can be negative. Suppose further that the profit per unit of item a and item b is \$5 and \$8, respectively.

In our simple problem, the aim is to maximize profit per day. This can be expressed mathematically by the following equation:

$$\text{Maximize } P = 5a + 8b$$

where a is the number of units of item a manufactured per day and b the number of units of item b, produced each day. From a quick inspection it should be obvious that the function above is linear. This equation, known as the objective function, is expressed in terms of the *decision variables* (i.e., quantities that we can control so as to maximize profits). The *constraints* in this problem concern available machining time in each department and are also linear in terms of our two decision variables as seen below.

$40a + 30b \leq 480$ (constraint on available time in department 1)

$24a + 32b \leq 480$ (constraint on available time in department 2)

$20a + 24b \leq 480$ (constraint on available time in department 3)

If it is assumed that only two products are being manufactured and that all machining time in departments 1, 2, and 3 is available solely for this purpose, we can formulate this problem as a linear programming (LP) problem, since the special characteristics involving linearity of the objective function and constraints are present here. Thus the problem can be written:

$$\text{Maximize } P = 5a + 8b$$

$$\text{subject to} \quad 40a + 30b \leq 480$$

$$24a + 32b \leq 480$$

$$20a + 24b \leq 480$$

$$a \geq 0, \quad b \geq 0$$

Our task now is to determine values of a and b that maximize profits and at the same time satisfy the linear constraints.

Numerous alternatives are available for the solution of this problem. Consider, for example, what would happen if we decided to produce only item a *or* item b.

Department	Item a Only	Item b Only
1	$40a \leq 480$, or $a \leq 12.0$	$30b \leq 480$, or $b \leq 16.0$
2	$24a \leq 480$, or $a \leq 20.0$	$32b \leq 480$, or $b \leq 15.0$
3	$20a \leq 480$, or $a \leq 24.0$	$24b \leq 480$, or $b \leq 20.0$
	No more than 12 units can be produced without violating the constraint on time of department 1	No more than 15 units can be produced without violating the constraint on time of department 2
	$P = 12$ units ($5/unit) = $60	$P = 15$ units ($8/unit) = $120

Other alternatives involving various combinations of item a and item b could also be proposed and evaluated. However, in larger problems enumerating all possible combinations of the decision variables could be quite time-consuming.

Systematic solution procedures are available for solving large LP problems with the aid of a digital computer, and one of the best known is the simplex method. The interested reader is encouraged to learn more about this very useful method.†

†The following references are suggested:

Bazaraa, M. S., and J. J. Jarvis, *Linear Programming and Network Flows* (New York: John Wiley & Sons, Inc., 1977).

Claycombe, W. W., and W. G. Sullivan, *Foundations of Mathematical Programming* (Reston, Va.: Reston Publishing Company, Inc., 1975).

Murty, K. G., *Linear and Combinatorial Programming* (New York: John Wiley & Sons, Inc., 1976).

Simmonnard, M., *Linear Programming* (Englewood Cliffs, N.J.: Prentice-Hall, Inc., 1966).

Strum, J. E., *Introduction to Linear Programming* (San Francisco: Holden-Day, Inc., 1972).

Zoints, S., *Linear and Integer Programming* (Englewood Cliffs, N.J.: Prentice-Hall, Inc., 1974).

Returning to our manufacturing problem, a solution can be discovered by first drawing a graph with units of item *a* along the ordinate and units of item *b* along the abscissa. If the constraint equations are then plotted on this graph, we would have Figure 17-1. The shaded area, called the *feasible region*, defines a convex polygon that contains the optimal solution to this problem. The feasible region is convex because a straight line connecting any two points in it will lie entirely within the region.

As one can easily verify, the feasible region permits these constraints to be satisfied:

$$a \geq 0, \quad b \geq 0$$

$$40a + 30b \leq 480$$

$$24a + 32b \leq 480$$

Thus any point satisfying all the constraints and nonnegativity conditions is a feasible solution to our problem. The last constraint (available time in department 3) does not lie in the shaded region because it is not a binding constraint. This means there is no way of using all available time in department 3 to produce items *a* and *b* without violating constraints on time in departments 1 and 2. If it were possible to schedule overtime operation in these two departments, the constraint on department 3 could become binding (or "active"), but this possibility is not being considered in the present problem. Thus we do not regard the time available in department 3 as a limited resource, and only two constraints are necessary in this problem.

Now that the feasible region for a solution has been defined, we must attempt to maximize our objective function by specifying the optimal number of units

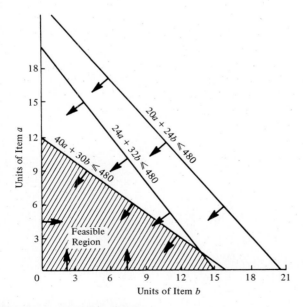

FIGURE 17-1   Feasible Region.

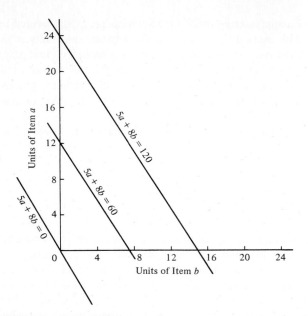

FIGURE 17-2   Objective Function.

of items a and b to manufacture. This is done by superimposing the objective function on Figure 17-1 and moving it as far as possible from the origin without leaving the feasible region. (We move away from the origin, since we are trying to maximize profits.) This can be understood readily by referring to Figures 17-2 and 17-3.

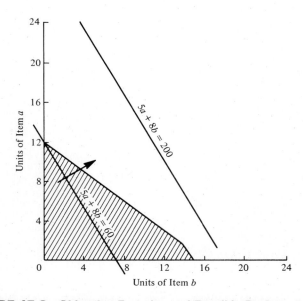

FIGURE 17-3   Objective Function and Feasible Region.

In Figure 17-2 the objective function is plotted. When $5a + 8b = P$, where P is any constant value, there is a direct relationship between a and b that can be used to plot the objective function. For example, suppose that $P = \$120$. When $a = 0$ and $b = 15$, the profit is $120. But when $a = 24$ and $b = 0$, profit will also be $120. Table 17-1 illustrates a few of the other combinations of a and b resulting in a profit of $120. To plot the objective function, $120 = 5a + 8b$, we could solve for $a = 24 - \frac{8}{5}b$, which is a straight line with a slope of $-\frac{8}{5}$ and an a-intercept of 24. This line is plotted in Figure 17-2. Also shown are other members of a family of objective functions with slopes of $-\frac{8}{5}$ and a-intercepts of $P/5$ appearing as a series of parallel lines.

Some values of P, however, will cause the decision variables to violate one or more constraints. That is, all or a part of the line formed by $5a + 8b = P$ may not lie in the feasible region defined by our constraint set. The combination of a and b that we want to determine is the one allowing the maximum value of P to occur in the equation $5a + 8b = P$ while this same equation lies in the feasible region at one or more points.

An example of too large a value of P is given in Figure 17-3, where $P = 200$. This graph also illustrates a situation in which the objective function lies in the feasible region but is not at its maximum possible value. In this case, the objective function, $5a + 8b = 60$, must move farther away from the origin to be maximized.

Finally, Figure 17-4 shows the maximum value of the objective function to be $5(0) + 8(15) = 120$ for $a = 0$ and $b = 15$. Note in Figure 17-4 that the objective function touches the feasible region at exactly one point. It is also clear that the feasible region depends only on the constraint equations and is unaffected by changes in the objective function. Moreover, the optimal solution is determined by the slope of the objective function, given a certain set of constraints. For example, suppose the objective function were $P = 9a + 3b$. The optimal solution would then be $a = 12$, $b = 0$, with a profit of $108.

The optimal solution to our illustrative production problem could also have been determined by making inequalities into equalities and then solving a set of simultaneous linear equations. In this case there are only two equations to deal with simultaneously, since we have two decision variables. But when several constraints are imposed on the problem, it is often difficult to determine *which two equations* to solve to find the maximum (or minimum) value of the objective function. This raises another interesting possibility for solving linear programming problems, in addition to the graphical method illustrated above.

TABLE 17-1 Combinations of a and b Giving $120 Profit

a	b	Profit (P)
10	8.75	$120
4.80	12	120
8	10	120
16	5	120

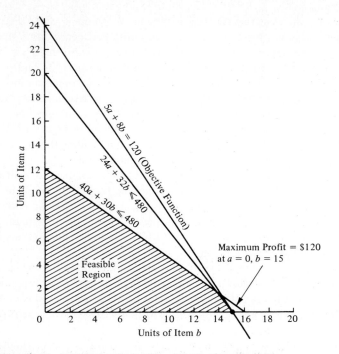

FIGURE 17-4 **Solution of the Manufacturing Problem.**

If we determine the points of intersection for every pair of constraint equations (equalities), it would be possible to find the point resulting in a maximum value of $P = 5a + 8b$. Suppose for the moment that we do not know that constraint 3 is nonbinding in the problem. All pairs of constraint equations and their points of intersection are shown below. There are $\binom{5}{2} = 5!/3!2! = 10$ points of intersection, since we have five constraints and two decision variables. Also, $\binom{5}{2}$ is a combinatorial term that can be written in terms of factorials:

$$\binom{5}{2} = \frac{5!}{2!(5-2)!}$$

(5! is read ''5 factorial'' and is equal to the product $5 \times 4 \times 3 \times 2 \times 1 = 120$; the other factorials are computed in similar fashion.)

Pair 1:*	$40a + 30b = 480$	$a = \frac{12}{7}$	$P = 118\frac{2}{7}$
	$24a + 32b = 480$	$b = \frac{96}{7}$	
Pair 2:	$40a + 30b = 480$	$a = -8$	$P = $ undefined
	$20a + 24b = 480$	$b = \frac{80}{3}$	
Pair 3:	$24a + 32b = 480$	$a = 60$	$P = $ undefined
	$20a + 24b = 480$	$b = -30$	

Pair 4:	$40a + 30b = 480$	$a = 0$	$P = 128$
	$a = 0$	$b = 16$	
Pair 5:*	$40a + 30b = 480$	$a = 12$	$P = 60$
	$b = 0$	$b = 0$	
Pair 6:*	$24a + 32b = 480$	$a = 0$	$P = 120$
	$a = 0$	$b = 15$	
Pair 7:	$24a + 32b = 480$	$a = 20$	$P = 100$
	$b = 0$	$b = 0$	
Pair 8:	$20a + 24b = 480$	$a = 0$	$P = 160$
	$a = 0$	$b = 20$	
Pair 9:	$20a + 24b = 480$	$a = 24$	$P = 120$
	$b = 0$	$b = 0$	
Pair 10:*	$a = 0$		$P = 0$
	$b = 0$		

The solution to each pair of equations must now be inserted into the original set of constraints to ensure that values of a and b are feasible (i.e., values of a and b do not violate the constraints). Recall that our constraints are

$$40a + 30b \leq 480 \quad \text{(constraint 1)}$$

$$24a + 32b \leq 480 \quad \text{(constraint 2)}$$

$$20a + 24b \leq 480 \quad \text{(constraint 3)}$$

$$a \geq 0 \quad \text{(constraint 4)}$$

$$b \geq 0 \quad \text{(constraint 5)}$$

The solution to pair 1 satisfies the three constraints (3, 4, and 5 above) not used to determine the point of intersection at $a = \frac{12}{7}$ and $b = \frac{96}{7}$. That is,

$$20(\tfrac{12}{7}) + 24(\tfrac{96}{7}) < 480$$

$$\tfrac{12}{7} > 0$$

$$\tfrac{96}{7} > 0$$

Solutions to pairs 2 and 3 are not permissible, since a and b must be nonnegative. After evaluating other points of intersection in the same manner, it is apparent that solutions to pairs 1, 5, 6, and 10 satisfy all five constraints. Asterisks have been placed by each of these solutions.

These four points are termed *basic feasible solutions* to our linear programming problem and lie at the vertices of the feasible region formed by the constraints. The optimal solution will be located at one of these vertices (a single point). It is also true that all points lying on or within the feasible region are feasible solutions to the problem. If the objective function is coincident with (parallel with) one of the binding constraints there are an infinite number of solutions resulting in a profit of P.

The enumeration of intersection points above illustrates that *the optimal solution is the basic feasible solution that maximizes* (or minimizes) *the objective function.* In our problem it is seen that the solution to pair 6 yields the maximum profit of \$120 at $a = 0$ and $b = 15$. Therefore, we would recommend that 15 units of item b be produced each day.

In making this recommendation to management, we could carry the analysis one step further and calculate idle time in each department:

Department 1: $480 - 40(0) - 30(15) = $ 30 minutes/day

Department 2: $480 - 24(0) - 32(15) = $ 0 minutes/day

Department 3: $480 - 20(0) - 24(15) = $ 120 minutes/day

We may now want to suggest that management consider the production of item c(a new product line), which would require machine work mainly in departments 1 and 3.

To complete this illustrative problem, suppose that management decides to add a new product line (item c) that requires 20 minutes of machining time in department 1 and 32 minutes of machining time in department 3. The constraints and corresponding feasible region for the problem would now be

$$
\begin{array}{lll}
(1) & 40a + 30b + 20c \le 480 & \text{Department 1 time} \\
(2) & 24a + 32b \phantom{{}+ 20c} \le 480 & \text{Department 2 time} \\
(3) & 20a + 24b + 32c \le 480 & \text{Department 3 time} \\
(4) & a \ge 0 & \left. \vphantom{\begin{array}{c}a\\b\\c\end{array}} \right\} \text{Nonnegativity} \\
(5) & b \ge 0 & \text{constraints on} \\
(6) & c \ge 0 & \text{each product line}
\end{array}
$$

The three-dimensional graph in Figure 17-5 shows the feasible region to be a polyhedron.

If the profit per unit of item c were \$8, the objective function is $P = 5a + 8b + 8c$. This is the shaded plane in the graph whose location depends on the value of P and whose slope is determined by the coefficients of the objective function. Because we are attempting to maximize P, the plane would move away from the origin until the optimal solution is reached.

It is difficult to determine from Figure 17-5 the point resulting in maximum profits. Apparently, the optimum solution lies in the b–c plane and is equal to 12 units of item b and 6 units of item c. The profit would be \$144, which exceeds our maximum profit of \$120 when only items a and b are manufactured.

The curious reader can verify that a maximum profit of \$144 per day results in this problem when $b = 12$ and $c = 6$. Instead of 10 pairs of equations that were presented earlier, a total of $\binom{6}{3} = 20$ sets of constraints would have to be evaluated. It is obvious that for larger problems, such enumerative search procedures can consume prohibitive amounts of time. Fortunately, the simplex method can greatly simplify the solution of large LP problems and is readily available for use on digital computers.

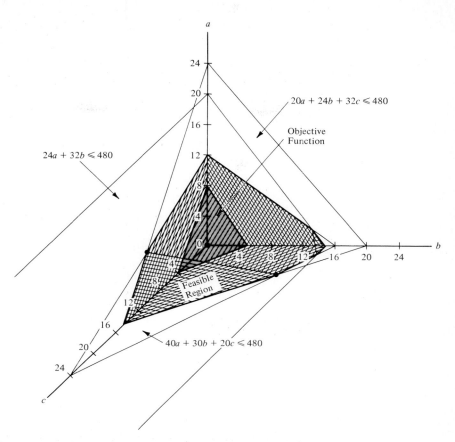

FIGURE 17-5 Solution Space for Three Decision Variables.

A GRAPHICAL SOLUTION TO A COST PROBLEM

A problem is presented here that illustrates how an economy study involving cost minimization can be formulated as a linear programming problem and solved graphically.

Suppose that the Ajax Furniture Company buys its lumber from two sources, companies G and H, and classifies the lumber according to three different grades, A, B, and C. The following matrix shows the expected proportion of lumber in each grade from each source. Board-feet requirements are also shown.

| Grade | Source | | Board Feet |
	G	H	Required
A	0.15	0.60	3,000
B	0.25	0.30	2,500
C	0.60	0.10	2,000

Company G charges $1 per board foot and H charges $1.50 per board foot. How much should be purchased from each supplier to satisfy requirements and minimize total cost?

We must first formulate our objective function and constraint equations. Let decision variable x_1 be the number of board feet purchased from G, and x_2 be the board feet from H. The objective is to

$$\text{Minimize } \$1.00x_1 + \$1.50x_2$$

$$\text{subject to } \quad 0.15x_1 + 0.60x_2 \geq 3{,}000$$

$$0.25x_1 + 0.30x_2 \geq 2{,}500$$

$$0.60x_1 + 0.10x_2 \geq 2{,}000$$

$$x_1 \geq 0, \quad x_2 \geq 0$$

The constraints and objective function are illustrated in Figure 17-6. Because we are minimizing costs, the optimal solution is the last point in the feasible region as the objective function moves toward the origin. This occurs at $x_1 = 5{,}716$ and $x_2 = 3{,}571$, with a minimum cost of $11,072.

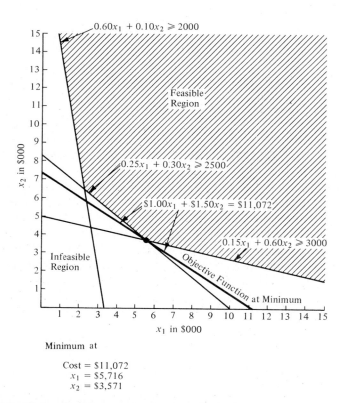

FIGURE 17-6   Solution of the Lumber Problem.

THE TRANSPORTATION PROBLEM

This section deals with a special type of linear programming problem, the *transportation problem*, in which there are *m* origins, each capable of producing and shipping a known amount of goods; and there are *n* destinations which require a specified shipment of goods. The penalty of shipping goods from an origin to a destination is given on a per unit basis (e.g., cost per ton-mile of goods shipped or miles over which a standard load is transported). The object is to determine the production shipping pattern that minimizes the total cost of transportation.

This problem can be simply illustrated as shown in Figure 17-7. As depicted in Figure 17-7, at plants A and B there are 6 and 8 truckloads, respectively, of cartons that are needed at warehouses C and D. Warehouse C requires 10 loads and warehouse D requires 4 loads of the cartons. The distances between the two plants and the two warehouses are as shown in Figure 17-7. How can the cartons at the plants be allocated between the two warehouses so as to minimize the loadmiles hauled? Table 17-2 shows how we might solve this problem by listing possible ways by which the required distribution might be achieved, and then computing the total mileage for each possibility. Using this approach, we can readily determine that by shipping 2 loads from plant A to warehouse C and 4 loads to warehouse D and by supplying the remaining requirements at warehouse C from plant B, a minimum mileage of 1,340 would be obtained.

This problem can also be represented in another manner, as shown in Figure 17-8. The availabilities and requirements are shown, and the transport distances are given in the small boxes.

Figure 17-9 shows how this transportation problem can be represented and solved graphically by computing the mileage resulting from the extreme con-

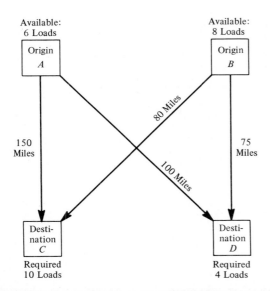

FIGURE 17-7 Linear Programming Problem: Transporting Available Loads to Required Destinations.

TABLE 17-2 Possible Combinations of Deliveries from Plants A and B to Warehouses C and D with Resulting Load-Miles

From:	A				B				
To:	C		D		C		D		
	Loads	(× 150) Load-Miles	Loads	(× 100) Load-Miles	Loads	(× 80) Load-Miles	Loads	(× 75) Load-Miles	Total Load-Miles
	6	900	0	—	4	320	4	300	1,520
	5	750	1	100	5	400	3	225	1,475
	4	600	2	200	6	480	2	150	1,430
	3	450	3	300	7	560	1	75	1,385
	2	300	4	400	8	640	0	—Min. →	1,340

ditions and connecting these values by straight lines, recognizing that the relationships are linear, and then adding the mileages to obtain the totals for those cases where the requirement of a total of 10 loads from the two plants to warehouse C is met.

It is apparent that this problem is very simple, because there are only five possible combinations of deliveries by which the required allocation of loads could be achieved. Therefore, obtaining a solution by the method used in Table 17-2 was not unduly laborious. However, it is equally clear that for problems more complex—and most real problems are considerably more complex—determining all the possible combinations and computing the necessary values manually would be prohibitively costly in time and money.

To / From	C	D	Available
A	150	100	6
B	80	75	8
Required	10	4	Totals 14

FIGURE 17-8 Matrix Representation of the Illustrative Problem Shown in Figure 17-7.

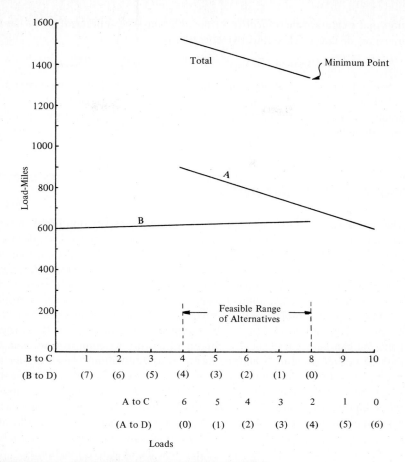

FIGURE 17-9 Graphical Solution of Example 17-1.

A mathematical statement of the situation represented in Figure 17-8 can be written from the information given. From the availabilities at each plant, expressed in the rows of the matrix,

$$X_{AC} + X_{AD} = 6$$

$$X_{BC} + X_{BD} = 8$$

where the X values indicate the allocations—X_{AC} is the number of loads assigned from plant A to warehouse C, and so on. Similarly,

$$X_{AC} + X_{BC} = 10$$

$$X_{AD} + X_{BD} = 4$$

These equations express the *constraints* imposed by the availabilities at the origins and the demands of the destinations. They represent the allocation model. To complete the mathematical statement of the problem, we can write an equation that expresses the *objective* of the solution—the minimization of the total transport distance. This objective equation is

$$\text{Minimize } 150X_{\text{AC}} + 100X_{\text{AD}} + 80X_{\text{BC}} + 75X_{\text{BD}}$$

Special computational procedures have been developed for solving large transportation problems. Their discussion is beyond the scope of this text, but references given in the footnote on page 538 include complete coverage of such procedures.

THE ASSIGNMENT PROBLEM

The *assignment problem* is a special case of the transportation problem. It can be viewed as a transportation problem with a demand of one at each destination and a supply of one at each origin (source), where all shipments must be either zero or one. A shipment of 1 indicates that a source is assigned to a destination. The solution involves assigning n sources to n destinations to minimize the total assignment cost. The assignment matrix must be a square $(n \times n)$ matrix.

Many realistic problems can be solved with the assignment algorithm. If, for example, there are n machines to be assigned to n locations such that materials handling can be accomplished at minimum cost to the company, the machines may be viewed as sources and the locations as destinations. The costs for three machines and three locations are given by a matrix such as that of Table 17-3. It can be seen that the cost of having machine U in location 2 is 22 units. In the solution there must be exactly one assignment in each column, and one assignment in each row—there must be exactly one machine for each location. The linear programming formulation of the assignment problem is straightforward. By the use of Table 17-3, we find that the objective function is

$$\text{Minimize } Z = 43x_{U1} + 22x_{U2} + 28x_{U3} + 40x_{V1} + 38x_{V2}$$

$$+ 37x_{V3} + 24x_{W1} + 25x_{W2} + 36x_{W3}$$

TABLE 17-3 Assignment Cost Matrix

		Location		
		1	2	3
Machine	U	43	22	28
	V	40	38	37
	W	24	25	36

where

$x_{ij} = 1$ if machine $i(i = U, V, W)$ is assigned to location $j(j = 1, 2, 3)$

$x_{ij} = 0$ if machine i is not assigned to location j

To avoid assigning two locations to one machine or two machines to a single location, a set of constraint equations could be written:

$$\left.\begin{array}{l} x_{U1} + x_{U2} + x_{U3} = 1 \\ x_{V1} + x_{V2} + x_{V3} = 1 \\ x_{W1} + x_{W2} + x_{W3} = 1 \end{array}\right\} \begin{array}{l}\text{requires that every machine} \\ \text{be assigned to exactly one location}\end{array}$$

$$\left.\begin{array}{l} x_{U1} + x_{V1} + x_{W1} = 1 \\ x_{U2} + x_{V2} + x_{W2} = 1 \\ x_{U3} + x_{V3} + x_{W3} = 1 \end{array}\right\} \begin{array}{l}\text{requires that each location be} \\ \text{serviced by one machine only}\end{array}$$

In general, the assignment problem can be summarized as follows.

Minimize $Z = \sum\limits_{i}^{n} \sum\limits_{j}^{n} c_{ij}x_{ij}$ (c_{ij} = materials-handling cost of assigning machine i to location j for a given production schedule)

subject to $\sum\limits_{i=1}^{n} x_{ij} = 1$

$$\sum\limits_{j=1}^{n} x_{ij} = 1$$

$$x_{ij} = 1 \text{ or } 0$$

For the simple problem above, assignments can be made by trial-and-error in an effort to discover the minimum cost. Unfortunately, for large problems, this approach would be impractical. Even when $n = 3$ (a small problem), there are six possible assignments for our machine–location problem, as follows:

Trial Number	Machine–Location Assignment	Cost
1	$U-1, V-2, W-3$	$117
2	$U-1, V-3, W-2$	105
3	$U-2, V-1, W-3$	98
4* (optimum)	$U-2, V-3, W-1$	83*
5	$U-3, V-1, W-2$	93
6	$U-3, V-2, W-1$	90

Thus the optimal assignment would be expressed as in Table 17-4 in matrix form.

In general, there are $n! = (n)(n-1)(n-2) \cdots (3)(2)(1)$ possible assignments to be considered. Obviously, when $n = 6$, it would be difficult to evaluate

TABLE 17-4   Optimal Assignment

Machine

$Z^* = 83$

all 720 possibilities. For large assignment problems, a special algorithm has been developed that considerably reduces solution time relative to trial-and-error enumeration methods. Discussion of this algorithm appears in most of the references listed in the footnote on page 538.

CAPITAL BUDGETING PROBLEMS

Linear programming is a useful technique for solving certain types of multiperiod *capital budgeting problems* when a firm is not able to implement all projects that increase its present worth. For example, constraints may exist on how much investment capital can be committed during each fiscal year, or interdependencies among projects may affect the extent to which projects can be successfully carried out during a specified planning horizon.

The final section of Chapter 6 dealt with the enumeration of mutually exclusive combinations of alternatives from sets of investment projects that can, themselves, be mutually exclusive, independent, and/or contingent. Suppose that the goal of a firm is to maximize its net present worth from the adoption of a capital budget that includes at least two mutually exclusive combinations. When the number of possible combinations becomes fairly large, manual methods that were described in Chapter 6 for determining the optimal investment plan tend to become quite complicated and time consuming. In such a situation, it is often worthwhile to consider linear programming as a solution procedure. The remainder of this section describes how simple capital budgeting problems can be formulated as LP problems. We hope that the reader will obtain some feeling of how more involved problems might also be modeled.

The objective function of the capital budgeting problem can be written as follows:

$$\text{Maximize net P.W.} = \sum_{j=1}^{n} B_j X_j$$

where B_j = net present worth of investment opportunity j during the planning period being considered

X_j = fraction of opportunity j that is implemented during the planning period (*Note*: In most problems of interest, X_j will be either 0 or 1); the X_j's are the decision variables

n = number of mutually exclusive combinations of alternatives under consideration

In computing the net P.W. of each mutually exclusive combination, a minimum attractive rate of return must be specified.

In view of the remaining notation defined as follows,

c_{ij} = cash outlay (e.g., initial investment or annual operating budget) required for opportunity j in time period t

C_t = maximum cash outlay that is permissible in time period t

there are typically two types of constraints present in capital budgeting problems:

1. Limitations on cash outlays for period t of the planning horizon

$$\sum_{j=1}^{n} c_{t_j} X_j \leq C_t$$

2. Interrelationships among investment opportunities. The following are examples:
   (a) If projects p, q, and r are mutually exclusive, then

$$X_p + X_q + X_r \leq 1$$

   (b) If project r can be undertaken only if project s has already been selected, then

$$X_r \leq X_s \quad \text{or} \quad X_r - X_s \leq 0$$

   (c) If projects u and v are mutually exclusive and project r is dependent (contingent) on the acceptance of u or v, then

$$X_u + X_v \leq 1$$
$$X_r \leq X_u + X_v$$

For a very simple illustration, consider the capital budgeting problem presented earlier in Example 6-13. The linear programming formulation of that particular problem is the following.

Maximize
$$13.4X_{B1} + 8.0X_{B2} - 1.3X_{C1} + 0.9X_{C2} + 9.0X_D$$

subject to

$$50X_{B1} + 30X_{b2} + 14X_{C1} + 15X_{C2} + 10X_D \leq 48 \qquad \text{(constraint on investment funds)}$$

$$X_{B1} + X_{B2} \leq 1 \qquad (B1, B2 \text{ mutually exclusive})$$

$$X_{C1} + X_{C2} \leq X_{B2} \qquad (C1 \text{ or } C_2 \text{ contingent on } B2)$$

$$X_D \leq X_{C1} \qquad (D \text{ contingent on } C1)$$

$$X_j = 0 \text{ or } 1 \qquad \text{(no fractional projects allowed)}$$

A problem such as the above could be solved quite readily by using the simplex method of linear programming if the last constraint ($X_j = 0$ or 1) were not present. With that constraint included, the problem is classified as a linear *integer* programming problem. There are many computer programs available for solving large linear integer programming problems.

As a second example, consider a three-period capital budgeting problem. Estimates of cash flows are as follows:

Investment Opportunity		Net Cash Flow ($000s), End of Year[a]:				Net P.W. ($000s) at 12%[b]
		0	1	2	3	
A1		− 225	150 (60)	150 (70)	150 (70)	+ 135.3
A2	mutually exclusive	− 290	200 (180)	180 (80)	160 (80)	+ 146.0
A3		− 370	210 (290)	200 (170)	200 (170	+ 119.3
B1	independent	− 600	100 (100)	400 (200)	500 (300)	+ 164.1
B2		− 1,200	500 (250)	600 (400)	600 (400)	+ 151.9
C1	mutually exclusive and dependent on acceptance of A1 or A2	− 160	70 (80)	70 (50)	70 (50)	+ 8.1
C2		− 200	90 (65)	80 (65)	60 (65)	− 13.1
C3		− 225	90 (100)	95 (60)	100 (70)	+ 2.3

[a]Estimates in parentheses are annual operating expenses (which have already been subtracted in determination of net cash flows).
[b]For example, net P.W. for $A1 = -225,000 + \$150,000(P/A, 12\%, 3) = +\$135,300$.

The M.A.R.R. is 12% and the ceiling on investment funds available is $1,200,000. In addition, there is a constraint on operating funds for support of the alternative selected, and it is $400,000 in year 1. From these constraints on funds outlays and the interrelationships among opportunities indicated above, we shall formulate this situation in terms of a linear integer programming problem.

First, the net present worth of each investment opportunity at 12% is calculated. The objective function then becomes

$$\text{Maximize net P.W.} = 135.3X_{A1} + 146.0X_{A2} + 119.3X_{A3} + 164.1X_{B1}$$
$$+ 151.9X_{B2} + 8.1X_{C1} - 13.1X_{C2} + 2.3X_{C3}$$

The budget constraints are the following.

Investment funds constraint:

$$225X_{A1} + 290X_{A2} + 370X_{A3} + 600X_{B1} + 1{,}200X_{B2}$$
$$+ 160X_{C1} + 200X_{C2} + 225X_{C3} \leq 1{,}200$$

First year's operating cost constraint:

$$60X_{A1} + 180X_{A2} + 290X_{A3} + 100X_{B1} + 250X_{B2}$$
$$+ 80X_{C1} + 65X_{C2} + 100X_{C3} \leq 400$$

Interrelationships among the investment opportunities give rise to these constraints on the problem:

$X_{A1} + X_{A2} + X_{A3} \leq 1$	A1, A2, A3 are mutually exclusive
$X_{B1} \leq 1$	
$X_{B2} \leq 1$	B1, B2 are independent
$X_{C1} + X_{C2} + X_{C3} \leq X_{A1} + X_{A2}$	accounts for dependence of C1, C2, C3 on A1 *or* A2

Finally, if all decision variables are required to be either 0 (not in optimal solution) or 1 (included in optimal solution), the last constraint on the problem would be written:

$$X_j = 0, 1 \quad \text{for } j = A1, A2, A3, B1, B2, C1, C2, C3$$

As one can see, a fairly simple problem such as this one would require an inordinate amount of time to solve by listing and evaluating all mutually exclusive combinations as suggested in Chapter 6. Consequently, it is recommended that a suitable computer program be utilized to obtain solutions for all but the most simple capital budgeting problems.

PROBLEMS

17-1   Determine the feasible region for the following LP problem:

$$\text{Minimize } C = 4x_1 + 6x_2$$
$$\text{subject to} \quad 2x_1 + 5x_2 \leq 10$$
$$3x_1 + 2x_2 \leq 6$$
$$x_1 \geq 0, \quad x_2 \geq 0$$

(a) Using graphical means, determine the optimal solution.
(b) If the objective function changes to $C = x_1 + 8x_2$, will there be a different optimal solution?

17-2   A small company binds books and has two bindings available. Binding A is a high-quality product that results in a profit of $1.80 per book. Binding B is a lower-quality product with a profit margin of $1.50 per book. If only the lower-quality binding were available, the company could bind 500 books each day. When binding A is requested, it requires 50% more time than does binding B. However, because of material shortages, only 350 books each day can be produced regardless of the type of binding. The high-quality binding requires a special glueing operation that has a maximum output of 250 books per day. Formulate this production problem as a linear programming problem to maximize profit and solve it graphically.

17-3   A person wishes to select a diet consisting of bread, butter, and/or milk, which has a minimum cost, but yields an adequate amount of vitamins A and B. The minimum vitamin requirements are 11 units of A and 10 units of B. The diet should not contain more than 13 units of A because more may be harmful. Furthermore, he likes milk and requires that the diet include at least 3 units of milk but is indifferent to the amount of butter and bread. A dietician has measured the vitamin contents and found, per unit of product, 1 unit of A and 3 of B in bread, 4 units of A and 1 of B in butter, 2 units of A and 2 of B in milk. The market price of 1 unit of bread, butter, and milk is 2, 9, and 1, cent, respectively. Formulate this problem as a linear programming problem. Be sure to specify the decision variables and the objective function. Solve this problem graphically.

17-4   A furniture manufacturer wants to determine how many tables, chairs, desks, and/or bookcases to make to optimize utilization of his available resources. These products use two different types of lumber, and he has on hand 1,500 board-feet of the first type and 1,000 board-feet of the second type. He has 800 man-hours of labor available for the total job. His sales forecast plus his backorders require him to make at least 40 tables, 130 chairs, 30 desks and no more than 10 bookcases. Each table, chair, desk, and bookcase requires 5, 1, 9, and 12 board-feet, respectively, of the first type of lumber; and 2, 3, 4, and 1 board-feet, respectively, of the second type. A table requires 3 man-hours to make, a chair requires 2 man-hours, a desk 5 man-hours, and a bookcase 10 man-hours. The manufacturer makes a total profit of $12 on a table, $5 on a chair, $15 on a desk, and $10 on a bookcase. Formulate this situation as a linear programming problems such that profits are maximized.

17-5   A company supplies aggregate from three quarries, A, B, and C, to five premix concrete plants. The following matrix gives the dollar transportation costs from these quarries to the plants, and the capacities and requirements per day. Set this up as a transportation problem and attempt to enumerate several feasible solutions. The least-cost solution is $90. How close did you get to this minimum cost?

Quarry	Plant					Capacity (units)
	1	2	3	4	5	
A	4	2	3	2	6	9
B	5	4	5	2	1	10
C	6	5	4	7	3	13
Requirements:	4	5	7	8	8	

17-6   A company has three factories, A, B, and C, from which it supplies a product to four company-operated retail stores, D, E, F, and G. The monthly capacities at the factories are 120, 140, and 130 units, respectively, and it has found that the four stores can sell 80, 90, 110, and 160 units per month, respectively, if they are available. The unit profits that result from the supply and sale are as follows:

From	To			
	D	E	F	G
A	$42	$48	$38	$36
B	40	49	52	39
C	38	36	45	44

Formulate this situation as an LP problem and try to enumerate some feasible solutions to it. The maximum profit is $18,310. How close did you get to this solution? (*Hint:* Add a row to account for unsatisfied demand. In the optimal solution, 20 units of D's demand is unmet and 30 units of G's demand is not satisfied.)

17-7 Four manufacturing plants must be built in four different geographic regions, with only one plant in each location. The costs of building each plant in the various locations are as follows, in millions of dollars:

Plant	Location			
	1	2	3	4
1	60	51	32	32
2	48	×	37	43
3	39	26	×	33
4	40	×	51	30

In this matrix an × indicates that, because of the need for highly specialized labor, it is impractical to assign a plant to a particular location. Formulate this as an assignment problem and generate, by trial and error, several feasible solutions to it.

17-8 Refer to Problem 6-36. Formulate this capital budgeting problem as a linear integer programming problem.

17-9 Refer to Problem 6-37. Set this problem up as a linear integer programming problem.

CHAPTER 18

Critical Path Economy

In carrying out many types of projects in industry and government, a large number of interdependent tasks must be planned, scheduled, and controlled. These component tasks may be interdependent with respect to their timing, manpower requirements, equipment needs, and so on. Examples of large and complex one-time projects include development of a prototype space vehicle and construction of a high-rise apartment building.

Typically, these jobs require many months or years for completion, and many of the component tasks can be completed in varying amounts of time and cost, depending on how different amounts of resources are assigned to them. Consequently, the total job will involve different amounts of time for completion, and will involve different costs, depending on how the component tasks are done. However, in most instances only a few of the tasks are critical in that they will affect the actual outcome in time and cost. Several *critical path procedures* have been developed that enable such projects to be managed in an effective and economical manner. In most cases the ultimate goal is to determine when each task must begin to assure that the overall project will be finished within a specified amount of time.

These procedures have been widely adopted by industry, but they have also been found useful in many nonindustrial activities, such as planning a national meeting of a technical society. In most cases they have an economic objective, and present economy studies and decisions are involved in their application.

The purpose of this chapter is to describe several project management techniques in which trade-offs between task timing and costs are possible.

THE GANTT CHART

Critical path methods are expansions of the old, well-known *Gantt charts*, which have been used since the early 1900s. Figure 18-1 is a simple Gantt chart that depicts the time schedule of the major activities that are involved in building a house. Such a chart shows the required activities, the time period over which each activity will extend, and the time at which each activity should start in order that each prerequisite activity will be completed, or have progressed sufficiently, so that the following activities can start. In addition, a time indicator usually is provided, which can be moved to indicate the current time. As work progresses, the activity bars are filled in so that at any time we can determine whether the projected schedule is being met. In Figure 18-1, for example, some portion of the exterior painting is behind schedule and the interior decorating is ahead of schedule.

As the normal Gantt, or bar, chart is used, it has two rather serious limitations. First, it does not clearly indicate details regarding the progress of activities. This is particularly bothersome for activities that extend over relatively long periods of time. For example, in Figure 18-1 the interior decorating is scheduled to extend over 5 weeks. Although the decoration as a whole—perhaps in man-hours—is ahead of schedule, there is not any certainty that it is being done in

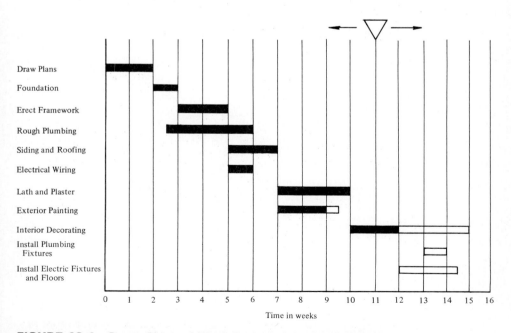

FIGURE 18-1   Gantt Chart of Work Tasks Involved in Building a House.

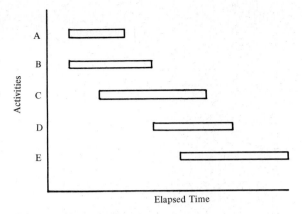

FIGURE 18-2   Activity–Elapsed-Time Chart.

the proper sequence so that the bathrooms will be decorated before the trim on the plumbing fixtures must be installed. The second, and more important, deficiency is that such charts do not give a clear indication as to what portions of any activity are specifically prerequisite to following activities or to dependent activities that may overlap.

These deficiencies of bar charts may be eliminated, to a large degree, by the procedures shown in Figures 18-2 and 18-3, wherein the activities are broken

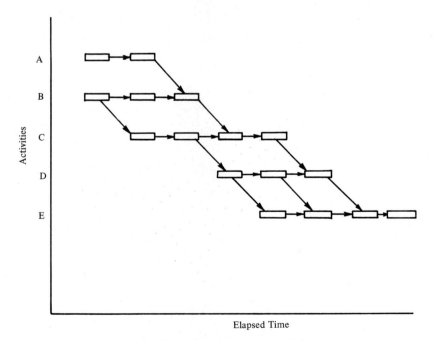

FIGURE 18-3   Activity–Elapsed-Time Chart, with Interdependency Shown by Arrows.

into subparts and the interdependence is shown by means of connecting arrows. These procedures lead directly to project network techniques.

ACTIVITY–EVENT NETWORKS

An activity–event network portrays the interrelationships between the activities and events that comprise a project or job, where an *event* is the start or end of a task and thus represents a point in time, and an *activity* is the work required to complete a specific event. Activities require time and resources for their accomplishment and represent recognizable parts of the project. In an activity–event network, as shown in Figure 18-4, events commonly are represented by circles or ovals and activities by arrows. Thus events are related to one another by activities. Figure 18-4 illustrates a simple activity-on-arrow diagram.

Events ordinarily are described by two- or three-word phrases, with the first word of the phrase being *start* or *complete*. Thus "start drawings," "start electrical design," "complete foundation," and "complete drawings" are typical descriptions. Because, except for initial events and a few other cases, an event represents the completion of an activity, most events utilize *complete* as the first word of the description. Activities are described by brief phrases, such as "prepare mechanical drawings" or "build cabinets."

The events are numbered 10, 20, 30, and so on when the network is first drawn. Because the passage of time progresses from left to right on a network, early events are given low numbers and later events are assigned higher numbers. If it becomes necessary to alter the network by adding more events, these can be assigned numbers, 11, 21, 31, and so on. Activities usually are referred to by showing the events that they connect: activity 10–20, activity 20–30, and so forth.

In activity–event networks, no events, except an initial one, can be started until the completion of all the activities that are connected to it by *incoming* arrows. Thus, in Figure 18-4, event 30 cannot occur until both activity 10–30 and activity 20–30 are completed. Similarly, no activity can start until the event from which it emerges has occurred.

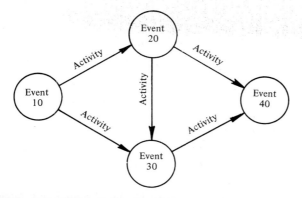

FIGURE 18-4 Activity–Event Chart.

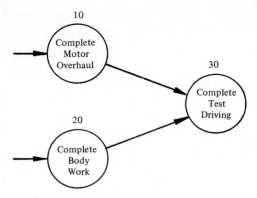

FIGURE 18-5 Activity—Event Sequence with the Final Event Dependent on Two Duplicate Activities (10—30 and 20—30).

Sometimes it is desirable to introduce a dummy activity into a network to have consistent logic. For example, suppose that the final steps in overhauling an automobile are as shown in Figure 18-5. Obviously, activities 10–30 and 20–30 are the same—testing the car—and both events 10 and 20 must be completed before the test driving can start. This type of situation can be resolved by use of a dummy activity, 10–20, as shown in Figure 18-6. A dummy activity requires no time and no resources.

Another important use of a dummy activity is to indicate that one of two independent events must precede the other. Thus, according to the logic depicted in Figure 18-5, the completions of the motor overhaul and the body work could occur simultaneously. But the use of the dummy activity in Figure 18-6 specifies that the completion of the motor overhaul (event 10) must precede the completion of the body work (event 20). Similarly, dummy activities are used in showing that one activity must precede another, as might be necessitated by the same piece of equipment being required for both activities. Although the dummy activity does not require any time, the resulting change in the sequence of the activities and/or events may extend the time required for a project and may alter the resource requirements.

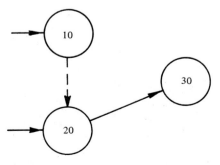

FIGURE 18-6 Use of "Dummy" Activity (10—20) to Avoid Redundant Activities Portrayed in Figure 18-5.

OTHER USEFUL METHODS
FOR MINIMIZING RESOURCE REQUIREMENTS

CPM AND PERT

Several variations of activity–event networks have been developed and are in widespread use. Probably the two most common are CPM (critical path method) and PERT (program evaluation and review technique). The du Pont Company developed CPM in the late 1950s as a technique for planning and controlling complex engineering projects with respect to their time and costs. About the same time, PERT was developed by the U.S. Navy as a means of managing the Polaris missile program. The successful completion of the Polaris program some 2 years ahead of the original forecast was attributed to use of the PERT technique.

CPM and PERT are based on the same concepts, the fundamental one being identification of a critical sequence of activities to be closely monitored and controlled in meeting a specified completion date. The main difference is the manner in which activity–event networks are graphically constructed.

CPM is applied, primarily, to projects that are more deterministic in nature, such as construction work, in which cost and time estimates can be predicted with considerable certainty due to the existence of past experience. PERT, on the other hand, tends to be applied to one-time projects, such as research and development work, in which intellectual effort and first-time prototype manufacturing are involved, so that time and cost estimates tend to be quite uncertain. Consequently, probabilistic methods may be employed in connection with PERT.

When times that will be required for activities must be based solely on subjective estimates, it is common practice to obtain these estimates from several persons who are familiar with, or are to be involved in, each activity. In such cases some of the persons will give quite optimistic estimates, while others anticipate many possible difficulties and will give quite pessimistic estimates. In some cases the "pessimistic," "optimistic," and "probable" time estimates for activities may be included on a PERT chart. Common practice is to compute a single time estimate using the relationship

$$t_{\text{est}} = \frac{t_0 + 4t_m + t_p}{6} \qquad (18\text{-}1)$$

where t_0 = optimistic time

t_m = most likely time

t_p = pessimistic time

Numerous variations of both CPM and PERT have been developed, so that in some cases the procedures are very similar. For further discussion in this chapter, CPM, with some modifications, is employed to illustrate the economy study aspects of these procedures.

CRITICAL PATHS

Assume that the network shown in Figure 18-7 represents activities and events that are required for carrying out a certain project. The numerals shown above

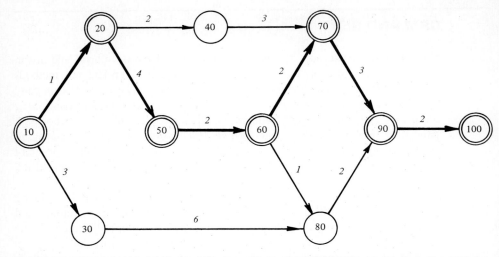

FIGURE 18-7 Typical Activity–Event Network with Critical Path Indicated by Heavy Arrows and Critical Events Indicated by Double Circles.

each activity arrow are the number of working days estimated for the completion of each activity, when normal procedures are used.† From examination of the diagram it is clear that there are four paths between the first event (10) and the final event (100). Each of these paths must be undertaken in completing the overall project. If the total time along each path is determined, the results are as follows:

1. For path 10–20–40–70–90–100: 11 days.
2. For path 10–20–50—60–70–90–100: 14 days.
3. For path 10–20–50–60–80–90–100: 12 days.
4. For path 10–30–80–90–100: 13 days.

It thus is evident that path 10–20–50–60–70–90–100 determines the minimum time in which the project can be completed, and it is called the *critical path*. Hence, the critical path is a series of connected activities which, if delayed, will cause the entire project to be delayed.

For a simple project such as this one, the critical path can be determined quite quickly by enumerating all paths and evaluating the time required to complete each. However, networks for most real projects are more complex, and a considerable number of paths may exist. Therefore, a more systematic method of analysis is desirable. Such a procedure for making the critical path calculations is demonstrated below for the network of Figure 18-7. This procedure would also apply to more complex networks.

In view of the estimated activity times calculated with Equation 18-1, the critical path of the network and other information useful to managing a large project can be determined by a two-phased procedure. The *forward pass* through a network commences at the initial node and proceeds toward the terminal node.

---

†Activity times are always shown in working periods—days, weeks, or months—not in calendar periods.

Conversely, the *backward pass* moves from the terminal node to the initial node. Each of these phases is now demonstrated for the simple network of Figure 18-7.

In the forward pass a number is determined at each node that represents the earliest occurrence of the respective event. Activities are described in terms of an $(i-j)$ index, such that $i < j$. For example, in Figure 18-7 the activity with duration of 4 days would be designated 20–50. Let ES_i denote the earliest start times for all activities emanating from event i, and ES_j represents the completion time. When $i = 1$ (the initial event), E_1 is defined to be zero. Further, d_{ij} denotes the duration of activity $i-j$.

To carry out the forward pass, the earliest start time for the jth event in the network is calculated with this relationship:

$$ES_j = maximum \text{ value of } [ES_i + d_{ij}] \text{ for all set of activities}$$
$$emanating\ from \text{ event } i \text{ that lead to event } j$$

This indicates that the earliest possible start time for a particular event is dependent on previous events plus the activity times of all preceding activities. When the relationship above is applied to the network of Figure 18-7, these early start times are obtained:

$$ES_{10} = 0$$
$$ES_{20} = ES_{10} + d_{10-20} = 0 + 1 = 1$$
$$ES_{30} = ES_{10} + d_{10-30} = 0 + 3 = 3$$
$$ES_{40} = ES_{20} + d_{20-40} = 1 + 2 = 3$$
$$ES_{50} = ES_{20} + d_{20-50} = 1 + 4 = 5$$
$$ES_{60} = ES_{50} + d_{20-60} = 5 + 2 = 7$$
$$ES_{70} = max\ [ES_{40} + d_{40-70}, ES_{60} + d_{60-70}]$$
$$= max\ [3 + 3, 7 + 2] = 9$$
$$ES_{80} = max\ [ES_{30} + d_{30-80}, ES_{60} + d_{60-80}]$$
$$= max\ [3 + 6, 7 + 1] = 9$$
$$ES_{90} = max\ [ES_{70} + d_{70-90}, ES_{80} + d_{80-90}]$$
$$= max\ [9 + 3, 9 + 2] = 12$$
$$ES_{100} = ES_{90} + d_{90-100} = 12 + 2 = 14$$

The earliest start time for event 100 (i.e., finished project) is, therefore, 14 days. This completes the forward pass through the network.

Now the backward pass is initiated at the terminal (end) event and the aim here is to compute the latest possible finish, or completion, times for all activities coming into each event. The latest finish time for event i is designated LF_i and for the terminal event $LF_i = ES_i$. As in the calculation of early start times, a simple relationship is utilized to determine the latest finish times for any given event i:

$LF_i = minimum$ value of $[LF_j - d_{ij}]$ for all sets of activities emanating from event i to event j (recall that $i < j$ in numbering events from left to right on the network)

To illustrate the use of this relationship, the latest finish times for events in Figure 18-7 are developed as follows:

$$LF_{100} = ES_{100} = 14$$

$$LF_{90} = LF_{100} - d_{90-100} = 14 - 2 = 12$$

$$LF_{80} = LF_{90} - d_{80-90} = 12 - 2 = 10$$

$$LF_{70} = LF_{90} - d_{70-90} = 12 - 3 = 9$$

$$LF_{60} = min [LF_{70} - d_{60-70}, LF_{80} - d_{60-80}]$$

$$= min [9 - 2, 10, - 1] = 7$$

$$LF_{50} = LF_{60} - d_{50-60} = 7 - 2 = 5$$

$$LF_{40} = LF_{70} - d_{40-70} = 9 - 3 = 6$$

$$LF_{30} = LF_{80} - d_{30-80} = 10 - 6 = 4$$

$$LF_{20} = min [LF_{40} - d_{20-40}, LF_{50} - d_{20-50}]$$

$$= min [6 - 2, 5 - 4] = 1$$

$$LF_{10} = min [LF_{20} - d_{10-20}, LF_{30} - d_{10-30}]$$

$$= min [1 - 1, 4 - 3] = 0$$

This completes the backward pass through the network.

Based on results of the forward and backward passes, the critical path activities can be easily identified. An $i–j$ activity is part of the critical path if three conditions are satisfied:

1. $ES_i = LF_i$.
2. $ES_j = LF_j$.
3. $ES_j - ES_i = LF_j - LF_i = d_{ij}$.

Hence these conditions indicate that there is no *slack time* between the earliest time an activity can start and its latest finish time. When the slack for an event is zero, it means that any delay in the occurrence of that event will cause delay in starting subsequent activities. On the other hand, when there is positive slack, the activity could be delayed by the amount indicated without causing any delay in the total project. Table 18-1 summarizes the ES_i and LF_i times and shows the amount of slack associated with each event.

If the events having zero slack are connected, as shown by heavy arrows in Figure 18-7, the resulting path (or paths) is critical; such paths determine the minimum time required for completing the project, and delays in any of the events on the critical path will result in extending the completion time of the project. In Figure 18-7 the events having zero slack are shown as double circles. Readers should check path 10–20–50–60–70–90–100 to assure themselves that the three conditions listed above for a critical path have indeed been satisfied.

TABLE 18-1   Slack Times for Each Event in Figure 18-7

Event i	ES	LF	Slack (days)
10	0	0	0
20	1	1	0
30	3	4	1
40	3	6	3
50	5	5	0
60	7	7	0
70	9	9	0
80	9	10	1
90	12	12	0
100	14	14	0

ECONOMY STUDY ASPECTS OF CPM AND PERT

Once a critical path is determined, it is apparent that in many cases economies can be obtained by shifting resources, or by changing procedures, so as to shorten the time required to complete a project. Getting the project completed and into operation at an earlier than predicted date can result in earlier and greater revenues. In construction projects, earlier completion may reduce or eliminate penalties for late completion or earn bonuses for early completion. Another possibility may be that money can be saved by reducing the allocation of resources to an activity on a noncritical path. However, reallocation of resources and changes in procedures usually involve some increases in costs. The added costs above normal costs by which shorter activity times can be achieved are frequently called *crash costs*. Obviously, there usually is a range of crash costs that must be considered. Thus the usual economy study problem of balancing costs versus revenues arises, and frequently alternative choices exist. It also is apparent that, because of the short time involved and the nature of the required expenditures, we are dealing with out-of-pocket costs in most instances, so that the economic analysis usually is one of present economy.

EXAMPLE 18-1

In the project depicted in Figure 18-7, it is found that a gross benefit of $150 per day can be obtained for each day saved, up to 2 days. Investigation reveals that the cost–time relationships for activities 20–50, 30–80, 50–60, 80–90, and 90–100 are as shown in Figure 18-8. What reallocation of resources should be made?

Solution

If 1 day's time were eliminated in the 20–50–60 path, the 10–20–50–60–70–90–100 path would be balanced with path 10–30–80–90–100; each would require 13

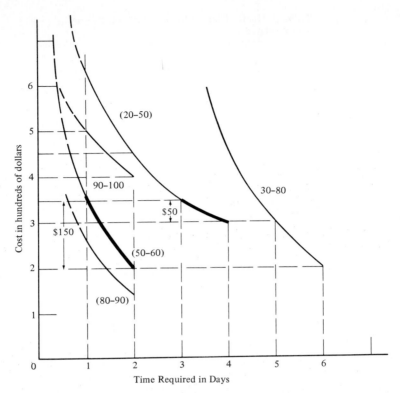

FIGURE 18-8 Cost–Time Relationships for Several Activities Shown in Figure 18-7.

days. There then would be no advantage in further reducing the time required for the activities of the first path, unless reduction also was made for the time of the activities of the second path. It is apparent that the first reduction should be made in the 10–20, the 20–50, or the 50–60 activities. Since activity 10–20 presently requires only 1 day, it is not likely that 1 day can be saved there. What we wish to do, therefore, is to reduce the time for either activity 20–50 or activity 50–60 at minimum cost.

Considering activities 20–50 and 50–60, it will be noticed that the slopes of the two curves, in Figure 18-8, make it apparent that reduction in time can be obtained at smaller cost in the case of activity 20–50 than for activity 50–60. This can be seen by examining the slopes of the two heavy-line segments. Further, the marginal cost of 1 day's time for activity 50–60 is $150, thus permitting no net benefit. On the other hand, because 1 day can be saved in connection with activity 20–50 at a cost of only $50, a net saving of $100 will be achieved. With a reduction of 1 day made in the time required for activity 20–50, the critical path network is as shown in Figure 18-9.

With the total time for path 10–20–50–60–70–90–100 now reduced to 13 days—the same as for path 10–30–80–90–100—a consideration of any further

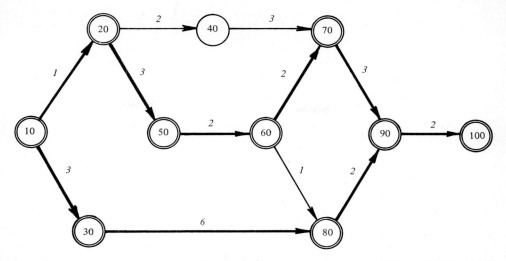

FIGURE 18-9 New Activity–Event Network, Resulting from Eliminating 1 Day for Activity 20–50, as Shown in Figure 18-7.

reduction in time would involve the time–cost relationships of the activities of both paths. Investigation revealed that only activities 80–90 and 90–100, in addition to those considered previously, offered any possibilities for net benefits, having marginal costs below $150. As shown in Figure 18-8, the marginal costs of each of these is $100. However, if the time for activity 30–80 is reduced, a concurrent reduction of 1 day would have to be made in activity 20–50 at a cost of $100 because both paths are critical. For this plan the total marginal cost would be $200, and thus it is not an economic solution. Because a reduction of 1 day in activity 90–100 affects both paths, and can be achieved for $100, this would be the alternative to select. ∎

OTHER APPLICATIONS OF CPM

Critical path networks can be extended to aid in the economic solutions of numerous problems. Example 18-2 illustrates such an application to a common problem of minimizing the cost that may be associated with possible variations in crew size.

EXAMPLE 18-2

The activities required on a small construction job, and their durations and manpower requirements are

Activity	Duration (days)	Workers per Day
10–20	1	1
20–30	2	3
10–40	4	3
30–50	3	2
40–50	3	1
10–60	4	3
50–70	2	3
60–70	3	2

It is desired to complete the job in the smallest number of days and, because of penalty costs associated with hiring and training workers, not to vary the work force any more than necessary from some minimum number. What will be the minimum time required for the job, and what is the minimum penalty cost for crew-size variation, assuming that the penalty costs are as shown below and that once an activity is started it must continue without interruption until completed?

Extra Workers	Cost per Day
1	$10
2	15
3	25
4	30
5	35

Solution

Figure 18-10 is the activity–event network for this job, and Figure 18-11 is a modified slack chart, giving the solution to Example 18-2. It will be noted that the left-hand portion of this figure is a modified slack chart with activities listed, rather than events, and with T_{ES} representing the earliest time at which an activity can start and T_{LF} the latest time at which it can be completed according to the constraints of the network. It is evident that the earliest completion time for the project will be 9 days, determined by the critical path 10–40–50–70.

Figure 18-11 is a slack-manning chart useful for this type of problem. The horizontal line opposite each activity shows the days during which it could occur—extending from T_{ES} to T_{LF}. Thus activity 20–30 can start on the second day and may be finished as late as the fourth day. For any activity,

$$\text{slack} = T_{LF} - \text{duration} - T_{ES}$$

Consequently, for activity 20–30 there is 1 day of slack. Since the activity requires 2 days for completion, and assuming that it will go forward to com-

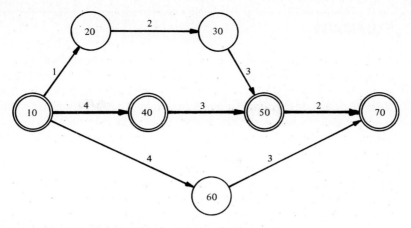

FIGURE 18-10 Activity Network for Example 18-2.

pletion without interruption when started, it could be accomplished either on the second and third days or on the third and fourth days. Where an activity has no slack, the space below the horizontal line on the chart has been filled in with crosshatching. The numerals above the horizontal lines show the number of men that will be required during each day for each activity that occurs.

Once such a chart is completed to this stage, the solution lies in trial-and-error adjustments to the manning table. This is done to cut off slack days from either the beginning or end of the horizontal lines so as to minimize crew-size variations. The days that are eliminated are shown crossed out. In this case, the minimum crew size is 4, and the total penalty cost is $125. ■

This type of manning chart can be used advantageously to allocate limited work crews or craft types to projects so as to achieve minimum cost and to avoid unforseen delays.

Activity	Dura-tion	T_{ES}	T_{LF}	Slack	Men	Days 1	2	3	4	5	6	7	8	9
10–20	1	0	2	1	1	1	1							
20–30	2	1	4	1	3		3	3	3					
10–40	4	0	4	0	3	3	3	3	3					
30–50	3	3	7	1	2				2	2	2	2		
40–50	3	4	7	0	1					1	1	1		
10–60	4	0	6	2	3	3	3	3	3	3	3			
50–70	2	7	9	0	3								3	3
60–70	3	4	9	2	2					2	2	2	2	2
Crew Size						4	6	9	6	6	6	5	5	5
Penalty Cost ($)						0	15	35	15	15	15	10	10	10

FIGURE 18-11 Slack Chart and Manning Table for Example 18-2.

SUMMARY

The applications of critical path networks that have been presented in this chapter show only a few of their uses. It is evident they can be of considerable assistance in solving several types of problems that involve economic decisions and that these decisions usually involve present economy considerations. Once the basic concepts and procedures of activity–event networks have been learned, there is little difficulty in applying the principles of present economy studies to specific problems.

The methods that have been discussed here for determining critical paths and slack times and for selecting the most economic alternatives are fairly routine and may readily be programmed for computers. A number of computer programs are available that permit quite complex networks to be solved with considerable ease. These greatly extend the usefulness of critical path methods.

PROBLEMS

18-1 Construct a project network consisting of activities on arrows that has 12 activities, numbered 10 through 120. The following relationships must be satisfied in the network.
(1) 10, 20, and 30 are the first activities of the project and are begun simultaneously.
(2) 10 and 20 must precede 40.
(3) 20 must precede 50 and 60.
(4) 30 and 60 must precede 70.
(5) 50 must precede 80 and 100.
(6) 40 and 80 must precede 90.
(7) 70 must precede 110 and 120.
(8) 90 must precede 120.
(9) 100, 110, and 120 are the terminal activities of the project.
Note that activities have been numbered here as opposed to events. Number or letter the nodes any way you desire as long as the relationships (network constraints) above are satisfied. You should find that two dummy activities are required.

18-2 For a minor construction project the activities and associated times are as follows:

Activity	Expected Time (days)
10–20	3
20–30	2
20–40	4
30–60	1
40–50	2
30–50	0
50–70	5
60–70	2
70–80	7

(a) Draw an activity–event diagram.

(b) Determine the critical path and minimum time for the project.

(c) Calculate and tabulate the earliest and latest times and the slack for each of the events.

18-3   The following activities and normal times are required for installing a large machine tool:

Activity	Normal Time (days)
00–10	2
10–20	4
10–30	3
10–40	2
20–60	2
30–50	4
40–50	3
50–70	3
60–80	5
70–80	2
80–90	1

(a) Draw the activity–event diagram.

(b) Determine the critical path and minimum time that will be required for the installation.

(c) Calculate the earliest and latest starting times and the slack for each event, and tabulate these values.

18-4   In Problem 18-3, assume that a bonus of $300 per day can be obtained for each day that the installation time can be reduced. Investigation shows that the times for activities 20–60, 30–50, and 70–80 can each be reduced, with the time–cost relationships being as portrayed by curves 50–60, 20–50, and 90–100, respectively, in Figure 18-8. Determine to what extent it would be economic to reduce the installation time and how it should be accomplished.

18-5   Refer to Figure 18-10.

(a) By using the forward pass and backward pass procedures described in the text, carry out the calculations required to determine the ES and LF times for each *event* and set up a table similar to that of Table 18-1.

(b) What is the critical path and minimum time required to complete the project?

(c) Check your answers with Figure 18-11 and indicate what differences exist between your table and the left side of Figure 18-11. Explain how you can easily convert from event-determined ES and LF times to T_{ES} and T_{LF} times for activities.

18-6   A company is planning a series of conferences for management personnel. These are to be held at a mountain-resort hotel in a sequence that will permit a series of interrelated decisions to be reached. Consequently, the seminar meetings can be considered to be activities leading to events (decisions). The following tabulation gives the duration of the seminars and the number of people who will be involved. It is desirable to know how to schedule the conferences so that a minimum total time will be required and so that the number of individual hotel rooms that will have to be reserved on any one night will be minimized. Assume that the personnel involved are nondependent.

Activity	Duration (days)	People
10–20	2	2
10–30	3	1
20–50	4	3
30–40	2	3
30–60	3	2
40–70	5	2
60–70	3	3
50–70	4	2
70–80	3	4

(a) Draw the activity–event diagram.
(b) Determine the minimum number of days that will be required to complete the conference series.
(c) Arrange the schedules so as to minimize the number of hotel rooms needed any night.

18-7 The activities, their durations, and manpower requirements on a small construction job are listed in the following table.

Activity	Duration (days)	People
10–20	1	1
10–40	4	3
10–60	4	3
20–60	2	3
30–50	3	2
40–50	3	1
50–70	2	3
60–70	3	2

It is desired to complete the job in the minimum time and, because of penalty costs associated with hiring and training workers, not to vary the size of the work force more than necessary from some minimum number.
(a) Draw the activity–event diagram.
(b) Determine the critical path and the minimum number of days for completing the job.
(c) Assuming the penalty costs for crew-size variation to be

Extra Men	Cost per Day
1	$ 1
2	4
3	9
4	16
5	25

arrange the schedule so as to minimize the penalty cost.

Glossary of Commonly Used Symbols and Terminology

ECONOMIC ANALYSIS METHODS AND COSTS

A.C.	Annual cost
A.W.	Annual worth
A.W.-C.	Annual worth–cost (same as A.C.)
C.R.	Capital recovery cost (annual cost of depreciation plus interest on investment)
E_k	Net expenses for kth compounding period
E.R.R.	External rate of return
E.R.R.R.	Explicit reinvestment rate of return
F.W.	Future worth
I.R.R.	Internal rate of return
M.A.R.R.	Minimum attractive rate of return
P.W.	Present worth
P.W.-C.	Present worth–cost
R_k	Net receipts (revenues) for kth compounding period
S_N	Net salvage value during compounding period N

COMPOUND INTEREST SYMBOLS

e	Reinvestment rate per interest period (as used for E.R.R. and E.R.R.R. methods)
f	Effective inflation rate per interest period
i	Effective interest rate per interest period
i'	Interest or rate of return (to be determined)
k	Index for compounding periods
r	Nominal interest rate per year
$\underline{r}$	Nominal interest rate per year, compounded continuously
N	Number of compounding periods
P	Present sum of money (present worth); the equivalent worth of one or more cash flows at a relative point in time called the present
F	Future sum of money (future worth); the equivalent worth of one or more cash flows at a relative point in time called the future
A	End-of-period cash flows (or equivalent end-of-period values) in a uniform series continuing for a specified number of periods
G	Uniform period-by-period increase or decrease in cash flows or amounts (the arithmetic gradient)
$\overline{P}$ or $\overline{F}$	Amount of money (or equivalent value) flowing continuously and uniformly during a given period
$\overline{A}$	Amount of money (or equivalent value) flowing continuously and uniformly during each and every period continuing for a specific number of periods

Functional Forms (Symbols) for Compound Interest Factors[a]

Functional Format (Symbol)	Name of Factor
ALL CASH FLOWS DISCRETE: END-OF-PERIOD COMPOUNDING	
$(F/P, i\%, N)$	Compound amount factor (single payment)
$(P/F, i\%, N)$	Present worth factor (single payment)
$(A/F, i\%, N)$	Sinking fund factor
$(A/P, i\%, N)$	Capital recovery factor
$(F/A, i\%, N)$	Compound amount factor (uniform series)
$(P/A, i\%, N)$	Present worth factor (uniform series)
$(A/G, i\%, N)$	Arithmetic gradient conversion factor (to uniform series)
$(P/G, i\%, N)$	Arithmetic gradient conversion factor (to present value)
ALL CASH FLOWS DISCRETE: CONTINUOUS COMPOUNDING	
$(F/P, \underline{r}\%, N)$	Continuous compounding: compound amount factor (single payment)
$(P/F, \underline{r}\%, N)$	Continuous compounding: present worth factor (single payment)

Functional Format (Symbol)	Name of Factor
$(A/F, \underline{r}\%, N)$	Continuous compounding: sinking fund factor
$(A/P, \underline{r}\%, N)$	Continuous compounding: capital recovery factor
$(F/A, \underline{r}\%, N)$	Continuous compounding: compound amount factor (uniform series)
$(P/A, \underline{r}\%, N)$	Continuous compounding: present worth factor (uniform series)

CONTINUOUS, UNIFORM CASH FLOWS: CONTINUOUS COMPOUNDING
(PAYMENTS DURING ONE PERIOD ONLY)

$(P/\overline{F}, \underline{r} \text{ or } i\%, N)$	Continuous compounding: present worth factor (single, continuous payment)
$(F/\overline{P}, \underline{r} \text{ or } i\%, N)$	Continuous compounding: compound amount factor (single, continuous payment)

CONTINUOUS, UNIFORM CASH FLOWS: CONTINUOUS COMPOUNDING
(PAYMENTS DURING A CONTINUOUS SERIES OF PERIODS)

$(\overline{A}/F, \underline{r} \text{ or } i\%, N)$	Continuous compounding: sinking fund factor (continuous, uniform payments)
$(\overline{A}/P, \underline{r} \text{ or } i\%, N)$	Continuous compounding: capital recovery factor (continuous, uniform payments)
$(F/\overline{A}, \underline{r} \text{ or } i\%, N)$	Continuous compounding: compound amount factor (continuous, uniform payments)
$(P/\overline{A}, \underline{r} \text{ or } i\%, N)$	Continuous compounding: present worth factor (continuous, uniform payments)

[a]From *American National Standard Publication ANSI Z94.5–1972*, published by the American Society of Mechanical Engineers, New York, except that use of $\underline{r}$ is added for continuous compounding.

TECHNICAL TERMS USED IN ENGINEERING ECONOMY†

Amortization—(a) (1) As applied to a capitalized asset, the distribution of the initial cost by periodic charges to operations as in depreciation. Most properly applies with indefinite life; (2) the reduction of a debt by either periodic or irregular payments; (b) a plan to pay off a financial obligation according to some prearranged program.

Annual Equivalent—(a) In *Time Value of Money* (q.v.), a uniform annual amount for a prescribed number of years that is equivalent in value to the present worth of any sequence of financial events for a given interest rate; (b) one of a sequence of equal end-of-year payments which would have the same financial effect when interest is considered as another payment or

†From *American National Standard Publication ANSI Z94.5–1972*, published by the American Society of Mechanical Engineers, New York.

sequence of payments which are not necessarily equal in amount or equally spaced in time.

Annuity—(a) An amount of money payable to a beneficiary at regular intervals for a prescribed period of time out of a fund reserved for that purpose; (b) a series of equal payments occurring at equal periods of time.

Annuity Factor—The function of interest rate and time that determines the amount of periodic annuity that may be paid out of a given fund.

Annuity Fund—A fund that is reserved for payment of annuities. The present worth of funds required to support future annuity payments.

Apportion—In accounting or budgeting, to assign a cost responsibility to a specific individual, organization unit, product, project, or order.

Average-Interest Method—A method of computing required return on investment based on the average book value of the asset during its life or during a specified study period.

Book Value—The recorded current value of an asset. First cost less accumulated depreciation, amortization, or depletion; (b) original cost of an asset less the accumulated depreciation; (c) the worth of a property as shown on the accounting records of a company. It is ordinarily taken to mean the original cost of the property less the amounts that have been charged as depreciation expense.

Breakeven Chart—A graphic representation of the relation between total income and total costs for various levels of production and sales indicating areas of profit and loss.

Breakeven Point (S)—(a) (1) In business operations, the rate of operations, output, or sales at which income is sufficient to equal operating cost, or operating cost plus additional obligations that may be specified; (2) the operating condition, such as output, at which two alternatives are equal in economy. (b) The percentage of capacity operation of a manufacturing plant at which income will just cover expenses.

Capacity Factor—(a) The ratio of average load to maximum capacity; (b) the ratio between average load and the total capacity of the apparatus, which is the optimum load; (c) the ratio of the average actual use to the available capacity.

Capital—(a) The financial resources involved in establishing and sustaining an enterprise or project (see *Investment* and *Working Capital*); (b) a term describing wealth which may be utilized to economic advantage. The form that this wealth takes may be as cash, land, equipment, patents, raw materials, finished product, etc.

Capital Recovery—(a) Charging periodically to operations amounts that will ultimately equal the amount of capital expenditure (see *Amortization, Depletion,* and *Depreciation*); (b) the replacement of the original cost of an asset plus interest; (c) the process of regaining the net investment in a project by means of revenue in excess of the costs from the project. (Usually implies amortization of principal plus interest on the diminishing unrecovered balance.)

Capital Recovery Factor—A factor used to calculate the sum of money required at the end of each of a series of periods to regain the net investment of a project plus the compounded interest on the unrecovered balance.

Capitalized Cost—(a) The present worth of a uniform series of periodic costs that continue for an indefinitely long time (hypothetically infinite). Not to be confused with a capitalized expenditure; (b) the value at the purchase date of the first life of the asset of all expenditures to be made in reference to this asset over an infinite period of time. This cost can also be regarded as the sum of capital which, if invested in a fund earning a stipulated interest rate, will be sufficient to provide for all payments required to maintain the asset in perpetual service.

Cash Flow—(a) The flowback of profit plus depreciation from a given project; (b) the actual dollars passing into and out of the treasury of a financial venture.

Common Costs—Costs which cannot be identified with a given output of products, operations, or services.

Compound Amount—The future worth of a sum invested (or loaned) at compound interest.

Compound Amount Factor—(a) The function of interest rate and time that determines the compound amount from a stated initial sum; (b) a factor which when multiplied by the single sum or uniform series of payments will give the future worth at compound interest of such single sum or series.

Compound Interest—(a) The type of interest that is periodically added to the amount of investment (or loan) so that subsequent interest is based on the cumulative amount; (b) the interest charges under the condition that interest is charged on any previous interest earned in any time period, as well as on the principal.

Compounding, Continuous—(a) A compound interest situation in which the compounding period is zero and the number of periods infinitely great. A mathematical concept that is practical for dealing with frequent compounding and small interest rates; (b) a mathematical procedure for evaluating compound interest factors based on a continuous interest function rather than discrete interest periods.

Compounding Period—The time interval between dates at which interest is paid and added to the amount of an investment or load. Designates frequency of compounding.

Decisions Under Certainty—Simple decisions that assume complete information and no uncertainty connected with the analysis of the decisions.

Decisions Under Risk—A decision problem in which the analyst elects to consider several possible futures, the probabilities of which can be estimated.

Decisions Under Uncertainty—A decision for which the analyst elects to consider several possible futures, the probabilities of which *cannot* be estimated.

Declining Balance Depreciation—Also known as *percent on diminishing value*. A method of computing depreciation in which the annual charge is a fixed percentage of the depreciated book value at the beginning of the year to which the depreciation applies.

Demand Factor—(a) The ratio of the maximum instantaneous production rate to the production rate for which the equipment was designed; (b) the ratio between the maximum power demand and the total connected load of the system.

Depletion—(a) A form of capital recovery applicable to extractive property (e.g., mines). Can be a unit-of-output basis the same as straight-line depreciation related to original or current appraisal of extent and value of deposit. (Known as *cost depletion*). Can also be a percentage of income received from extractions (known as *percentage depletion*). (b) A lessening of the value of an asset due to a decrease in the quantity available. It is similar to depreciation except that it refers to such natural resources as coal, oil, and timber in forests.

Depreciated Book Value—The first cost of the capitalized asset minus the accumulation of annual depreciation cost charges.

Depreciation—(a) (1) Decline in value of a capitalized asset; (2) a form of capital recovery applicable to a property with two or more years' life span, in which an appropriate portion of the asset's value is periodically charged to current operations. (b) The loss of value because of obsolescence or due to attrition. In accounting, depreciation is the allocation of this loss of value according to some plan.

Development Cost—The sum of all the costs incurred by an inventor or sponsor of a project up to the time that the project is accepted by those who will promote it.

Discounted Cash Flow—(a) The present worth of a sequence in time of sums of money when the sequence is considered as a flow of cash into and/or out of an economic unit; (b) an investment analysis which compares the present worth of projected receipts and disbursements occurring at designated future times in order to estimate the rate of return from the investment or project.

Earning Value—The present worth of an income producer's probable future net earnings, as prognosticated on the basis of recent and present expense and earnings and the business outlook.

Economic Return—The profit derived from a project or business enterprise without consideration of obligations to financial contributors and claims of others based on profit.

Economy—The cost or profit situation regarding a practical enterprise or project, as in *economy study, engineering economy, project economy*.

Effective Interest—The true value of interest rate computed by equations for compound interest for a 1-year period.

Endowment—A fund established for the support of some project or succession of donations or financial obligations.

Endowment Method—As applied to economy study, a comparison of alternatives based on the present worth of the anticipated financial events.

Engineering Economy—(a) The application of engineering or mathematical analysis and synthesis to economic decisions; (b) a body of knowledge and techniques concerned with the evaluation of the worth of commodities and services relative to their cost; (c) the economic analysis of engineering alternatives.

Estimate—The true magnitude as closely as it can be determined by the exercise of sound judgment based on approximate computations and is not to be confused with offhand approximations that are little better than outright guesses.

Expected Return—The profit anticipated from a venture.

Expected Yield—The ratio expected return/investment, usually expressed as a percentage on an annual basis.

First Cost—The initial cost of a capitalized property, including transportation, installation, preparation for service, and other related initial expenditures.

Future Worth—(a) The equivalent value at a designated future date based on *time value of money*; (b) the monetary sum, at a given future time, which is equivalent to one or more sums at given earlier times when interest is compounded at a given rate.

Going-Concern Value—The difference between the value of a property as it stands possessed of its going elements and the value of the property alone as it would stand at completion of construction as a bare or inert assembly of physical parts.

Good-Will Value—That element of value which inheres in the fixed and favorable consideration of customers arising from an established well-known and well-conducted business.

Incremental Cost—The additional cost that will be incurred as the result of increasing the output one more unit. Conversely, it can be defined as the cost that will not be incurred if the output is reduced one unit. More technically, it is the variation in output resulting from a unit change in input. It is known as the marginal cost.

In-Place Value—A value of a physical property—market value plus costs of transportation to site and installation.

Intangibles—(a) In economy studies, conditions, or economy factors that cannot be readily evaluated in *quantitative* terms as in money; (b) in accounting, the assets that cannot be reliably evaluated (e.g., *goodwill*).

Interest—(a) (1) financial share in a project or enterprise; (2) periodic compensation for the lending of money; (3) in economy study, synonymous with *required return*, expected profit, or *charge* for the use of capital. (b) The cost for the use of capital. Sometimes referred to as the *Time Value of Money (q.v.)*.

Interest Rate—The ratio of the interest payment to the principal for a given unit of time and is usually expressed as a percentage of the principal.

Interest Rate, Effective—An interest rate for a stated period (per year unless otherwise specified) that is the equivalent of a smaller rate of interest that is more frequently compounded.

Interest Rate, Nominal—The customary type of interest rate designation on an annual basis without consideration of compounding periods. The usual basis for computing periodic interest payments.

Investment—(a) As applied to an enterprise as a whole, the cost (or present value) if all the properties and funds necessary to establish and maintain the enterprise as a going concern. The *capital* tied up in the enterprise or project. (b) Any expenditure which has substantial and enduring value (at least two years' anticipated life) and which is therefore capitalized.

Investor's Method—(see *Discounted Cash Flow*)

Irreducibles—(a) A term that may be used for the class of intangible conditions or economy factors that can only be *qualitatively* appraised (e.g., ethical considerations); (b) matters that cannot readily be reduced to estimated money receipts and disbursements.

Life—(a) Economic: that period of time after which a machine or facility should be discarded or replaced because of its excessive costs or reduced profitability. The economic impairment may be absolute or relative. (b) Physical: that period of time after which a machine or facility can no longer be repaired in order to perform its design function properly. (c) Service: the period of time that a machine or facility will satisfactorily perform its function without major overhaul.

Load Factor—(a) A ratio that applies to physical plant or equipment: average loan/maximum demand, usually expressed as a percentage. Equivalent to percent of capacity operation if facilities just accommodate the maximum demand. (b) Is defined as the ratio of average load to maximum load.

MAPI Method—(a) A procedure for replacement analysis sponsored by the Machinery and Allied Products Institute. (b) A method of capital investment analysis which has been formulated by the Machinery and Allied Products Institute. This method uses a fixed format and provides charts and graphs to facilitate calculations. A prominent feature of this method is that it explicitly includes obsolescence.

Market Value—(see *Salvage Value*)

Marginal Analysis—An economic concept concerned with those elements of costs and revenue which are associated directly with a specific course of action, normally using available current costs and revenue as a base and usually independent of traditional accounting allocation procedures.

Marginal Cost—(a) The cost of one additional unit of production, activity, or service; (b) the rate of change of cost with production or output.

Matheson Formula—A title for the formula used for *Declining Balance Depreciation* (*q.v.*)

Nominal Interest—The number employed loosely to describe the annual interest rate.

Obsolescence—(a) The condition of being out of date. A loss of value occasioned by new developments which place the older property at a competitive disadvantage. A factor in depreciation. (b) A decrease in the value of an asset brought about by the development of new and more economical methods, processes, and/or machinery. (c) The loss of usefulness or worth of a product or facility as a result of the appearance of better and/or more economical products, methods, or facilities.

Payoff Period—(a) Regarding an investment, the number of years (or months) required for the related profit or savings in operating cost to equal the amount of said investment; (b) the period of time at which a machine, facility, or other investment has produced sufficient net revenue to recover its investment costs.

Perpetual Endowment—An endowment with hypothetically infinite life. (See *Capitalized Cost* and *Endowment*.)

Present Worth—(a) The equivalent value at the present, based on *time value of money*. (b) (1) The monetary sum which is equivalent to a future sum(s) when interest is compounded at a given rate. (2) The discounted value of future sums.

Present Worth Factor—(a) A mathematical expression also known as the present value of an annuity of one; (b) one of a set of mathematical formulas used to facilitate calculation of present worth in economic analyses involving compound interest.

Profitability Index—The rate of return in an economy study or investment decision when calculated by the *Discounted Cash Flow Method* or the *Investor's Method* (*q.v.*).

Promotion Cost—The sum of all the expenses found to be necessary to arrange for the financing and organizing of the business unit which will build and operate the project.

Rate of Return—(a) The interest rate at which the present worth of the cash flows on a project is zero; (b) the interest rate earned by an investment.

Replacement Policy—A set of decision rules (usually optimal) for the replacement of facilities that wear out, deteriorate, or fail over a period of time. Replacement models are generally concerned with weighing the increasing operating costs (and possibly decreasing revenues) associated with aging equipment against the net proceeds from alternative equipment.

Replacement Study—An economic analysis involving the comparison of an existing facility and a facility proposed to supplant the existing facility.

Required Return—The *minimum* return or profit necessary to justify an investment. Often termed *interest, expected* return or profit, or *charge for the use of capital*. It is the minimum acceptable percentage, no more and no less.

Required Yield—The ratio of required return over amount of investment, usually expressed as a percentage on an annual basis.

Retirement of Debt—The termination of a debt obligation by appropriate settlement with lender—understood to be in full amount unless partial settlement is specified.

Salvage Value—(a) The cost recovered or which could be recovered from a used property when removed, sold, or scrapped. A factor in appraisal of property value and in computing depreciation. (b) The market value of a machine or facility at any point in time. Normally, an estimate of an asset's net market value at the end of its estimated life.

Sensitivity—The relative magnitude of the change in one or more elements of an engineering economy problem that will reverse a decision among alternatives.

Simple Interest—(a) Interest that is not compounded—is not added to the income-producing investment or loan; (b) the interest charges under the condition that interest in any time is only charged on the principal.

Sinking Fund—(a) A fund accumulated by period deposits and reserved exclusively for a specific purpose, such as retirement of a debt or replacement of a property; (b) a fund created by making periodic deposits (usually equal) at compound interest in order to accumulate a given sum at a given future time for some specific purpose.

Sinking Fund Depreciation—(a) A method of computing depreciation in which the periodic amount is presumed to be deposited in a *sinking fund* that earns interest at a specified rate. Sinking fund may be real but is usually hypothetical. (b) A method of depreciation where a fixed sum of money is regularly deposited at compound interest in a real or imaginary fund in order to accumulate an amount equal to the total depreciation of an asset at the end of the asset's estimated life. The sinking fund depreciation in any year equals the sinking fund deposit plus interest in the sinking fund balance.

Sinking Fund Factor—(a) The function of interest rate and time that determines the cumulative amount of a sinking fund resulting from specified periodic deposits. Future worth per unit of uniform periodic amounts. (b) The mathematical formulas used to facilitate sinking fund calculations.

Straight Line Depreciation—Method of depreciation whereby the amount to be recovered (written off) is spread uniformly over the estimated life of the asset in terms of time periods or units of output. May be designated *percent of initial value*.

Study Period—In economy study, the length of time that is presumed to be covered in the schedule of events and appraisal of results. Often the anticipated life of the project under consideration, but a shorter time may be more appropriate for decision making.

Sum-of-Digits Method—Also known as sum-of-the-years'-digits method. A method of computing depreciation in which the amount for any year is based on the ratio: (years of remaining life)$/(1 + 2 + 3 + \cdots + N)$, being the total anticipated life.

Sunk Cost—(a) The unrecovered balance of an investment. It is a cost, already paid, that is not relevant to the decision concerning the future that is being made. Capital already invested that for some reason cannot be retrieved. (b) A past cost which has no relevance with respect to future receipts and disbursements of a facility undergoing an engineering economy study. This concept implies that since a past outlay is the same regardless of the alternative selected, it should not influence the choice between alternatives.

Tangibles—Things that can be *quantitatively* measured or valued, such as items of cost and physical assets.

Time Value of Money—(a) The cumulative effect of elapsed time on the money value of an event, based on the earning power of equivalent invested funds (see *Future Worth* and *Present Worth*). (b) The expected interest rate that capital should or will earn.

Traceable Costs—Costs which can be identified with a given product, operation, or service.

Valuation or Appraisal—The art of estimating the fair-exchange value of specific properties.

Working Capital—(a) That portion of investment represented by *current assets* (assets that are not capitalized) less the *current liabilities*. The capital necessary to sustain operations. (b) Those funds that are required to make the enterprise or project a going concern.

Yield—The ratio of return or profit over the associated investment, expressed as a percentage or decimal usually on an annual basis.

Selected References

American Telephone and Telegraph Company, Engineering Department, *Engineering Economy*, 3rd ed. New York: American Telephone and Telegraph Co., 1977.

Au, T. and T. P. Au, *Engineering Economics for Capital Investment Analysis*. Boston: Allyn and Bacon, Inc., 1983.

Barish, N. N., and S. Kaplan, *Economic Analysis for Engineering and Managerial Decision Making*. New York: McGraw-Hill Book Company, 1978.

Bierman, H., Jr., and S. Smidt, *The Capital Budgeting Decision*, 5th ed. New York: Macmillan Publishing Co., Inc., 1980.

Blank, L. T., and A. J. Tarquin, *Engineering Economy*, 2nd ed., New York: McGraw-Hill Book Company, 1983.

Bussey, Lynn E. *The Economic Analysis of Industrial Projects*. Englewood Cliffs, N.J.: Prentice-Hall, Inc., 1978.

Canada, J. R., and J. A. White, *Capital Investment Decision Analysis for Management and Engineering*. Englewood Cliffs, N.J.: Prentice-Hall, Inc., 1980.

Chemical Engineering. Published biweekly by McGraw-Hill, Inc., New York

Cochrane, J. L., and M. Zeleny, *Multiple Criteria Decision Making*. Columbia, S.C.: University of South Carolina, 1973.

Collier, C. A. and W. B. Ledbetter, *Engineering Cost Analysis*. New York: Harper and Row, Publishers, 1982.

Engineering Economist, The. A quarterly journal jointly published by the Engineering Economy Division of the American Society for Engineering Education and the Institute of Industrial Engineers. Published by IIE, Norcross, Ga.

Engineering News-Record. Published by McGraw-Hill, Inc., New York; refer to the most recent issue.

ENGLISH, J. M., ed. *Cost Effectiveness: Economic Evaluation of Engineering Systems.* New York: John Wiley & Sons, Inc., 1968.

FABRYCKY, W. J., and G. J. THUESEN *Economic Decision Analysis*, 2nd ed. Englewood Cliffs, N.J.: Prentice-Hall, Inc. 1980.

FLEISCHER, G. A., *Risk and Uncertainty: Non-deterministic Decision Making in Engineering Economy.* Publication EE-75-1, Norcross, Ga. Institute of Industrial Engineers, 1975.

GRANT, E. L., W. G. IRESON, and R. S. LEAVENWORTH, *Principles of Engineering Economy*, 7th ed. New York: John Wiley & Sons, Inc., 1982.

HAPPEL, J., and D. JORDAN, *Chemical Process Economics*, 2nd ed. New York: Marcel Dekker, Inc., 1975.

Harvard Business Review, Published bimonthly by the Harvard University Press, Boston.

Industrial Engineering. A monthly magazine published by the Institute of Industrial Engineers, Norcross, Ga.

JELEN, F. C., and J. H. BLACK, *Cost and Optimization Engineering*, 2nd ed. New York: McGraw-Hill Book Company, 1983.

JEYNES, P. H., *Profitability and Economic Choice.* Ames, Iowa: Iowa State University Press, 1968.

JONES, B. W., *Inflation in Engineering Economic Analysis.* New York: John Wiley & Sons, Inc., 1982.

KENNEY, R. L., and H. RAIFFA, *Decisions with Multiple Objectives: Preferences and Value Tradeoffs.* New York: John Wiley & Sons, Inc. 1976.

LASSER, J. K., *Your Income Tax.* New York: Simon and Schuster, Inc. (see latest edition).

Machinery and Allied Products Institute, *MAPI Replacement Manual.* Washington, D.C.: Machinery and Allied Products Institute, 1950.

MALLIK, A. K., *Engineering Economy with Computer Applications.* Mahomet, Ill.: Engineering Technology, Inc., 1979.

MAO, JAMES, *Quantitative Analysis of Financial Decisions.* New York: Macmillan Publishing Co., Inc., 1969.

MAYER, R. R., *Capital Expenditure Analysis for Managers and Engineers.* Prospect Heights, Ill.: Waveland Press, Inc., 1978.

MERRETT, A. J., and A. SYKES. *The Finance and Analysis of Capital Projects.* New York: John Wiley & Sons, Inc., 1963.

MISHAN, E. J., *Cost–Benefit Analysis.* New York: Praeger Publishers, Inc., 1976.

MORRIS, W. T., *Engineering Economic Analysis.* Reston, Va.: Reston Publishing Company, Inc., 1976.

NEWNAN, DONALD G., *Engineering Economic Analysis*, rev. ed. San Jose, Calif.: Engineering Press, 1980.

OAKFORD, R. V., *Capital Budgeting: A Quantitative Evaluation of Investment Alternatives.* New York: John Wiley & Sons, Inc. 1970.

OSO, J. B. "The proper role of the tax-adjusted cost of capital in present value studies," *The Engineering Economist*, Vol. 24, No. 1 (Fall 1978), pp. 1–12.

OSTWALD, P. F., *Cost Estimating for Engineering and Management.* Englewood Cliffs, N.J.: Prentice-Hall, Inc., 1974.

PETERS, M. S., and K. D. TIMMERHAUS, *Plant Design and Economics for Chemical Engineers*, 2nd ed. New York: McGraw-Hill Book Company, 1968.

REISMAN, A., *Managerial and Engineering Economics*. Boston: Allyn and Bacon, Inc., 1971.

RIGGS, J. L., *Engineering Economics*. New York: McGraw-Hill Book Company, 1977.

ROSE, L. M., *Engineering Investment Decisions: Planning Under Uncertainty*. Amsterdam: Elsevier Scientific Publishing Co., 1976.

SMITH, G. W., *Engineering Economy: The Analysis of Capital Expenditures*, 2nd ed. Ames, Iowa: The Iowa State University Press, 1973.

STEVENS, G. T., *Economic and Financial Analysis of Capital Investment*. New York: John Wiley & Sons, Inc., 1979.

STUART, R. D., *Cost Estimating*. New York: John Wiley & Sons, Inc., 1982.

SULLIVAN, W. G., and W. W. CLAYCOMBE *Fundamentals of Forecasting*. Reston, Va: Reston Publishing Company, Inc., 1977.

TAYLOR, G. A., *Managerial and Engineering Economy*, 3rd ed. New York: Van Nostrand Reinhold Company, 1980.

TERBORGH, GEORGE, *Business Investment Management*. Washington, D.C.: Machinery and Allied Products Institute, 1967.

THUESEN, H. G., W. J. FABRYCKY, and G. J. THUESEN, *Engineering Economy*, 5th ed. Englewood Cliffs, N.J.: Prentice-Hall, Inc., 1977.

VANHORNE, J. C., *Financial Management and Policy*, 5th ed. Englewood Cliffs, N.J.: Prentice-Hall, Inc., 1980.

WARD, T. L., and W. G. SULLIVAN, "Equivalence of the present worth and revenue requirements methods of capital investment analysis," *AIIE Transactions*, Vol. 13, No.1 (March 1981), pp. 29–40.

WEINGARTNER, H. M., *Mathematical Programming and the Analysis of Capital Budgeting Problems*. Englewood Cliffs, N.J.: Prentice-Hall, Inc., 1975.

WHITE, J. A., M. H. AGEE, and K. E. CASE, *Principles of Engineering Economic Analysis*. New York: John Wiley & Sons, Inc., 1977.

WOODS, D. R., *Financial Decision Making in the Process Industry*. Englewood Cliffs, N.J.: Prentice-Hall, Inc., 1975.

Computer Program for Use in Engineering Economy Studies

GENERAL DESCRIPTION

This appendix presents and demonstrates a computer program, written in the BASIC programming language and suitable for personal computers, that has wide applicability to many problems and homework exercises in this book. The program is written in a manner that permits users to "talk" with the computer; that is, it is a conversational computer program. This program is a powerful computational aid that offers these capabilities:

1. A choice of any one or all of six methods for computing a measure of merit in an engineering economy study. These methods are identified in the program listing.
2. A set of instructions to the user that can be utilized at the discretion of the analyst. Instructions comprise statements 40 through 460 of the program listing.
3. A choice of discrete or continuous compounding of a periodic rate of interest (e.g., an effective or nominal interest rate per year).
4. The ability to conduct sensitivity studies and correct data entry errors with a minimum of effort.

5. The calculation of various measures of merit at negative rates of interest that sometimes arise in inflation-related problems.
6. The ability to deal with negative time periods that occur when capital investment costs are spread over several years. Measures of merit are computed relative to end-of-year 0, which is usually defined as the start of commercial operation of the venture.

PROGRAM LISTING*

```
20     DIM C(100)
30     REM : ************* INPUT DATA *************
40     PRINT "DO YOU WANT TO SEE INSTRUCTIONS FOR DATA
       INPUT (TYPE YES OR NO)";
50     INPUT Q1$
60     IF Q1$="NO" THEN 480
70     PRINT " INSTRUCTIONS FOR ENTERING CASH FLOW
       DATA"
75     PRINT
80     PRINT " THIS PROGRAM GIVES THE USER THE OPTION
       OF CALCULATING ANY OF"
90     PRINT "THE FOLLOWING MEASURES OF MERIT :
       INTERNAL RATE OF RETURN ,"
100    PRINT "PRESENT WORTH, FUTURE WORTH, ANNUAL
       WORTH, SIMPLE PAYBACK, OR"
110    PRINT "DISCOUNTED PAYBACK PERIOD."
120    PRINT " END OF PERIOD CASH FLOWS ARE ENTERED IN
       THE FOLLOWING MANNER :"
121    PRINT "    C,J1,J2 "
130    PRINT " WHERE C = CASH FLOW FOR EACH YEAR FROM
       THE YEAR J1 THROUGH J2."
140    PRINT " ALL CALCULATIONS ARE BASED ON YEAR 0
       BEING THE PRESENT PERIOD"
150    PRINT "(IF NEGATIVE YEARS ARE ENTERED, THE
       PRESENT WORTH IS BASED ON"
160    PRINT "PERIOD 0 THAT WAS ENTERED BY THE USER.)"
170    PRINT " THE NUMBER OF DIFFERENT CASH FLOW VALUES
       ASKED FOR IN THE"
180    PRINT "PROGRAM IS THE NUMBER OF DIFFERENT CASH
       FLOW SEQUENCES THAT YOU WILL"
190    PRINT "ENTER, FOR EXAMPLE :"
200    PRINT "    CASH FLOW    PERIOD"
210    PRINT "      10000         0"
220    PRINT "        0           1"
230    PRINT "      10000         2"
```

*Each statement comprises one line of the BASIC program (it may appear here as two lines due to narrow margins in the book).

```
240    PRINT "          10000          3"
250    PRINT "          10000          4"
260    PRINT "FOR THE ABOVE CASH FLOWS, THE NUMBER OF
       DIFFERENT CASH FLOW SEQUENCES"
270    PRINT "IS 3 AND THE CASH FLOWS SHOULD BE ENTERED
       AS FOLLOWS :"
280    PRINT "10000,0,0"
290    PRINT " 0,1,1"
300    PRINT "10000,2,4"
310    PRINT " ALL CASH FLOWS THAT OCCUR AT THE END OF
       A PARTICULAR TIME"
320    PRINT "PERIOD MUST BE ENTERED AS A NET AMOUNT,
       FOR EXAMPLE, A -10000"
330    PRINT "AT THE END OF YEAR 2 AND A +2000 AT THE
       END OF YEAR 2 WOULD BE"
340    PRINT "ENTERED AS -8000,2,2 ,"
350    PRINT " A NET CASH FLOW AMOUNT IS ENTERED IN
       THIS MANNER FOR EACH TIME"
360    PRINT "PERIOD IN THE LIFE OF THE ALTERNATIVE
       BEING CONSIDERED,"
370    PRINT "E.G. A PROJECT HAVING A $10000
       EXPENDITURE AT END OF YEAR 0"
380    PRINT "          FOLLOWED BY SAVINGS OF $3000/YEAR AT
       END OF YEAR 1 TO 6 AND"
390    PRINT "          A FINAL SALVAGE VALUE OF $2000 AT
       END OF YEAR 6 WOULD BE "
400    PRINT "          ENTERED AS FOLLOWS :"
410    PRINT "-10000,0,0"
420    PRINT "   3000,1,5"
430    PRINT "   5000,6,6"
440    PRINT " THE USER IS LIMITED TO END-OF-PERIOD
       CASH FLOW CONVENTION BUT"
450    PRINT "HAS THE OPTION OF SELECTING DISCRETE OR
       CONTINUOUS COMPOUNDING,"
460    PRINT "MULTIPLE INTERNAL RATES OF RETURN CANNOT
       BE IDENTIFIED,"
461    PRINT
462    PRINT
463    PRINT
464    PRINT
470    PRINT "PLEASE ANSWER THE FOLLOWING QUESTIONS"
480    PRINT "HOW MANY DIFFERENT CASH FLOW VALUES ARE
       THERE?"
490    INPUT W
500    PRINT "ENTER CASH FLOW, FIRST PERIOD, LAST
       PERIOD"
510    FOR I = 1 TO W
```

```
520     INPUT C1,J1,J2
530     IF I = 1 THEN 1730
540     FOR Z = J1 TO J2
550     C(Z-S) = C1
560     NEXT Z
570     NEXT I
580     PRINT
590     PRINT " YEAR     CASH FLOW"
600     FOR I = S TO J2
610     I8 = 0
620     PRINT USING 630,I,C(I-S)
630     :####     ####,###.##
640     NEXT I
650     PRINT "WANT TO MAKE ANY CORRECTIONS OR ADDITIONS
        (Y,N)";
660     INPUT A$
670     IF A$ = "N" THEN 740
680     IF A$ = "NO" THEN 740
690     PRINT "ENTER YEAR (FOR CORRECTION OR ADDITION) ";
700     INPUT K
710     PRINT "ENTER CASH FLOW FOR ABOVE YEAR";
720     INPUT C(K-S)
730     GO TO 650
740     REM : ******** ASK FOR TRANSACTION TYPE ********
750     PRINT "WHICH METHOD WOULD YOU LIKE TO USE? (TYPE
        PW, FW, AW, IRR, SPP, DPP, TABLE,? OR END)"
760     INPUT A$
770     IF A$ <> "?" THEN 870
780     PRINT " PRESENT WORTH =============> (PW)"
790     PRINT " FUTURE WORTH ==============> (FW)"
800     PRINT " ANNUAL WORTH ==============> (AW)"
810     PRINT " INTERNAL RATE OF RETURN ===> (IRR)"
820     PRINT " SIMPLE PAYBACK PERIOD =====> (SPP)"
830     PRINT " DISCOUNTED PAYBACK PERIOD => (DPP)"
840     PRINT " SEE THE TABLE AGAIN =======> (TABLE)"
850     PRINT " END THE PROGRAM ===========> (END)"
860     INPUT A$
870     IF A$ = "PW" THEN 980
880     IF A$ = "FW" THEN 980
890     IF A$ = "AW" THEN 980
900     IF A$ = "DPP" THEN 980
910     IF A$ = "IRR" THEN 1100
920     IF A$ = "SPP" THEN 1480
930     IF A$ = "END" THEN 2030
940     IF A$ = "TABLE" THEN 580
950     PRINT "PLEASE RETYPE"
960     GO TO 860
```

```
970    REM : ********* PWORTH *********
980    PRINT "ENTER INTEREST RATE AS DECIMAL , FOR
       EXAMPLE ENTER 10% AS .1)"
990    INPUT I3
1000   PRINT "DO YOU WANT CONTINUOUS (CON) OR DISCRETE
       (DIS) COMPOUNDING";
1010   INPUT B$
1020   IF B$ = "CON" THEN 1750
1030   IF A$ = "DPP" THEN 1590
1040   GOSUB 1680
1050   IF A$ = "FW" THEN 1770
1060   IF A$ = "AW" THEN 1790
1070   PRINT "=========> ";A$;" = ";P
1080   GO TO 1830
1090   REM : ********* RATE *********
1100   E = 0.001
1110   E = E/100
1120   I1 = 0
1130   I3 = 0
1140   I2 = 1
1150   GO SUB 1680
1160   IF P <> 0 THEN 1200
1170   A = 0
1180   B = 0
1190   GO TO 1460
1200   D = 0.1
1210   Z1 = ABS (P-C(0))
1220   IF ABS (C(0)) < Z1 THEN 1250
1230   PRINT "NO INTEREST RATE CAN BE FOUND "
1240   GO TO 1830
1250   P1 = P
1260   FOR I3 = I1 TO I2 STEP D
1270   GO SUB 1680
1280   V = SGN(P1)*SGN(P)
1290   IF V < 0 THEN 1320
1300   P1 = P
1310   NEXT I3
1320   IF D<= E THEN 1370
1330   I1 = I3-D
1340   I2 = I3
1350   D = D/10
1360   GO TO 1260
1370   PRINT "DO YOU WANT DISCRETE (DIS) OR CONTINUOUS
       (CON) INTEREST RATE"
1380   INPUT A$
1390   A = I3-D
1400   B = I3
1410   IF A$ = "DIS" THEN 1440
```

```
1420  A = LN (1+A)
1430  B = LN (1+B)
1440  A = A*100
1450  B = B*100
1460  PRINT "=========> IRR IS BETWEEN ";A;" AND
      ";B;"%"
1470  GO TO 1830
1480  REM : ********* SIMPLE PAYBACK PERIOD *********
1490  P = 0
1500  FOR I = S TO J2
1510  P = C(I-S) + P
1520  IF P >= 0 THEN 1560
1530  NEXT I
1540  PRINT "=========> SIMPLE PAYBACK PERIOD > YEAR
      (I.E.NEVER) ";I
1550  GO TO 1830
1560  PRINT "=========> SIMPLE PAYBACK PERIOD = YEAR
      ";I
1570  GO TO 1830
1580  REM : ******* DISCOUNTED PAYBACK PERIOD *******
1590  P = 0
1600  FOR I = S TO J2
1610  P = P + C(I-S)/(1+I3)**(I)
1620  IF P >= 0 THEN 1660
1630  NEXT I
1640  PRINT "=========> DISCOUNTED PAYBACK PERIOD >
      YEAR ";I
1650  GO TO 1830
1660  PRINT "=========> DISCOUNTED PAYBACK PERIOD =
      YEAR ";I
1670  GO TO 1830
1680  P = 0
1690  FOR I = S TO J2
1700  P = P+C(I-S)/((I3+1)**I)
1710  NEXT I
1720  RETURN
1730  S = J1
1740  GO TO 540
1750  I3 = EXP(I3)-1
1760  GO TO 1030
1770  P = P * (1+I3)**J2
1780  GO TO 1060
1790  IF S=0 GO TO 1810
1800  P = P*(1+I)**(0-S)
1810  P = (P*I3*(I3+1)**(J2-S))/(((I3+1)**(J2-S))-1)
1820  GO TO 1070
1830  PRINT "_____"
1840  PRINT "WANT ANOTHER RUN ? (Y,N) ";
```

```
1850  INPUT A$
1860  IF A$ = "N" THEN 2030
1870  PRINT "TYPE NEW,OLD OR CHANGE (TYPE '?' FOR
      EXPLANATION)";
1880  INPUT A$
1890  IF A$ <> "?" THEN 1940
1900  PRINT "TO ENTER COMPLETE NEW CASH FLOW TYPE
      ========> NEW"
1910  PRINT "TO KEEP OLD CASH FLOWS WITH NO CHANGE
      TYPE ==> OLD"
1920  PRINT "TO MAKE SOME CHANGES ON OLD CASH FLOWS
      TYPE => CHANGE"
1930  INPUT A$
1940  IF A$ = "NEW" THEN 480
1950  IF A$ = "OLD" THEN 750
1960  IF A$ = "CHANGE" THEN 1990
1970  PRINT "PLEASE RETYPE"
1980  GO TO 1930
1990  PRINT "WOULD YOU LIKE TO SEE THE OLD CASH FLOW
      TABLE (YES OR NO)";
2000  INPUT A$
2010  IF A$ = "NO" THEN 690
2020  GO TO 580
2030  END
```

EXAMPLE PROBLEMS

Several examples from Chapters 4 and 5 are presented to illustrate the versatility of the computer program. It should be noted that after-tax cash flows of Chapter 10 and 11 could also be reduced to a desired measure of merit with this program.

Example 5-2 With Three Different Cash Flow Values

```
HOW MANY DIFFERENT CASH FLOW VALUES
 ?3
ENTER CASH FLOW, FIRST PERIOD, LAST PERIOD
 ?-25000,0,0
 ?8000,1,4
 ?13000,5,5

YEAR    CASH FLOW
   0    -25,000.00
   1      8,000.00
   2      8,000.00
   3      8,000.00
   4      8,000.00
   5     13,000.00
```

```
WANT TO MAKE ANY CORRECTIONS OR ADDITIONS (Y,N)
?N
WHAT WOULD YOU LIKE TO DO (TYPE
PW,FW,AW,IRR,SPP,DPP,TABLE,? OR END)
 ?PW
ENTER INTEREST RATE AS DECIMAL , FOR EXAMPLE
ENTER 10% AS .1)
 ?.2
DO YOU WANT CONTINUOUS (CON) OR DISCRETE (DIS)
COMPOUNDING ?DIS
=========> PW = 934.285
```

Example 5-2 With Cash Flow Values for Each Individual Year

```
HOW MANY DIFFERENT CASH FLOW VALUES
 ?6
ENTER CASH FLOW, FIRST PERIOD, LAST PERIOD
 ?-25000,0,0
 ?8000,1,1
 ?8000,2,2
 ?8000,3,3
 ?8000,4,4
 ?13000,5,5

YEAR      CASH FLOW
  0      -25,000.00
  1        8,000.00
  2        8,000.00
  3        8,000.00
  4        8,000.00
  5       13,000.00

WANT TO MAKE ANY CORRECTIONS OR ADDITIONS (Y,N)
 ?N
WHAT WOULD YOU LIKE TO DO (TYPE
PW,FW,AW,IRR,SPP,DPP,TABLE,? OR END)
 ?PW
ENTER INTEREST RATE AS DECIMAL , FOR EXAMPLE
ENTER 10% AS .1)
 ?.2
DO YOU WANT CONTINUOUS (CON) OR DISCRETE (DIS)
COMPOUNDING ?DIS
=========> PW = 934.285
```

Example 5-2 With Continuous Compounding

```
--------------------------------------------------
WANT ANOTHER RUN ? (Y,N) ?Y
TYPE NEW,OLD OR CHANGE (TYPE '?' FOR
 EXPLANATION) ?OLD
WHAT WOULD YOU LIKE TO DO (TYPE
 PW,FW,AW,IRR,SPP,DPP,TABLE,? OR END)
 ?PW
ENTER INTEREST RATE AS DECIMAL , FOR EXAMPLE
 ENTER 10% AS ,1)
 ?,2
DO YOU WANT CONTINUOUS (CON) OR DISCRETE (DIS)
 COMPOUNDING ?CON
==========> PW = -320,036
```

Example 5-4, 5-6, and 5-8 (In Order)

```
--------------------------------------------------
WHAT ANOTHER RUN ? (Y,N) ?Y
TYPE NEW,OLD OR CHANGE (TYPE '?' FOR
 EXPLANATION) ??
TO ENTER COMPLETE NEW CASH FLOW TYPE ========>
 NEW
TO KEEP OLD CASH FLOWS WITH NO CHANGE TYPE ==>
 OLD
TO MAKE SOME CHANGES ON OLD CASH FLOWS TYPE =>
 CHANGE
 ?OLD
WHAT WOULD YOU LIKE TO DO (TYPE
 PW,FW,AW,IRR,SPP,DPP,TABLE,? OR END)
 ?AW
ENTER INTEREST RATE AS DECIMAL , FOR EXAMPLE
 ENTER 10% AS ,1)
 ?,2
DO YOU WANT CONTINUOUS (CON) OR DISCRETE (DIS)
 COMPOUNDING ?DIS
==========> AW = 312,406
--------------------------------------------------
WANT ANOTHER RUN ? (Y,N) ?Y
TYPE NEW,OLD OR CHANGE (TYPE '?' FOR
 EXPLANATION) ?OLD
WHAT WOULD YOU LIKE TO DO (TYPE
 PW,FW,AW,IRR,SPP,DPP,TABLE,? OR END)
 ?FW
```

```
ENTER INTEREST RATE AS DECIMAL , FOR EXAMPLE
 ENTER 10% AS .1)
 ?.2
DO YOU WANT CONTINUOUS (CON) OR DISCRETE (DIS)
 COMPOUNDING ?DIS
==========> FW = 2324.8
--------------------------------------------------
WANT ANOTHER RUN ? (Y,N) ?Y
TYPE NEW,OLD OR CHANGE (TYPE '?' FOR
 EXPLANATION) ?OLD
WHAT WOULD YOU LIKE TO DO (TYPE
 PW,FW,AW,IRR,SPP,DPP,TABLE,? OR END)
 ?IRR
DO YOU WANT DISCRETE (DIS) OR CONTINUOUS (CON)
 INTEREST RATE
 ?DIS
==========> IRR IS BETWEEN 21.577 AND 21.578 %
```

Example 4-12 To Illustrate a Negative Interest Rate

```
--------------------------------------------------
WANT ANOTHER RUN ? (Y,N) ?Y
TYPE NEW,OLD OR CHANGE (TYPE '?' FOR
 EXPLANATION) ?EXAMPLE 4-12
PLEASE RETYPE
 ?NEW
HOW MANY DIFFERENT CASH FLOW VALUES
 ?1
ENTER CASH FLOW, FIRST PERIOD, LAST PERIOD
 ?600,1,15

YEAR      CASH FLOW
  1         600.00
  2         600.00
  3         600.00
  .            .
  .            .
  .            .
 14         600.00
 15         600.00

WANT TO MAKE ANY CORRECTIONS OR ADDITIONS (Y,N)
 ?N
WHAT WOULD YOU LIKE TO DO (TYPE
 PW,FW,AW,IRR,SPP,DPP,TABLE,? OR END)
 ?PW
```

```
ENTER INTEREST RATE AS DECIMAL , FOR EXAMPLE
 ENTER 10% AS .1)
 ?-0.0175
DO YOU WANT CONTINUOUS (CON) OR DISCRETE (DIS)
 COMPOUNDING ?DIS
=========> PW = 10395.5
```

Example 5-19 Industrial Case Study

```
---------------------------------------------------
WANT ANOTHER RUN ? (Y,N) ?Y
TYPE NEW,OLD OR CHANGE (TYPE '?' FOR
 EXPLANATION) ?NEW
HOW MANY DIFFERENT CASH FLOW VALUES
 ?17
ENTER CASH FLOW, FIRST PERIOD, LAST PERIOD
 ?-40000,0,0
 ?-522100,1,1
 ?-777000,2,2
 ?-46400,3,3
 ?373600,4,4
 ?385800,5,5
 ?265800,6,6
 ?385800,7,7
 ?385800,8,8
 ?385800,9,9
 ?385800,10,10
 ?265800,11,11
 ?385800,12,12
 ?385800,13,13
 ?385800,14,14
 ?385800,15,15
 ?471200,16,16

YEAR        CASH FLOW
  0         -40,000.00
  1        -522,100.00
  2        -777,000.00
  3         -46,400.00
  4         373,600.00
  5         385,800.00
  6         265,800.00
  7         385,800.00
  8         385,800.00
  9         385,800.00
 10         385,800.00
 11         265,800.00
```

```
     12          385,800.00
     13          385,800.00
     14          385,800.00
     15          385,000.00
     16          471,200.00
WANT TO MAKE ANY CORRECTIONS OR ADDITIONS (Y,N)
 ?N
WHAT WOULD YOU LIKE TO DO (TYPE
 PW,FW,AW,IRR,SPP,DPP,TABLE,? OR END)
 ?IRR
DO YOU WANT DISCRETE (DIS) OR CONTINUOUS (CON)
 INTEREST RATE
 ?DIS
==========> IRR IS BETWEEN 18.342 AND 18.343 %
--------------------------------------------------
WANT ANOTHER RUN ? (Y,N) ?Y
TYPE NEW,OLD OR CHANGE (TYPE '?' FOR
 EXPLANATION) ?OLD
WHAT WOULD YOU LIKE TO DO (TYPE
 PW,FW,AW,IRR,SPP,DPP,TABLE,? OR END)
 ?PW
ENTER INTEREST RATE AS DECIMAL , FOR EXAMPLE
 ENTER 10% AS .1)
 ?.2
DO YOU WANT CONTINUOUS (CON) OR DISCRETE (DIS)
 COMPOUNDING ?DIS
==========> PW = -87137.3
--------------------------------------------------
```

Example 5-19 to Illustrate Sensitivity Analysis: What is the Effect of Not Recovering Working Capital of $85,400 at the End of the Project?

```
--------------------------------------------------
WANT ANOTHER RUN ? (Y,N) ?Y
TYPE NEW,OLD OR CHANGE (TYPE '?' FOR
 EXPLANATION) ?CHANGE
WOULD YOU LIKE TO SEE THE OLD CASH FLOW TABLE
 (YES OR NO) ?NO
ENTER YEAR (FOR CORRECTION OR ADDITION) ?16
ENTER CASH FLOW FOR ABOVE YEAR ?385800
WANT TO MAKE ANY CORRECTIONS OR ADDITIONS (Y,N)
 ?N
WHAT WOULD YOU LIKE TO DO (TYPE
 PW,FW,AW,IRR,SPP,DPP,TABLE,? OR END)
 ?IRR
```

```
DO YOU WANT DISCRETE (DIS) OR CONTINUOUS (CON)
 INTEREST RATE
 ?DIS
==========> IRR IS BETWEEN 18.24 AND 18.241 %
--------------------------------------------------
WANT ANOTHER RUN ? (Y,N) ?YES
TYPE NEW,OLD OR CHANGE (TYPE '?' FOR
 EXPLANATION) ?OLD
WHAT WOULD YOU LIKE TO DO (TYPE
 PW,FW,AW,IRR,SPP,DPP,TABLE,? OR END)
 ?PW
ENTER INTEREST RATE AS DECIMAL , FOR EXAMPLE
 ENTER 10% AS .1)
 ?.2
DO YOU WANT CONTINUOUS (CON) OR DISCRETE (DIS)
 COMPOUNDING ?DIS
==========> PW = -91756.4
--------------------------------------------------
```

The MAPI Method for Replacement and General Investment Analysis†

This appendix is intended to provide a brief summary of the main provisions, strengths, and weaknesses of the current (third) version of the MAPI method and to illustrate its use for a typical project analysis. The MAPI method or system for investment analysis has been evolved through several major works over the past 20 years by George Terborgh, Research Director of the Machinery and Allied Products Institute. The first version, which applied only to replacement problems, was described in *Dynamic Equipment Policy* and the *MAPI Replacement Manual*, both published in 1950. The second, and more-general-purpose MAPI system was published in *Business Investment Policy* in 1958. The current (third) version was summarized in 1967 in *Business Investment Management*. All were published by the Machinery and Allied Products Institute, Washington, D.C.

Basically, the MAPI method provides a series of charts and forms to facilitate investment analysis computation. One of its main features is the inclusion of provision for consideration of obsolescence and deterioration, assumed to affect operating results as a linear function of time. The MAPI charts provide for ease in determining the percentage retention value and are computed for various

---

†This discussion of the MAPI method was prepared primarily by Jack Turvaville, Professor of Industrial Engineering, Tennessee Technological University, and is used with his permission. It is based largely on pp. 149–158 of *Business Investment Management*, by George Terborgh, Machinery and Allied Products Institute, 1967, and reproduced by permission of the publisher.

service lives, salvage values, and tax write-off methods. They are available for straight line, double declining balance, sum-of-the-years'-digits, and current expense depreciation methods for either a 1-year or longer-than-1-year comparison period—a total of eight charts. Figures D-1 and D-2 illustrate two such charts as given for a wide range of service lives and salvage value ratios. Assumptions on which the charts are based include 50% income tax rate, a 25%–75% debt-to-equity capital ratio, an average debt capital interest rate of 3% (after-tax), and an after-tax equity return of 10%.

It should be recognized that allowance for deterioration and obsolescence is built into the chart retention values. According to Terborgh, the computation

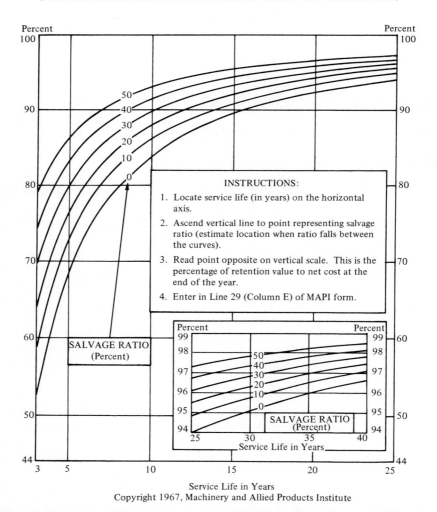

MAPI CHART No. 1A
(ONE-YEAR COMPARISON PERIOD AND SUM-OF-DIGITS TAX DEPRECIATION)

Copyright 1967, Machinery and Allied Products Institute

FIGURE D-1 Example of a Machinery and Allied Products Institute (MAPI) Chart.

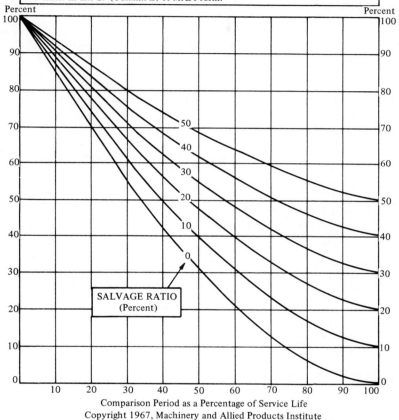

MAPI CHART No. 1B

(LONGER THAN ONE-YEAR COMPARISON PERIODS
AND SUM-OF-DIGITS TAX DEPRECIATION)

INSTRUCTIONS:
1. Locate on horizontal axis percentage which comparison period is of service life.
2. Ascend vertical line to point representing salvage ratio (estimate location when ratio falls between the curves).
3. Read point opposite on vertical scale. This is the percentage of retention value to net cost at end of comparison period.
4. Enter in line 29 (Column E) of MAPI form.

SALVAGE RATIO
(Percent)

Comparison Period as a Percentage of Service Life

Copyright 1967, Machinery and Allied Products Institute

FIGURE D-2   Another Example of a MAPI Chart.

for the retention values projects a stream of pretax earnings over the estimated service life that conforms in shape to the projection pattern and in size to the following requirements: (1) It includes the income tax, payable at the specified rate, on the excess of the earnings over the deductions provided by tax depreciation and interest; (2) the remainder of the after-tax earnings suffices (with terminal salvage, if any) to permit a full recovery of the investment over the service life and to provide a return throughout on the unrecovered balance of equity at the prescribed after-tax return rate and on the unrecovered balance of debt at the prescribed interest rate.

MAPI SUMMARY FORM
(AVERAGING SHORTCUT)

PROJECT_____

ALTERNATIVE_____

COMPARISON PERIOD (YEARS)   (P)   1

ASSUMED OPERATING RATE OF PROJECT (HOURS PER YEAR)   1,200

I. OPERATING ADVANTAGE
(<u>NEXT-YEAR</u> FOR A 1-YEAR COMPARISON PERIOD,* ANNUAL AVERAGES FOR LONGER PERIODS)

A. EFFECT OF PROJECT ON REVENUE

		INCREASE	DECREASE	
1	FROM CHANGE IN QUALITY OF PRODUCTS	$	$	1
2	FROM CHANGE IN VOLUME OF OUTPUT			2
3	TOTAL	$ X	$ Y	3

B. EFFECT ON OPERATING COSTS

		INCREASE	DECREASE	
4	DIRECT LABOR	$ 900	$	4
5	INDIRECT LABOR	150		5
6	FRINGE BENEFITS	190		6
7	MAINTENANCE	200		7
8	TOOLING	80		8
9	MATERIALS AND SUPPLIES		16,800	9
10	INSPECTION			10
11	ASSEMBLY			11
12	SCRAP AND REWORK			12
13	DOWN TIME			13
14	POWER	40		14
15	FLOOR SPACE		1,000	15
16	PROPERTY TAXES AND INSURANCE	320		16
17	SUBCONTRACTING			17
18	INVENTORY		1,100	18
19	SAFETY			19
20	FLEXIBILITY			20
21	OTHER			21
22	TOTAL	$ 1,880 Y	$ 18,900 X	22

C. COMBINED EFFECT

23	NET INCREASE IN REVENUE (3X−3Y)	$	23
24	NET DECREASE IN OPERATING COSTS (22X−22Y)	$ 17,020	24
25	ANNUAL OPERATING ADVANTAGE (23 + 24)	$ 17,020	25

*Next year means the first year of project operation. For projects with a significant break-in period, use performance after break-in.

FIGURE D-3   MAPI Summary Form, Sheet 1 (with Entries for Example Project).

The use of the MAPI method is consummated through a standard form as shown in Figures D-3 and D-4 with example amounts entered therein. The example amounts are based on the following example project taken from p. 156 of Terborgh's *Business Investment Management.*†

An analyst desires to investigate whether it would be more economical for the company to make its own corrugated containers. He finds that to do this the company will have to purchase a large box machine and a box stitcher at a combined cost of $29,800.

†Published by Machinery and Allied Products Institute, 1967.

II. INVESTMENT AND RETURN

A. INITIAL INVESTMENT

26	INSTALLED COST OF PROJECT	$ 29,000			
	MINUS INITIAL TAX BENEFIT OF	$ 2,100	(Net Cost)	$ 27,700	26
27	INVESTMENT IN ALTERNATIVE				
	CAPITAL ADDITIONS MINUS INITIAL TAX BENEFIT	$			
	PLUS: DISPOSAL VALUE OF ASSETS RETIRED				
	BY PROJECT*	$ 4,000		$ 4,000	27
28	INITIAL NET INVESTMENT (26—27)			23,700	28

B. TERMINAL INVESTMENT

29 RETENTION VALUE OF PROJECT AT END OF COMPARISON PERIOD
(ESTIMATE FOR ASSETS, IF ANY, THAT CANNOT BE DEPRECIATED OR EXPENSED, FOR OTHERS, ESTIMATE OR USE MAPI CHARTS.)

Item or Group	Installed Cost, Minus Initial Tax Benefit (Net Cost) A	Service Life (Years) B	Disposal Value, End of Life (Percent of Net Cost) C	MAPI Chart Number D	Chart Percent-age E	Retention Value $\left(\dfrac{A \times E}{100}\right)$ F
Box Machine and Stitcher	$ 27,700	13	10	1A	89.4	$ 24,760

	ESTIMATED FROM CHARTS (TOTAL OF COL. F)	$ 24,760			
	PLUS: OTHERWISE ESTIMATED	$	$ 24,760	29	
30	DISPOSAL VALUE OF ALTERNATIVE AT END OF PERIOD *		$ 4,000	30	
31	TERMINAL NET INVESTMENT (29—30)		$ 20,760	31	

C. RETURN

32	AVERAGE NET CAPITAL CONSUMPTION $\left(\dfrac{28-31}{P}\right)$	$ 2,940	32
33	AVERAGE NET INVESTMENT $\left(\dfrac{28+31}{2}\right)$	$ 22,230	33
34	BEFORE-TAX RETURN $\left(\dfrac{25-32}{33} \times 100\right)$	% 63.3	34
35	INCREASE IN DEPRECIATION AND INTEREST DEDUCTIONS	$ 4,190	35
36	TAXABLE OPERATING ADVANTAGE (25—35)	$ 12,830	36
37	INCREASE IN INCOME TAX (36 X TAX RATE)	$ 6,415	37
38	AFTER-TAX OPERATING ADVANTAGE (25—37)	$ 10,605	38
39	AVAILABLE FOR RETURN ON INVESTMENT (38—32)	$ 7,665	39
40	AFTER-TAX RETURN $\left(\dfrac{39}{33} \times 100\right)$	% 34.5	40

*After terminal tax adjustments.

Copyright 1967, Machinery and Allied Products Institute

FIGURE D-4   MAPI Summary Form, Sheet 2 (with Entries for Example Project).

It is estimated that direct labor cost will be increased by $900 a year, indirect labor (supervision) by $50, and fringe benefits by $190. Maintenance will be higher by $200, tool costs by $80, power consumption by $40, and property taxes and insurance by $320. On the other hand, there will be a saving of $16,800 in the cost of purchased materials, $1,100 in inventory carrying costs (other than floor space), and $1,000 in floor space. The net cost reduction is therefore $17,020.

In addition to this operating advantage, the equipment will permit a reduction

THE MAPI METHOD FOR REPLACEMENT
AND GENERAL INVESTMENT ANALYSIS

of $4,000 in inventory investment. After consulting with operating officials and others, the analyst comes up with the following stipulations:

Comparison period	1 year
Project operating rate	1,200 hours
Service life	13 years
Terminal salvage ratio	10 percent of net cost
Tax depreciation method	sum-of-the-years'-digits
Tax rate	50%
Debt ratio	30%
Debt interest rate	5%
Investment credit	7%

The results shown on line 40 of Figure D-4 indicate an after-tax return of 34.5% for the project above.

In reviewing the application of the MAPI method to the example project, it should be observed that all the considerations given to the determination of the operating advantage would be necessary regardless of the technique used in developing a measure of comparison. It must be further noted that line 29 of Figure D-4 states that an estimate of the retention value does not have to come from the MAPI charts. If the analyst has a better estimate, he can use it.

An after-tax cash flow diagram using the information from the MAPI forms in Figures D-3 and D-4 would be

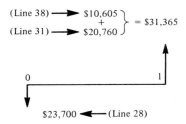

Upon solving, $-\$23,700 + \$31,365(P/F, i\%, 1) = 0$, and from interest tables, i can be interpolated to be 32.4%. This is close to the 34.5% return by the MAPI method. Generally, if the MAPI assumptions are not violated substantially, the return calculated by the MAPI method is a good approximation of the rate of return calculated by the discounted cash flow method. The MAPI method can be applied to problems involving a comparison period of more than 1 year in a manner very similar to the example above.

To most people trained in engineering economy principles, the use of a formula approach such as the MAPI method is too rigid. The assumptions built into the method can be inappropriate and cause error in the analysis to result.†
However, the MAPI method does provide for obsolescence and deterioration, offers an excellent checklist, will provide consistent results, and is relatively easy to use.

---

†The Alfred V. Bodine/SME Award is given to the author of the best paper submitted each year on the topic of "machine tool justification and its relationship to productivity." A concern underlying the award is that manufacturing productivity may be blunted by formula-based equipment replacement methods such as MAPI.

Interest and Annuity Tables for Discrete Compounding

(For various values of i from $\frac{1}{2}\%$ to 50%)

i = interest rate per period (usually 1 year)

N = number of periods

$$(F/P, i\%, N) = (1 + i)^N$$

$$(P/F, i\%, N) = \frac{1}{(1 + i)^N}$$

$$(F/A, i\%, N) = \frac{(1 + i)^N - 1}{i}$$

$$(P/A, i\%, N) = \frac{(1 + i)^N - 1}{i(1 + i)^N}$$

$$(A/F, i\%, N) = \frac{i}{(1 + i)^N - 1}$$

$$(A/P, i\%, N) = \frac{i(1 + i)^N}{(1 + i)^N - 1} \quad \longleftarrow \quad \text{Amortization}$$

$$(P/G, i\%, N) = \frac{1}{i}\left[\frac{(1 + i)^N - 1}{i(1 + i)^N} - \frac{N}{(1 + i)^N}\right]$$

$$(A/G, i\%, N) = \frac{1}{i} - \frac{N}{(1 + i)^N - 1}$$

Interest tables for selected integer-valued interest rates have been included in Appendixes E and F. The following computer program, written in BASIC, can be used on most personal computers to obtain interest tables for discrete cash flows and discrete or continuous compounding of integer or fractional interest rates expressed in decimal form.

LISTING

```
10     PRINT "THIS PROGRAM COMPUTES INTEREST TABLES FOR
       DISCRETE (D)"
20     PRINT "COMPOUNDING AND CONTINUOUS (C)
       COMPOUNDING, END-OF-"
30     PRINT "PERIOD CASH FLOW CONVENTION IS ASSUMED,
       ENTER 'I' PER"
40     PRINT "PERIOD AS A DECIMAL FOR DISCRETE
       COMPOUNDING AND 'R' PER PERIOD AS"
50     PRINT "A DECIMAL FOR CONTINUOUS COMPOUNDING, 'N'
       IS THE NUMBER"
60     PRINT "OF COMPOUNDING PERIODS,"
70     PRINT
80     PRINT "ENTER VALUE OF INTEREST RATE AS A
       DECIMAL ";
90     INPUT I
100    I3 = I
110    PRINT "DISCRETE (D) OR CONTINUOUS (C)
       COMPOUNDING ";
120    INPUT B$
130    IF B$ = "D" THEN 150
140    I = EXP(I)-1
150    PRINT
160    PRINT "ENTER STARTING AND ENDING VALUES OF N"
170    PRINT "FOR TABULATIONS,";
180    INPUT Q,R
190    PRINT
200    PRINT
210    PRINT "   N      F/P    P/F    P/A    A/P    A/F";
220    PRINT "    F/A"
230    PRINT " === ===== ===== ===== ===== ====";
240    PRINT "= ====="
250    FOR N = Q TO R
260    A = (1+I)^N
270    C = 1/A
280    D = (A-1)/(I*A)
290    E = 1/D
300    F = E-I
310    G = (A-1)/I
```

```
320    PRINT USING 330, N,A,C,D,E,F,G
330    :### #####.#### #####.#### #####.#### #####.####
       #####.#### #####.####
340    NEXT N
350    PRINT
360    PRINT
370    PRINT "  N      P/G       A/G"
380    PRINT " ===    =====     ====="
390    FOR N = Q TO R
400    A = (1 + I)^N
410    P = ((A-1)/(I*A)-N/A)/I
420    T = 1/I - N/(A-1)
430    PRINT USING 440,N,P,T
440    :### #####.#### #####.####
450    NEXT N
460    PRINT
470    PRINT
480    PRINT "DO YOU WANT TO RUN FOR ANOTHER VALUE OF
       INTEREST RATE (Y OR N)";
490    INPUT S$
500    IF S$ = "Y" THEN 70
510    END
```

EXAMPLES

```
ENTER VALUE OF INTEREST RATE AS A DECIMAL ?.1
DISCRETE (D) OR CONTINUOUS (C) COMPOUNDING ?D

ENTER STARTING AND ENDING VALUES OF N
FOR TABULATIONS. ?1,10
```

N	F/P	P/F	P/A	A/P	A/F
===	=====	=====	=====	=====	=====
1	1.1000	0.9091	0.9091	1.1000	1.0000
2	1.2100	0.8264	1.7355	0.5762	0.4762
3	1.3310	0.7513	2.4869	0.4021	0.3021
4	1.4641	0.6830	3.1699	0.3155	0.2155
5	1.6105	0.6209	3.7908	0.2638	0.1638
6	1.7716	0.5645	4.3553	0.2296	0.1296
7	1.9487	0.5132	4.8684	0.2054	0.1054
8	2.1436	0.4665	5.3349	0.1874	0.0874
9	2.3579	0.4241	5.7590	0.1736	0.0736
10	2.5937	0.3855	6.1446	0.1627	0.0627

```
    N      F/A       P/G       A/G
   ===    =====     =====     =====
    1    1.0000    0.0000    0.0000
    2    2.1000    0.8264    0.4762
    3    3.3100    2.3291    0.9366
    4    4.6410    4.3781    1.3812
    5    6.1051    6.8618    1.8101
    6    7.7156    9.6842    2.2236
    7    9.4872   12.7631    2.6216
    8   11.4359   16.0287    3.0045
    9   13.5795   19.4214    3.3724
   10   15.9374   22.8913    3.7255
```

DO YOU WANT TO RUN FOR ANOTHER VALUE OF INTEREST
RATE (Y OR N) ?Y

ENTER VALUE OF INTEREST RATE AS A DECIMAL ?.5
DISCRETE (D) OR CONTINUOUS (C) COMPOUNDING ?D

ENTER STARTING AND ENDING VALUES OF N
FOR TABULATIONS. ?20,25

```
    N        F/P        P/F       P/A       A/P       A/F
   ===      =====      =====     =====     =====     =====
   20     3325.2567   0.0003    1.9994    0.5002    0.0002
   21     4987.8850   0.0002    1.9996    0.5001    0.0001
   22     7481.8276   0.0001    1.9997    0.5001    0.0001
   23    11222.7414   0.0001    1.9998    0.5000    0.0000
   24    16834.1120   0.0001    1.9999    0.5000    0.0000
   25    25251.1680   0.0000    1.9999    0.5000    0.0000
```

```
    N        F/A        P/G       A/G
   ===      =====      =====     =====
   20     6648.5134   3.9868    1.9940
   21     9973.7700   3.9908    1.9958
   22    14961.6552   3.9936    1.9971
   23    22443.4828   3.9955    1.9980
   24    33666.2240   3.9969    1.9986
   25    50500.3360   3.9979    1.9990
```

DO YOU WANT TO RUN FOR ANOTHER VALUE OF INTEREST
RATE (Y OR N) ?Y

ENTER VALUE OF INTEREST RATE AS A DECIMAL ?.3
DISCRETE (D) OR CONTINUOUS (C) COMPOUNDING ?D

ENTER STARTING AND ENDING VALUES OF N
FOR TABULATIONS. ?20,25

**INTEREST AND ANNUITY TABLES
FOR DISCRETE COMPOUNDING**

N	F/P	P/F	P/A	A/P	A/F
===	=====	=====	=====	=====	=====
20	190.0496	0.0053	3.3158	0.3016	0.0016
21	247.0645	0.0040	3.3198	0.3012	0.0012
22	321.1838	0.0031	3.3230	0.3009	0.0009
23	417.5390	0.0024	3.3254	0.3007	0.0007
24	542.8007	0.0018	3.3272	0.3006	0.0006
25	705.6409	0.0014	3.3286	0.3004	0.0004

N	F/A	P/G	A/G
===	=====	=====	=====
20	630.1654	10.7019	3.2275
21	820.2150	10.7828	3.2480
22	1067.2795	10.8482	3.2646
23	1388.4633	10.9009	3.2781
24	1806.0023	10.9433	3.2890
25	2348.8029	10.9773	3.2979

DO YOU WANT TO RUN FOR ANOTHER VALUE OF INTEREST
RATE (Y OR N) ?N

**INTEREST AND ANNUITY TABLES
FOR DISCRETE COMPOUNDING**

TABLE E-1 Discrete Compounding; *i* = 1/2%

	SINGLE PAYMENT		UNIFORM SERIES				
	Compound Amount Factor	Present Worth Factor	Compound Amount Factor	Present Worth Factor	Sinking Fund Factor	Capital Recovery Factor	
N	To find *F* Given *P* F/P	To find *P* Given *F* P/F	To find *F* Given *A* F/A	To find *P* Given *A* P/A	To find *A* Given *F* A/F	To find *A* Given *P* A/P	*N*
1	1.0050	0.9950	1.0000	0.9950	1.0000	1.0050	1
2	1.0100	0.9901	2.0050	1.9851	0.4988	0.5038	2
3	1.0151	0.9851	3.0150	2.9702	0.3317	0.3367	3
4	1.0202	0.9802	4.0301	3.9505	0.2481	0.2531	4
5	1.0253	0.9754	5.0502	4.9259	0.1980	0.2030	5
6	1.0304	0.9705	6.0755	5.8964	0.1646	0.1696	6
7	1.0355	0.9657	7.1059	6.8621	0.1407	0.1457	7
8	1.0407	0.9609	8.1414	7.8229	0.1228	0.1278	8
9	1.0459	0.9561	9.1821	8.7790	0.1089	0.1139	9
10	1.0511	0.9513	10.2280	9.7304	0.0978	0.1028	10
11	1.0564	0.9466	11.2791	10.6770	0.0887	0.0937	11
12	1.0617	0.9419	12.3355	11.6189	0.0811	0.0861	12
13	1.0670	0.9372	13.3972	12.5561	0.0746	0.0796	13
14	1.0723	0.9326	14.4642	13.4887	0.0691	0.0741	14
15	1.0777	0.9279	15.5365	14.4166	0.0644	0.0694	15
16	1.0831	0.9233	16.6142	15.3399	0.0602	0.0652	16
17	1.0885	0.9187	17.6973	16.2586	0.0565	0.0615	17
18	1.0939	0.9141	18.7857	17.1727	0.0532	0.0582	18
19	1.0994	0.9096	19.8797	18.0823	0.0503	0.0553	19
20	1.1049	0.9051	20.9791	18.9874	0.0477	0.0527	20
21	1.1104	0.9006	22.0839	19.8879	0.0453	0.0503	21
22	1.1160	0.8961	23.1944	20.7840	0.0431	0.0481	22
23	1.1216	0.8916	24.3103	21.6756	0.0411	0.0461	23
24	1.1272	0.8872	25.4319	22.5628	0.0393	0.0443	24
25	1.1328	0.8828	26.5590	23.4456	0.0377	0.0427	25
26	1.1385	0.8784	27.6918	24.3240	0.0361	0.0411	26
27	1.1442	0.8740	28.8303	25.1980	0.0347	0.0397	27
28	1.1499	0.8697	29.9744	26.0676	0.0334	0.0384	28
29	1.1556	0.8653	31.1243	26.9330	0.0321	0.0371	29
30	1.1614	0.8610	32.2799	27.7940	0.0310	0.0360	30
35	1.1907	0.8398	38.1453	32.0353	0.0262	0.0312	35
40	1.2208	0.8191	44.1587	36.1721	0.0226	0.0276	40
45	1.2516	0.7990	50.3240	40.2071	0.0199	0.0249	45
50	1.2832	0.7793	56.6450	44.1427	0.0177	0.0227	50
55	1.3156	0.7601	63.1256	47.9813	0.0158	0.0208	55
60	1.3488	0.7414	69.7698	51.7254	0.0143	0.0193	60
65	1.3829	0.7231	76.5818	55.3773	0.0131	0.0181	65
70	1.4178	0.7053	83.5658	58.9393	0.0120	0.0170	70
75	1.4536	0.6879	90.7262	62.4135	0.0110	0.0160	75
80	1.4903	0.6710	98.0674	65.8022	0.0102	0.0152	80
85	1.5280	0.6545	105.594	69.107	0.0095	0.0145	85
90	1.5666	0.6383	113.311	72.331	0.0088	0.0138	90
95	1.6061	0.6226	121.22	75.475	0.0082	0.0132	95
100	1.6467	0.6073	129.33	78.542	0.0077	0.0127	100
∞				200.0		0.0050	∞

TABLE E-2   Discrete Compounding; $i = 1\%$

	SINGLE PAYMENT		UNIFORM SERIES				
	Compound Amount Factor	Present Worth Factor	Compound Amount Factor	Present Worth Factor	Sinking Fund Factor	Capital Recovery Factor	
N	To find F Given P F/P	To find P Given F P/F	To find F Given A F/A	To find P Given A P/A	To find A Given F A/F	To find A Given P A/P	N
1	1.0100	0.9901	1.0000	0.9901	1.0000	1.0100	1
2	1.0201	0.9803	2.0100	1.9704	0.4975	0.5075	2
3	1.0303	0.9706	3.0301	2.9410	0.3300	0.3400	3
4	1.0406	0.9610	4.0604	3.9020	0.2463	0.2563	4
5	1.0510	0.9515	5.1010	4.8534	0.1960	0.2060	5
6	1.0615	0.9420	6.1520	5.7955	0.1625	0.1725	6
7	1.0721	0.9327	7.2135	6.7282	0.1386	0.1486	7
8	1.0829	0.9235	8.2857	7.6517	0.1207	0.1307	8
9	1.0937	0.9143	9.3685	8.5660	0.1067	0.1167	9
10	1.1046	0.9053	10.4622	9.4713	0.0956	0.1056	10
11	1.1157	0.8963	11.5668	10.3676	0.0865	0.0965	11
12	1.1268	0.8874	12.6825	11.2551	0.0788	0.0888	12
13	1.1381	0.8787	13.8093	12.1337	0.0724	0.0824	13
14	1.1495	0.8700	14.9474	13.0037	0.0669	0.0769	14
15	1.1610	0.8613	16.0969	13.8650	0.0621	0.0721	15
16	1.1726	0.8528	17.2578	14.7178	0.0579	0.0679	16
17	1.1843	0.8444	18.4304	15.5622	0.0543	0.0643	17
18	1.1961	0.8360	19.6147	16.3982	0.0510	0.0610	18
19	1.2081	0.8277	20.8109	17.2260	0.0481	0.0581	19
20	1.2202	0.8195	22.0190	18.0455	0.0454	0.0554	20
21	1.2324	0.8114	23.2391	18.8570	0.0430	0.0530	21
22	1.2447	0.8034	24.4715	19.6603	0.0409	0.0509	22
23	1.2572	0.7954	25.7162	20.4558	0.0389	0.0489	23
24	1.2697	0.7876	26.9734	21.2434	0.0371	0.0471	24
25	1.2824	0.7798	28.2431	22.0231	0.0354	0.0454	25
26	1.2953	0.7720	29.5256	22.7952	0.0339	0.0439	26
27	1.3082	0.7644	30.8208	23.5596	0.0324	0.0424	27
28	1.3213	0.7568	32.1290	24.3164	0.0311	0.0411	28
29	1.3345	0.7493	33.4503	25.0657	0.0299	0.0399	29
30	1.3478	0.7419	34.7848	25.8077	0.0287	0.0387	30
35	1.4166	0.7059	41.6602	29.4085	0.0240	0.0340	35
40	1.4889	0.6717	48.8863	32.8346	0.0205	0.0305	40
45	1.5648	0.6391	56.4809	36.0945	0.0177	0.0277	45
50	1.6446	0.6080	64.4630	39.1961	0.0155	0.0255	50
55	1.7285	0.5785	72.8523	42.1471	0.0137	0.0237	55
60	1.8167	0.5505	81.6695	44.9550	0.0122	0.0222	60
65	1.9094	0.5237	90.9364	47.6265	0.0110	0.0210	65
70	2.0068	0.4983	100.676	50.1684	0.0099	0.0199	70
75	2.1091	0.4741	110.912	52.5870	0.0090	0.0190	75
80	2.2167	0.4511	121.671	54.8881	0.0082	0.0182	80
85	2.3298	0.4292	132.979	57.0776	0.0075	0.0175	85
90	2.4486	0.4084	144.86	59.161	0.0069	0.0169	90
95	2.5735	0.3886	157.35	61.143	0.0064	0.0164	95
100	2.7048	0.3697	170.48	63.029	0.0059	0.0159	100
∞				100.000		0.0100	∞

TABLE E-3 Discrete Compounding; $i = 1\ 1/2\%$

	SINGLE PAYMENT		UNIFORM SERIES				
	Compound Amount Factor	Present Worth Factor	Compound Amount Factor	Present Worth Factor	Sinking Fund Factor	Capital Recovery Factor	
N	To find F Given P F/P	To find P Given F P/F	To find F Given A F/A	To find P Given A P/A	To find A Given F A/F	To find A Given P A/P	N
1	1.0150	0.9852	1.0000	0.9852	1.0000	1.0150	1
2	1.0302	0.9707	2.0150	1.9559	0.4963	0.5113	2
3	1.0457	0.9563	3.0452	2.9122	0.3284	0.3434	3
4	1.0614	0.9422	4.0909	3.8544	0.2444	0.2594	4
5	1.0773	0.9283	5.1523	4.7826	0.1941	0.2091	5
6	1.0934	0.9145	6.2295	5.6972	0.1605	0.1755	6
7	1.1098	0.9010	7.3230	6.5982	0.1366	0.1516	7
8	1.1265	0.8877	8.4328	7.4859	0.1186	0.1336	8
9	1.1434	0.8746	9.5593	8.3605	0.1046	0.1196	9
10	1.1605	0.8617	10.7027	9.2222	0.0934	0.1084	10
11	1.1779	0.8489	11.8632	10.0711	0.0843	0.0993	11
12	1.1956	0.8364	13.0412	10.9075	0.0767	0.0917	12
13	1.2136	0.8240	14.2368	11.7315	0.0702	0.0852	13
14	1.2318	0.8118	15.4504	12.5434	0.0647	0.0797	14
15	1.2502	0.7999	16.6821	13.3432	0.0599	0.0749	15
16	1.2690	0.7880	17.9323	14.1312	0.0558	0.0708	16
17	1.2880	0.7764	19.2013	14.9076	0.0521	0.0671	17
18	1.3073	0.7649	20.4893	15.6725	0.0488	0.0638	18
19	1.3270	0.7536	21.7967	16.4261	0.0459	0.0609	19
20	1.3469	0.7425	23.1236	17.1686	0.0432	0.0582	20
21	1.3671	0.7315	24.4705	17.9001	0.0409	0.0559	21
22	1.3876	0.7207	25.8375	18.6208	0.0387	0.0537	22
23	1.4084	0.7100	27.2251	19.3308	0.0367	0.0517	23
24	1.4295	0.6995	28.6335	20.0304	0.0349	0.0499	24
25	1.4509	0.6892	30.0630	20.7196	0.0333	0.0483	25
26	1.4727	0.6790	31.5139	21.3986	0.0317	0.0467	26
27	1.4948	0.6690	32.9866	22.0676	0.0303	0.0453	27
28	1.5172	0.6591	34.4814	22.7267	0.0290	0.0440	28
29	1.5400	0.6494	35.9986	23.3761	0.0278	0.0428	29
30	1.5631	0.6398	37.5386	24.0158	0.0266	0.0416	30
35	1.6839	0.5939	45.5920	27.0756	0.0219	0.0369	35
40	1.8140	0.5513	54.2678	29.9158	0.0184	0.0334	40
45	1.9542	0.5117	63.6141	32.5523	0.0157	0.0307	45
50	2.1052	0.4750	73.6827	34.9997	0.0136	0.0286	50
55	2.2679	0.4409	84.5294	37.2714	0.0118	0.0268	55
60	2.4432	0.4093	96.2145	39.3802	0.0104	0.0254	60
65	2.6320	0.3799	108.803	41.3378	0.0092	0.0242	65
70	2.8355	0.3527	122.364	43.1548	0.0082	0.0232	70
75	3.0546	0.3274	136.97	44.8416	0.0073	0.0223	75
80	3.2907	0.3039	152.71	46.4073	0.0065	0.0215	80
85	3.5450	0.2821	169.66	47.8607	0.0059	0.0209	85
90	3.8189	0.2619	187.93	49.2098	0.0053	0.0203	90
95	4.1141	0.2431	207.61	50.4622	0.0048	0.0198	95
100	4.4320	0.2256	228.80	51.6247	0.0044	0.0194	100
∞				66.667		0.0150	∞

TABLE E-4 Discrete Compounding; $i = 2\%$

	SINGLE PAYMENT		UNIFORM SERIES				
	Compound Amount Factor	Present Worth Factor	Compound Amount Factor	Present Worth Factor	Sinking Fund Factor	Capital Recovery Factor	
N	To find F Given P F/P	To find P Given F P/F	To find F Given A F/A	To find P Given A P/A	To find A Given F A/F	To find A Given P A/P	N
1	1.0200	0.9804	1.0000	0.9804	1.0000	1.0200	1
2	1.0404	0.9612	2.0200	1.9416	0.4950	0.5150	2
3	1.0612	0.9423	3.0604	2.8839	0.3268	0.3468	3
4	1.0824	0.9238	4.1216	3.8077	0.2426	0.2626	4
5	1.1041	0.9057	5.2040	4.7135	0.1922	0.2122	5
6	1.1262	0.8880	6.3081	5.6014	0.1585	0.1785	6
7	1.1487	0.8706	7.4343	6.4720	0.1345	0.1545	7
8	1.1717	0.8535	8.5830	7.3255	0.1165	0.1365	8
9	1.1951	0.8368	9.7546	8.1622	0.1025	0.1225	9
10	1.2190	0.8203	10.9497	8.9826	0.0913	0.1113	10
11	1.2434	0.8043	12.1687	9.7868	0.0822	0.1022	11
12	1.2682	0.7885	13.4121	10.5753	0.0746	0.0946	12
13	1.2936	0.7730	14.6803	11.3484	0.0681	0.0881	13
14	1.3195	0.7579	15.9739	12.1062	0.0626	0.0826	14
15	1.3459	0.7430	17.2934	12.8493	0.0578	0.0778	15
16	1.3728	0.7284	18.6393	13.5777	0.0537	0.0737	16
17	1.4002	0.7142	20.0121	14.2919	0.0500	0.0700	17
18	1.4282	0.7002	21.4123	14.9920	0.0467	0.0667	18
19	1.4568	0.6864	22.8405	15.6785	0.0438	0.0638	19
20	1.4859	0.6730	24.2974	16.3514	0.0412	0.0612	20
21	1.5157	0.6598	25.7833	17.0112	0.0388	0.0588	21
22	1.5460	0.6468	27.2990	17.6580	0.0366	0.0566	22
23	1.5769	0.6342	28.8449	18.2922	0.0347	0.0547	23
24	1.6084	0.6217	30.4218	18.9139	0.0329	0.0529	24
25	1.6406	0.6095	32.0303	19.5234	0.0312	0.0512	25
26	1.6734	0.5976	33.6709	20.1210	0.0297	0.0497	26
27	1.7069	0.5859	35.3443	20.7069	0.0283	0.0483	27
28	1.7410	0.5744	37.0512	21.2813	0.0270	0.0470	28
29	1.7758	0.5631	38.7922	21.8444	0.0258	0.0458	29
30	1.8114	0.5521	40.5681	22.3964	0.0246	0.0446	30
35	1.9999	0.5000	49.9944	24.9986	0.0200	0.0400	35
40	2.2080	0.4529	60.4019	27.3555	0.0166	0.0366	40
45	2.4379	0.4102	71.8927	29.4902	0.0139	0.0339	45
50	2.6916	0.3715	84.5793	31.4236	0.0118	0.0318	50
55	2.9717	0.3365	98.5864	33.1748	0.0101	0.0301	55
60	3.2810	0.3048	114.051	34.7609	0.0088	0.0288	60
65	3.6225	0.2761	131.126	36.1975	0.0076	0.0276	65
70	3.9996	0.2500	149.978	37.4986	0.0067	0.0267	70
75	4.4158	0.2265	170.792	38.6771	0.0059	0.0259	75
80	4.8754	0.2051	193.772	39.7445	0.0052	0.0252	80
85	5.3829	0.1858	219.144	40.7113	0.0046	0.0246	85
90	5.9431	0.1683	247.16	41.5869	0.0040	0.0240	90
95	6.5617	0.1524	278.08	42.3800	0.0036	0.0236	95
100	7.2446	0.1380	312.23	43.0983	0.0032	0.0232	100
∞				50.0000		0.0200	∞

TABLE E-5 Discrete Compounding; $i = 3\%$

	SINGLE PAYMENT		UNIFORM SERIES				
	Compound Amount Factor	Present Worth Factor	Compound Amount Factor	Present Worth Factor	Sinking Fund Factor	Capital Recovery Factor	
N	To find F Given P F/P	To find P Given F P/F	To find F Given A F/A	To find P Given A P/A	To find A Given F A/F	To find A Given P A/P	N
1	1.0300	0.9709	1.0000	0.9709	1.0000	1.0300	1
2	1.0609	0.9426	2.0300	1.9135	0.4926	0.5226	2
3	1.0927	0.9151	3.0909	2.8286	0.3235	0.3535	3
4	1.1255	0.8885	4.1836	3.7171	0.2390	0.2690	4
5	1.1593	0.8626	5.3091	4.5797	0.1884	0.2184	5
6	1.1941	0.8375	6.4684	5.4172	0.1546	0.1846	6
7	1.2299	0.8131	7.6625	6.2303	0.1305	0.1605	7
8	1.2668	0.7894	8.8923	7.0197	0.1125	0.1425	8
9	1.3048	0.7664	10.1591	7.7861	0.0984	0.1284	9
10	1.3439	0.7441	11.4639	8.5302	0.0872	0.1172	10
11	1.3842	0.7224	12.8078	9.2526	0.0781	0.1081	11
12	1.4258	0.7014	14.1920	9.9540	0.0705	0.1005	12
13	1.4685	0.6810	15.6178	10.6349	0.0640	0.0940	13
14	1.5126	0.6611	17.0863	11.2961	0.0585	0.0885	14
15	1.5580	0.6419	18.5989	11.9379	0.0538	0.0838	15
16	1.6047	0.6232	20.1569	12.5611	0.0496	0.0796	16
17	1.6528	0.6050	21.7616	13.1661	0.0460	0.0760	17
18	1.7024	0.5874	23.4144	13.7535	0.0427	0.0727	18
19	1.7535	0.5703	25.1168	14.3238	0.0398	0.0698	19
20	1.8061	0.5537	26.8703	14.8775	0.0372	0.0672	20
21	1.8603	0.5375	28.6765	15.4150	0.0349	0.0649	21
22	1.9161	0.5219	30.5367	15.9369	0.0327	0.0627	22
23	1.9736	0.5067	32.4528	16.4436	0.0308	0.0608	23
24	2.0328	0.4919	34.4264	16.9355	0.0290	0.0590	24
25	2.0938	0.4776	36.4592	17.4131	0.0274	0.0574	25
26	2.1566	0.4637	38.5530	17.8768	0.0259	0.0559	26
27	2.2213	0.4502	40.7096	18.3270	0.0246	0.0546	27
28	2.2879	0.4371	42.9309	18.7641	0.0233	0.0533	28
29	2.3566	0.4243	45.2188	19.1884	0.0221	0.0521	29
30	2.4273	0.4120	47.5754	19.6004	0.0210	0.0510	30
35	2.8139	0.3554	60.4620	21.4872	0.0165	0.0465	35
40	3.2620	0.3066	75.4012	23.1148	0.0133	0.0433	40
45	3.7816	0.2644	92.7197	24.5187	0.0108	0.0408	45
50	4.3839	0.2281	112.797	25.7298	0.0089	0.0389	50
55	5.0821	0.1968	136.071	26.7744	0.0073	0.0373	55
60	5.8916	0.1697	163.053	27.6756	0.0061	0.0361	60
65	6.8300	0.1464	194.332	28.4529	0.0051	0.0351	65
70	7.9178	0.1263	230.594	29.1234	0.0043	0.0343	70
75	9.1789	0.1089	272.630	29.7018	0.0037	0.0337	75
80	10.6409	0.0940	321.362	30.2008	0.0031	0.0331	80
85	12.3357	0.0811	377.856	30.6311	0.0026	0.0326	85
90	14.3004	0.0699	443.35	31.0024	0.0023	0.0323	90
95	16.5781	0.0603	519.27	31.3227	0.0019	0.0319	95
100	19.2186	0.0520	607.29	31.5989	0.0016	0.0316	100
∞				33.3333		0.0300	∞

TABLE E-6 Discrete Compounding; $i = 5\%$

	SINGLE PAYMENT		UNIFORM SERIES				
	Compound Amount Factor	Present Worth Factor	Compound Amount Factor	Present Worth Factor	Sinking Fund Factor	Capital Recovery Factor	
N	To find F Given P F/P	To find P Given F P/F	To find F Given A F/A	To find P Given A P/A	To find A Given F A/F	To find A Given P A/P	N
1	1.0500	0.9524	1.0000	0.9524	1.0000	1.0500	1
2	1.1025	0.9070	2.0500	1.8594	0.4878	0.5378	2
3	1.1576	0.8638	3.1525	2.7232	0.3172	0.3672	3
4	1.2155	0.8227	4.3101	3.5460	0.2320	0.2820	4
5	1.2763	0.7835	5.5256	4.3295	0.1810	0.2310	5
6	1.3401	0.7462	6.8019	5.0757	0.1470	0.1970	6
7	1.4071	0.7107	8.1420	5.7864	0.1228	0.1728	7
8	1.4775	0.6768	9.5491	6.4632	0.1047	0.1547	8
9	1.5513	0.6446	11.0266	7.1078	0.0907	0.1407	9
10	1.6289	0.6139	12.5779	7.7217	0.0795	0.1295	10
11	1.7103	0.5847	14.2068	8.3064	0.0704	0.1204	11
12	1.7959	0.5568	15.9171	8.8633	0.0628	0.1128	12
13	1.8856	0.5303	17.7130	9.3936	0.0565	0.1065	13
14	1.9799	0.5051	19.5986	9.8986	0.0510	0.1010	14
15	2.0789	0.4810	21.5786	10.3797	0.0463	0.0963	15
16	2.1829	0.4581	23.6575	10.8378	0.0423	0.0923	16
17	2.2920	0.4363	25.8404	11.2741	0.0387	0.0887	17
18	2.4066	0.4155	28.1324	11.6896	0.0355	0.0855	18
19	2.5269	0.3957	30.5390	12.0853	0.0327	0.0827	19
20	2.6533	0.3769	33.0659	12.4622	0.0302	0.0802	20
21	2.7860	0.3589	35.7192	12.8212	0.0280	0.0780	21
22	2.9253	0.3418	38.5052	13.1630	0.0260	0.0760	22
23	3.0715	0.3256	41.4305	13.4886	0.0241	0.0741	23
24	3.2251	0.3101	44.5020	13.7986	0.0225	0.0725	24
25	3.3864	0.2953	47.7271	14.0939	0.0210	0.0710	25
26	3.5557	0.2812	51.1134	14.3752	0.0196	0.0696	26
27	3.7335	0.2678	54.6691	14.6430	0.0183	0.0683	27
28	3.9201	0.2551	58.4026	14.8981	0.0171	0.0671	28
29	4.1161	0.2429	62.3227	15.1411	0.0160	0.0660	29
30	4.3219	0.2314	66.4388	15.3725	0.0151	0.0651	30
35	5.5160	0.1813	90.3203	16.3742	0.0111	0.0611	35
40	7.0400	0.1420	120.800	17.1591	0.0083	0.0583	40
45	8.9850	0.1113	159.700	17.7741	0.0063	0.0563	45
50	11.4674	0.0872	209.348	18.2559	0.0048	0.0548	50
55	14.6356	0.0683	272.713	18.6335	0.0037	0.0537	55
60	18.6792	0.0535	353.584	18.9293	0.0028	0.0528	60
65	23.8399	0.0419	456.798	19.1611	0.0022	0.0522	65
70	30.4264	0.0329	588.528	19.3427	0.0017	0.0517	70
75	38.8327	0.0258	756.653	19.4850	0.0013	0.0513	75
80	49.5614	0.0202	971.228	19.5965	0.0010	0.0510	80
85	63.2543	0.0158	1245.09	19.6838	0.0008	0.0508	85
90	80.7303	0.0124	1594.61	19.7523	0.0006	0.0506	90
95	103.035	0.0097	2040.69	19.8059	0.0005	0.0505	95
100	131.501	0.0076	2610.02	19.8479	0.0004	0.0504	100
∞				20.0000		0.0500	∞

TABLE E-7   Discrete Compounding; _i_ = 8%

	SINGLE PAYMENT		UNIFORM SERIES				
	Compound Amount Factor	Present Worth Factor	Compound Amount Factor	Present Worth Factor	Sinking Fund Factor	Capital Recovery Factor	
_N_	To find _F_ Given _P_ _F/P_	To find _P_ Given _F_ _P/F_	To find _F_ Given _A_ _F/A_	To find _P_ Given _A_ _P/A_	To find _A_ Given _F_ _A/F_	To find _A_ Given _P_ _A/P_	_N_
1	1.0800	0.9259	1.0000	0.9259	1.0000	1.0800	1
2	1.1664	0.8573	2.0800	1.7833	0.4808	0.5608	2
3	1.2597	0.7938	3.2464	2.5771	0.3080	0.3880	3
4	1.3605	0.7350	4.5061	3.3121	0.2219	0.3019	4
5	1.4693	0.6806	5.8666	3.9927	0.1705	0.2505	5
6	1.5869	0.6302	7.3359	4.6229	0.1363	0.2163	6
7	1.7138	0.5835	8.9228	5.2064	0.1121	0.1921	7
8	1.8509	0.5403	10.6366	5.7466	0.0940	0.1740	8
9	1.9990	0.5002	12.4876	6.2469	0.0801	0.1601	9
10	2.1589	0.4632	14.4866	6.7101	0.0690	0.1490	10
11	2.3316	0.4289	16.6455	7.1390	0.0601	0.1401	11
12	2.5182	0.3971	18.9771	7.5361	0.0527	0.1327	12
13	2.7196	0.3677	21.4953	7.9038	0.0465	0.1265	13
14	2.9372	0.3405	24.2149	8.2442	0.0413	0.1213	14
15	3.1722	0.3152	27.1521	8.5595	0.0368	0.1168	15
16	3.4259	0.2919	30.3243	8.8514	0.0330	0.1130	16
17	3.7000	0.2703	33.7502	9.1216	0.0296	0.1096	17
18	3.9960	0.2502	37.4502	9.3719	0.0267	0.1067	18
19	4.3157	0.2317	41.4463	9.6036	0.0241	0.1041	19
20	4.6610	0.2145	45.7620	9.8181	0.0219	0.1019	20
21	5.0338	0.1987	50.4229	10.0168	0.0198	0.0998	21
22	5.4365	0.1839	55.4567	10.2007	0.0180	0.0980	22
23	5.8715	0.1703	60.8933	10.3711	0.0164	0.0964	23
24	6.3412	0.1577	66.7647	10.5288	0.0150	0.0950	24
25	6.8485	0.1460	73.1059	10.6748	0.0137	0.0937	25
26	7.3964	0.1352	79.9544	10.8100	0.0125	0.0925	26
27	7.9881	0.1252	87.3507	10.9352	0.0114	0.0914	27
28	8.6271	0.1159	95.3388	11.0511	0.0105	0.0905	28
29	9.3173	0.1073	103.966	11.1584	0.0096	0.0896	29
30	10.0627	0.0994	113.283	11.2578	0.0088	0.0888	30
35	14.7853	0.0676	172.317	11.6546	0.0058	0.0858	35
40	21.7245	0.0460	259.056	11.9246	0.0039	0.0839	40
45	31.9204	0.0313	386.506	12.1084	0.0026	0.0826	45
50	46.9016	0.0213	573.770	12.2335	0.0017	0.0817	50
55	68.9138	0.0145	848.923	12.3186	0.0012	0.0812	55
60	101.257	0.0099	1253.21	12.3766	0.0008	0.0808	60
65	148.780	0.0067	1847.25	12.4160	0.0005	0.0805	65
70	218.606	0.0046	2720.08	12.4428	0.0004	0.0804	70
75	321.204	0.0031	4002.55	12.4611	0.0002	0.0802	75
80	471.955	0.0021	5886.93	12.4735	0.0002	0.0802	80
85	693.456	0.0014	8655.71	12.4820	0.0001	0.0801	85
90	1018.92	0.0010	12723.9	12.4877	_a_	0.0801	90
95	1497.12	0.0007	18071.5	12.4917	_a_	0.0801	95
100	2199.76	0.0005	27484.5	12.4943	_a_	0.0800	100
∞				12.5000		0.0800	∞

_a_Less than 0.0001.

TABLE E-8 Discrete Compounding; $i = 10\%$

	SINGLE PAYMENT		UNIFORM SERIES				
	Compound Amount Factor	Present Worth Factor	Compound Amount Factor	Present Worth Factor	Sinking Fund Factor	Capital Recovery Factor	
N	To find F Given P F/P	To find P Given F P/F	To find F Given A F/A	To find P Given A P/A	To find A Given F A/F	To find A Given P A/P	N
1	1.1000	0.9091	1.0000	0.9091	1.0000	1.1000	1
2	1.2100	0.8264	2.1000	1.7355	0.4762	0.5762	2
3	1.3310	0.7513	3.3100	2.4869	0.3021	0.4021	3
4	1.4641	0.6830	4.6410	3.1699	0.2155	0.3155	4
5	1.6105	0.6209	6.1051	3.7908	0.1638	0.2638	5
6	1.7716	0.5645	7.7156	4.3553	0.1296	0.2296	6
7	1.9487	0.5132	9.4872	4.8684	0.1054	0.2054	7
8	2.1436	0.4665	11.4359	5.3349	0.0874	0.1874	8
9	2.3579	0.4241	13.5795	5.7590	0.0736	0.1736	9
10	2.5937	0.3855	15.9374	6.1446	0.0627	0.1627	10
11	2.8531	0.3505	18.5312	6.4951	0.0540	0.1540	11
12	3.1384	0.3186	21.3843	6.8137	0.0468	0.1468	12
13	3.4523	0.2897	24.5227	7.1034	0.0408	0.1408	13
14	3.7975	0.2633	27.9750	7.3667	0.0357	0.1357	14
15	4.1772	0.2394	31.7725	7.6061	0.0315	0.1315	15
16	4.5950	0.2176	35.9497	7.8237	0.0278	0.1278	16
17	5.0545	0.1978	40.5447	8.0216	0.0247	0.1247	17
18	5.5599	0.1799	45.5992	8.2014	0.0219	0.1219	18
19	6.1159	0.1635	51.1591	8.3649	0.0195	0.1195	19
20	6.7275	0.1486	57.2750	8.5136	0.0175	0.1175	20
21	7.4002	0.1351	64.0025	8.6487	0.0156	0.1156	21
22	8.1403	0.1228	71.4027	8.7715	0.0140	0.1140	22
23	8.9543	0.1117	79.5430	8.8832	0.0126	0.1126	23
24	9.8497	0.1015	88.4973	8.9847	0.0113	0.1113	24
25	10.8347	0.0923	98.3470	9.0770	0.0102	0.1102	25
26	11.9182	0.0839	109.182	9.1609	0.0092	0.1092	26
27	13.1100	0.0763	121.100	9.2372	0.0083	0.1083	27
28	14.4210	0.0693	134.210	9.3066	0.0075	0.1075	28
29	15.8631	0.0630	148.631	9.3696	0.0067	0.1067	29
30	17.4494	0.0573	164.494	9.4269	0.0061	0.1061	30
35	28.1024	0.0356	271.024	9.6442	0.0037	0.1037	35
40	45.2592	0.0221	442.592	9.7791	0.0023	0.1023	40
45	72.8904	0.0137	718.905	9.8628	0.0014	0.1014	45
50	117.391	0.0085	1163.91	9.9148	0.0009	0.1009	50
55	189.059	0.0053	1880.59	9.9471	0.0005	0.1005	55
60	304.481	0.0033	3034.81	9.9672	0.0003	0.1003	60
65	490.370	0.0020	4893.71	9.9796	0.0002	0.1002	65
70	789.746	0.0013	7887.47	9.9873	0.0001	0.1001	70
75	1271.89	0.0008	12708.9	9.9921	a	0.1001	75
80	2048.40	0.0005	20474.0	9.9951	a	0.1000	80
85	3298.97	0.0003	32979.7	9.9970	a	0.1000	85
90	5313.02	0.0002	53120.2	9.9981	a	0.1000	90
95	8556.67	0.0001	85556.7	9.9988	a	0.1000	95
100	13780.6	a	137796	9.9993	a	0.1000	100
∞				10.0000		0.1000	∞

aLess than 0.0001.

	SINGLE PAYMENT		UNIFORM SERIES				
	Compound Amount Factor	Present Worth Factor	Compound Amount Factor	Present Worth Factor	Sinking Fund Factor	Capital Recovery Factor	
N	To find F Given P F/P	To find P Given F P/F	To find F Given A F/A	To find P Given A P/A	To find A Given F A/F	To find A Given P A/P	N
1	1.1200	0.8929	1.0000	0.8929	1.0000	1.1200	1
2	1.2544	0.7972	2.1200	1.6901	0.4717	0.5917	2
3	1.4049	0.7118	3.3744	2.4018	0.2963	0.4163	3
4	1.5735	0.6355	4.7793	3.0373	0.2092	0.3292	4
5	1.7623	0.5674	6.3528	3.6048	0.1574	0.2774	5
6	1.9738	0.5066	8.1152	4.1114	0.1232	0.2432	6
7	2.2107	0.4523	10.0890	4.5638	0.0991	0.2191	7
8	2.4760	0.4039	12.2997	4.9676	0.0813	0.2013	8
9	2.7731	0.3606	14.7757	5.3282	0.0677	0.1877	9
10	3.1058	0.3220	17.5487	5.6502	0.0570	0.1770	10
11	3.4785	0.2875	20.6546	5.9377	0.0484	0.1684	11
12	3.8960	0.2567	24.1331	6.1944	0.0414	0.1614	12
13	4.3635	0.2292	28.0291	6.4235	0.0357	0.1557	13
14	4.8871	0.2046	32.3926	6.6282	0.0309	0.1509	14
15	5.4736	0.1827	37.2797	6.8109	0.0268	0.1468	15
16	6.1304	0.1631	42.7533	6.9740	0.0234	0.1434	16
17	6.8660	0.1456	48.8837	7.1196	0.0205	0.1405	17
18	7.6900	0.1300	55.7497	7.2497	0.0179	0.1379	18
19	8.6128	0.1161	63.4397	7.3658	0.0158	0.1358	19
20	9.6463	0.1037	72.0524	7.4694	0.0139	0.1339	20
21	10.8038	0.0926	81.6987	7.5620	0.0122	0.1322	21
22	12.1003	0.0826	92.5026	7.6446	0.0108	0.1308	22
23	13.5523	0.0738	104.603	7.7184	0.0096	0.1296	23
24	15.1786	0.0659	118.155	7.7843	0.0085	0.1285	24
25	17.0001	0.0588	133.334	7.8431	0.0075	0.1275	25
26	19.0401	0.0525	150.334	7.8957	0.0067	0.1267	26
27	21.3249	0.0469	169.374	7.9426	0.0059	01.259	27
28	23.8839	0.0419	190.699	7.9844	0.0052	0.1252	28
29	26.7499	0.0374	214.583	8.0218	0.0047	0.1247	29
30	29.9599	0.0334	241.333	8.0552	0.0041	0.1241	30
35	52.7996	0.0189	431.663	8.1755	0.0023	0.1223	35
40	93.0509	0.0107	767.091	8.2438	0.0013	0.1213	40
45	163.988	0.0061	1358.23	8.2825	0.0007	0.1207	45
50	289.002	0.0035	2400.02	8.3045	0.0004	0.1204	50
55	509.320	0.0020	4236.00	8.3170	0.0002	0.1202	55
60	897.596	0.0011	7471.63	8.3240	0.0001	0.1201	60
65	1581.87	0.0006	13173.9	8.3281	a	0.1201	65
70	2787.80	0.0004	23223.3	8.3303	a	0.1200	70
75	4913.05	0.0002	40933.8	8.3316	a	0.1200	75
80	8658.47	0.0001	72145.6	8.3324	a	0.1200	80
∞				8.333		0.1200	∞

^{a}Less than 0.0001.

	SINGLE PAYMENT		UNIFORM SERIES				
	Compound Amount Factor	Present Worth Factor	Compound Amount Factor	Present Worth Factor	Sinking Fund Factor	Capital Recovery Factor	
N	To find F Given P F/P	To find P Given F P/F	To find F Given A F/A	To find P Given A P/A	To find A Given F A/F	To find A Given P A/P	N
1	1.1500	0.8696	1.0000	0.8696	1.0000	1.1500	1
2	1.3225	0.7561	2.1500	1.6257	0.4651	0.6151	2
3	1.5209	0.6575	3.4725	2.2832	0.2880	0.4380	3
4	1.7490	0.5718	4.9934	2.8550	0.2003	0.3503	4
5	2.0114	0.4972	6.7424	3.3522	0.1483	0.2983	5
6	2.3131	0.4323	8.7537	3.7845	0.1142	0.2642	6
7	2.6600	0.3759	11.0668	4.1604	0.0904	0.2404	7
8	3.0590	0.3269	13.7268	4.4873	0.0729	0.2229	8
9	3.5179	0.2843	16.7858	4.7716	0.0596	0.2096	9
10	4.0456	0.2472	20.3037	5.0188	0.0493	0.1993	10
11	4.6524	0.2149	24.3493	5.2337	0.0411	0.1911	11
12	5.3502	0.1869	29.0017	5.4206	0.0345	0.1845	12
13	6.1528	0.1625	34.3519	5.5831	0.0291	0.1791	13
14	7.0757	0.1413	40.5047	5.7245	0.0247	0.1747	14
15	8.1371	0.1229	47.5804	5.8474	0.0210	0.1710	15
16	9.3576	0.1069	55.7175	5.9542	0.0179	0.1679	16
17	10.7613	0.0929	65.0751	6.0472	0.0154	0.1654	17
18	12.3755	0.0808	75.8363	6.1280	0.0132	0.1632	18
19	14.2318	0.0703	88.2118	6.1982	0.0113	0.1613	19
20	16.3665	0.0611	102.444	6.2593	0.0098	0.1598	20
21	18.8215	0.0531	118.810	6.3125	0.0084	0.1584	21
22	21.6447	0.0462	137.632	6.3587	0.0073	0.1573	22
23	24.8915	0.0402	159.276	6.3988	0.0063	0.1563	23
24	28.6252	0.0349	184.168	6.4338	0.0054	0.1554	24
25	32.9189	0.0304	212.793	6.4641	0.0047	0.1547	25
26	37.8568	0.0264	245.712	6.4906	0.0041	0.1541	26
27	43.5353	0.0230	283.569	6.5135	0.0035	0.1535	27
28	50.0656	0.0200	327.104	6.5335	0.0031	0.1531	28
29	57.5754	0.0174	377.170	6.5509	0.0027	0.1527	29
30	66.2118	0.0151	434.745	6.5660	0.0023	0.1523	30
35	133.176	0.0075	881.170	6.6166	0.0011	0.1511	35
40	267.863	0.0037	1779.09	6.6418	0.0006	0.1506	40
45	538.769	0.0019	3585.13	6.6543	0.0003	0.1503	45
50	1083.66	0.0009	7217.71	6.6605	0.0001	0.1501	50
55	2179.62	0.0005	14524.1	6.6636	a	0.1501	55
60	4384.00	0.0002	29220.0	6.6651	a	0.1500	60
65	8817.78	0.0001	58778.5	6.6659	a	0.1500	65
70	17735.7	a	118231	6.6663	a	0.1500	70
75	35672.8	a	237812	6.6665	a	0.1500	75
80	71750.8	a	478332	6.6666	a	0.1500	80
∞				6.667		0.1500	∞

aLess than 0.0001.

TABLE E-11 Discrete Compounding; i = 20%

	SINGLE PAYMENT		UNIFORM SERIES				
	Compound Amount Factor	Present Worth Factor	Compound Amount Factor	Present Worth Factor	Sinking Fund Factor	Capital Recovery Factor	
N	To find F Given P F/P	To find P Given F P/F	To find F Given A F/A	To find P Given A P/A	To find A Given F A/F	To find A Given P A/P	N
1	1.2000	0.8333	1.0000	0.8333	1.0000	1.2000	1
2	1.4400	0.6944	2.2000	1.5278	0.4545	0.6545	2
3	1.7280	0.5787	3.6400	2.1065	0.2747	0.4747	3
4	2.0736	0.4823	5.3680	2.5887	0.1863	0.3863	4
5	2.4883	0.4019	7.4416	2.9906	0.1344	0.3344	5
6	2.9860	0.3349	9.9299	3.3255	0.1007	0.3007	6
7	3.5832	0.2791	12.9159	3.6046	0.0774	0.2774	7
8	4.2998	0.2326	16.4991	3.8372	0.0606	0.2606	8
9	5.1598	0.1938	20.7989	4.0310	0.0481	0.2481	9
10	6.1917	0.1615	25.9587	4.1925	0.0385	0.2385	10
11	7.4301	0.1346	32.1504	4.3271	0.0311	0.2311	11
12	8.9161	0.1122	39.5805	4.4392	0.0253	0.2253	12
13	10.6993	0.0935	48.4966	4.5327	0.0206	0.2206	13
14	12.8392	0.0779	59.1959	4.6106	0.0169	0.2169	14
15	15.4070	0.0649	72.0351	4.6755	0.0139	0.2139	15
16	18.4884	0.0541	87.4421	4.7296	0.0114	0.2114	16
17	22.1861	0.0451	105.931	4.7746	0.0094	0.2094	17
18	26.6233	0.0376	128.117	4.8122	0.0078	0.2078	18
19	31.9480	0.0313	154.740	4.8435	0.0065	0.2065	19
20	38.3376	0.0261	186.688	4.8696	0.0054	0.2054	20
21	46.0051	0.0217	225.026	4.8913	0.0044	0.2044	21
22	55.2061	0.0181	271.031	4.9094	0.0037	0.2037	22
23	66.2474	0.0151	326.237	4.9245	0.0031	0.2031	23
24	79.4968	0.0126	392.484	4.9371	0.0025	0.2025	24
25	95.3962	0.0105	471.981	4.9476	0.0021	0.2021	25
26	114.475	0.0087	567.377	4.9563	0.0018	0.2018	26
27	137.371	0.0073	681.853	4.9636	0.0015	0.2015	27
28	164.845	0.0061	819.223	4.9697	0.0012	0.2012	28
29	197.814	0.0051	984.068	4.9747	0.0010	0.2010	29
30	237.376	0.0042	1181.88	4.9789	0.0008	0.2008	30
35	590.668	0.0017	2948.34	4.9915	0.0003	0.2003	35
40	1469.77	0.0007	7343.85	4.9966	0.0001	0.2001	40
45	3657.26	0.0003	18281.3	4.9986	a	0.2001	45
50	9100.43	0.0001	45497.2	4.9995	a	0.2000	50
55	22644.8	a	113219	4.9998	a	0.2000	55
60	56347.5	a	281732	4.9999	a	0.2000	60
∞				5.0000		0.2000	∞

a Less than 0.0001.

TABLE E-12 Discrete Compounding; $i = 25\%$

	SINGLE PAYMENT		UNIFORM SERIES				
	Compound Amount Factor	Present Worth Factor	Compound Amount Factor	Present Worth Factor	Sinking Fund Factor	Capital Recovery Factor	
N	To find F Given P F/P	To find P Given F P/F	To find F Given A F/A	To find P Given A P/A	To find A Given F A/F	To find A Given P A/P	N
1	1.2500	0.8000	1.0000	0.8000	1.0000	1.2500	1
2	1.5625	0.6400	2.2500	1.4400	0.4444	0.6944	2
3	1.9531	0.5120	3.8125	1.9520	0.2623	0.5123	3
4	2.4414	0.4096	5.7656	2.3616	0.1734	0.4234	4
5	3.0518	0.3277	8.2070	2.6893	0.1218	0.3718	5
6	3.8147	0.2621	11.2588	2.9514	0.0888	0.3388	6
7	4.7684	0.2097	15.0735	3.1611	0.0663	0.3163	7
8	5.9605	0.1678	19.8419	3.3289	0.0504	0.3004	8
9	7.4506	0.1342	25.8023	3.4631	0.0388	0.2888	9
10	9.3132	0.1074	33.2529	3.5705	0.0301	0.2801	10
11	11.6415	0.0859	42.5661	3.6564	0.0235	0.2735	11
12	14.5519	0.0687	54.2077	3.7251	0.0184	0.2684	12
13	18.1899	0.0550	68.7596	3.7801	0.0145	0.2645	13
14	22.7374	0.0440	86.9495	3.8241	0.0115	0.2615	14
15	28.4217	0.0352	109.687	3.8593	0.0091	0.2591	15
16	35.5271	0.0281	138.109	3.8874	0.0072	0.2572	16
17	44.4089	0.0225	173.636	3.9099	0.0058	0.2558	17
18	55.5112	0.0180	218.045	3.9279	0.0046	0.2546	18
19	69.3889	0.0144	273.556	3.9424	0.0037	0.2537	19
20	86.7362	0.0115	342.945	3.9539	0.0029	0.2529	20
21	108.420	0.0092	429.681	3.9631	0.0023	0.2523	21
22	135.525	0.0074	538.101	3.9705	0.0019	0.2519	22
23	169.407	0.0059	673.626	3.9764	0.0015	0.2515	23
24	211.758	0.0047	843.033	3.9811	0.0012	0.2512	24
25	264.698	0.0038	1054.79	3.9849	0.0009	0.2509	25
26	330.872	0.0030	1319.49	3.9879	0.0008	0.2508	26
27	413.590	0.0024	1650.36	3.9903	0.0006	0.2506	27
28	516.988	0.0019	2063.95	3.9923	0.0005	0.2505	28
29	646.235	0.0015	2580.94	3.9938	0.0004	0.2504	29
30	807.794	0.0012	3227.17	3.9950	0.0003	0.2503	30
35	2465.19	0.0004	9856.76	3.9984	0.0001	0.2501	35
40	7523.16	0.0001	30088.7	3.9995	a	0.2500	40
45	22958.9	a	91831.5	3.9998	a	0.2500	45
50	70064.9	a	280256	3.9999	a	0.2500	50
∞				4.0000		0.2500	∞

aLess than 0.0001.

TABLE E-13 Discrete Compounding; $i = 30\%$

	SINGLE PAYMENT		UNIFORM SERIES				
	Compound Amount Factor	Present Worth Factor	Compound Amount Factor	Present Worth Factor	Sinking Fund Factor	Capital Recovery Factor	
N	To find F Given P F/P	To find P Given F P/F	To find F Given A F/A	To find P Given A P/A	To find A Given F A/F	To find A Given P A/P	N
1	1.3000	0.7692	1.000	0.769	1.0000	1.3000	1
2	1.6900	0.5917	2.300	1.361	0.4348	0.7348	2
3	2.1970	0.4552	3.990	1.816	0.2506	0.5506	3
4	2.8561	0.3501	6.187	2.166	0.1616	0.4616	4
5	3.7129	0.2693	9.043	2.436	0.1106	0.4106	5
6	4.8268	0.2072	12.756	2.643	0.0784	0.3784	6
7	6.2749	0.1594	17.583	2.802	0.0569	0.3569	7
8	8.1573	0.1226	23.858	2.925	0.0419	0.3419	8
9	10.604	0.0943	32.015	3.019	0.0312	0.3312	9
10	13.786	0.0725	42.619	3.092	0.0235	0.3235	10
11	17.922	0.0558	56.405	3.147	0.0177	0.3177	11
12	23.298	0.0429	74.327	3.190	0.0135	0.3135	12
13	30.287	0.0330	97.625	3.223	0.0102	0.3102	13
14	39.374	0.0254	127.91	3.249	0.0078	0.3078	14
15	51.186	0.0195	167.29	3.268	0.0060	0.3060	15
16	66.542	0.0150	218.47	3.283	0.0046	0.3046	16
17	86.504	0.0116	285.01	3.295	0.0035	0.3035	17
18	112.46	0.0089	371.52	3.304	0.0027	0.3027	18
19	146.19	0.0068	483.97	3.311	0.0021	0.3021	19
20	190.05	0.0053	630.16	3.316	0.0016	0.3016	20
21	247.06	0.0040	820.21	3.320	0.0012	0.3012	21
22	321.18	0.0031	1067.3	3.323	0.0009	0.3009	22
23	417.54	0.0024	1388.5	3.325	0.0007	0.3007	23
24	542.80	0.0018	1806.0	3.327	0.0005	0.3005	24
25	705.64	0.0014	2348.8	3.329	0.0004	0.3004	25
26	917.33	0.0011	3054.4	3.330	0.0003	0.3003	26
27	1192.5	0.0008	3971.8	3.331	0.0003	0.3003	27
28	1550.3	0.0006	5164.3	3.331	0.0002	0.3002	28
29	2015.4	0.0005	6714.6	3.332	0.0002	0.3002	29
30	2620.0	0.0004	8730.0	3.332	0.0001	0.3001	30
31	3406.0	0.0003	11350.	3.332	a	0.3001	31
32	4427.8	0.0002	14756.	3.333	a	0.3001	32
33	5756.1	0.0002	19184.	3.333	a	0.3001	33
34	7483.0	0.0001	24940.	3.333	a	0.3000	34
35	9727.8	0.0001	32423.	3.333	a	0.3000	35
∞				3.333		0.3000	∞

aLess than 0.0001.

TABLE E-14   Discrete Compounding; $i = 40\%$

	SINGLE PAYMENT		UNIFORM SERIES				
	Compound Amount Factor	Present Worth Factor	Compound Amount Factor	Present Worth Factor	Sinking Fund Factor	Capital Recovery Factor	
N	To find F Given P F/P	To find P Given F P/F	To find F Given A F/A	To find P Given A P/A	To find A Given F A/F	To find A Given P A/P	N
1	1.4000	0.7143	1.000	0.714	1.000	1.4000	1
2	1.9600	0.5102	2.400	1.224	0.4167	0.8167	2
3	2.7440	0.3644	4.360	1.589	0.2294	0.6294	3
4	3.8416	0.2603	7.104	1.849	0.1408	0.5408	4
5	5.3782	0.1859	10.946	2.035	0.0934	0.4914	5
6	7.5295	0.1328	16.324	2.168	0.0613	0.4613	6
7	10.541	0.0949	23.853	2.263	0.0419	0.4419	7
8	14.758	0.0678	34.395	2.331	0.0291	0.4291	8
9	20.661	0.0484	49.153	2.379	0.0203	0.4203	9
10	28.925	0.0346	69.814	2.414	0.0143	0.4143	10
11	40.496	0.0247	98.739	2.438	0.0101	0.4101	11
12	56.694	0.0176	139.23	2.456	0.0072	0.4072	12
13	79.371	0.0126	195.93	2.469	0.0051	0.4051	13
14	111.12	0.0090	275.30	2.478	0.0036	0.4036	14
15	155.57	0.0064	386.42	2.484	0.0026	0.4026	15
16	217.80	0.0046	541.99	2.489	0.0018	0.4019	16
17	304.91	0.0033	759.78	2.492	0.0013	0.4013	17
18	426.88	0.0023	1064.7	2.494	0.0009	0.4009	18
19	597.63	0.0017	1491.6	2.496	0.0007	0.4007	19
20	836.68	0.0012	2089.2	2.497	0.0005	0.4005	20
21	1171.4	0.0009	2925.9	2.498	0.0003	0.4003	21
22	1639.9	0.0006	4097.2	2.498	0.0002	0.4002	22
23	2295.9	0.0004	5737.1	2.499	0.0002	0.4002	23
24	3214.2	0.0003	8033.0	2.499	0.0001	0.4001	24
25	4499.9	0.0002	11247.	2.499	a	0.4001	25
26	6299.8	0.0002	15747.	2.500	a	0.4001	26
27	8819.8	0.0001	22047.	2.500	a	0.4000	27
28	12348.	0.0001	30867.	2.500	a	0.4000	28
29	17287.	0.0001	43214.	2.500	a	0.4000	29
30	24201.	a	60501.	2.500	a	0.4000	30
∞				2.500		0.4000	∞

aLess than 0.0001.

	SINGLE PAYMENT		UNIFORM SERIES				
	Compound Amount Factor	Present Worth Factor	Compound Amount Factor	Present Worth Factor	Sinking Fund Factor	Capital Recovery Factor	
N	To find F Given P F/P	To find P Given F P/F	To find F Given A F/A	To find P Given A P/A	To find A Given F A/F	To find A Given P A/P	N
1	1.5000	0.6667	1.000	0.667	1.0000	1.5000	1
2	2.2500	0.4444	2.500	1.111	0.4000	0.9000	2
3	3.3750	0.2963	4.750	1.407	0.2101	0.7105	3
4	5.0625	0.1975	8.125	1.605	0.1231	0.6231	4
5	7.5938	0.1317	13.188	1.737	0.0758	0.5758	5
6	11.391	0.0878	20.781	1.824	0.0481	0.5481	6
7	17.086	0.0585	32.172	1.883	0.0311	0.5311	7
8	25.629	0.0390	49.258	1.922	0.0203	0.5203	8
9	38.443	0.0260	74.887	1.948	0.0134	0.5134	9
10	57.665	0.0173	113.33	1.965	0.0088	0.5088	10
11	86.498	0.0116	171.00	1.977	0.0059	0.5059	11
12	129.75	0.0077	257.49	1.985	0.0039	0.5039	12
13	194.62	0.0051	387.24	1.990	0.0026	0.5026	13
14	291.93	0.0034	581.86	1.993	0.0017	0.5017	14
15	437.89	0.0023	873.79	1.995	0.0011	0.5011	15
16	656.84	0.0015	1311.7	1.997	0.0008	0.5008	16
17	985.26	0.0010	1968.5	1.998	0.0005	0.5005	17
18	1477.9	0.0007	2953.8	1.999	0.0003	0.5003	18
19	2216.8	0.0005	4431.7	1.999	0.0002	0.5002	19
20	3325.3	0.0003	6648.5	1.999	0.0002	0.5002	20
21	4987.9	0.0002	9973.8	2.000	0.0001	0.5001	21
22	7481.8	0.0001	14962.	2.000	a	0.5001	22
23	11223.	0.0001	22443.	2.000	a	0.5000	23
24	16834.	0.0001	33666.	2.000	a	0.5000	24
25	25251.	a	50500.	2.000	a	0.5000	25
∞				2.000		0.5000	∞

TABLE E-16 Gradient to Present Worth Conversion Factor for Discrete Compounding (to Find P, Given G)

$$(P/G, i\%, N) = \frac{1}{i}\left[\frac{(1+i)^N - 1}{i(1+i)^N} - \frac{N}{(1+i)^N}\right]$$

N	1%	2%	5%	8%	10%	12%	15%	20%	25%	30%	50%	N
1	0.00	0.00	0.00	0.00	0.00	0.00	0.00	0.00	0.00	0.00	0.00	1
2	0.98	0.96	0.91	0.86	0.83	0.80	0.76	0.69	0.64	0.59	0.44	2
3	2.92	2.85	2.63	2.45	2.33	2.22	2.07	1.85	1.66	1.50	1.04	3
4	5.80	5.62	5.10	4.65	4.38	4.13	3.79	3.30	2.89	2.55	1.63	4
5	9.61	9.24	8.24	7.37	6.86	6.40	5.78	4.91	4.20	3.63	2.16	5
6	14.32	13.68	11.97	10.52	9.68	8.93	7.94	6.58	5.51	4.67	2.60	6
7	19.92	18.90	16.23	14.02	12.76	11.64	10.19	8.26	6.77	5.62	2.95	7
8	26.38	24.88	20.97	17.81	16.03	14.47	12.48	9.88	7.95	6.48	3.22	8
9	33.69	31.57	26.13	21.81	19.42	17.36	14.75	11.43	9.02	7.23	3.43	9
10	41.84	38.95	31.65	25.98	22.89	20.25	16.98	12.89	9.99	7.89	3.58	10
11	50.80	47.00	37.50	30.27	26.40	23.13	19.13	14.23	10.85	8.45	3.70	11
12	60.57	55.67	43.62	34.63	29.90	25.95	21.18	15.47	11.60	8.92	3.78	12
15	94.48	85.20	63.29	47.89	40.15	33.92	26.69	18.51	13.33	9.92	3.92	15
20	165.46	144.60	98.49	69.09	55.41	44.97	33.58	21.74	14.89	10.70	3.99	20
25	252.89	214.26	134.23	87.80	67.70	53.10	38.03	23.43	15.56	10.98	4.00	25
30	355.00	291.72	168.62	103.46	77.08	58.78	40.75	24.26	15.83	11.07	—	30
35	470.15	374.88	200.58	116.09	83.99	62.61	42.36	24.66	15.94	11.10	—	35
40	596.85	461.99	229.55	126.04	88.95	65.12	43.28	24.85	15.98	11.11	—	40
45	733.70	551.56	255.31	133.73	92.45	66.73	43.81	24.93	15.99	—	—	45
50	879.41	642.36	277.91	139.59	94.89	67.76	44.10	24.97	16.00	—	—	50
60	1192.80	823.70	314.34	147.30	97.70	68.81	44.34	24.99	—	—	—	60
70	1528.64	999.83	340.84	151.53	98.99	69.21	44.42	—	—	—	—	70
80	1879.87	1166.79	359.65	153.80	99.56	69.36	44.47	—	—	—	—	80
90	2240.55	1322.17	372.75	154.99	99.81	—	—	—	—	—	—	90
100	2605.76	1464.75	381.75	155.61	99.92	—	—	—	—	—	—	100

$$(A/G, i\%, N) = \frac{1}{i} - \frac{N}{(1+i)^N - 1}$$

N	1%	2%	5%	8%	10%	12%	15%	20%	25%	30%	50%	N
1	0.0001	0.0000	0.00	0.0000	0.0000	0.0000	0.0000	0.0000	0.0000	0.00	0.00	1
2	0.4974	0.4950	0.49	0.4808	0.4762	0.4717	0.4651	0.4545	0.4444	0.43	0.40	2
3	0.9932	0.9868	0.97	0.9487	0.9366	0.9246	0.9071	0.8791	0.8525	0.83	0.74	3
4	1.4874	1.4752	1.44	1.4040	1.3812	1.3589	1.3263	1.2742	1.2249	1.18	1.02	4
5	1.9799	1.9604	1.90	1.8465	1.8101	1.7746	1.7228	1.6405	1.5631	1.49	1.24	5
6	2.4708	2.4422	2.36	2.2763	2.2236	2.1720	2.0972	1.9788	1.8683	1.77	1.42	6
7	2.9600	2.9208	2.81	2.6937	2.6216	2.5515	2.4498	2.2902	2.1424	2.01	1.56	7
8	3.4476	3.3961	3.24	3.0985	3.0045	2.9131	2.7813	2.5756	2.3872	2.22	1.68	8
9	3.9335	3.8680	3.68	3.4910	3.3724	3.2574	3.0922	2.8364	2.6048	2.40	1.76	9
10	4.4177	4.3367	4.10	3.8713	3.7255	3.5847	3.3832	3.0739	2.7971	2.55	1.82	10
11	4.9003	4.8021	4.51	4.2395	4.0641	3.8953	3.6549	3.2893	2.9663	2.68	1.87	11
12	5.3813	5.2642	4.92	4.5957	4.3884	4.1897	3.9082	3.4841	3.1145	2.80	1.91	12
15	6.8141	6.6309	6.10	5.5945	5.2789	4.9803	4.5650	3.9588	3.4530	3.03	1.97	15
20	9.1692	8.8433	7.90	7.0369	6.5081	6.0202	5.3651	4.4643	3.7667	3.23	1.99	20
25	11.4829	10.9744	9.52	8.2254	7.4580	6.7708	5.8834	4.7352	3.9052	3.30	2.00	25
30	13.7555	13.0251	10.97	9.1897	8.1762	7.2974	6.2066	4.8731	3.9628	3.32	—	30
35	15.9869	14.9961	12.25	9.9611	8.7086	7.6577	6.4019	4.9406	3.9858	3.33	—	35
40	18.1774	16.8885	13.38	10.5699	9.0962	7.8988	6.5168	4.9728	3.9947	3.33	—	40
45	20.3271	18.7033	14.36	11.0447	9.3740	8.0572	6.5830	4.9877	3.9980	—	—	45
50	22.4362	20.4420	15.22	11.4107	9.5704	8.1597	6.6205	4.9945	3.9993	—	—	50
60	26.5331	23.6961	16.61	11.9015	9.8023	8.2664	6.6530	4.9989	—	—	—	60
70	30.4701	26.6632	17.62	12.1783	9.9113	8.3082	6.6627	—	—	—	—	70
80	34.2490	29.3572	18.35	12.3301	9.9609	8.3241	6.6656	—	—	—	—	80
90	37.8723	31.7929	18.87	12.4116	9.9831	—	—	—	—	—	—	90
100	41.3424	33.9863	19.23	12.4545	9.9927	—	—	—	—	—	—	100

Interest and Annuity Tables for Continuous Compounding

(For various values of r from 1% to 25%)

r = nominal interest rate per year, compounded continuously

N = number of compounding periods

$$(F/P, r\%, N) = e^{rN}$$

$$(P/F, r\%, N) = e^{-rN} = \frac{1}{e^{rN}}$$

$$(F/A, r\%, N) = \frac{e^{rN} - 1}{e^{r} - 1}$$

$$(P/A, r\%, N) = \frac{e^{rN} - 1}{e^{rN}(e^{r} - 1)}$$

$$(F/\overline{A}, r\%, N) = \frac{e^{rN} - 1}{r}$$

$$(P/\overline{A}, r\%, N) = \frac{e^{rN} - 1}{re^{rN}}$$

TABLE F-1 Continuous Compounding; $r = 1\%$

	Discrete Flows				Continuous Flows		
	SINGLE PAYMENT		UNIFORM SERIES		UNIFORM SERIES		
	Compound Amount Factor	Present Worth Factor	Compound Amount Factor	Present Worth Factor	Compound Amount Factor	Present Worth Factor	
N	To find F Given P F/P	To find P Given F P/F	To find F Given A F/A	To find P Given A P/A	To find F Given $\overline{A}$ $F/\overline{A}$	To find P Given $\overline{A}$ $P/\overline{A}$	N
1	1.0101	0.9900	1.0000	0.9900	1.0050	0.9950	1
2	1.0202	0.9802	2.0101	1.9703	2.0201	1.9801	2
3	1.0305	0.9704	3.0303	2.9407	3.0455	2.9554	3
4	1.0408	0.9608	4.0607	3.9015	4.0811	3.9211	4
5	1.0513	0.9512	5.1015	4.8527	5.1271	4.8771	5
6	1.0618	0.9418	6.1528	5.7945	6.1837	5.8235	6
7	1.0725	0.9324	7.2146	6.7269	7.2508	6.7606	7
8	1.0833	0.9231	8.2871	7.6500	8.3287	7.6884	8
9	1.0942	0.9139	9.3704	8.5639	9.4174	8.6069	9
10	1.1052	0.9048	10.4646	9.4688	10.5171	9.5163	10
11	1.1163	0.8958	11.5698	10.3646	11.6278	10.4166	11
12	1.1275	0.8869	12.6860	11.2515	12.7497	11.3080	12
13	1.1388	0.8781	13.8135	12.1296	13.8828	12.1905	13
14	1.1503	0.8694	14.9524	12.9990	15.0274	13.0642	14
15	1.1618	0.8607	16.1026	13.8597	16.1834	13.9292	15
16	1.1735	0.8521	17.2645	14.7118	17.3511	14.7856	16
17	1.1853	0.8437	18.4380	15.5555	18.5305	15.6335	17
18	1.1972	0.8353	19.6233	16.3908	19.7217	16.4730	18
19	1.2092	0.8270	20.8205	17.2177	20.9250	17.3041	19
20	1.2214	0.8187	22.0298	18.0365	22.1403	18.1269	20
21	1.2337	0.8106	23.2512	18.8470	23.3678	18.9416	21
22	1.2461	0.8025	24.4849	19.6496	24.6077	19.7481	22
23	1.2586	0.7945	25.7309	20.4441	25.8600	20.5466	23
24	1.2712	0.7866	26.9895	21.2307	27.1249	21.3372	24
25	1.2840	0.7788	28.2608	22.0095	28.4025	22.1199	25
26	1.2969	0.7711	29.5448	22.7806	29.6930	22.8948	26
27	1.3100	0.7634	30.8417	23.5439	30.9964	23.6621	27
28	1.3231	0.7558	32.1517	24.2997	32.3130	24.4216	28
29	1.3364	0.7483	33.4748	25.0480	33.6427	25.1736	29
30	1.3499	0.7408	34.8113	25.7888	34.9859	25.9182	30
35	1.4191	0.7047	41.6976	29.3838	41.9068	29.5312	35
40	1.4918	0.6703	48.9370	32.8034	49.1825	32.9680	40
45	1.5683	0.6376	56.5476	36.0563	56.8312	36.2372	45
50	1.6487	0.6065	64.5483	39.1505	64.8721	39.3469	50
55	1.7333	0.5769	72.9593	42.0939	73.3253	42.3050	55
60	1.8221	0.5488	81.8015	44.8936	82.2119	45.1188	60
65	1.9155	0.5220	91.0971	47.5569	91.5541	47.7954	65
70	2.0138	0.4966	100.869	50.0902	101.375	50.3415	70
75	2.1170	0.4724	111.143	52.5000	111.700	52.7633	75
80	2.2255	0.4493	121.942	54.7923	122.554	55.0671	80
85	2.3396	0.4274	133.296	56.9727	133.965	57.2585	85
90	2.4596	0.4066	145.232	59.0468	145.960	59.3430	90
95	2.5857	0.3867	157.780	61.0198	158.571	61.3259	95
100	2.7183	0.3679	170.971	62.8965	171.828	63.2121	100

TABLE F-2 Continuous Compounding; $r = 2\%$

	Discrete Flows				Continuous Flows		
	SINGLE PAYMENT		UNIFORM SERIES		UNIFORM SERIES		
	Compound Amount Factor	Present Worth Factor	Compound Amount Factor	Present Worth Factor	Compound Amount Factor	Present Worth Factor	
N	To find F Given P F/P	To find P Given F P/F	To find F Given A F/A	To find P Given A P/A	To find F Given $\overline{A}$ $F/\overline{A}$	To find P Given $\overline{A}$ $P/\overline{A}$	N
1	1.0202	0.9802	1.0000	0.9802	1.0101	0.9901	1
2	1.0408	0.9608	2.0202	1.9410	2.0405	1.9605	2
3	1.0618	0.9418	3.0610	2.8828	3.0918	2.9118	3
4	1.0833	0.9231	4.1228	3.8059	4.1644	3.8442	4
5	1.1052	0.9048	5.2061	4.7107	5.2585	4.7581	5
6	1.1275	0.8869	6.3113	5.5976	6.3748	5.6540	6
7	1.1503	0.8694	7.4388	6.4670	7.5137	6.5321	7
8	1.1735	0.8521	8.5891	7.3191	8.6755	7.3928	8
9	1.1972	0.8353	9.7626	8.1544	9.8609	8.2365	9
10	1.2214	0.8187	10.9598	8.9731	11.0701	9.0635	10
11	1.2461	0.8025	12.1812	9.7756	12.3038	9.8741	11
12	1.2712	0.7866	13.4273	10.5623	13.5625	10.6686	12
13	1.2969	0.7711	14.6985	11.3333	14.8465	11.4474	13
14	1.3231	0.7558	15.9955	12.0891	16.1565	12.2108	14
15	1.3499	0.7408	17.3186	12.8299	17.4929	12.9591	15
16	1.3771	0.7261	18.6685	13.5561	18.8564	13.6925	16
17	1.4049	0.7118	20.0456	14.2678	20.2474	14.4115	17
18	1.4333	0.6977	21.4505	14.9655	21.6665	15.1162	18
19	1.4623	0.6839	22.8839	15.6494	23.1142	15.8069	19
20	1.4918	0.6703	24.3461	16.3197	24.5912	16.4840	20
21	1.5220	0.6570	25.8380	16.9768	26.0981	17.1477	21
22	1.5527	0.6440	27.3599	17.6208	27.6354	17.7982	22
23	1.5841	0.6313	28.9126	18.2521	29.2037	18.4358	23
24	1.6161	0.6188	30.4967	18.8709	30.8037	19.0608	24
25	1.6487	0.6065	32.1128	19.4774	32.4361	19.6735	25
26	1.6820	0.5945	33.7615	20.0719	34.1014	20.2740	26
27	1.7160	0.5827	35.4435	20.6547	35.8003	20.8626	27
28	1.7507	0.5712	37.1595	21.2259	37.5336	21.4395	28
29	1.7860	0.5599	38.9102	21.7858	39.3019	22.0051	29
30	1.8221	0.5488	40.6962	22.3346	41.1059	22.5594	30
35	2.0138	0.4966	50.1824	24.9199	50.6876	25.1707	35
40	2.2255	0.4493	60.6663	27.2591	61.2770	27.5336	40
45	2.4596	0.4066	72.2528	29.3758	72.9802	29.6715	45
50	2.7183	0.3679	85.0578	31.2910	85.9141	31.6060	50
55	3.0042	0.3329	99.2096	33.0240	100.208	33.3564	55
60	3.3201	0.3012	114.850	34.5921	116.006	34.9403	60
65	3.6693	0.2725	132.135	36.0109	133.465	36.3734	65
70	4.0552	0.2466	151.238	37.2947	152.760	37.6702	70
75	4.4817	0.2231	172.349	38.4564	174.084	38.8435	75
80	4.9530	0.2019	195.682	39.5075	197.652	39.9052	80
85	5.4739	0.1827	221.468	40.4585	223.697	40.8658	85
90	6.0496	0.1653	249.966	41.3191	252.482	41.7351	90
95	6.6859	0.1496	281.461	42.0978	284.295	42.5216	95
100	7.3891	0.1353	316.269	42.8023	319.453	43.2332	100

TABLE F-3 Continuous Compounding: $r = 3\%$

	Discrete Flows				Continuous Flows		
	SINGLE PAYMENT		UNIFORM SERIES		UNIFORM SERIES		
	Compound Amount Factor	Present Worth Factor	Compound Amount Factor	Present Worth Factor	Compound Amount Factor	Present Worth Factor	
	To find F Given P	To find P Given F	To find F Given A	To find P Given A	To find F Given $\overline{A}$	To find P Given $\overline{A}$	
N	F/P	P/F	F/A	P/A	$F/\overline{A}$	$P/\overline{A}$	N
1	1.0305	0.9704	1.0000	0.9704	1.0152	0.9851	1
2	1.0618	0.9418	2.0305	1.9122	2.0612	1.9412	2
3	1.0942	0.9139	3.0923	2.8261	3.1391	2.8690	3
4	1.1275	0.8869	4.1865	3.7131	4.2499	3.7693	4
5	1.1618	0.8607	5.3140	4.5738	5.3945	4.6431	5
6	1.1972	0.8353	6.4758	5.4090	6.5739	5.4910	6
7	1.2337	0.8106	7.6730	6.2196	7.7893	6.3139	7
8	1.2712	0.7866	8.9067	7.0063	9.0416	7.1124	8
9	1.3100	0.7634	10.1779	7.7696	10.3321	7.8874	9
10	1.3499	0.7408	11.4879	8.5105	11.6620	8.6394	10
11	1.3910	0.7189	12.8378	9.2294	13.0323	9.3692	11
12	1.4333	0.6977	14.2287	9.9271	14.4443	10.0775	12
13	1.4770	0.6771	15.6621	10.6041	15.8994	10.7648	13
14	1.5220	0.6570	17.1390	11.2612	17.3987	11.4318	14
15	1.5683	0.6376	18.6610	11.8988	18.9437	12.0791	15
16	1.6161	0.6188	20.2293	12.5176	20.5358	12.7072	16
17	1.6653	0.6005	21.8454	13.1181	22.1764	13.3168	17
18	1.7160	0.5827	23.5107	13.7008	23.8669	13.9084	18
19	1.7683	0.5665	25.2267	14.2663	25.6089	14.4825	19
20	1.8221	0.5488	26.9950	14.8151	27.4040	15.0396	20
21	1.8776	0.5326	28.8171	15.3477	29.2537	15.5803	21
22	1.9348	0.5169	30.6947	15.8646	31.1597	16.1050	22
23	1.9937	0.5016	32.6295	16.3662	33.1239	16.6141	23
24	2.0544	0.4868	34.6232	16.8529	35.1478	17.1083	24
25	2.1170	0.4724	36.6776	17.3253	37.2333	17.5878	25
26	2.1815	0.4584	38.7946	17.7837	39.3824	18.0531	26
27	2.2479	0.4449	40.9761	18.2285	41.5969	18.5047	27
28	2.3164	0.4317	43.2240	18.6603	43.8789	18.9430	28
29	2.3869	0.4190	45.5404	19.0792	46.2304	19.3683	29
30	2.4596	0.4066	47.9273	19.4858	48.6534	19.7810	30
35	2.8577	0.3499	60.9975	21.3453	61.9217	21.6687	35
40	3.3201	0.3012	76.1830	22.9459	77.3372	23.2935	40
45	3.8574	0.2592	93.8260	24.3235	95.2475	24.6920	45
50	4.4817	0.2231	114.324	25.5092	116.056	25.8957	50
55	5.2070	0.1920	138.140	26.5297	140.233	26.9317	55
60	6.0496	0.1653	165.809	27.4081	168.322	27.8234	60
65	7.0287	0.1423	197.957	28.1641	200.956	28.5909	65
70	8.1662	0.1225	235.307	28.8149	238.872	29.2515	70
75	9.4877	0.1054	278.702	29.3750	282.924	29.8200	75
80	11.0232	0.0907	329.119	29.8570	334.106	30.3094	80
85	12.8071	0.0781	387.696	30.2720	393.570	30.7306	85
90	14.8797	0.0672	455.753	30.6291	462.658	31.0931	90
95	17.2878	0.0578	534.823	30.9365	542.926	31.4052	95
100	20.0855	0.0498	626.690	31.2010	636.185	31.6738	100

TABLE F-4 Continuous Compounding; $r = 5\%$

	Discrete Flows				Continuous Flows		
	SINGLE PAYMENT		UNIFORM SERIES		UNIFORM SERIES		
	Compound Amount Factor	Present Worth Factor	Compound Amount Factor	Present Worth Factor	Compound Amount Factor	Present Worth Factor	
N	To find F Given P F/P	To find P Given F P/F	To find F Given A F/A	To find P Given A P/A	To find F Given $\overline{A}$ $F/\overline{A}$	To find P Given $\overline{A}$ $P/\overline{A}$	N
1	1.0513	0.9512	1.0000	0.9512	1.0254	0.9754	1
2	1.1052	0.9048	2.0513	1.8561	2.1034	1.9033	2
3	1.1618	0.8607	3.1564	2.7168	3.2367	2.7858	3
4	1.2214	0.8187	4.3183	3.5355	4.4281	3.6254	4
5	1.2840	0.7788	5.5397	4.3143	5.6805	4.4240	5
6	1.3499	0.7408	6.8237	5.0551	6.9972	5.1836	6
7	1.4191	0.7047	8.1736	5.7598	8.3814	5.9062	7
8	1.4918	0.6703	9.5926	6.4301	9.8365	6.5936	8
9	1.5683	0.6376	11.0845	7.0678	11.3662	7.2474	9
10	1.6487	0.6065	12.6528	7.6743	12.9744	7.8694	10
11	1.7333	0.5759	14.3015	8.2512	14.6651	8.4610	11
12	1.8221	0.5488	16.0347	8.8001	16.4424	9.0238	12
13	1.9155	0.5220	17.8569	9.3221	18.3108	9.5591	13
14	2.0138	0.4966	19.7724	9.8187	20.2751	10.0683	14
15	2.1170	0.4724	21.7862	10.2911	22.3400	10.5527	15
16	2.2255	0.4493	23.9032	10.7404	24.5108	11.0134	16
17	2.3396	0.4274	26.1287	11.1678	26.7929	11.4517	17
18	2.4596	0.4066	28.4683	11.5744	29.1921	11.8686	18
19	2.5857	0.3867	30.9279	11.9611	31.7142	12.2652	19
20	2.7183	0.3679	33.5137	12.3290	34.3656	12.6424	20
21	2.8577	0.3499	36.2319	12.6789	37.1530	13.0012	21
22	3.0042	0.3329	39.0896	13.0118	40.0833	13.3426	22
23	3.1582	0.3166	42.0938	13.3284	43.1639	13.6673	23
24	3.3201	0.3012	45.2519	13.6296	46.4023	13.9761	24
25	3.4903	0.2865	48.5721	13.9161	49.8069	14.2699	25
26	3.6693	0.2725	52.0624	14.1887	53.3859	14.5494	26
27	3.8574	0.2592	55.7317	14.4479	57.1485	14.8152	27
28	4.0552	0.2466	59.5891	14.6945	61.1040	15.0681	28
29	4.2631	0.2346	63.6443	14.9291	65.2623	15.3086	29
30	4.4817	0.2231	67.9074	15.1522	69.6338	15.5374	30
35	5.7546	0.1738	92.7346	16.1149	95.0921	16.5245	35
40	7.3891	0.1353	124.613	16.8646	127.781	17.2933	40
45	9.4877	0.1054	165.546	17.4484	169.755	17.8920	45
50	12.1825	0.0821	218.105	17.9032	223.650	18.3583	50
55	15.6426	0.0639	285.592	18.2573	292.853	18.7214	55
60	20.0855	0.0498	372.247	18.5331	381.711	19.0043	60
65	25.7903	0.0388	483.515	18.7479	495.807	19.2245	65
70	33.1155	0.0302	626.385	18.9152	642.309	19.3961	70
75	42.5211	0.0235	809.834	19.0455	830.422	19.5296	75
80	54.5981	0.0183	1045.39	19.1469	1071.963	19.6337	80
85	70.1054	0.0143	1347.84	19.2260	1382.108	19.7147	85
90	90.0171	0.0111	1736.20	19.2875	1780.342	19.7778	90
95	115.584	0.0087	2234.87	19.3354	2291.686	19.8270	95
100	148.413	0.0067	2875.17	19.3727	2948.263	19.8652	100

TABLE F-5 Continuous Compounding; $r = 8\%$

	Discrete Flows				Continuous Flows		
	SINGLE PAYMENT		UNIFORM SERIES		UNIFORM SERIES		
	Compound Amount Factor	Present Worth Factor	Compound Amount Factor	Present Worth Factor	Compound Amount Factor	Present Worth Factor	
N	To find F Given F F/P	To find P Given P P/F	To find F Given A F/A	To find P Given A P/A	To find F Given $\overline{A}$ $F/\overline{A}$	To find P Given $\overline{A}$ $P/\overline{A}$	N
1	1.0833	0.9231	1.0000	0.9231	1.0411	0.9610	1
2	1.1735	0.8521	2.0833	1.7753	2.1689	1.8482	2
3	1.2712	0.7866	3.2568	2.5619	3.3906	2.6672	3
4	1.3771	0.7261	4.5280	3.2880	4.7141	3.4231	4
5	1.4918	0.6703	5.9052	3.9584	6.1478	4.1210	5
6	1.6161	0.6188	7.3970	4.5771	7.7009	4.7652	6
7	1.7507	0.5712	9.0131	5.1483	9.3834	5.3599	7
8	1.8965	0.5273	10.7637	5.6756	11.2060	5.9088	8
9	2.0544	0.4868	12.6602	6.1624	13.1804	6.4156	9
10	2.2255	0.4493	14.7147	6.6117	15.3193	6.8834	10
11	2.4109	0.4148	16.9402	7.0265	17.6362	7.3152	11
12	2.6117	0.3829	19.3511	7.4094	20.1462	7.7138	12
13	2.8292	0.3535	21.9628	7.7629	22.8652	8.0818	13
14	3.0649	0.3263	24.7920	8.0891	25.8107	8.4215	14
15	3.3201	0.3012	27.8569	8.3903	29.0015	8.7351	15
16	3.5966	0.2780	31.1770	8.6684	32.4580	9.0245	16
17	3.8962	0.2567	34.7736	8.9250	36.2024	9.2917	17
18	4.2207	0.2369	38.6698	9.1620	40.2587	9.5384	18
19	4.5722	0.2187	42.8905	9.3807	44.6528	9.7661	19
20	4.9530	0.2019	47.4627	9.5826	49.4129	9.9763	20
21	5.3656	0.1864	52.4158	9.7689	54.5694	10.1703	21
22	5.8124	0.1720	57.7813	9.9410	60.1555	10.3494	22
23	6.2965	0.1588	63.5938	10.0998	66.2067	10.5148	23
24	6.8120	0.1466	69.8903	10.2464	72.7620	10.6674	24
25	7.3891	0.1353	76.7113	10.3817	79.8632	10.8083	25
26	8.0045	0.1249	84.1003	10.5067	87.5559	10.9384	26
27	8.6711	0.1153	92.1048	10.6220	95.8892	11.0584	27
28	9.3933	0.1065	100.776	10.7285	104.917	11.1693	28
29	10.1757	0.0983	110.169	10.8267	114.696	11.2716	29
30	11.0232	0.0907	120.345	10.9174	125.290	11.3660	30
35	16.4446	0.0608	185.439	11.2765	193.058	11.7399	35
40	24.5325	0.0408	282.547	11.5172	294.157	11.9905	40
45	36.5982	0.0273	427.416	11.6786	444.978	12.1585	45
50	54.5982	0.0183	643.535	11.7868	669.977	12.2711	50
55	81.4509	0.0123	965.947	11.8593	1005.64	12.3465	55
60	121.510	0.0082	1446.93	11.9079	1506.38	12.3971	60
65	181.272	0.0055	2164.47	11.9404	2253.40	12.4310	65
70	270.426	0.0037	3234.91	11.9623	3367.83	12.4538	70
75	403.429	0.0025	4831.83	11.9769	5030.36	12.4690	75
80	601.845	0.0017	7214.15	11.9867	7510.56	12.4792	80
85	897.847	0.0011	10768.1	11.9933	11210.6	12.4861	85
90	1339.43	0.0007	16070.1	11.9977	16730.4	12.4907	90
95	1998.20	0.0005	23979.7	12.0007	24964.9	12.4937	95
100	2980.96	0.0003	35779.3	12.0026	37249.5	12.4958	100

TABLE F-6 Continuous Compounding; $r = 10\%$

	SINGLE PAYMENT		UNIFORM SERIES		UNIFORM SERIES		
	Compound Amount Factor	Present Worth Factor	Compound Amount Factor	Present Worth Factor	Compound Amount Factor	Present Worth Factor	
	To find F Given P	To find P Given F	To find F Given A	To find P Given A	To find F Given $\overline{A}$	To find P Given $\overline{A}$	
N	F/P	P/F	F/A	P/A	$F/\overline{A}$	$P/\overline{A}$	N
1	1.1052	0.9048	1.0000	0.9048	1.0517	0.9516	1
2	1.2214	0.8187	2.1052	1.7236	2.2140	1.8127	2
3	1.3499	0.7408	3.3266	2.4644	3.4986	2.5918	3
4	1.4918	0.6703	4.6764	3.1347	4.9182	3.2968	4
5	1.6487	0.6065	6.1683	3.7412	6.4872	3.9347	5
6	1.8221	0.5488	7.8170	4.2900	8.2212	4.5119	6
7	2.0138	0.4966	9.6391	4.7866	10.1375	5.0341	7
8	2.2255	0.4493	11.6528	5.2360	12.2554	5.5067	8
9	2.4596	0.4066	13.8784	5.6425	14.5960	5.9343	9
10	2.7183	0.3679	16.3380	6.0104	17.1828	6.3212	10
11	3.0042	0.3329	19.0563	6.3433	20.0417	6.6713	11
12	3.3201	0.3012	22.0604	6.6445	23.2012	6.9881	12
13	3.6693	0.2725	25.3806	6.9170	26.6930	7.2747	13
14	4.0552	0.2466	29.0499	7.1636	30.5520	7.5340	14
15	4.4817	0.2231	33.1051	7.3867	34.8169	7.7687	15
16	4.9530	0.2019	37.5867	7.5886	39.5303	7.9810	16
17	5.4739	0.1827	42.5398	7.7713	44.7395	8.1732	17
18	6.0496	0.1653	48.0137	7.9366	50.4965	8.3470	18
19	6.6859	0.1496	54.0634	8.0862	56.8589	8.5043	19
20	7.3891	0.1353	60.7493	8.2215	63.8906	8.6466	20
21	8.1662	0.1225	68.1383	8.3440	71.6617	8.7754	21
22	9.0250	0.1108	76.3045	8.4548	80.2501	8.8920	22
23	9.9742	0.1003	85.3295	8.5550	89.7418	8.9974	23
24	11.0232	0.0907	95.3037	8.6458	100.232	9.0928	24
25	12.1825	0.0821	106.327	8.7278	111.825	9.1791	25
26	13.4637	0.0743	118.509	8.8021	124.637	9.2573	26
27	14.8797	0.0672	131.973	8.8693	138.797	9.3279	27
28	16.4446	0.0608	146.853	8.9301	154.446	9.3919	28
29	18.1741	0.0550	163.298	8.9852	171.741	9.4498	29
30	20.0855	0.0498	181.472	9.0349	190.855	9.5021	30
35	33.1155	0.0302	305.364	9.2212	321.154	9.6980	35
40	54.5981	0.0183	509.629	9.3342	535.982	9.8168	40
45	90.0171	0.0111	846.404	9.4027	890.171	9.8889	45
50	148.413	0.0067	1401.65	9.4443	1474.13	9.9326	50
55	244.692	0.0041	2317.10	9.4695	2436.92	9.9591	55
60	403.429	0.0025	3826.43	9.4848	4024.29	9.9752	60
65	665.142	0.0015	6314.88	9.4940	6641.42	9.9850	65
70	1096.63	0.0009	10417.6	9.4997	10956.3	9.9909	70
75	1808.04	0.0006	17182.0	9.5031	18070.7	9.9945	75
80	2980.96	0.0003	28334.4	9.5051	29799.6	9.9966	80
85	4914.77	0.0002	46721.7	9.5064	49137.7	9.9980	85
90	8103.08	0.0001	77037.3	9.5072	81020.8	9.9988	90
95	13359.7	a	127019	9.5076	133587	9.9993	95
100	22026.5	a	209425	9.5079	220255	9.9995	100

Discrete Flows (SINGLE PAYMENT, UNIFORM SERIES); Continuous Flows (UNIFORM SERIES)

[a]Less than 0.0001.

TABLE F-7   Continuous Compounding: _r_ = 12%

	Discrete Flows				Continuous Flows		
	SINGLE PAYMENT		UNIFORM SERIES		UNIFORM SERIES		
	Compound Amount Factor	Present Worth Factor	Compound Amount Factor	Present Worth Factor	Compound Amount Factor	Present Worth Factor	
_N_	To find _F_ Given _P_ _F/P_	To find _P_ Given _F_ _P/F_	To find _F_ Given _A_ _F/A_	To find _P_ Given _A_ _P/A_	To find _F_ Given $\overline{A}$ $F/\overline{A}$	To find _P_ Given $\overline{A}$ $P/\overline{A}$	_N_
1	1.1275	0.8869	1.0000	0.8869	1.0625	0.9423	1
2	1.2712	0.7866	2.1275	1.6735	2.2604	1.7781	2
3	1.4333	0.6977	3.3987	2.3712	3.6111	2.5194	3
4	1.6161	0.6188	4.8321	2.9900	5.1340	3.1768	4
5	1.8221	0.5488	6.4481	3.5388	6.8510	3.7599	5
6	2.0544	0.4868	8.2703	4.0256	8.7869	4.2771	6
7	2.3164	0.4317	10.3247	4.4573	10.9679	4.7357	7
8	2.6117	0.3829	12.6411	4.8402	13.4308	5.1426	8
9	2.9447	0.3396	15.2528	5.1798	16.2057	5.5034	9
10	3.3201	0.3012	18.1974	5.4810	19.3343	5.8234	10
11	3.7434	0.2671	21.5176	5.7481	22.8618	6.1072	11
12	4.2207	0.2369	25.2610	5.9850	26.8391	6.3589	12
13	4.7588	0.2101	29.4817	6.1952	31.3235	6.5822	13
14	5.3656	0.1864	34.2405	6.3815	36.3796	6.7802	14
15	6.0496	0.1653	39.6061	6.5468	42.0804	6.9558	15
16	6.8210	0.1466	45.6557	6.6934	48.5080	7.1116	16
17	7.6906	0.1300	52.4767	6.8235	55.7551	7.2498	17
18	8.6711	0.1153	60.1673	6.9388	63.9261	7.3723	18
19	9.7767	0.1023	68.8384	7.0411	73.1390	7.4810	19
20	11.0232	0.0907	78.6151	7.1318	83.5265	7.5774	20
21	12.4286	0.0805	89.6383	7.2123	95.2383	7.6628	21
22	14.0132	0.0714	102.067	7.2836	108.443	7.7387	22
23	15.7998	0.0633	116.080	7.3469	123.332	7.8059	23
24	17.8143	0.0561	131.880	7.4030	140.119	7.8655	24
25	20.0855	0.0498	149.694	7.4528	159.046	7.9184	25
26	22.6464	0.0442	169.780	7.4970	180.386	7.9654	26
27	25.5337	0.0392	192.426	7.5362	204.448	8.0070	27
28	28.7892	0.0347	217.960	7.5709	231.577	8.0439	28
29	32.4597	0.0308	246.749	7.6017	262.164	8.0766	29
30	36.5982	0.0273	279.209	7.6290	296.652	8.1056	30
35	66.6863	0.0150	515.200	7.7257	547.386	8.2084	35
40	121.510	0.0082	945.203	7.7788	1004.25	8.2648	40
45	221.406	0.0045	1728.72	7.8079	1836.72	8.2957	45
50	403.429	0.0025	3156.38	7.8239	3353.57	8.3127	50
55	735.095	0.0014	5757.75	7.8327	6117.46	8.3220	55
60	1339.43	0.0007	10497.8	7.8375	11153.6	8.3271	60
65	2440.60	0.0004	19134.6	7.8401	20330.0	8.3299	65
70	4447.07	0.0002	34872.0	7.8416	37050.6	8.3315	70
75	8103.08	0.0001	63547.3	7.8424	67517.4	8.3323	75
80	4764.8	_a_	115797	7.8428	123032	8.3328	80

_a_Less than 0.0001.

TABLE F-8 Continuous Compounding; $r = 15\%$

	Discrete Flows				Continuous Flows		
	SINGLE PAYMENT		UNIFORM SERIES		UNIFORM SERIES		
	Compound Amount Factor	Present Worth Factor	Compound Amount Factor	Present Worth Factor	Compound Amount Factor	Present Worth Factor	
N	To find F Given P F/P	To find P Given F P/F	To find F Given A F/A	To find P Given A P/A	To find F Given $\overline{A}$ $F/\overline{A}$	To find P Given $\overline{A}$ $P/\overline{A}$	N
1	1.1618	0.8607	1.0000	0.8607	1.0789	0.9286	1
2	1.3499	0.7408	2.1618	1.6015	2.3324	1.7279	2
3	1.5683	0.6376	3.5117	2.2392	3.7887	2.4158	3
4	1.8221	0.5488	5.0800	2.7880	5.4808	3.0079	4
5	2.1170	0.4724	6.9021	3.2603	7.4467	3.5176	5
6	2.4596	0.4066	9.0191	3.6669	9.7307	3.9562	6
7	2.8577	0.3499	11.4787	4.0168	12.3843	4.3337	7
8	3.3201	0.3012	14.3364	4.3180	15.4674	4.6587	8
9	3.8574	0.2592	17.6565	4.5773	19.0495	4.9384	9
10	4.4817	0.2231	21.5139	4.8004	23.2113	5.1791	10
11	5.2070	0.1920	25.9956	4.9925	28.0465	5.3863	11
12	6.0496	0.1653	31.2026	5.1578	33.6643	5.5647	12
13	7.0287	0.1423	37.2522	5.3000	40.1913	5.7182	13
14	8.1662	0.1225	44.2809	5.4225	47.7745	5.8503	14
15	9.4877	0.1054	52.4471	5.5279	56.5849	5.9640	15
16	11.0232	0.0907	61.9348	5.6186	66.8212	6.0619	16
17	12.8071	0.0781	72.9580	5.6967	78.7140	6.1461	17
18	14.8797	0.0672	85.7651	5.7639	92.5315	6.2186	18
19	17.2878	0.0578	100.645	5.8217	108.585	6.2810	19
20	20.0855	0.0498	117.933	5.8715	127.237	6.3348	20
21	23.3361	0.0429	138.018	5.9144	148.907	6.3810	21
22	27.1126	0.0369	161.354	5.9513	174.084	6.4208	22
23	31.5004	0.0317	188.467	5.9830	203.336	6.4550	23
24	36.5982	0.0273	219.967	6.0103	237.322	6.4845	24
25	42.5211	0.0235	256.565	6.0338	276.807	6.5099	25
26	49.4024	0.0202	299.087	6.0541	322.683	6.5317	26
27	57.3975	0.0174	348.489	6.0715	375.983	6.5505	27
28	66.6863	0.0150	405.886	6.0865	437.909	6.5667	28
29	77.4785	0.0129	472.573	6.0994	509.856	6.5806	29
30	90.0171	0.0111	550.051	6.1105	593.448	6.5926	30
35	190.566	0.0052	1171.36	6.1467	1263.78	6.6317	35
40	403.429	0.0025	2486.67	6.1638	2682.86	6.6501	40
45	854.059	0.0012	5271.19	6.1719	5687.06	6.6589	45
50	1808.04	0.0006	11166.0	6.1757	12046.9	6.6630	50
55	3827.63	0.0003	23645.3	6.1775	25510.8	6.6649	55
60	8103.08	0.0001	50064.1	6.1784	54013.9	6.6658	60
65	17154.2	a	105993	6.1788	114355	6.6663	65
70	36315.5	a	224393	6.1790	242097	6.6665	70
75	76879.9	a	475047	6.1791	512526	6.6666	75
80	162755	a	1005680	6.1791	1085030	6.6666	80

aLess than 0.0001.

TABLE F-9   Continuous Compounding; $r = 20\%$

	Discrete Flows				Continuous Flows		
	SINGLE PAYMENT		UNIFORM SERIES		UNIFORM SERIES		
	Compound Amount Factor	Present Worth Factor	Compound Amount Factor	Present Worth Factor	Compound Amount Factor	Present Worth Factor	
N	To find F Given P F/P	To find P Given F P/F	To find F Given A F/A	To find P Given A P/A	To find F Given $\overline{A}$ $F/\overline{A}$	To find P Given $\overline{A}$ $P/\overline{A}$	N
1	1.2214	0.8187	1.0000	0.8187	1.1070	0.9063	1
2	1.4918	0.6703	2.2214	1.4891	2.4591	1.6484	2
3	1.8221	0.5488	3.7132	2.0379	4.1106	2.2559	3
4	2.2255	0.4493	5.5353	2.4872	6.1277	2.7534	4
5	2.7183	0.3679	7.7609	2.8551	8.5914	3.1606	5
6	3.3201	0.3012	10.4792	3.1563	11.6006	3.4940	6
7	4.0552	0.2466	13.7993	3.4029	15.2760	3.7670	7
8	4.9530	0.2019	17.8545	3.6048	19.7652	3.9905	8
9	6.0496	0.1653	22.8075	3.7701	25.2482	4.1735	9
10	7.3891	0.1353	28.8572	3.9054	31.9453	4.3233	10
11	9.0250	0.1108	36.2462	4.0162	40.1251	4.4460	11
12	11.0232	0.0907	45.2712	4.1069	50.1159	4.5464	12
13	13.4637	0.0743	56.2944	4.1812	62.3187	4.6286	13
14	16.4446	0.0608	69.7581	4.2420	77.2232	4.6959	14
15	20.0855	0.0498	86.2028	4.2918	95.4277	4.7511	15
16	24.5325	0.0408	106.288	4.3325	117.663	4.7962	16
17	29.9641	0.0334	130.821	4.3659	144.820	4.8331	17
18	36.5982	0.0273	160.785	4.3932	177.991	4.8634	18
19	44.7012	0.0224	197.383	4.4156	218.506	4.8881	19
20	54.5981	0.0183	242.084	4.4339	267.991	4.9084	20
21	66.6863	0.0150	296.682	4.4489	328.432	4.9250	21
22	81.4509	0.0123	363.369	4.4612	402.254	4.9386	22
23	99.4843	0.0101	444.820	4.4713	492.422	4.9497	23
24	121.510	0.0082	544.304	4.4795	602.552	4.9589	24
25	148.413	0.0067	665.814	4.4862	737.066	4.9663	25
26	181.272	0.0055	814.227	4.4917	901.361	4.9724	26
27	221.406	0.0045	995.500	4.4963	1102.03	4.9774	27
28	270.426	0.0037	1216.91	4.5000	1347.13	4.9815	28
29	330.299	0.0030	1487.33	4.5030	1646.50	4.9849	29
30	403.429	0.0025	1817.63	4.5055	2012.14	4.9876	30
35	1096.63	0.0009	4948.60	4.5125	5478.17	4.9954	35
40	2980.96	0.0003	13459.4	4.5151	14899.8	4.9983	40
45	8103.08	0.0001	36594.3	4.5161	40510.4	4.9994	45
50	22026.5	a	99481.4	4.5165	110127	4.9998	50
55	59874.1	a	270426	4.5166	299366	4.9999	55
60	162755	a	735103	4.5166	813769	5.0000	60

aLess than 0.0001.

	Discrete Flows				Continuous Flows		
	SINGLE PAYMENT		UNIFORM SERIES		UNIFORM SERIES		
	Compound Amount Factor	Present Worth Factor	Compound Amount Factor	Present Worth Factor	Compound Amount Factor	Present Worth Factor	
N	To find F Given P F/P	To find P Given F P/F	To find F Given A F/A	To find P Given A P/A	To find F Given $\overline{A}$ $F/\overline{A}$	To find P Given $\overline{A}$ $P/\overline{A}$	N
1	1.2840	0.7788	1.0000	0.7788	1.1361	0.8848	1
2	1.6487	0.6065	2.2840	1.3853	2.5949	1.5739	2
3	2.1170	0.4724	3.9327	1.8577	4.4680	2.1105	3
4	2.7183	0.3679	6.0497	2.2256	6.8731	2.5285	4
5	3.4903	0.2865	8.7680	2.5121	9.9614	2.8540	5
6	4.4817	0.2231	12.2584	2.7352	13.9268	3.1075	6
7	5.7546	0.1738	16.7401	2.9090	19.0184	3.3049	7
8	7.3891	0.1353	22.4947	3.0443	25.5562	3.4587	8
9	9.4877	0.1054	29.8837	3.1497	33.9509	3.5784	9
10	12.1825	0.0821	39.3715	3.2318	44.7300	3.6717	10
11	15.6426	0.0639	51.5539	3.2957	58.5705	3.7443	11
12	20.0855	0.0498	67.1966	3.3455	76.3421	3.8009	12
13	25.7903	0.0388	87.2821	3.3843	99.1614	3.8449	13
14	33.1155	0.0302	113.073	3.4145	128.462	3.8792	14
15	42.5211	0.0235	146.188	3.4380	166.084	3.9059	15
16	54.5982	0.0183	188.709	3.4563	214.393	3.9267	16
17	70.1054	0.0143	243.307	3.4706	276.422	3.9429	17
18	90.0171	0.0111	313.413	3.4817	356.068	3.9556	18
19	115.584	0.0087	403.430	3.4904	458.337	3.9654	19
20	148.413	0.0067	519.014	3.4971	589.653	3.9730	20
21	190.566	0.0052	667.427	3.5023	758.265	3.9790	21
22	244.692	0.0041	857.993	3.5064	974.768	3.9837	22
23	314.191	0.0032	1102.69	3.5096	1252.76	3.9873	23
24	403.429	0.0025	1416.88	3.5121	1609.72	3.9901	24
25	518.013	0.0019	1820.30	3.5140	2068.05	3.9923	25
26	665.142	0.0015	2338.31	3.5155	2656.57	3.9940	26
27	854.059	0.0012	3003.46	3.5167	3412.23	3.9953	27
28	1096.63	0.0009	3857.52	3.5176	4382.53	3.9964	28
29	1408.10	0.0007	4954.15	3.5183	5628.42	3.9972	29
30	1808.04	0.0006	6362.26	3.5189	7228.17	3.9978	30
35	6310.69	0.0002	22215.2	3.5203	25238.8	3.9994	35
40	22026.5	a	77547.5	3.5207	88101.9	3.9998	40
45	76879.9	a	270676	3.5208	307516	3.9999	45
50	268337	a	944762	3.5208	1073350	4.0000	50

^{a}Less than 0.0001.

The Standardized Normal Distribution Function

Table on page 643 from Ronald E. Walpole and Raymond H. Myers, *Probability and Statistics for Engineers and Scientists,* 2nd ed. New York: Macmillan, 1978, p. 513.

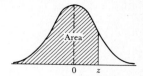

Area

0   z

z	0.00	0.01	0.02	0.03	0.04	0.05	0.06	0.07	0.08	0.09
−3.4	0.0003	0.0003	0.0003	0.0003	0.0003	0.0003	0.0003	0.0003	0.0003	0.0002
−3.3	0.0005	0.0005	0.0005	0.0004	0.0004	0.0004	0.0004	0.0004	0.0004	0.0003
−3.2	0.0007	0.0007	0.0006	0.0006	0.0006	0.0006	0.0006	0.0005	0.0005	0.0005
−3.1	0.0010	0.0009	0.0009	0.0009	0.0008	0.0008	0.0008	0.0008	0.0007	0.0007
−3.0	0.0013	0.0013	0.0013	0.0012	0.0012	0.0011	0.0011	0.0011	0.0010	0.0010
−2.9	0.0019	0.0018	0.0017	0.0017	0.0016	0.0016	0.0015	0.0015	0.0014	0.0014
−2.8	0.0026	0.0025	0.0024	0.0023	0.0023	0.0022	0.0021	0.0021	0.0020	0.0019
−2.7	0.0035	0.0034	0.0033	0.0032	0.0031	0.0030	0.0029	0.0028	0.0027	0.0026
−2.6	0.0047	0.0045	0.0044	0.0043	0.0041	0.0040	0.0039	0.0038	0.0037	0.0036
−2.5	0.0062	0.0060	0.0059	0.0057	0.0055	0.0054	0.0052	0.0051	0.0049	0.0048
−2.4	0.0082	0.0080	0.0078	0.0075	0.0073	0.0071	0.0069	0.0068	0.0066	0.0064
−2.3	0.0107	0.0104	0.0102	0.0099	0.0096	0.0094	0.0091	0.0089	0.0087	0.0084
−2.2	0.0139	0.0136	0.0132	0.0129	0.0125	0.0122	0.0119	0.0116	0.0113	0.0110
−2.1	0.0179	0.0174	0.0170	0.0166	0.0162	0.0158	0.0154	0.0150	0.0146	0.0143
−2.0	0.0228	0.0222	0.0217	0.0212	0.0207	0.0202	0.0197	0.0192	0.0188	0.0183
−1.9	0.0287	0.0281	0.0274	0.0268	0.0262	0.0256	0.0250	0.0244	0.0239	0.0233
−1.8	0.0359	0.0352	0.0344	0.0336	0.0329	0.0322	0.0314	0.0307	0.0301	0.0294
−1.7	0.0446	0.0436	0.0427	0.0418	0.0409	0.0401	0.0392	0.0384	0.0375	0.0367
−1.6	0.0548	0.0537	0.0526	0.0516	0.0505	0.0495	0.0485	0.0475	0.0465	0.0455
−1.5	0.0668	0.0655	0.0643	0.0630	0.0618	0.0606	0.0594	0.0582	0.0571	0.0559
−1.4	0.0808	0.0793	0.0778	0.0764	0.0749	0.0735	0.0722	0.0708	0.0694	0.0681
−1.3	0.0968	0.0951	0.0934	0.0918	0.0901	0.0885	0.0869	0.0853	0.0838	0.0823
−1.2	0.1151	0.1131	0.1112	0.1093	0.1075	0.1056	0.1038	0.1020	0.1003	0.0985
−1.1	0.1357	0.1335	0.1314	0.1292	0.1271	0.1251	0.1230	0.1210	0.1190	0.1170
−1.0	0.1587	0.1562	0.1539	0.1515	0.1492	0.1469	0.1446	0.1423	0.1401	0.1379
−0.9	0.1841	0.1814	0.1788	0.1762	0.1736	0.1711	0.1685	0.1660	0.1635	0.1611
−0.8	0.2119	0.2090	0.2061	0.2033	0.2005	0.1977	0.1949	0.1922	0.1894	0.1867
−0.7	0.2420	0.2389	0.2358	0.2327	0.2296	0.2266	0.2236	0.2206	0.2177	0.2148
−0.6	0.2743	0.2709	0.2676	0.2643	0.2611	0.2578	0.2546	0.2514	0.2483	0.2451
−0.5	0.3085	0.3050	0.3015	0.2981	0.2946	0.2912	0.2877	0.2843	0.2810	0.2776
−0.4	0.3446	0.3409	0.3372	0.3336	0.3300	0.3264	0.3228	0.3192	0.3156	0.3121
−0.3	0.3821	0.3783	0.3745	0.3707	0.3669	0.3632	0.3594	0.3557	0.3520	0.3483
−0.2	0.4207	0.4168	0.4129	0.4090	0.4052	0.4013	0.3974	0.3936	0.3897	0.3859
−0.1	0.4602	0.4562	0.4522	0.4483	0.4443	0.4404	0.4364	0.4325	0.4286	0.4247
−0.0	0.5000	0.4960	0.4920	0.4880	0.4840	0.4801	0.4761	0.4721	0.4681	0.4641
0.0	0.5000	0.5040	0.5080	0.5120	0.5160	0.5199	0.5239	0.5279	0.5319	0.5359
0.1	0.5398	0.5438	0.5478	0.5517	0.5557	0.5596	0.5636	0.5675	0.5714	0.5753
0.2	0.5793	0.5832	0.5871	0.5910	0.5948	0.5987	0.6026	0.6064	0.6103	0.6141
0.3	0.6179	0.6217	0.6255	0.6293	0.6331	0.6368	0.6406	0.6443	0.6480	0.6517
0.4	0.6554	0.6591	0.6628	0.6664	0.6700	0.6736	0.6772	0.6808	0.6844	0.6879
0.5	0.6915	0.6950	0.6985	0.7019	0.7054	0.7088	0.7123	0.7157	0.7190	0.7224
0.6	0.7257	0.7291	0.7324	0.7357	0.7389	0.7422	0.7454	0.7486	0.7517	0.7549
0.7	0.7580	0.7611	0.7642	0.7673	0.7704	0.7734	0.7764	0.7794	0.7823	0.7852
0.8	0.7881	0.7910	0.7939	0.7967	0.7995	0.8023	0.8051	0.8078	0.8106	0.8133
0.9	0.8159	0.8186	0.8212	0.8238	0.8264	0.8289	0.8315	0.8340	0.8365	0.8389
1.0	0.8413	0.8438	0.8461	0.8485	0.8508	0.8531	0.8554	0.8577	0.8599	0.8621
1.1	0.8643	0.8665	0.8686	0.8708	0.8729	0.8749	0.8770	0.8790	0.8810	0.8830
1.2	0.8849	0.8869	0.8888	0.8907	0.8925	0.8944	0.8962	0.8980	0.8997	0.9015
1.3	0.9032	0.9049	0.9066	0.9082	0.9099	0.9115	0.9131	0.9147	0.9162	0.9177
1.4	0.9192	0.9207	0.9222	0.9236	0.9251	0.9265	0.9278	0.9292	0.9306	0.9319
1.5	0.9332	0.9345	0.9357	0.9370	0.9382	0.9394	0.9406	0.9418	0.9429	0.9441
1.6	0.9452	0.9463	0.9474	0.9484	0.9495	0.9505	0.9515	0.9525	0.9535	0.9545
1.7	0.9554	0.9564	0.9573	0.9582	0.9591	0.9599	0.9608	0.9616	0.9625	0.9633
1.8	0.9641	0.9649	0.9656	0.9664	0.9671	0.9678	0.9686	0.9693	0.9699	0.9706
1.9	0.9713	0.9719	0.9726	0.9732	0.9738	0.9744	0.9750	0.9756	0.9761	0.9767
2.0	0.9772	0.9778	0.9783	0.9788	0.9793	0.9798	0.9803	0.9808	0.9812	0.9817
2.1	0.9821	0.9826	0.9830	0.9834	0.9838	0.9842	0.9846	0.9850	0.9854	0.9857
2.2	0.9861	0.9864	0.9868	0.9871	0.9875	0.9878	0.9881	0.9884	0.9887	0.9890
2.3	0.9893	0.9896	0.9898	0.9901	0.9904	0.9906	0.9909	0.9911	0.9913	0.9916
2.4	0.9918	0.9920	0.9922	0.9925	0.9927	0.9929	0.9931	0.9932	0.9934	0.9936
2.5	0.9938	0.9940	0.9941	0.9943	0.9945	0.9946	0.9948	0.9949	0.9951	0.9952
2.6	0.9953	0.9955	0.9956	0.9957	0.9959	0.9960	0.9961	0.9962	0.9963	0.9964
2.7	0.9965	0.9966	0.9967	0.9968	0.9969	0.9970	0.9971	0.9972	0.9973	0.9974
2.8	0.9974	0.9975	0.9976	0.9977	0.9977	0.9978	0.9979	0.9979	0.9980	0.9981
2.9	0.9981	0.9982	0.9982	0.9983	0.9984	0.9984	0.9985	0.9985	0.9986	0.9986
3.0	0.9987	0.9987	0.9987	0.9988	0.9988	0.9989	0.9989	0.9989	0.9990	0.9990
3.1	0.9990	0.9991	0.9991	0.9991	0.9992	0.9992	0.9992	0.9992	0.9993	0.9993
3.2	0.9993	0.9993	0.9994	0.9994	0.9994	0.9994	0.9994	0.9995	0.9995	0.9995
3.3	0.9995	0.9995	0.9995	0.9996	0.9996	0.9996	0.9996	0.9996	0.9996	0.9997
3.4	0.9997	0.9997	0.9997	0.9997	0.9997	0.9997	0.9997	0.9997	0.9997	0.9998

A Computer Program for Performing After-Tax Present Worth Studies

GENERAL DESCRIPTION

This BASIC computer program is a computational tool for analyzing the net present worth of after-tax economy problems of Chapters 10 and 11. It is designed for actual-dollar cash flow studies that *include* differential escalation/ inflation or that *ignore* the effects of such price changes. Furthermore, the program is written to accept Accelerated Cost Recovery System (ACRS) percentages for a specified recovery period. It is an interactive program suitable for personal (mini-) computers and, as such, contains instructions to the user regarding its required data inputs and key assumptions. Typical printout for the program is provided in Chapter 10.

PROGRAM LISTING*

```
00010    REM: ***************************************
00027    REM: PROGRAM INFORMATION
00028    REM: ***************************************
00040    REM: WHILE TAKING INFLATION INTO
         CONSIDERATION, THIS PROGRAM
00050    REM: PERFORMS AN ACTUAL DOLLAR, AFTER-TAX
         ANALYSIS OF AN
00060    REM: INVESTMENT. SINCE THE PROGRAM PROMPTS THE
         USER FOR
00070    REM: INPUT INFORMATION, THE PROGRAM IS QUITE
         FLEXIBLE AND
00080    REM: EASILY UNDERSTOOD. IF NECESSARY, THE USER
         CAN PROVIDE
00090    REM: A DIFFERENT ANNUAL ESCALATION RATE FOR
         UTILITY, LABOR,
00100    REM: AND MATERIAL COSTS.
00150    REM:
00190    REM: ***************************************
00200    REM: DEFINITION OF MAJOR VARIABLES
00210    REM: ***************************************
00220    REM: T4: INCREMENTAL INCOME TAX RATE
00230    REM: T3: TAX RATE ON CAPITAL GAIN OR LOSS
00240    REM: G1: GENERAL INFLATION RATE
00250    REM: M1: AFTER-TAX, COMBINED DISCOUNT RATE
00260    REM: T5: INVESTMENT TAX CREDIT PERCENTAGE
00270    REM: P1: PERIOD OF STUDY IN YEARS
00280    REM: L4: USEFUL LIFE OF INVESTMENT IN YEARS
00290    REM: L5: TAX LIFE OF INVESTMENT IN YEARS
00300    REM: F1: FIRST COST OF INVESTMENT
00320    REM: S1: SALVAGE VALUE OF INVESTMENT IN ACTUAL
         DOLLARS
00330    REM: U1: ANNUAL COST OF UTILITIES IN REAL
         DOLLARS
00340    REM: A6: ANNUAL COST OF UTILITIES IN ACTUAL
         DOLLARS
00350    REM: E3: EFFECTIVE ANNUAL ESCALATION RATE FOR
         UTILITIES
00360    REM: L1: ANNUAL COST OF LABOR IN REAL DOLLARS
00370    REM: A2: ANNUAL COST OF LABOR IN ACTUAL
         DOLLARS
00380    REM: E1: EFFECTIVE ANNUAL ESCALATION RATE FOR
         LABOR
```

*Each statement comprises one line of the BASIC program (it may appear here as two lines due to narrow margins in the book).

```
00390   REM: M2: ANNUAL COST OF MATERIALS IN REAL
        DOLLARS
00400   REM: A4: ANNUAL COST OF MATERIALS IN ACTUAL
        DOLLARS
00410   REM: E2: EFFECTIVE ANNUAL ESCALATION RATE FOR
        MATERIALS
00420   REM: R1: ANNUAL REVENUES IN REAL DOLLARS
00430   REM: L2: ANNUAL LEASE COST IN REAL DOLLARS
00440   REM: L3: LIFE OF THE LEASE IN YEARS
00450   REM: A5: ANNUAL REVENUES IN ACTUAL DOLLARS
00460   REM: A3: LEASE COST ADJUSTED FOR RENEGOTIATION
00470   REM: I1: INVESTMENT TAX CREDIT
00480   REM: D2: DEPRECIATION OF THE INVESTMENT SUMMED
        OVER ITS LIFE
00490   REM: B1: BOOK VALUE OF THE INVESTMENT
00500   REM: C1: TAX ON THE CAPITAL GAIN OR LOSS
00510   REM: B2: BEFORE TAX CASH FLOW
00520   REM: D1: DEPRECIATION IN ACTUAL DOLLARS
00530   REM: T2: TAXABLE INCOME IN ACTUAL DOLLARS
00540   REM: T1: INCOME TAX IN ACTUAL DOLLARS
00550   REM: P2: PRESENT VALUE OF THE CASH FLOW IN
        THAT YEAR
00560   REM: N1: NET PRESENT VALUE OF THE INVESTMENT
00570   REM: E9: ESCALATION RATE FOR REVENUES
00580   REM: ******************************************
00590   REM: SETTING ALL VARIABLES EQUAL TO ZERO
00600   REM: ******************************************
00610   DIM A1(100), A6(100), A2(100), A4(100),
        P2(100)
00620   DIM B2(100), D1(100), T2(100), T1(100)
00630   DIM A7(100), A3(100), A5(100), S1(100)
01100   REM: ******************************************
01110   REM:    COLLECTING INPUT INFORMATION
01120   REM: ******************************************
00001   PRINT
00002   PRINT
00010   PRINT "THIS PROGRAM ALLOWS YOU TO CONDUCT AN
        ACTUAL DOLLAR ANALYSIS OF AN"
00011   PRINT "INVESTMENT ALTERNATIVE ON AN AFTER-TAX
        BASIS WHEN INFLATION/ESCALATION"
00012   PRINT "OF REVENUES AND COSTS ARE INVOLVED. THE
        REFERENCE YEAR FOR ALL CASH"
00013   PRINT "FLOWS , EXCEPT SALVAGE VALUE , IS THE
        TIME AT WHICH THE INITIAL"
00014   PRINT "INVESTMENT IS MADE (I.E. TIME 0) .
        THESE ARE TERMED 'REAL DOLLARS' IN"
00015   PRINT "THE PROMPTS BELOW . MULTIPLE
        REPLACEMENTS OF AN ALTERNATIVE ARE"
```

A COMPUTER PROGRAM FOR PERFORMING AFTER-TAX
PRESENT WORTH STUDIES

```
00016    PRINT "PERMITTED WHEN THE STUDY PERIOD IS
         LONGER THAN USEFUL LIFE. CONVERSELY"
00017    PRINT "THE STUDY PERIOD CAN BE SHORTER THAN
         USEFUL LIFE. SALVAGE VALUE AT THE"
00018    PRINT "END OF THE STUDY PERIOD IS SPECIFIED BY
         THE ANALYST."
00019    PRINT
00020    PRINT
00021    PRINT "* *************************************"
00022    PRINT "*                                     *"
00023    PRINT "* NOTE: ENTER ALL PERCENTAGES AND     *"
         PRINT "*       RATES IN DECIMAL FORM. FOR    *"
00024    PRINT "*       EXAMPLE, 10% SHOULD BE        *"
         PRINT "*       ENTERED IN THIS PROGRAM       *"
         PRINT "*       AS 0.10                       *"
00025    PRINT "*                                     *"
00026    PRINT " *************************************"
01381    PRINT
01382    PRINT
01390    PRINT
01400    PRINT
01410    PRINT
01420    PRINT
01430    PRINT "                  INVESTMENT"
01440    PRINT "                  =========="
01450    PRINT
01460    PRINT
01470    PRINT " INPUT INITIAL COST OF INVESTMENT ";
01480    INPUT F1
01490    PRINT
01500    PRINT " INPUT INVESTMENT TAX CREDIT PERCENTAGE
         IN DECIMAL FORM ";
01510    INPUT T5
01520    PRINT
01530    PRINT " INPUT THE USEFUL LIFE (IN YEARS) OF
         THE INVESTMENT ";
01540    INPUT L4
01550    PRINT
01560    PRINT " HOW LONG IS THE STUDY PERIOD";
01570    INPUT P1
01580    PRINT
01590    PRINT " INPUT THE SALVAGE VALUE OF THE
         INVESTMENT IN ACTUAL DOLLARS FOR "
01600    J = L4
01610    IF J > P1 GO TO 1640
01620    PRINT "YEAR ";J;" ";
01630    INPUT S1(J)
01640    IF J = P1 GO TO 1690
```

```
01650   J = J + L4
01660   IF J <= P1 GO TO 1610
01670   J = P1
01680   GO TO 1610
01690   PRINT
01700   PRINT
01710   PRINT
01720   PRINT
01730   PRINT "          DEPRECIATION"
01740   PRINT "          ============"
01750   PRINT
01760   PRINT
01770   PRINT " INPUT ACRS TAX LIFE OF INVESTMENT IN
        YEARS ";
01780   INPUT L5
01790   PRINT
01800   PRINT " INPUT THE APPROPRIATE ACRS RECOVERY
        PERCENTAGE IN DECIMAL FORM"
01810   FOR J = 1 TO L5
01820   PRINT "YEAR ";J;" ";
01830   INPUT A1(J)
01840   REM: *****************************************
01850   REM:    CHECKS IF A1 RECOVERY PERCENTAGES SUM
               TO ONE
01860   REM: *****************************************
01870   S2 = S2 + A1(J)
01880   NEXT J
01890   IF S2 <= 0.9999 THEN 1920
01900   IF S2 >= 1.0001 THEN 1920
01910   GO TO 1990
01920   PRINT "THE A1 RECOVERY PERCENTAGES DO NOT SUM
        TO ONE. TRY AGAIN."
01930   S2 = 0.0
01940   GO TO 1760
01950   REM:
01960   REM: *****************************************
01970   REM:    COLLECTING INPUT INFORMATION
01980   REM: *****************************************
01990   PRINT
02000   PRINT
02010   PRINT
02020   PRINT
02030   PRINT "      REVENUES AND EXPENSES"
02040   PRINT "      ===================="
02050   PRINT
02060   PRINT
02070   PRINT " INPUT THE ANNUAL REVENUES IN REAL
        DOLLARS ";
```

```
02080   INPUT R1
02090   PRINT
02100   PRINT " INPUT EFFECTIVE ANNUAL ESCALATION RATE
        FOR REVENUES IN DECIMAL FORM ";
02110   INPUT E9
02120   PRINT
02130   PRINT " INPUT THE ANNUAL COST OF UTILITIES IN
        REAL DOLLARS ";
02140   INPUT U1
02150   PRINT
02160   PRINT " INPUT EFFECTIVE ANNUAL ESCALATION RATE
        FOR UTILITIES IN DECIMAL FORM ";
02170   INPUT E3
02180   PRINT
02190   PRINT " INPUT THE ANNUAL COST OF LABOR IN REAL
        DOLLARS ";
02200   INPUT L1
02210   PRINT
02220   PRINT " INPUT EFFECTIVE ANNUAL ESCALATION RATE
        FOR LABOR IN DECIMAL FORM ";
02230   INPUT E1
02240   PRINT
02250   PRINT " INPUT THE ANNUAL COST OF MATERIALS IN
        REAL DOLLARS ";
02260   INPUT M2
02270   PRINT
02280   PRINT " INPUT EFFECTIVE ANNUAL ESCALATION RATE
        FOR MATERIALS IN DECIMAL FORM ";
02290   INPUT E2
02300   PRINT
02310   PRINT " INPUT THE ANNUAL LEASE COST IN ACTUAL
        DOLLARS ";
02320   INPUT L2
02330   PRINT
02340   PRINT " INPUT NUMBER OF YEARS BEFORE THE LEASE
        HAS TO BE RENEGOTIATED ";
02350   PRINT
02360   PRINT
02370   PRINT
02380   PRINT
02390   PRINT "          OTHER INFORMATION"
02400   PRINT "          =================="
02410   PRINT
02420   PRINT
02430   INPUT L3
02440   PRINT
02450   PRINT
```

```
02460    PRINT " INPUT INCREMENTAL INCOME TAX RATE IN
         DECIMAL FORM ";
02470    INPUT T4
02480    PRINT
02490    PRINT " INPUT TAX RATE ON CAPITAL GAIN OR LOSS
         IN DECIMAL FORM ";
02500    INPUT T3
02510    PRINT
02520    PRINT " INPUT INFLATION RATE FOR REPLACEMENT
         INVESTMENTS IN DECIMAL FORM ";
02530    INPUT G1
02540    PRINT
02550    PRINT " INPUT THE M.A.R.R. FOR AFTER-TAX
         ANALYSIS IN DECIMAL FORM ";
02560    INPUT M1
02570    REM:
02580    REM: ******************************************
02590    REM:    CALCULATING THE LEASE COST FOR EACH
                 YEAR
02600    REM: ******************************************
02610    A3(1) = -L2
02620    FOR L = 2 TO P1
02630        I2 = L-1
02640        K1 = K1+1
02650        IF K1 > = L3 GO TO 2680
02660        A3(L) = A3(I2)
02670        GO TO 2700
02680        A3(L) = -L2*(1+G1)**I2
02690        K1 = 0
02700    NEXT L
02710    REM:
02720    REM: ******************************************
02730    REM:    CALCULATING THE DEPRECIATION
02740    REM: ******************************************
02750    N3 = N3 + 1
02760    FOR I = 1 TO L4
02770        N2 = N2 + 1
02780        N4 = N3*L4
02790        K2 = N4 + I
02800        A1(K2) = A1(I)
02810          D1(N2)=-(F1*(1+G1)**(N4-L4))*A1(N2)
02820          IF N2 = P1 GO TO 2850
02830        IF N2 <> N4 THEN 3000
02840        IF N2 >= P1 THEN 3000
02850          N6 = N4-L4+1
02860    FOR K = N6 TO N2
02870          D2 = D2 - D1(K)
```

```
02880    NEXT K
02890        B1 = F1*(1+G1)**(N4-L4) - D2
02900        D2 = 0.0
02910        IF S1(N2) >= F1 THEN 2940
02920        IF S1(N2) <= B1 THEN 2940
02930        GO TO 2980
02940        REM:
02950        T2(N2) = S1(N2) - B1
02960        T1(N2) = -T2(N2) * T3
02970        GO TO 3000
02980        T2(N2) = S1(N2) - B1
02990        T1(N2) = -T2(N2) * T4
03000    NEXT I
03010    IF N2 < P1 GO TO 2750
03020    REM: *****************************************
03030    REM:    CALCULATING OTHER COSTS
03040    REM: *****************************************
03050    FOR K = 1 TO P1
03060    A6(K) = -U1 * (1+E3)**K
03070    A2(K) = -L1 * (1+E1)**K
03080    A4(K) = -M2 * (1+E2)**K
03090    A5(K) = R1 * (1+E9)**K
03100        B2(K) = A5(K)+A6(K)+A2(K)+A4(K)+A3(K)
03110    T2(K) = B2(K) + D1(K) + T2(K)
03120    T1(K) = -(B2(K) + D1(K))*T4 + T1(K)
03130    A7(K) = B2(K) + T1(K) + S1(K)
03140    NEXT K
03150    REM: *****************************************
03160    REM:    ADJUSTING THE REPURCHASING PRICE
03170    REM: *****************************************
03180    I1 = T5*F1
03190    F2 = -F1
03200    Y1 = F2 + I1
03210    J2 = J2 + 1
03220    FOR K = 1 TO L4
03230        J1 = J1 + 1
03240        J3 = J2*L4
03250        IF J1 <> J3 GO TO 3310
03260        IF J1 = P1 GO TO 3310
03270        A5(J1) = A5(J1) + F2*(1+G1)**J3+S1(J3)
03280        B2(J1) = B2(J1) + F2*(1+G1)**J3+S1(J3)
03290        T1(J1) = T1(J1) + I1*(1+G1)**J3
03300        A7(J1) = A7(J1) + Y1*(1+G1)**J3
03310        IF J1 <> P1 GO TO 3340
03320        A5(J1) = A5(J1) + S1(J1)
03330        B2(J1) = B2(J1) + S1(J1)
03340    NEXT K
03350    IF J1 < P1 GO TO 3210
```

```
03360    REM: ****************************************
03370    REM:    CALCULATING THE NET PRESENT VALUE
03380    REM: ****************************************
03390    FOR I = 1 TO P1
03400       P2(I) = A7(I)*(1+M1)**-I
03410       N5 = N5 + P2(I)
03420    NEXT I
03430    N1 = Y1 +N5
03440    REM: ****************************************
03450    REM:  PRINTING HEADINGS AND OUTPUT
03460    REM: ****************************************
03470    PRINT
03480    PRINT
03490    PRINT
03500    PRINT
03510    PRINT
03520    PRINT
03530    PRINT "      ACTUAL DOLLAR ANALYSIS"
03540    PRINT
03550    PRINT
03560    PRINT "    UTILITY   LABOR    MATERIAL    LEASE"
03570    PRINT "YR  COST      COST     COST        COST
03580    PRINT"__  -------  -----   --------    -----
         REVENUE    BTCF"
         -------    ----"
03590    PRINT USING 3600 , X9,X9,X9,X9,X9,F2,F2
03600    :## ######.## ######.## ######.## ######.##
         ######.## ######.##
03610    FOR N = 1 TO P1
03620    PRINT USING 3600 , N,A6(N),A2(N),A4(N),A3(N),
         A5(N),B2(N)
03630    NEXT N
03640    PRINT
03650    PRINT
03660    PRINT
03670    PRINT "             TAXABLE INCOME      PRESENT"
03680    PRINT "YR BTCF DEPR INCOME  TAXES  ATCF VALUE"
03690    PRINT "__ ____ __   ------- ------ ---- -------"
03700    PRINT USING 3600 , X9,F2,X9,X9,I1,Y1,Y1
03710    FOR J = 1 TO P1
03720    PRINT USING 3600 , J,B2(J),D1(J),T2(J),T1(J),
         A7(J),P2(J)
03730    NEXT J
03740    PRINT "                              ---------"
03750    PRINT USING 3760, N1
03760    :       NET PRESENT VALUE        ######.##
03770    PRINT
03780    IF N1 > 0.0 GO TO 3810
```

A COMPUTER PROGRAM FOR PERFORMING AFTER-TAX
PRESENT WORTH STUDIES

```
03790    PRINT "YOU HAVE A LOSER. THE NET PRESENT VALUE
         OF THE INVESTMENT IS NEGATIVE."
03800    GO TO 3820
03810    PRINT"YOU HAVE A WINNER. THE NET PRESENT VALUE
         OF THE INVESTMENT IS POSITIVE."
03820    PRINT
03830    PRINT "WOULD YOU LIKE TO EVALUATE ANOTHER
         INVESTMENT (YES=1, NO=2)";
03840    INPUT I9
03850    IF I9 = 1 GO TO 1420
03860    END
```

Answers to Even-Numbered Problems

1.16	Assets = $3,800, Net Income = $850
2-10	**(a)** $D^* = 300$ units/year (minimum loss)
2-12	**(a)** 70% (or $700,000) **(b)** 61.7% **(c)** 70%
2-14	Maximum profit = $0
3-2	Select aluminum.
3-4	Choose Fiber Y.
3-6	Choose charter service.
3-8	**(a)** Offset is cheaper for 25 copies **(b)** $N = 15$
3-10	**(a)** Produce 500 castings per hour. **(b)** If production costs rise by 42% or more, produce 100 castings per hour.
3-12	Choose aluminum.
3-14	First method, profit is $109.70. Second method, profit is $92.50.
3-16	**(a)** Hourly wage = $5.85
4-2	$I = $240 for one year; $1,920 for eight years
4-4	$I = $19,950 - $10,500 = $9,450
4-6	$A = $2,755.29
4-8	$A = $280.28

4-10 $A = \$4,425$

4-12 $N = 36.27$ years so age is 58.27 years.

4-14 $P = \$6,757.80$, $A = \$681.59$

4-16 $i' = 11\%$ per year

4-18 $F = \$3,500$

4-20 $P = \$153,034$

4-22 $A = \$189.63$

4-24 $P = \$28,655.40$

4-26 $Q = \$607.70$

4-28 $G = \$1,236.85$

4-30 $P = \$4,684.65$

4-32 $\$268,307$

4-34   (a) $F = \$6,340.64$ (b) $F = \$2,655.86$

4-36 $i' = 5.1\%$

4-38 $i' = 40.9\%/\text{yr.}$

4-40 $i' = 26.7\%/\text{yr.}$

4-42 $N = 88.4$ months

4-44 $N = 48$ months

4-46 $A = \$583.95$

4-48 $A = \$558.62$

4-50 $i' = 1.32\%/\text{yr.}$

4-52 $A = \$26,250.30$

4-54 $P = \$4,653.50$

4-56 $P = \$4,836$

4-58 $P = \$601.04$

4-60   (a) $\$13,094.20$ (b) $\$18,914.08$ (c) $\underline{r} = 9.19\%$

4-62   (a) $\$5,909.70$ (b) $\$353,475$

4-64   (a) $\$3.3 \times 10^6$ (b) $\$40,260.60$ (c) $\$9,252.34$

5-2   Present expenditure $= \$26,441$

5-4   (a) Select process R (b) C.R. Amount $= \$503.80$

5-6   (a) 14% (b) 0%   (c) 19.6%
       (d) 0%   (e) 33%   (f) 5.5%

5-8   (a) 7 years, 4 years   (b) 7 years, never

5-12   (a) I.R.R. $= 6.8\%$ for A; 9.3% for B
       E.R.R. $= 7.4\%$ for A; 8.7% for B
       E.R.R.R. $= 7.2\%$ for A; 8.9% for B
       (b) P.W of A $= -\$405.09$, P.W. of B $= \$321.22$
       A.W. of A $= -\$87.63$, A.W. of B $= \$69.68$

5-14 $i' = 21.4\%$

5-20   P.W. $= \$630.50$, F.W. $= \$1,677.14$, A.W. $= \$151.55$, I.R.R. $= 24.9\%$,
       E.R.R. $= 19.8\%$, E.R.R.R. $= 24.2\%$, $\theta = 4$ years, $\theta' = 6$ years

5-22 A.W. = $20,728 so make the investment.

5-24 (a) F.W. = −$1,128 so do not produce the new toy.
(b) Discounted payback = never.

5-26 $i' = 17.3\%$; install the steam pipes.

5-28 P.W. = $399,094; build the new paint shed.

5-30 (a) P.W. = $450,970
(b) Minimum selling price is $0.43 per pound.

5-32 Multiple I.R.R. exist at 4% and 32%. The E.R.R. is 7.6%.

5-34 (a) $i' = 0.5\%$ and 28.8%
(b) Total cash inflow is less than outflow so do not pursue the project.

6-2 I.R.R. of A → B = 30.2%, I.R. R. of B → C = 15.1%, etc. Select alternative C.

6-4 (a) F.W. of alternative C = −$706,246. Choose C.
(b) I.R.R. of A → C = 23%, etc. Select C.

6-6 A.W. of machine A = $129.56. Select A.

6-8 (a) A.W. of apartment house = $32,006. Choose the apartment house.
(b) I.R.R. of D.S. → A.H. = 25%, etc. Select the apartment house.

6-10 Recommend motor A.

6-12 (a) A.C. of QWIK rotor motor = $2,232. Select it.
(b) See text for repeatability assumptions.

6-14 I.R.R. of C → E = 20.2%; therefore recommend packaging equipment E.

6-16 (a) Select the copper piping system.
(b) Choose the plastic piping system.

6-18 Choose plan B (uninvested cash of $32,000 can earn 12% per year).

6-20 E.R.R. of A → C = 12.1%, so select C.

6-22 P = $482,635 (breakeven purchase price).

6-24 I.R.R. of II → I = 9.7% and I.R.R. of I → III = 12%. Recommend alternative III and check for multiple I.R.R.'s (there are none).

6-26 Capitalized worth of plan A (−$147,000) is less than that of plan B (−$170,900).

6-28 Select plan A.

6-30 Choose the pumping plan (P.W. = −$2,647,721).

6-32 Recommend alternative 1 and alternative 2.

6-34 Only alternative A should be chosen.

6-36 Selecting proposals B1, B2, and C2 and D maximizes P.W. ($25,600).

7-2 Select alternative 1 initially. To reverse this, a salvage value $2,050 would be needed for alternative 2.

7-4 (a) $N = 7.3$ years
(b) Select alternative 1, assuming salvage values remain unchanged.

7-6 The breakeven point is 8.2¢ per kWhr.

7-8 Annual savings is the most sensitive factor; the least sensitive is residual value.

7-10 Build the 4-lane bridge now.

7-14 (a) Select W   (b) Select W   (c) Select U   (d) Select W

7-16 Recommend pump N

7-18   Expected value = 1,350 cubic yards; the variance is 66,500 (cu. yds)^2.

7-20   Pr (Salvage Value $\geq$ 171) = 0.788.

7-22   Alternative E minimizes equivalent annual cost ($19,561).

7-24   The expected E.R.R.R. (for instance) is 21.5%. This project does not meet the 25% return criterion.

7-26   E (N.P.W.) = \$2,477.40, V (N.P.W.) = 1.1×10^6, Pr (N.P.W. $\geq$ 0) = 0.99.

7-28   E (N.P.W.) = $-$\$27,600, V (N.P.W.) = 404.7×10^6

8-4   \$871,749

8-6   \$232,400 (total cost)

8-8   \$262,224

8-10   Unit selling price = \$1.63

8-12   9 years

8-14   PW-C of A = \$388,777; PW-C of B = \$369,110.

8-16   Wait 2 years to purchase.

8-18   Amount to deposit = \$43,753.

8-20   By trial and error, N = 5 years

8-22   (a) i' = 15.9%
           (b) escalation rate = 7.85% per year

8-24   Escalation rate = 2.77% per year

8-26   F.W. = \$21,742

8-28   P.W. = $-$\$10,869

8-30   Maximum justifiable retrofit cost = 393.83×10^6

9-8   (a) BV_5 = \$42,857   (b) BV_5 = \$32,571
           (c) BV_5 = \$27,760   (d) BV_5 = \$18,000

9-10   (a) BV_4 = \$1,100   (b) BV_4 = \$1,181   (c) BV_4 = \$569.53
           (d) BV_4 = \$788.90   (e) BV_4 = \$144

9-12   CR Amount = \$1,052

9-14   (a) $-$\$1,660   (b) $-$\$481.82   (c) \$220

9-16   (a) \$17,200; \$14,000   (b) \$5,900

9-18   (a) d = \$1,500   (b) \$4,000

9-20   (a) \$56,203   (b) \$45,483

10-2   (a) \$869,850   (b) \$44.95%
           (c) \$1,065,150   (d) \$4,480,150

10-4   Investment in property has the highest after-tax return 1 year later.

10-6   (a) Jones has the greater after-tax income
           (b) 8.8%

10-8   (a) \$918   (b) 14.55%

10-10 P.W. of A.C.R.S. = \$15,711; P.W. of D.D.B. = \$15,227

10-12 (a) P.W. = $-$\$788; do not purchase.
           (b) P.W. = $-$\$742; do not purchase.

10-14 The unknown quantity is \$68,000.

10-16 $27,717

10-18 E.R.R. = 16.1%

10-20 Extra costs could be as much as $1,251 per year.

10-22 **(a)** Recommend press B   **(b)** It is preferable to buy.

10-24 **(a)** $i' = 63\%$ for B → A   **(b)** $i' = 67\%$ for B → A.

10-26 P.W. = $-$121,722$; A.W. = $-$32,135$

10-28 Select machine A because I.R.R. of A → B < 0

11-2 Keep the old machine

11-4 Select plan B

11-6 Coterminate at 8 years, choose plan III

11-8 **(a)** Select the improved machine.
(b) Keep the present machine for 2 more years.

11-10 The replacement is justified (A.W. = $-$2,523$).

11-12 Keep the present machine.

11-14 **(a)** $N^* = 3$ years   **(b)** $N^* = 3$ years

11-16 $N^* = 5$ years (A.C. = $5,333)

11-18 **(a)** Retain existing facilities.
(b) Recommend the new installation.

11-20 Keep the defender for at least another year.

11-22 The defender should be retained.

11-24 $i' = 14.6\%$ (after taxes)

12-4 $56,839

12-6 Retain the steel pier.

12-8 Private vehicles should pay 28¢ (round to 30¢).

12-10 Select plans A, B and C.

12-12 Plan 2 should be chosen.

12-14 Projects A and C would be recommended.

13-8 Select alternative B (levelized revenue requirement = $17,407).

13-10 The price should be $0.019/kWhr

13-12 **(a)** Retain old plume ($\overline{RR}$ = $10,100)
(b) Retain old plume ($\overline{RR}$ = $11,841)

13-14 Elect to install full capacity now ($\overline{RR}$ = $162,615)

14-4 Total annual cost = $605,000

14-6 $i' = 6.5\%$ per 6 months

14-8 $75,000 > CR cost; buy Small company.

14-10 The purchase should be made.

14-12 Total return on original owners' equity is greater for Plan I.

14-14 $K'_a = 9.24\%$

14-16 Recommend leasing to minimize the present worth of costs ($-$6,034).

14-18 Borrowed funds: P.W. = $7,114, I.R.R. = ∞; Equity Funds: P.W. = $4,582,
I.R.R. = 31%

14-20 $1,624

14-22 Select projects B, D and A.

15-2 Annual setup costs = annual holding costs

15-4 (a) Q drops to $Q/\sqrt{4}$.   (b) Q is reduced to $Q/\sqrt{3}$. (c) Q is reduced.

15-6 2,000 cases

15-8 2,357 fixtures

15-10 Order size should be 501

15-12 Manufacture the parts in your company.

15-14 Group replacement should occur every third period.

16-2 The product may not have sales appeal.

16-4 Esteem value often provides the market appeal for an item when use value cannot.

17-2 Number of books with binding A = 250 and with binding B = 100.

17-6 Maximum profit is $18,310

17-8 Maximize $10.75\,B1 + 6.45\,B2 - 1.85\,C1 + 0.19\,C2 + 8.22\,D$, subject to $C1 + C2 \leq 1$, $C1 + C2 \leq B2$, and $D \leq 1$.

18-2 (b) Critical path is 10–20–40–50–70–80. Minimum time = 21 days.

18-4 Reduce activity 20–60 by 1 day; also reduce either 70–80 or 30–50 by 1 day. Total savings would be $300.

18-6 Minimum time = 13 days

INDEX

Combined interest rate, 292, 295
Common stock, 472
Competition, 24
Compound interest, 69
Compounding
 continuous, with continuous cash flows, 111
 continuous, with discrete cash flows, 108
 discrete, 75
 more often than cash flows, 101
 more often than yearly, 99
 less often than cash flows, 103
Computer program
 for calculating equivalent worths and internal rate of return, 143, 590
 for calculating interest tables, 610
 for calculating after-tax cash flows, 378–81, 644
 for Monte Carlo simulation, 254–58
Conductor, electrical, economic size, 516
Consumer goods, 21
Consumer Price Index, 288, 291
Contingent projects, 206, 552
Continuous compounding
 continuous cash flows, interest formulas for, 111
 discrete payments, interest formulas for, 108
Corporation, 471
Cost
 book, 44, 398
 capital recovery, 133
 capitalized, 157
 cash, 44
 crash, 567
 decision, 38, 534
 direct labor, 14
 direct material, 14
 elements of, 14
 first, 37, 282
 fixed, 27, 37, 158
 incremental, 34, 37, 38, 41
 indexes, 280
 investment, 37, 157, 282
 marginal, 38, 405
 opportunity, 42–43, 67, 431
 overhead, 14–16, 283, 286
 sunk, 41, 396
 unit, 32, 41
 variable, 27, 37, 158
Cost accounting, 14
Cost concepts, 37, 533
Cost estimating, 276
Cost indexes, 280–82
Cost of capital, 67, 450, 476, 479
Cost recovery percentages (A.C.R.S.), 328–29

Cost reduction program, 154, 529
Cost–time tradeoffs, CPM, 567
Cost–volume relationships, 27
Coterminated assumption, 176, 183, 186
 replacement alternatives, 408
Critical path, 563–64
Critical path method (CPM), 563
Cycle time, 55

Data sources for cost estimating, 278–80
Debt capital, 8, 449, 472
Decision analysis, multiple objectives in, 260
Decision criteria, supplemental, 153, 260
Decision-making process, 8
Decision rules, complete uncertainty, 235–37
Declining balance method, 321–23
Defender (old asset), 393
 economic life, 405
Deferment period, breakeven, 227
Deferred annuities, 82
Deferred investment
 alternatives, 199, 461
 example of, 201
Deflation, 300
Demand, 23–24, 26, 200, 518
Depletion, 334
Depletion allowances, effect of on after-tax results, 382
Depreciation, 44, 313
 accounting for, 335
 A.C.R.S. percentages, 328–29
 as capital source, 17, 477
 changes in price levels, 317
 declining balance method, 321
 differences from other costs, 316
 functional, 317
 illustration of various methods in after-tax analysis, 357
 machine-hour method, 327
 physical, 317
 purposes of, 315
 recovery, 390–91
 requirements for selection of method, 317–18
 service output method, 326
 sinking fund, 326
 straight line method, 320
 sum-of-the-years'-digits method, 323
 switchover option, 323
 types of, 317
 useful life guidelines, 319
Depreciation accounting
 methods of, 318
 multiple asset, 333